A TREATISE ON DINITROGEN FIXATION
Section III

A TREATISE ON DINITROGEN FIXATION

Section III: Biology

GENERAL EDITOR

R. W. F. Hardy

SECTION III EDITOR

W. S. Silver

A WILEY-INTERSCIENCE PUBLICATION

JOHN WILEY & SONS, New York • London • Sydney • Toronto

Copyright © 1977 by John Wiley & Sons, Inc.

All rights reserved. Published simultaneously in Canada.

No part of this book may be reproduced by any means, nor transmitted, not translated into a machine language without the written permission of the publisher.

Library of Congress Cataloging in Publication Data:
Main entry under title:

A Treatise on dinitrogen fixation.

   Section 3 edited by W. S. Silver, section 4 by A. H. Gibson.
   "A Wiley-Interscience publication."
   Includes indexes.
   CONTENTS: [2] Section 3, biology.—[3] Section 4, agronomy and ecology.
   1. Nitrogen—Fixation. 2. Micro-organisms, Nitrogen-fixing. 3. Legumes. I. Hardy, Ralph W. F., 1934- II. Silver, Warren S. III. Gibson, A. H.

QR89.7.T73   589'.7'04133   76-15278
ISBN 0-471-35138-5 (Section 3)

Printed in the United States of America

10 9 8 7 6 5 4 3 2 1

ERRATA FOR:

A TREATISE ON DINITROGEN FIXATION - SECTION III: BIOLOGY

Edited by R. W. F. Hardy and W. S. Silver

Following page 385 legends for Plates 8.2 and 8.3 should read as follows:

Plate 8.2. a, b. Drawings of infection threads in Vicia faba roots published by Ward in 1887 (10). Note the similarity of his Plate 13 with Plate 8.1g, and the vesicle in the right-hand thread with Plate 8.1f. c. Diagram of a Pisum sativum nodule showing two vascular connections with the root. Note that the root endodermis is continuous between the nodule traces indicating that the nodule arose in the cortex. a.- root stele, e - root endodermis, n - nodule endodermis, v - nodule vascular trace and ve - its associated bundle endodermis, m - nodule meristem, d - differentiating zone, b - bacteroid zone, db - degenerate zone. After Bond (192).

Plate 8.3. Drawings of nodule development in Pisum sativum taken from sections, by Libbenga and Harkes (188). In 8.1a the infection thread has initiated meristematic activity in the inner cortex opposite a protoxylem point, while still growing through the outer cortex. b. The thread has reached the meristematic zone in the inner cortex. c. The thread has grown into the nodule initial with many cortex cells now dividing and some penetrated by the threads. d. The zone of dividing cells is now larger, with older cells in the initial now expanding as they cease to divide, and the meristematic region has become localized at the edge of the initial away from the root stele.

Following page 400 legend for Plate 8.10 should include:

Insert. Cowpea bacteroid showing condensed nucleoid.

			wrong	right
Page 400	para 2	line 18	lenticullular	lenticellular
Page 400(b)	para 1	line 9	single	singly
Page 400(b)	para 1	line 11	connection	connecting
Page 400(b)	para 1	line 11	bacterios	bacteroid
Page 415	para 4	line 1	85%	83%
Page 428	para 4	line 12	mucleoids	nucleoids
Page 429	para 3	line 4	cub	club
Page 450	para 4	line 8	closed	cold
Page 452	table	line 8	add sugar	add agar
Page 459	Ref. No. 44		initial stages of Clovers...	initial stages of infection of Clovers...

(over)

(continued)

Figure for Section 8.4.2.3 Page 454.

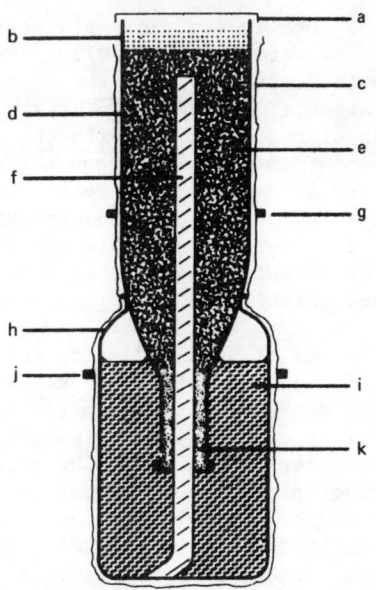

MODIFIED LEONARD BOTTLE-JAR ASSEMBLY (360)

a: glass or plastic cover (removed after seedling emergence),
b: dry gravel mulch (added after seedling emergence, may be
replaced by paraffined sand), c: water resistant paper bag
(covering assembly during sterilization, remaining to cover
junction of bottle and jar), d: inverted bottle with bottom
cut off, e: coarse river sand, f: wick, g: rubber band,
h: jar, i: nutrient solution (½ strength), j: rubber band,
k: cotton or glass wool.

Contributors

SECTION III: BIOLOGY

J. H. Becking, Institute for Atomic Sciences in Agriculture, Wageningen, The Netherlands

F. J. Bergersen, Division of Plant Industry, CSIRO Canberra, Australia

B. E. Caldwell, Department of Crop Science, North Carolina State University, Raleigh, North Carolina, U. S. A.

Peter Dart, ICRISAT, Hyderabad, India

T. A. LaRue, Prairie Regional Laboratory, National Research Council, Saskatoon, Saskatchewan, Canada

J. W. Millbank, Botany Department, Imperial College, London, England

C. A. Parker, Department of Soil Science and Plant Nutrition, Institute of Agriculture, University of Western Australia, Nedlands, Perth, Australia

J. S. Pate, Botany Department, University of Western Australia, Nedlands, Australia

E. A. Schwinghamer, Division of Plant Industry, CSIRO, Canberra, Australia

W. S. Silver, Department of Biology, University of South Florida, Tampa, Florida, U. S. A.

W. D. P. Stewart, Department of Biological Sciences, University of Dundee, Dundee, Scotland

Stanley Streicher, Department of Biology, Massachusetts Institute of Technology, Cambridge, Massachusetts, U. S. A.

R. C. Valentine, Plant Growth Laboratory and Agronomy and Range Science Department, University of California, Davis, California, U. S. A.

H. G. Vest, Department of Horticulture, Michigan State University, East Lansing, Michigan, U. S. A.

J. M. Vincent, Department of Microbiology, University of Sydney, Sydney, Australia

Preface

During the last quarter of the twentieth century, the world's population will increase from 4 billion to 6 or 7 billion, and the demand for food will more than double. Provision of an adequate supply of fixed nitrogen is central to the successful meeting of this food challenge. The supplementation of biological dinitrogen fixation with 40 million tons of fertilizer nitrogen at a cost of 8-10 billion U. S. dollars in 1975, and the estimated need for a supplement equivalent to 160 million tons of fertilizer nitrogen in 2000 A.D. indicates the magnitude of the problem. Moreover, the growing awareness of environmental quality and limitations of nonrenewable resources may introduce additional constraints.

Ignited by the extraction of the dinitrogen-fixing enzyme from a bacterial cell and abiological fixation of dinitrogen under ambient conditions, research on dinitrogen fixation entered a new log phase of growth about 15 years ago, and this remarkable phase continues to grow in both accomplishments and person power. Our objective is to provide the first comprehensive interdisciplinary reference work on dinitrogen fixation combining these recent advances with earlier work. Section I of this work on Inorganic and Physical Chemistry is edited by Frank Bottomley, Section II on Biochemistry is edited by Richard C. Burns, Section III on Biology is edited by Warren S. Silver, and Section IV on Agronomy and Ecology is edited by Alan H. Gibson, with Ralph W. F. Hardy as general editor for all sections. We hope that this coverage will facilitate research to explore biological and abiological dinitrogen fixation and to develop technologies for provision of fixed nitrogen.

Biological dinitrogen fixation has contributed to production in natural and agricultural habitats from an early stage in the development of living matter on earth. Until the late 1880s, none of the causal agents had been identified positively, although early writing, as far back as 5000 years ago, testify to the appreciation of legumes as contributors to soil fertility. Since the pioneering studies of Boussingault,

Hellriegal, Winogradsky, and Beijerinck, a very considerable understanding of the various diazotrophic forms, the factors affecting their activity, and their contribution to the nitrogen economy of many habitats, has been achieved. However, this understanding is far from complete, and as we stand at the threshold of an era of intensive investigation in all aspects of dinitrogen fixation, it is timely that existing knowledge be reviewed and collated.

Two major reasons can be given for the necessity to intensify our studies of dinitrogen fixation. The first is the increasing need to understand the earth's biological environment, to understand the inputs and outputs from all habitats, and to appreciate how perturbation of any elements of existing ecosystems will influence the behavior of all component parts. Developed and developing soils, lakes, rivers, tundras, and oceans have been and are receiving close attention, and with the availability of new techniques, additional diazotrophic forms are being identified. Many of these diazotrophs are free-living procaryotes (mainly bacteria or blue-green algae), but increasingly, facultative, and more particularly, associative symbiotic systems are being described from many habitats (e.g., marine, tundra, rhizosphere of cereals and dicotyledonous species). In many ways the associative symbiotic systems are essentially unknown with regard to our understanding of the diazotroph involved, of the factors involved in the association, and of the nitrogen contribution to the host or to the habitat.

The principal need for a deeper understanding of biological dinitrogen fixation is the urgent requirement to increase agricultural production for a seemingly ever-expanding human population. Although all forms of biological dinitrogen fixation ultimately contribute to such production, increases in the short term are likely to depend on the greater and more efficient utilization of legume-Rhizobium associations. The grain and pulse legumes are foremost in current research approaches to increasing protein production, although pasture and forage legumes, both in temperate areas and in the tropics, and green manure legumes, make significant contributions to animal production and the soil's nitrogen supply, respectively. The problems of increasing food supply are exacerbated by difficulties in producing nitrogenous fertilizers. In the short term the high capital cost of building fertilizer plants prevents many countries from

Preface

producing the quantities of fertilizer required, whereas in the long term there is grave concern about the supplies of natural gas required to manufacture these fertilizers. In addition, the inefficiency of uses of fertilizer nitrogen by agricultural crops, potential pollution of groundwater by unused nitrogen, denitrification loss of fertilizer nitrogen, possible destructive effects of denitrification products on atmospheric ozone, and transportation, storage and application costs for fertilizer nitrogen are limitations of our current technology. The nature and the magnitude of these problems require that intensive efforts be made to increase biological nitrogen fixation, not only from legumes but from blue-green algae and other procaryotes, either in the free-living state or in some form of symbiosis with a macrosymbiont driven by solar energy.

Because of the current importance of the legumes, as well as for convenience, there is a broad division of the subject into the legume and "nonlegume" systems. The latter comprises the bacterial and blue-green algal diazotrophs that function either in the free-living state or as microsymbionts associated with lower plants or higher plants, other than legumes. Section III, the biological section, is devoted to a consideration of the broader aspects of dinitrogen fixation — the organisms involved, pertinent physiological processes of the microbe and the host plant, and where appropriate, certain phytogeographical facets. The current importance of the leguminous symbiosis is reflected in the six chapters devoted to this topic, and three chapters on the genetic aspects of dinitrogen fixation attest to its increasing importance as a possible approach to potentiating and extending the process of dinitrogen fixation by genetic manipulation.

Fred, Baldwin, and McCoy, in introducing their monumental monograph "Root Nodule Bacteria and Leguminous Plants" in 1932, expressed their concern with the volume of material at their disposal, and the adequacy of its communication, by quoting W. J. Humphrey: "Many investigations are lost for years, if not forever, in the jungle of journals and the tangle of tongues." How much more relevant such comment is today! The contributing authors have not attempted to provide a complete citation of all references; rather they have concentrated on citing more recent work and landmark studies in particular areas, and through such references, providing access for interested readers to earlier studies. This, we believe, has allowed them to provide completeness of cover of their topics, to present current thinking on the subject, and to indicate their thoughts on the most profitable areas for future investigation.

R. W. F. Hardy
W. S. Silver

Wilmington, Delaware
Tampa, Florida
June 1976

SPECIAL NOTE

The *Plant and Soil, Special Volume* which is referred to in many chapters is catalogued in some libraries as *Biological Nitrogen Fixation in Natural and Agricultural Habitats*, T. A. Lie and E. G. Mulder, Eds., Martinus Nijhoff, The Hague, 1971.

Contents

SECTION III: BIOLOGY

Chapter 1	Perspectives in Biological Dinitrogen Fixation C. A. Parker	3
Chapter 2	The Bacteria T. A. LaRue	19
Chapter 3	Blue-Green Algae W. D. P. Stewart	63
Chapter 4	Lower Plant Associations J. W. Millbank	125
Chapter 5	Foliar Associations in Higher Plants W. S. Silver	153
Chapter 6	Dinitrogen-fixing Associations in Higher Plants other than Legumes J. H. Becking	185
Chapter 7	Rhizobium: General Microbiology J. M. Vincent	277
Chapter 8	Infection and Development of Leguminous Nodules Peter Dart	367
Chapter 9	Functional Biology of Dinitrogen Fixation by Legumes J. S. Pate	473
Chapter 10	Physiological Chemistry of Dinitrogen Fixation by Legumes F. J. Bergersen	519

Chapter 11	Genetic Aspects of Nodulation and Dinitrogen Fixation by Legumes: The Macrosymbiont B. E. Caldwell, H. G. Vest	557
Chapter 12	Genetic Aspects of Nodulation and Dinitrogen Fixation by Legumes: The Microsymbiont E. A. Schwinghamer	577
Chapter 13	The Genetic Basis of Dinitrogen Fixation in _Klebsiella pneumoniae_ Stanley Streicher, R. C. Valentine	623

SUBJECT INDEX 657

TAXONOMIC INDEX 665

A TREATISE ON DINITROGEN FIXATION

Section III

SECTION III
Biology

CHAPTER 1

Perspectives in Biological Dinitrogen Fixation

C. A. PARKER

Department of Soil Science and Plant Nutrition
Institute of Agriculture
University of Western Australia
Nedlands, Perth, Australia

1.1. Retrospect, 3
1.2. The Nitrogen Economy of the Biosphere, 6
    1.2.1. Sources of Nitrogen, 6
        1.2.1.1. Rainfall, 6
        1.2.1.2. Autotrophic Microorganisms, 7
        1.2.1.3. Heterotrophic Bacteria, 7
        1.2.1.4. Symbiotic Associations, 8
    1.2.2. Mechanisms of Nitrogen Loss, 9
        1.2.2.1. Denitrification, 9
        1.2.2.2. Fire, 9
    1.2.3. The Nitrogen Balance, 9
1.3. Major Determinants of N_2 Fixation, 10
    1.3.1. Sources of Energy, 10
    1.3.2. Availability of Nitrogen in Soil, 10
    1.3.3. Oxygen: Specific Inhibitor of N_2 Fixation, 11
1.4. Concluding Remarks, 12
1.5. Questions in Need of Answers, 13
1.6. References, 14

1.1 RETROSPECT

The aim of this chapter is to place biological N_2 fixation in perspective as a natural phenomenon and as an essential part of the ecosystem. It concludes by asking questions that beg answers.

Excellent books and reviews are available on the subject of dinitrogen (N_2) fixation, so that readers requiring a more detailed historical treatment are referred to these (1-8).

Professor Perry Wilson kindly sent me a copy of Aulie's historical article on "Boussingault and the Nitrogen Cycle" (9). It makes a good starting point, for Boussingault's long engagement (1836-1876) with the problem of the source of nitrogen in plants which will strike a sympathetic chord in many contemporary minds. Boussingault's field evidence and biological observations with legumes were so much at variance with his careful and controlled chemical evidence, that he finally abandoned his earlier ideas about N_2 utilization by legumes, and came to view the soil as a chemical system capable of supplying the nitrogen needs of all plants.

The enigma of leguminous behavior was solved when Hellriegel and Wilfarth proved N_2 fixation by nodulated legumes in 1887, the year of Boussingault's death. This was followed in 1888 by the isolation of the first nodule bacteria by Beijerinck (1). Winogradsky reported in 1893 the isolation of a soil bacterium, Clostridium pasteurianum, which fixed N_2 in the test tube (3). Beijerinck isolated the aerobic diazotroph Azotobacter in 1901. Contradictory reports appeared in the years 1889-1928 about N_2 fixation in pure cultures of blue-green algae (2).

In 1932 Fred, Baldwin, and McCoy published their monograph (1) which collated, reviewed, and analyzed existing information on legume symbiosis (including a section of other N_2-fixing plants). They achieved a biological synthesis that had a profound influence on subsequent theory and practice. The monograph published by P. W. Wilson in 1945 achieved a similar but more biochemical synthesis and provided a blueprint for future research strategy.

For the next two decades Wilson and Burris with their colleagues and students, employing the new technique of mass spectrometry with $^{15}N_2$ to trace reactions, dominated the scene on the biochemistry of N_2 fixation. In a series of brilliantly conceived experiments, based on considerations of enzyme theory and the likely pathways of N_2 fixation (10) they traced the pathway of fixation through ammonia to glutamic acid and, thence, to other amino acids.

Finally they obtained enriched juvenile ammonia directly from Clostridium in 1951, and in 1953 achieved the same result with Azotobacter (11).

A new impetus to biochemical studies occurred when extracts capable of fixing N_2 were prepared from Clostridium by Carnahan, Mortenson, Mower, and Castle in 1960 (12, 13). Cell-free extracts of other diazotrophs were soon prepared (14), and it became possible to investigate the nature of "nitrogenase," uncomplicated by growth, nutrition, or respiration. The enzyme itself has now been purified and and resolved into two components (15). Exciting recent research shows that "nitrogenase" activity can be transferred from N_2-fixing Klebsiella to nonnitrogen-fixing mutants (16, 17) and to E. coli (18).

Meanwhile, through the use of $^{15}N_2$, a large number of diazotrophs were added to the list (19, 20) while others were confirmed or removed (21). Symbiotic systems have been confirmed and extended (8, 22-26) and some challenged (27). We can now be confident that N_2-fixing symbioses are present in certain members of the Coriariales, Rosales, Myricales, Fagales, Casuarinales, and Rhamnales, as well as Leguminales. All have bacterial root-nodule endophytes (8, 22).

The herbaceous angiosperm Gunnera (Haloragidaceae), which carries blue-green algae in stem cortical tissues (24), fixes appreciable amounts of N_2 (26). Associations with blue-green algae as the diazotrophs are widespread, involving lichens, liverworts, ferns, cycads, as well as higher plants (24, 23). Associations between blue-green algae and some protozoa, e.g., Cyanophora paradoxa (24), and green algae (23) are known but the function of the blue-green algae is poorly understood.

Confirmation of ammonia as the key intermediate in N_2 fixation had important ecological and biochemical implications. Wilson and Burris (10) and Bayliss (28) had calculated from thermodynamic data that in the reduction of N_2 to NH_4^+, using glucose as the source of electrons, energy might become available and not be required from external sources to "drive" the reaction.

What, then, was to prevent asymbiotic diazotrophs from exploiting their enzymic advantage in competition for organic carbon? Again, since N_2 could be considered as a major alternative oxidant, the question of the relationship between O_2 and N_2 in aerobic diazotrophs was reviewed (29). Evidence that these gases may in fact compete for substrate electrons was reported by Parker and Scutt (30) and supported by Dalton and Postgate (31).

In legumes also, ammonia was shown by Kennedy (32, 33) to be the key intermediate in N_2 fixation, and this was confirmed by Bergersen (34). Kennedy also implicated bacteroids as the site of nitrogen fixation (33, 35-37), but it remained for Bergersen and Turner (38), and Koch, Evans, and Russell (39) to demonstrate this proposal by obtaining fixation in bacteroids separated from nodule tissues. Glucose and/or fructose were found to be the likely substates for bacteroids in Ornithopus and Lupinus nodules (40).

Recognition of the ability of nitrogenase to reduce substances with $C \equiv C$, $C \equiv N$ or $N \equiv N$ structures began with the report on N_2O reduction by Mozen and Burris in 1954 (41). A wealth of chemical information for comparison with the inorganic complexes of dinitrogen has followed (see II.4). Perhaps most useful was that showing that "acetylene and N_2 reduction are analogous processes, catalysed by the same enzyme system" (42), which led to the proposal by Hardy and Knight for the use of acetylene reduction as a sensative assay for nitrogenase and its widespread application in many laboratory and field studies (see IV.12).

1.2 THE NITROGEN ECONOMY OF THE BIOSPHERE

1.2.1 Sources of Nitrogen

An excellent discussion on the bio-geochemistry of nitrogen is given by Alexander (43), who considers sources, sinks, states, and amounts of nitrogen on Earth (also see IV.1). "To maintain a steady-state level of nitrogen in the biosphere, losses must equal gains, and the leakage from the biosphere is apparently made up by the unique organisms that bring about nitrogen fixation..." (43).

1.2.1.1 Rainfall

Allison (44) lists from many parts of the world rainfall nitrogen values ranging from 1.8-22 kg/ha·yr. In northern Australia Wetselaar and Hutton (45) conducted careful experiments to estimate the annual accession of rainfall N to the soil, and to judge its origin. The annual average rainfall at the site is 925 mm, the area is subject to frequent electric storms, and is remote from industrial or urban activity. They found annual increments of nitrate plus ammonium nitrogen amounting to 1.5 kg/ha, the concentration of ions becoming lower as the season progressed. The authors concluded from consideration of the ionic balance that neither the ocean nor the atmosphere could be major contributors, and that "most of the material in the rain water is part of a terrestrial cycle and not a true accession."

Thus it appears that rainfall makes minor contributions of "new" fixed nitrogen to the biosphere.

1.2.1.2 Autotrophic Microorganisms

N_2 fixation in aquatic environments by photosynthetic bacteria (46, 47) or by blue-green algae (48) has been estimated to vary from a few kilograms to 60-70 kg/ha in paddy soils. In arid areas algal crusts may contribute meaningful amounts of nitrogen to the ecosystem (62, 48). Marine gains from N_2 fixation are about one-third of the terrestrial gains (43).

1.2.1.3 Heterotrophic Bacteria

Certain bacteria in the genera Clostridium, Bacillus, Klebsiella, Azotobacter, Derxia, Beijerinckia, and Mycobacterium fix N_2, either anerobically or more efficiently at low pO_2 (49). Their activity can be measured using $^{15}N_2$ or acetylene, but their numbers are likely to be grossly underestimated by laboratory counting methods (50, 51).

Nitrogen accumulation in unsupplemented soils growing grass without legumes has been reported (52, 53). Recent evidence indicates fixation of N_2 in the rhizosphere of grass (54, 55) and herbs (56), the quantities ranging from a few kilograms up to 90 kg nitrogen/ha·yr. The microbial agents appear to be embedded in the mucigel on the root surface (55). This looser type of symbiosis, termed associative by Hardy and Holsten (88), may be of considerable importance in tropical soils.

The importance of N_2 fixation in the phylloplane (57) is more difficult to assess. High pO_2 would exclude most N_2-fixing bacteria from activity, and aerobic N_2 fixers would be relatively inefficient. However, Jones (58) has estimated combined soil and phyllosphere fixation rates of 53-120 g nitrogen/ha·day, in Douglas Fir forest. For comparison, these data extrapolate to give values of about 8-18 kg nitrogen/ha·yr.

1.2.1.4 Symbiotic Associations

The root-nodule associations known to fix N_2 belong in 8 orders of Angiosperms (after Hutchinson, ref. 59), and in the family Cycadaceae of the gymnosperms. Of these the Leguminales house eubacteria; Gunnera and the gymnosperms house blue-green algae. The others contain antinomycetes in their root nodules (8, 22) with exception of Trema which is nodulated by Rhizobium (87).

Nitrogen fixation by pasture legumes is well documented, and varies from around 40 to several hundred kg N/ha·yr (60, 61). For leguminous shrubs and trees, as for all the other root-nodulated plants the evidence is harder to obtain. Bond (8) cites evidence for fixation excluding legumes of around 60 kg N/ha·yr, and this seems a likely figure for woody legumes. Gunnera was estimated to contribute around 60 kg N/ha·yr to its particular ecosystem in New Zealand (26).

The importance of symbiotic associations with blue-green algae is less well understood. Lichens are undoubtedly important in harsh environments (24) for their nitrogen and carbon contributions to bare surfaces, and in arid climates (62). Azolla has been suggested as a green manuring plant in rice fields by Tuzimura et al. in Japan (63).

Double symbiosis, where legumes support both N_2-fixing and mycorrhizal associations at the same time, has been reported (64, 65). This has important biological and practical implications, and may well apply to other N_2-fixing systems.

On land it appears that root nodule associations with higher plants are of major importance in maintaining soil N in shrub and forest communities. Even here, possible contributions from asymbiotes should not be ignored. In grasslands asymbiotic N_2 fixation must be of sufficient magnitude to maintain the nitrogen balance against losses by denitrification and fire (see III.1.2.2.).

1.2.2 Mechanisms of Nitrogen Loss

1.2.2.1 Denitrification

Organic N eventually appears as ammonia some of which may be volatilized, while some is oxidized to nitrate. Where the biological demand for O_2 exceeds the rate of supply, which may easily happen in wet soil or stagnant water in the presence of readily decomposable substrate, nitrate nitrogen is used as an electron acceptor by facultatively anaerobic bacteria. N_2 and N_2O are formed (66) and fixed N is lost.

The rate and extent of denitrification are influenced by many factors such as the amount and type of organic matter, pH, temperature, and moisture - all factors influencing oxygen demand relative to oxygen supply. Experiments show losses from soil in pots may be as high as 40% of the added N (66). Marine denitrification may be greater than the increment gained from N_2 fixation (43).

1.2.2.2 Fire

When grassland is burned, over 90% of the organic nitrogen in the grass is lost to the atmosphere (67). Similar losses take place in forest fires, where it has also been shown that about 95% of the organic nitrogen is liberated as N_2 (68). As everyone knows fires can be started by either natural causes, such as lightning, or by man. Lightning fires are sporadic and infrequent, and because of accumulation of material they are very destructive. While man-made fires may be largely accidental, fire has been used intelligently for 20,000 yr or more, making a profound impact on plant communities (69, 70). For example, much of the world's grassland is thought to have been won from forest and shrubland through the agency of man-manipulated fire (71).

Vast quantities of fixed nitrogen must have been lost to the atmosphere over the ages, because of fire, losses which have been offset by biological N_2 fixation.

1.2.3 The Nitrogen Balance

Nearly 80 yr after Boussingault's death, Allison (72) titled a review "The Enigma of Soil Nitrogen Balance

Sheets." Ten years later he was not much more confident about such calculations (73). On a global scale, taking the oceans and the land together, good estimates of true gains or losses of fixed N to the biosphere cannot be made.

Accurate measurements of N gains or losses are difficult to make in soil, even under uniformly growing pasture, because of variability and possible nitrogen movement in the soil profile. Measurements are even more difficult under trees and shrubs where greater depths of soil are being exploited, and the sampling problems are truly formidable. Newer techniques such as the acetylene reduction assay should provide better answers to these measurement problems.

It is, therefore, easy to see that if the "steady state" is being maintained against losses by fire and denitrification, biological N_2 fixation has possibly been underestimated in the past, especially on grasslands.

1.3 MAJOR DETERMINANTS OF N_2 FIXATION

1.3.1 Sources of Energy

The use of light energy for N_2 fixation on open land surfaces, or in aquatic habitats, is presumably limited mainly by the availability of other nutrients such as phosphate or iron in aquatic or humid habitats, or by water itself in arid areas and on rock surfaces. The amount of organic carbon available to diazotrophs as an energy source depends ultimately on the amount of photosynthetic product. It may be available directly in phototrophs (74) as sugars in nodules (40), as organic exudates from the roots (75), or as dead organic material. The total increment of dry organic matter varies with climate and soil and may range from a few hundred kg/ha·yr in semiarid regions to 50 tons/ha·yr with sugar cane (76). Production in a vigorous young tropical rain forest would possibly be much greater.

1.3.2 Availability of Nitrogen in Soil

Available N depresses N_2 fixation in asymbiotic diazotrophs, and in legumes and other N_2-fixing angiosperms (23). The carbon-nitrogen ratio in the plant sap is important for effective symbiosis (3),

Major Determinants of N_2 Fixation

probably because the plant and the bacterium are in competition for soluble carbohydrate. Where available soil nitrogen is high, plants may absorb sufficient nitrogen for growth, the carbohydrate being used by the plant for the synthesis of protein and structural complexes.

In the case of asymbiotic N_2 fixation, the diazotrophs are in competition for organic carbon with other soil microorganisms. To gain competitive advantage available nitrogen should be low, relative to carbon, and oxygen should be undersupplied (see III.1.3.3).

"Soils gain organic carbon, for the most part, from the growth and decomposition of plants in situ, and the living green plant is a powerful ecological factor in microbial processes" (52). While the plant adds organic carbon to the soil, as exudates (75) or as senescent tissues, it also takes up available nitrogen from the soil. At the same time, by root respiration or occupation of soil channels, oxygen may be depleted. These conditions are likely to favor heterotrophic diazotrophs, and can be found in grasslands. It is not at all the same thing to add ground straw to soil and incubate the mixture in the laboratory!

1.3.3 Oxygen: Specific Inhibitor of N_2 Fixation

Oxygen inhibits the nitrogenase system in both anaerobes and aerobes (14). It precludes the growth of anaerobes such as Clostridium, and it inhibits N_2 fixation in Azotobacter specifically and possibly competitively (30, 31).

Whether O_2 respiration is a protective mechanism for nitrogenase (49) or not, the fact remains that in reducing O_2 the bacteria, while increasing ATP content, are losing electrons that might have been used to reduce N_2. Nor is it antithetical that O_2 inhibits nitrogenase directly, it is necessary for aerobic diazotrophs, and at the same time competes with N_2 for substrate electrons. Gases such as H_2, CO, C_2H_2, N_2O, all inhibit N_2 fixation [there is one dissenting report (77) concerning the competitive role of H_2) and are important as biochemical tools for elucidating the nature of N_2 fixation. Oxygen inhibition is an overriding biological reality: With N_2, oxygen makes up most of the terrestrial atmosphere. The relationship

1.4 CONCLUDING REMARKS

There seems to be a discrepancy in the efficiency of asymbiotic diazotrophs in N_2 fixation, compared with the legume symbiosis. Taking 40 mg N_2 fixed/g mannitol consumed by <u>Azotobacter</u> (31) and calculating the N_2-fixing cost of a legume on that basis, indicates a requirement for about 40% of the total photosynthate. This is unlikely, as Gibson (78) found no significant difference in the growth of legumes using either fixed nitrogen or N_2 under controlled conditions. Yet at 40 mg N/g mannitol, <u>Azotobacter</u> has achieved extraordinary efficiency. This amount of N represents 0.25 g protein, which, at a proportion of one-third protein, would represent a conversion of 75% of the mannitol to dry cell material. The discrepancy, then, probably lies in the largely growth-bound nature of N_2 fixation by asymbiotic bacteria, and in fluctuating environmental factors such as moisture and pO_2.

In the legume, growth of the nodule bacteria is obviously controlled; the bacteria do not continue to divide, nor do they quickly die. In our experiments with lupins, especially if these are well nourished, nodule tissues may remain functional until seed setting. Thus the original bacterial cells, secure in the enclosing sacs, have been fixing N_2 for 120 days. May this not be envisioned as controlled N_2 respiration (30)? Bayliss' (28) thermochemical equations show the possibility that 600-700 mg N_2 could be fixed per gram glucose metabolised. With legumes, oxygen consumption by the bacteroids would reduce this.

There is no known difference, in principle, between N_2 fixation by asymbionts and by symbionts, nor can there be serious doubt that nodule symbionts evolved from asymbionts (79). These systems thus represent the two ends of an evolutionary scale. The procaryotes brought with them the unique properties of "nitrogenase" (49), thereby liberating the plant from dependence on soil nitrogen. The procaryotes, in turn, obtained greater security from oxygen excess, moisture stress, and the intense competition of other microorganisms.

There is a possibility that intermediate stages, not showing visible nodules, might go unnoticed (79). Because of this, and observations made by Stevenson (80), we examined a number of non-nodulated plants which grow strongly in nitrogen-poor soils (Parker, Oakley, Williams, unpublished). Whole plants were dug and exposed in a field within minutes after disturbing to $^{15}N_2$ with a gas mixture of 95% $^{15}N_2$-enriched N_2, 5%; O_2, 10%; argon, 85%. Exposure was for 24-48 hours. No enrichment was found in the leaves, stems or roots of the following plants: Pinus radiata D. Don; Eucalyptus calophylla R. Br.; Ricinus communis, L; Banksia menziesii R. Br.; Rubus fruiticosus; Trichinium manglesii, Lindl.; Arctotheca nivea, Hoffm.; Scaevola crassifolia, Labill.; Myoporum insulare R. Br.; Helianthus annus, L.; Cakile maritima Scop.; Carpobrotus aequilateralis (Haw.) N. E. Br.; Anthericum divaricatum, Jacq.; Cryptostemma calendulaceum, L. R. Br.; Stirlingia latifolia, (R. Br.) Steud.; Watsonia bulbillifera, Mathews and L. Boluf.; Perlargonium drummondii, Turcz.; Oenothera drummondii, Hook.; Hypochoeris glabra, L. Small enrichments were sometimes obtained with cores of soil growing grass. Pots of annual ryegrass (L. rigidum) and rice (Oryza sativa L.) were substantially enriched after exposure for 10 days, under conditions in which sunlight was excluded from the soil surface.

It seems at this stage of knowledge that symbiotic systems, particularly in root-nodulated plants, are much more efficient than asymbiotic heterotrophs in terms of N_2 fixed per unit of photosynthate produced. Less highly developed systems, which do not show themselves as visible nodules, deserve further investigation.

1.5 QUESTIONS IN NEED OF ANSWERS

Since these introductory remarks are intended to preface the chapters of Part III of the treatise in which the biological aspects are discussed, it is not unreasonable to raise some biological questions which need to be answered.

1. All symbiotic diazotrophs in Angiosperms are housed in root nodules (excepting, perhaps, Gunnera). Are there less obvious systems which have not been observed?

2. Few plant families have succeeded in establishing nodule symbiosis. What is the evolutionary significance in the fact that most of these are woody families (81), which are clustered around the Rosales in Hutchinson's evolutionary scheme (59)?
3. The non-legumes have actinomycetes in their root nodules with one known exception whereas the legumes have eubacteria. Within the legumes, however, the Caesalpiniaceae are largely unnodulated, while the Papilionaceae and Mimosaceae are mostly nodulated (82). Does this mean that the evolution of these associations is geologically rather late (83)?
4. If blue-green algal cells were the origin of certain plant organelles (84, 85), why didn't plants "take over" the nitrogenase system as well? Did the primitive blue-green algae possess nitrogenase? Could they have gained it later by transfer from another bacterium?

In the chapters which follow it will be noted that some of the questions are to varying degrees answered, some are perhaps unanswerable, and most assuredly many equally intriguing new questions are raised!

1.6 REFERENCES

1. Fred, E. B., I. L. Baldwin, and E. M. McCoy, Root Nodule Bacteria and Leguminous Plants, Wisconsin University Studies in Science, (5), 1932.
2. Waksman, S. A., Principles of Soil Microbiology, 2nd ed., Williams & Wilkins, Baltimore, Md., 1932.
3. Wilson, P. W., The Biochemistry of Symbiotic Nitrogen Fixation, University of Wisconsin Press, 1940.
4. Allen, E. K., and O. N. Allen, Bact. Rev., 14, 273 (1950).
5. Jensen, H. L., in Soil Nitrogen, W. V. Bartholomew and F. E. Clark, Eds., Am. Soc. Agron., Madison, Wis. pp. 436-480, 1965.
6. Nutman, P. S., Biol. Rev., 31, 109 (1956).
7. Nutman, P. S., Proc. Roy. Soc. Ser. B, 172, 417 (1969).
8. Bond, G., Ann. Rev. Plant Physiol., 18, 107 (1967).

References

9. Aulie, R. P., *Proc. Am. Phil. Soc.*, **114**, 435 (1970).
10. Wilson, P. W., and R. H. Burris, *Bact. Rev.*, **11**, 41 (1947).
11. Wilson, P. W., *Proc. Roy. Soc. Ser. B.*, **172**, 319 (1969).
12. Carnahan, J. E., L. E. Mortenson, H. F. Mower, and J. E. Castle, *Biochim. Biophys. Acta*, **38**, 188 (1960a).
13. Carnahan, J. E., L. E. Mortenson, H. F. Mower, and J. E. Castle, *Biochim. Biophys. Acta*, **44**, 520 (1960b).
14. Burris, R. H., in *The Chemistry and Biochemistry of Nitrogen Fixation*, J. R. Postgate, Ed., Plenum Press, London/New York, pp. 105-160, 1971.
15. Hardy, R. W. F., and R. C. Burns, in *Iron-Sulphur Proteins*, W. Lovenberg, Ed., Academic Press, New York, p. 65, 1973.
16. Streicher, S., E. Gurney, and R. C. Valentine, *Proc. Nat. Acad. Sci. U. S.*, **68**, 1174 (1971).
17. Dixon, R. A., and J. R. Postgate, *Nature*, **234**, 47 (1971).
18. Dixon, R. A., and J. R. Postgate, *Nature*, **237**, 102 (1972).
19. Wilson, P. W., and R. H. Burris, *Ann. Rev. Microbiol.*, **7**, 415 (1953).
20. Stewart, W. D. P., *Proc. Roy. Soc. Ser. B.*, **172**, 368 (1969).
21. Hill, S., and J. R. Postgate, *J. Gen. Microbiol.*, **58**, 277 (1969).
22. Becking, J. H., *Plant Soil*, **32**, 611 (1970).
23. Stewart, W. D. P., *Nitrogen Fixation in Plants*, University of London Press, 1966.
24. Scott, G. D., *Plant Symbiosis*, Edw. Arnold, 1969.
25. Morrison, T. M., *Nature*, **189**, 945 (1961).
26. Silvester, W. B., and D. R. Smith, *Nature*, **224**, 1231 (1969).
27. Silver, W. S., in *The Chemistry and Biochemistry of Nitrogen Fixation*, J. R. Postgate, Ed., Plenum Press, London, pp. 245-281, 1971.
28. Bayliss, N. S., *Aust. J. Biol. Sci.*, **9**, 364 (1956).
29. Parker, C. A., *Nature*, **173**, 780 (1954).
30. Parker, C. A., and P. B. Scutt, *Biochim. Biophys. Acta*, **38**, 230 (1960).
31. Dalton, H., and J. R. Postgate, *J. Gen. Microbiol.*, **54**, 463 (1969).
32. Kennedy, I. R., *Proc. Sec. Austral. Pl. Nutr. Conf.*, Perth, paper A(d) 7, 1964.

33. Kennedy, I. R., *Primary Products of Symbiotic Nitrogen Fixation*, Ph. D. Thesis, University of Western Australia, 1965.
34. Bergersen, F. J., *Aust. J. Biol. Sci.*, **18**, 1 (1965).
35. Kennedy, I. R., C. A. Parker, and D. K. Kidby, *Biochim. Biophys. Acta*, **130**, 517 (1966).
36. Kennedy, I. R., *Biochim. Biophys. Acta*, **130**, 285 (1966a).
37. Kennedy, I. R., *Biochim. Biophys. Acta*, **130**, 295 (1966b).
38. Bergersen, F. J., and G. L. Turner, *Biochim. Biophys. Acta*, **141**, 507 (1967).
39. Koch, B., H. J. Evans, and S. Russell, *Plant Physiol.*, **42**, 466 (1967).
40. Kidby, D. K., *Carbon Metabolism in Legume Root Nodules*, Ph. D. Thesis, University of Western Australia, 1967.
41. Mozen, M. M., and R. H. Burris, *Biochim. Biophys. Acta*, **14**, 577 (1954).
42. Dilworth, M. J., *Biochim. Biophys. Acta*, **127**, 285 (1966).
43. Alexander, M., *Microbial Ecology*, John Wiley and Sons, New York, pp. 419-429, 1971.
44. Allison, F. E., in *Soil Nitrogen*, W. V. Bartholomew and F. E. Clark, Eds., Am. Soc. Agron., Madison, Wis., pp. 573-606, 1965.
45. Wetselaar, R., and J. T. Hutton, *Aust. J. Agr. Res.*, **14**, 319 (1963).
46. Okuda, A., M. Yamaguchi, M. Kobayashi, and T. Katayama, *Soil Sci. Plant Nutr.*, **7**, 146 (1961).
47. Kobayashi, M., T. Katayama, and A. Okuda, *Soil Sci. Plant Nutri.*, **11**, 16 (1965).
48. Stewart, W. D. P., *Plant Soil*, **32**, 555 (1970).
49. Postgate, J. R., *The Chemistry and Biochemistry of Nitrogen Fixation*, J. R. Postgate, Ed., Plenum Press, London/New York, pp. 161-190, 1971.
50. Parker, C. A., *Aust. J. Agr. Res.*, **5**, 90 (1954).
51. Parker, C. A., *Aust. J. Agr. Res.*, **6**, 388 (1955).
52. Parker, C. A., *J. Soil Sci.*, **8**, 48 (1957).
53. Jenkinson, D. S., *Rothamsted Exp. Sta. Rep. for 1970*, Part 2, 113 (1971).
54. Kass, D. L., M. Drosdoff, and M. Alexander, *Soil Sci. Soc. Am. Proc.*, **35**, 286 (1971).
55. Dobereiner, J., J. M. Day, and P. J. Dart, *J. Gen. Microbiol.*, **71**, 103 (1972).
56. Harris, D., and P. J. Dart, *Soil Biol. Biochem.*, **5**, 277 (1973).

References

57. Ruinen, J., *Plant Soil*, 22, 375 (1965).
58. Jones, K., *Ann. Bot.*, 34, 239 (1970).
59. Hutchinson, J., *Evolution and Phylogeny of Flowering Plants*, Academic Press, London/New York, 1969.
60. Dawson, R. C., *Plant Soil*, 32, 655 (1970).
61. Henzell, E. F., and D. O. Norris, in *A Review of Nitrogen in the Tropics with Particular Reference to Pastures*, Cwlth. Bur. Pastures and Field Crops, Bull. 46, 1 (1962).
62. Rogers, R. W., R. T. Lange, and D. J. D. Nicholas, *Nature*, 209, 96 (1966).
63. Tuzimura, K., F. Ikeda, and K. Tsukamoto, *Jap. Soil Fert. J.*, 28, 275 (1957).
64. Ross, J. P., *Phytopathology*, 61, 1400 (1971).
65. Bowen, G. D., and A. D. Rovira, in *Proc. Easter School Agr. Sci.*, 15th Nottingham, W. J. Whittington, pp. 170-198 1968.
66. Broadbent, F. E., and F. E. Clark, in *Soil Nitrogen*, W. V. Bartholomew and F. E. Clark, Eds., Am. Soc. Agron., Madison, Wis., pp. 344-359, 1965.
67. Norman, M. J. T., and R. Wetselaar, *J. Aust. Inst. Agr. Sci.*, 26, 272 (1960).
68. Vines, R. G., L. Gibson, A. B. Hatch, N. K. King, D. A. MacArthur, D. R. Packham, and R. J. Taylor, *On the Nature, Properties, and Behavior of Bushfire Smoke*, C.S.I.R.O., Aust. Div. Appl. Chem. Tech. Paper No. 1, 32 pp (1971).
69. Smith, A. G., in *Studies in the Vegetational History of the British Isles*, D. Walker and R. G. West, Eds., Cambridge University Press, London, 1970.
70. Hallam, S. J., *Fire and Hearth: A Study of Aboriginal Usage and European Usurpation in South-Western Australia*, Aust. Inst. Aboriginal Studies, Canberra, 1975.
71. Barnard, C., and O. H. Frankel, in *Grasses and Grasslands*, C. Barnard, Ed., MacMillan and Co., London, New York, pp. 1-12, 1964.
72. Allison, F. E., in *Advances in Agronomy*, A. G. Norman, Ed., Academic Press, New York, Vol. 2, pp. 213-250, 1955.
73. Allison, F. E., in *Advances in Agronomy*, A. G. Norman, Ed., Academic Press, New York/London, Vol. 18, pp. 219-258, 1966.

74. Cox, R. M., and P. Fay, Proc. Roy. Soc. Ser. B., 172, 357 (1969).
75. Rovira, A. D., in Ecology of Soil Borne Pathogens, K. F. Baker and W. C. Snyder, Eds., John Murray London, pp.170-186, 1965.
76. Barnes, A. C., The Sugar Cane, Interscience, New York, 1964.
77. Parker, C. A., and M. J. Dilworth, Biochim. Biophys. Acta, 69, 152 (1963).
78. Gibson, A. H., Aust. J. Biol. Sci., 19, 499 (1966).
79. Parker, C. A., Nature, 179, 593 (1957).
80. Stevenson, G., Ann. Bot. N. S., 23, 622 (1959).
81. Bond, G., J. T. MacConnell, and A. H. McCallum, Ann. Bot. N. S., 20, 501 (1956).
82. Allen, E. K., and O. N. Allen, Recent Advances in Botany, University of Toronto Press, Vol. 1, pp.585-588, 1961.
83. Parker, C. A., in Festskrift til Hans Lauritz Jensen, Gadgaard Nielsens Bogtrykkeri, Lemvig, Denmark, pp.107-116, 1968.
84. Sagan, L., J. Theor. Biol., 14, 225 (1967).
85. Raven, P. H., Science, 169, 641 (1970).
86. Almon, L., Zentbl. Bakt., 2 Abt., 87, 289 (1933).
87. Trinick, M. J., Nature (London), 224, 459 (1973).
88. Hardy, R. W. F. and R. D. Holsten in The Aquatic Environment: Microbial Transformations and Water Management Problems, L. J. Guarraia and R. K. Ballentine, Eds., U. S. Government Printing Office, Publ. No. EPA 436/G-73-008, Washington, D. C., pp.89-132, 1973.

CHAPTER 2

The Bacteria[1]

T. A. LA RUE

Prairie Regional Laboratory
National Research Council
Saskatoon, Saskatchewan
Canada

2.1. Introduction, 20
    2.1.1. History, 20
    2.1.2. Description of Dinitrogen-fixing Bacteria, 21
        2.1.2.1. Hydrogen and Hydrogenase, 26
2.2. Methods of Growing Bacteria, 27
    2.2.1. Batch Cultures, 27
    2.2.2. Continuous Cultures, 28
    2.2.3. Synchronous Cultures, 29
    2.2.4. Mixed Cultures, 29
2.3. Physiology, 31
    2.3.1. Carbon, 31
    2.3.2. Energetics, 33
    2.3.3. Combined Nitrogen, 35
        2.3.3.1. Ammonia, 35
        2.3.3.2. Nitrate and Nitrogen Oxides, 36
        2.3.3.3. Amines, 37
        2.3.3.4. Hydrazine and Diimide, 38
    2.3.4. The Gas Phase, 39
        2.3.4.1. Nitrogen, 39
        2.3.4.2. Oxygen, 41
        2.3.4.3. Hydrogen, 45
        2.3.4.4. Carbon Monoxide, 45
        2.3.4.5. Acetylene, 46
        2.3.4.6. Inert Gases, 46
    2.3.5. pH, 47

*NRCC No. 14477

    2.3.6. Temperature, 47
    2.3.7. Minerals, 47
        2.3.7.1. Molybdenum, 47
        2.3.7.2. Iron, 49
        2.3.7.3. Other Elements, 49
        2.3.7.4. Osmotic Pressure and Ionic
                  Strength, 51
2.4. References, 52

2.1 INTRODUCTION

2.1.1 History[1]

That microbes fixed dinitrogen (N_2) was suggested in 1862 when Jodin reported "microderms" in nutrient containing only carbon and minerals. Later Berthelot observed an increase in total nitrogen in soil, which he attributed, with no good evidence, to bacteria. Winogradsky first demonstrated nonsymbiotic N_2 fixation; in 1895 (6) he isolated Clostridium pastorianum (later C. pasteurianum), capable of growth in the absence of fixed nitrogen.

In 1901 Beijerinck (7) discovered the aerobic N_2-fixing bacteria: Azotobacter chroococcum, from soil, and A. agilis from water. Shortly after Lipmann (8) isolated A. vinelandii. Since then, scores of putative N_2-fixing organisms have been reported, in the most diverse groups of microbes and in the most exotic environments. However, nearly all physiological studies have been conducted on these first four organisms.

Much of the research on N_2 fixation before 1930 is of little value, for many researchers assumed that isolation and growth of a microbe in nitrogen-free nutrient eliminated the possibility of contamination with non-N_2-fixing forms. In fact, the thick slime of most diazotrophs makes purification difficult. Probably many of the pleomorphic forms assigned to Azotobacter were due to contaminating organisms.

The use of inadequate nutrient media frequently made it impossible to obtain meaningful physiological data, for growth was very slow. In 1930, Bortels (9)

---

[1]Early studies of dinitrogen fixation are documented by Wilson (1, 2), Winogradsky (3), Waksman (4), and Burk (5).

demonstrated that molybdenum was necessary for growth of Azotobacter on N_2. After this was established, adequate nutrient media could be formulated, and data obtained from short-term experiments.

Present knowledge of the physiology of N_2 fixation rests on the work of Burk, who coined the term "nitrogenase" for the protein catalyzing the conversion of N_2 to a form in which it could be incorporated into cellular nitrogen. Burk demonstrated that under controlled conditions nitrogen gain was proportional to oxygen uptake by Azotobacter. Using a micro-respirometer to estimate fixation, he characterized the effects of physical and chemical factors on N_2 fixation. He resorted to whole cells to estimate the kinetic characteristics of nitrogenase, for he was unable to obtain N_2 fixation in cell-free preparations. Not the least of his contributions was the Lineweaver-Burk formulation for enzyme kinetics. Burk's studies remain an example of how basic biochemical information may be obtained using intact cells.

2.1.2 Description of Dinitrogen-fixing Bacteria

Tables 2.1 and 2.2 list the bacteria for which there is good evidence for N_2 fixation. Acetylene reduction or $^{15}N_2$ assays have not been published for some species, but their rapid growth and efficient nitrogen gain in nitrogen-free media leave little doubt that they possess this capacity.

The number of bacterial species capable of fixing N_2 is small. Moreover not every strain of a species may have this capacity. Some of the diazotrophs are aerobic, some facultative, some anaerobic. They may be capable of heterotrophic or photosynthetic growth. Collectively, they are capable of metabolizing a wide range of carbon compounds. Some species can grow at pH 3, others at pH 10. Their habitats are the soil, the rhizoplane, the phylloplane, sand, ground waters, estuaries, seas, mud, rotting vegetable matter, and animal intestines. No thermophilic N_2-fixing bacterium has yet been found; with this exception there is no niche where at least one N_2-fixing species may not survive.

TABLE 2.1. Dinitrogen-fixing Bacteria

	Description	Synonyms	Reference for N_2 Fixing Ability
PSEUDOMODALES			
Thiorhodaceae (Chromatiaceae)	10		
Thiocapsa roseopersicina	10		202
Chromatium minutissimum	10		
C. vinosum	10		
C. strain D			12
Athiorhodaceae (Rhodospirillaceae)	10		
Rhodospirillum rubrum			13
Rhodopseudomonas gelatinosa			14
R. capsulata			14
R. spheroides			14
R. palustris			14
HYPHOMICROBIALES			
Rhodomicrobium vanieli	10		15
Chlorobacteriaceae	10		
Chlorobium thiosulfatophilum	10	"Chlorobacterium sp"	13
C. limicola			
Spirillaceae	10		
Desulfovibrio desulfuricans	17		18
D. gigas			18, 19
D. vulgaris			19
Desulfotomaculum orientis	20		18
D. ruminis	20		18

EUBACTERIALES

Azotobacteraceae
Azotobacter chroococcum
A. agilis   21, 22, 23
                                             7   Azotococcus agilis (25)
                                             7   Azomonas agilis (26, 11)

A. vinelandii   8   24
A. beijerinckii   8, 21
A. macrocytogenes   22, 27   Azomonas macrocytogenes
                                                        (26, 11)

A. insignis   22, 28   Azomonas insignis (26, 11)
                                                        Azotococcus insigne (25)

A. miscellum   29, 30
A. paspali   30, 31
Beijerinckia indica   11, 21, 32, 203   B. lacticogenes (25)
                                                                A. indicum (33)
                                                                B. mobilum (35)

B. mobilis   11, 32, 34, 203
B. fluminesis   11, 32, 24, 203   B. acida (37)
B. derxii   11, 32, 35, 36,   B. congensis (32)
                                             203

Derxia gummosa   22, 38   39

Enterobacteriaceae
Klebsiella pneumoniae   10   K. aerogenes (40)   42
                                                        Aerobacter aerogenes (41)
Klebsiella pneumoniae N4-B   43   Achromobacter sp (44)   43, 45
(anaerogenic)

K. rubiacearum   46   Mycobacterium rubiacearum   46
Escherichia intermedia   10, 47   47
Enterobacter cloacae   48, 49, 50   49, 50
E. aerogenes   47   47, 204
E. agglomerans   205   205

                                                                                            (Continued)

TABLE 2.1. (Continued)

	Description	Synonyms	Reference for N_2 Fixing Ability
Corynebacteriaceae			
Corynebacterium autotrophicum	206		207
Bacillaceae			
Bacillus polymyxa	10, 51, 52		56, 57
B. macerans	53, 54		57
Clostridium pasteurianum	53	Clostridium sp (55)	58
C. aceticum		C. pastorianum (6, 51)	58
C. acetobutyricum		Plectridium aceticum	58
C. beijerinckii			58
C. butyricum			58
C. butylicum			58
C. felsinium			58
C. kluyverii		Terminosporus kluyverii	58
C. lactoacetophilum			58
C. madisonii			58
C. pectinovorum		Plectridium pectinovorum	58
		Granulobacter pectinovorum	
C. tetanomorphum		Plectridium tetanomorphum	58

Introduction

TABLE 2.2. Dinitrogen-Fixing Bacteria of Uncertain Classification

	Description Reference	Reference for N_2-Fixing Ability
Methylosinus trichosporium[a]	59	60
Mycobacter flavum[b]	61	62
Pseudomonas azotogensis[c]	63	62
Methanobacterium omelianski[d]	64, 65	66
Enterobacter-like sp[e]	67, 204	67, 204
Bacillus circulans[f]	47	47
Methane oxidizing bacterium[g]	208	208
Chlorospeudomonas ethylicum[h]	209	16
"Curved rod"[i]	210	210

[a] Pseudomonas methanitrificans. Gram-negative rod, obligate aerobe, oxidizes methane, forms exospore.
[b] Not a mycobacterium according to Bergey's scheme. Most likely a coryneform bacterium (68).
[c] Not a pseudomonad (22). May not be a N_2-fixer (88).
[d] A symbiotic association of two bacteria (65).
[e] Associated with wood rotting fungi. Taxonomic schemes for the Enterobacteriaceae were developed largely from clinical isolates of animal or human origin. The taxonomic keys may not be applicable to isolates from plant, soil or marine environments.
[f] Tentative classification, similar to Bergey's description.
[g] The bacterium is unique among dinitrogen fixers in that it may oxidize ethylene, thereby interfering with the acetylene reduction assay.
[h] A mixed culture containing C. limicola (109).
[i] Found in gut of marine shipworm Teredora malleolus.

Apart from their ability to fix N_2, there are no common characteristics to unite these bacteria. All can produce a hydrogenase, some produce spores or cysts, most have an extracellular capsule, many store poly-β-hydroxybutric acid, all are either anaerobic or sensitive to high oxygen tension, all can grow in simple media, requiring few or no growth factors. However, none of these traits is exclusive to N_2-fixing bacteria.

It is almost certain that additional N_2-fixing bacteria will be found. Most N_2-fixing bacteria discovered to date were isolated on minimal media lacking fixed nitrogen. There may be N_2-fixing bacteria that are capable of fixing only a portion of their nitrogen, or some that require complex growth factors or an obligatory symbiosis with another bacterium. Because fixation is common in the photosynthetic bacteria, one may predict that it will often be observed as new species are isolated. Nitrogenase is probably present in the recently described Pelodictyon sp. which has a hydrogen-evolving enzyme repressed by ammonium (69). Among the heterotrophs it is unlikely that many widely distributed species remain to be discovered. However, the recently developed acetylene-ethylene assay (see IV.12) will permit screening of leaf and soil samples, and additional species of N_2-fixing bacteria living in loose association with specific plants may be discovered.

2.1.2.1 Hydrogen and Hydrogenase

Since Burris and Wilson (70) discovered that hydrogen is an inhibitor of symbiotic N_2-fixation, the study of hydrogenase and hydrogen evolution has been an intimate part of N_2-fixation research. Hydrogenase is found in all asymbiotic diazotrophs, but possession of this enzyme is not restricted to N_2-fixing bacteria. Hydrogenases from different bacteria have dissimilar physical properties (71). In Azotobacter, the enzyme is inducible, being formed only during growth on N_2. The evolution of hydrogen by hydrogenase does not require ATP, and is inhibited by carbon monoxide. The azotobacter enzyme is particulate, and its close association with succinate oxidation suggests that it is bound to a cell membrane (72). Because of its particulate nature and extreme oxygen sensitivity, hydrogenase has been difficult to purify (211).

The hydrogenase of Azotobacter shows a low rate of exchange of hydrogen with deuterium from deuterium oxide; the hydrogenases of many other bacteria catalyze such an exchange.

Nitrogenase, in the absence of other substrates, will reduce protons to hydrogen. This reaction requires ATP, is not inhibited by CO, and has not been shown to be reversible. The nitrogenase-catalyzed reduction of N_2 is inhibited by high concentrations of hydrogen. In clostridia, which contain hydrogenase and ferredoxin, hydrogen may serve as a reductant for N_2 fixation. Thus hydrogen has several possible roles: product, inhibitor, and electron donor. It may be that the in vivo function of hydrogen evolution by nitrogenase is to provide reductant for oxygen by the adjacent hydrogenase and, therefore, decrease the oxygen concentration in the vicinity of nitrogenase. Although Azotobacter nitrogenase will evolve hydrogen in the absence of other substrates, the intact bacteria does not evolve hydrogen. The most likely explanation for this phenomenon is that the nitrogenase in Azotobacter is in a nonaqueous environment where protons are unavailable.

In the absence of ammonia, or a readily assimilated nitrogen source, strains of purple non-sulfur, sulfur, and green-sulfur bacteria produce molecular hydrogen in the light from organic growth substrates or inorganic electron donors. The hydrogen-evolving system is repressed by ammonium salts and its activity reversibly inhibited by N_2. The system is probably nitrogenase which, in the presence of photochemically produced electrons and ATP, and in the absence of N_2, evolves hydrogen.

2.2 METHODS OF GROWING BACTERIA

Ideally, microbes should be studied in an environment similar to their natural one. Throughout his studies (3), Winogradsky preferred to study Azotobacter in soil cultures or on solid silica gel, augmented with small amounts of alcohols or organic acids. In contrast, current practice which requires large amounts of cells, obtained rapidly and inexpensively, has led to the growth of bacteria in agitated liquid cultures, using high concentrations of rapidly metabolized substrates.

2.2.1 Batch Cultures

It is often assumed that the cell metabolism remains constant during the exponential phase. This is not so. The metabolism of microbes depends on their environment, and during the exponential phase the medium is changing rapidly in a complex manner. With

Azotobacter, it is probable that the decrease in dissolved oxygen is particularly critical. The metabolism of the cells changes, too, in response to the environment. Though some cellular features remain relatively constant, others are transient. It is not surprising to find that the activity of an enzyme varies from batch to batch despite the use of a "defined" medium.

A few reports on changes in N_2-fixing bacteria during the growth cycle illustrate the changing nature of a batch culture. Hydrogenase activity is closely associated with N_2 fixation, yet this relationship varied with culture age in B. polymyxa (56). While N_2 fixation increased during the exponential phase, hydrogenase activity was maximum at the "midlog" stage. Change in enzyme activity during the cycle may depend on unexpected factors; $^{15}N_2$ incorporation was maximal during the midlog phase if pyruvate was the carbon substrate but was less dependent on age when mannitol was the carbon source.

A. vinelandii may have more internal membrane in late exponential growth than in the early phase (73). Bergersen (74) noted that in A. vinelandii in batch culture the ratio $C_2H_4:NH_3$ was close to the theoretical 3:1 during the early part of the growth cycle but fell sharply during the later part.

The photosynthetic N_2-fixing bacteria are slow growers, but the generation time of most other N_2-fixing bacteria is in the range of 2-8 hours; thus the entire growth cycle will be at most a few days. The study of stationary phase cultures can yield little information on metabolic processes occurring during growth. Much early work on possible intermediates in N_2 fixation is based on the identity of nitrogen compounds in old cultures. In a review of the topic, Wilson and Burris (75) were too charitable in dismissing claims based on cultures as much as 30 days old.

Batch cultures will continue to be used for their convenience and economy, but the limitations of the method should be realized.

2.2.2 Continuous Cultures

The azotobacters are favored subjects for chemostat cultures, and Hill et al., (76) have reviewed their use in continuous culture to study morphology, enzyme induction, and cytochrome production. A principal advantage of continuous fermentation is that it produces large amounts of uniform cells with consistent properties. This feature was exemplified by Munson and Burris (77), who used the method to study nitrogenase from R. rubrum. In 49 extracts of batch-grown cells, 24 extracts had no activity. When R. rubrum was grown in continuous culture, nitrogenase was found in each of 14 extracts. The cultures could be maintained for up to 30 days.

Dalton and Postgate reported chemostat cultures of A. chroococcum which were apparently limited by nitrogen, despite adequate aeration (78). The components of the medium were in excess and oxygen was present; solubility considerations made it unlikely that dissolved dinitrogen was lacking. Therefore the authors proposed that the cells were limited by their ability to fix N_2, a status they called "N_2-limited." A similar situation was found by Daesch and Mortenson (79) with C. pasteurianum. Chemostat cultures contained twice the amount of nitrogenase required to maintain growth. The inherent limitation to N_2 fixation may be the supply of energy or reductant.

2.2.3 Synchronous Cultures

Kurz and LaRue (80) synchronized A. vinelandii by adding a fresh carbon-limited nutrient to a culture at regular 3-hour intervals. After synchrony was achieved, the bacteria were dividing shortly before the addition of fresh medium. Nitrogenase was synthesized within 1 hour after the addition, and its activity remained constant throughout the cell cycle. Ammonium salts prevented nitrogenase synthesis whenever they were added, and promoted a rapid disappearance of nitrogenase activity. Nitrate or glutamine had no immediate effect on nitrogenase synthesis or activity; synthesis stopped in the cell cycle following that in which fixed nitrogen was added.

2.2.4 Mixed Cultures

The growth of a bacterium in pure culture is an unnatural phenomenon; a species in a natural environment partakes in complex interactions with many other species. For example, Chlorobacteriaceae may occur in nature growing in a symbiotic association with non-photosynthetic bacteria (81). These "consortia" are composed of a large rod-shaped motile bacterium coated with green or brown Chlorobium species. The inner bacterium confers motility on the complex. Whether the consortium is capable of fixing N_2, and the nature of the contribution of the inner bacterium to such a symbiosis, remains to be investigated.

Jensen (82) has reviewed earlier work on mixed populations containing a N_2-fixing bacterium, and we shall note a few recent studies.

Rice and Paul (83) obtained a model system by adding straw to water-logged soil. Two layers of microbial activity developed, aerobes at the surface that partly degraded the straw and released metabolites, and anaerobes, predominantly C. butylicum, which used the metabolites to support N_2-fixation. The addition of pectin to soil plates stimulated the appearance of azotobacter colonies (84). Azotobacter could not degrade pectin, but presumably grew on galacturonic acid and galactose released by pectinolytic microbes.

Dommergues and Mutaftschiev (85) grew strains of B. indica and B. fluminensis in static cultures with Lipomyces starkeyi. L. starkeyi is an acid-tolerant yeast found in tropical soils like those that harbor Beijerinckia. In mixed cultures there was a 1.4 to 8-fold increase in fixed nitrogen compared to pure cultures of Beijerinckia. The authors suggested that the yeast produced some factor that stimulated the N_2-fixing bacteria. However, the addition of concentrated filtrate from a culture of L. starkeyi had only a slight effect on fixation by axenic B. indica, so it is unlikely that the synergism observed was due to growth factors produced by yeast. Fixation by the Azotobacteraceae is inhibited by excess oxygen, and the yeast likely promoted fixation by lowering the oxygen content of the medium (see III.5.1.2.2).

Okuda and Kobayashi (86) found that R. capsulatus was able to grow in the presence of oxygen when associated with A. vinelandii. Dinitrogen fixation was 3-12 times higher than by either bacterium alone. Probably in this association the Azotobacter lowered the oxygen tension of the medium sufficiently to allow the photosynthetic organism to survive. Okuda et al. (87) similarily observed N_2 fixation even in the dark in a mixture of R. capsulatus and B. megaterium in a

Physiology

medium containing glycerol, which is not metabolized by R. capsulatus. It is likely that the synergism was mediated when the Bacillus produced pyruvate as well as lowered the oxygen tension.

Line and Loutit (88) observed that a clostridial strain from New Zealand soils could fix N_2 and support growth of aerobes in a seemingly aerobic environment. High acetylene reduction activity was noted in a mixed culture with Pseudomonas azotogensis, a putative N_2 fixer. They suggested that in cases of apparent fixation by aerobes, a check for contaminating anaerobes be made.

In summary, a non-N_2-fixing microbe can promote fixation in two ways. It may provide a substrate not otherwise available to the N_2-fixer, or it may create an environment favorable for fixation. Of course, not all associations are beneficial to the N_2-fixing organism. Chan et al. (89) have demonstrated that even low numbers of pseudomonads can prevent the growth of A. chroococcum by lowering the pH below that where the organism could grow.

2.3 PHYSIOLOGY

2.3.1 Carbon

Detailed lists of carbon sources utilized are found in some of the published descriptions of N_2-fixing bacteria (Table 2.1). The Thiorhodaceae and Athiorhodaceae may use a carbon source as electron acceptor, and organic acids (usually malic or succinic) or alcohols (ethanol) are preferred. Among the aerobic diazotrophs, the Azotobacter can grow on the widest variety of carbon compounds. They are generally able to utilize many mono- and di-saccharides, organic acids, alcohols and aromatic compounds (21, 90). A. chroococcum is unique in its ability to grow and fix N_2 on starch. The Beijerinckia and Derxia are restricted to acids and a smaller range of sugars, and simple alcohols; M. flavum can use only a few alcohols and organic acids.

Carbon sources are generally evaluated using dinitrogen as N source, but additional carbon sources may support growth under non-N_2-fixing conditions. A. chroococcum can grow on glycollic acid if provided with ammonia, but this carbon source will not support growth on N_2 or nitrate (91). Presumably that substrate can provide carbon and energy for growth, but is too oxidized to provide reductant for N_2 fixation or nitrate reduction.

Azotobacter generally oxidizes carbon substrates completely, and acid is seldom formed. A. chroococcum, A. vinelandii, and A. beijerinckii produce acid from arabinose and xylose (92); Beijerinckia growing in an alkaline medium produces acetic acid from glucose (93).

The ability to utilize benzoic acid and shikimic acid distinguishes the soil organisms A. chroococcum, A. vinelandii, and A. beijerinckii from the aquatic A. agilis. The use of appropriate carbon substrates is a convenient aid to the identification or selective isolation of Azotobacter species (94, 95).

Some of the N_2-fixing clostridia will grow on hemicellulose and pectin, but the ability of these substrates to support N_2 fixation has not been determined. No N_2-fixing bacterium is known to degrade lignin; and the unidentified N_2 fixer in marine shipworms is unique in utilizing cellulose (210).

The substrates usually used for growing bacteria in the laboratory, such as sucrose or mannitol, are unlikely to arise in soil in any quantity. The natural carbon sources of the soil N_2-fixing bacteria are more likely to be ethanol, butanol, phenols, acetate, butyrate, and others arising from the incomplete degradation of plant residues by other organisms. Generally these compounds do not support as rapid a growth as do sugars. They may also favor a different morphology of the microbe; alcohols or β-hydroxybutyrate promote the formation of cysts in Azotobacter (96), while sugars do not. It is, therefore, unwise to assume that microbes behave in soil as they do in typical laboratory nutrients.

Azotobacter, Beijerinckia, and Derxia frequently accumulate large amounts (up to 70% by weight) of poly-β-hydroxybutyrate. In batch cultures polymer synthesis starts when the dissolved oxygen concentration approaches zero. In chemostats, polymer forms during oxygen-limited steady state growth. It has been proposed (97, 98) that the polymer is not only a readily available reserve of carbon, but that its synthesis serves a regulatory purpose, providing an electron "sink" for excess reducing power. In A. vinelandii, the polymer is degraded during encystment (98).

Physiology

It was realized early that some carbon compounds were more efficient in supporting N_2 fixation than others; with aerobes, the efficiency of carbon utilization is markedly affected by the oxygen tension. Parker and Scutt (99) found that 3 times as much N_2 was fixed per unit glucose at 0.04 atm O_2 as at 0.2 atm. Similarily, Schmidt-Lorenz and Rippel-Baldes (100) reported that A. chroococcum fixed 20 mg N_2/g glucose consumed at 0.02 atm oxygen, but only 10 mg N_2/g glucose at 0.4 atm. Because few investigators defined the oxygen tension in their experiments, it is impossible to make meaningful comparisons between efficiencies reported for different compounds or in different laboratories. However, it is likely that in soil the efficiency of N_2 fixation by Azotobacter is greater than has been estimated from laboratory experiments in aerated flasks.

The efficiencies of carbon compounds for support of N_2 fixation by clostridia are frequently expressed as mg N_2 fixed/g carbon supplied. Since the fermentation does not completely metabolize the carbon substrate, such figures underestimate the efficiency of anaerobic N_2 fixation.

2.3.2 Energetics

The requirement for ATP by nitrogenase in vitro varies with the conditions and ratio of Fe protein to Mo-Fe protein. Recent experiments indicate that the requirement is not less than four ATP molecules per 2 electrons (101), i.e., 12 ATP molecules per molecule of N_2 reduced (see II.3 and II.4).

Whether energy was necessary for N_2 fixation in vivo had been a matter of contention since Burk (102) proposed that N_2 fixation occurred with the liberation of energy, and that such energy was available for use in general metabolism. As a thermodynamic model, biological N_2 fixation may be regarded as the oxidation of glucose by dinitrogen. Wilson and Burris (1, 103) examined the thermodynamics of such a reaction and concluded that N_2 fixation required energy. However, Bayliss, who assumed different concentrations of reactants, calculated that N_2 fixation liberated energy (104).

Before it was realized that nitrogenase required ATP, Senez (105) described experiments with batch cultures of Desulfovibrio desulfuricans in which the growth substrate, lactic acid, was used at the same rate whether the bacteria were grown on ammonia or dinitrogen. However, the cells grown on dinitrogen yielded less organic matter per mole of substrate utilized. Senez suggested that N_2 fixation was limiting growth and that surplus lactate metabolism led to energy being dissipated as heat. In the light of present knowledge, it is likely that some of the lactate was utilized to provide energy for N_2 fixation.

Daesch and Mortenson (79) grew C. pasteurianum in continuous culture with sucrose as carbon source. Whether dinitrogen or $(NH_4)_2SO_4$ was the nitrogen source, the ratio of the end carbon products, acetate and butyrate, was the same. Therefore the authors assumed that metabolism of sucrose, and hence the ATP production, was the same in both cases. At all growth rates, N_2-fixing cells utilized more sucrose than did cells growing on fixed nitrogen. When dinitrogen was replaced by argon, sucrose catabolism of N_2-fixing cells decreased, showing that the increased carbon utilization in part reflected an energy demand for N_2-fixation. Calculations indicated that whole cells used about 20 moles of ATP for the fixation of 1 mole of N_2 to 2 moles of NH_3. An indeterminable part of this was probably wasted by the nitrogenase mediated production of hydrogen, which utilizes 4 moles of ATP/mole H_2 in vitro

Dalton and Postgate (106) grew A. chroococcum in continuous culture with N_2 or NH_3, at different dilution rates. The efficiency of carbon utilization for N fixation increased with larger dilution rates. This they ascribed to a decrease in respiration requirement to protect nitrogenase. Extrapolating to infinite growth rates, where the need for respiratory protection would be minimal, they obtained a requirement of 4-5 moles of ATP/mole N_2 reduced.

Two assumptions were made. The first was that the carbon metabolism of Azotobacter was the same on either nitrogen source. The possibility that N_2 fixation would change the pathways of carbon utilization was not examined. Recently Kurz and LaRue (106) found that the specific activity of citric acid cycle enzymes in A. vinelandii was different in cells grown on N_2 or fixed nitrogen. The second assumption was that the estimate of ATP consumed was based on a widely accepted value of $Y_{ATP}=10.5$ (107), i.e., 1 mg of bacteria synthesized represented 10.5 mmoles ATP consumed. Nagai et al. (108) reported that Y_{ATP} for A. vinelandii was much less than 10.5, and moreover varied with the dilution rate in chemostat cultures. If Y_{ATP} for Azotobacter is smaller than 10.5 then the value of 4-5 moles ATP/mole N_2 reduced in vivo is an underestimate.

In A. vinelandii the P/O ratio has been variously reported as 1 (109), 2 (110), or 3 (111). The different results may reflect the dissimilar experimental procedures used by the investigators. The same P/O ratio was obtained in the presence of Ar or N_2 in the gas phase, or with a mutant lacking nitrogenase. Therefore the pattern of oxidative phosphorylation was independent of N_2 fixation (111). Both components of nitrogenase can be synthesized even when oxidative phosphorylation is disrupted (112).

2.3.3 Combined Nitrogen

2.3.3.1 Ammonia

Ammonia is the preferred nitrogen source for all N_2-fixing bacteria, and provides more rapid growth than any other N source. Several metabolic pathways for ammonia incorporation have been proposed (113). Azotobacter and Clostridium growing on dinitrogen or nitrate will assimilate ammonia without lag (1). The addition of ammonia to a N_2-fixing chemostat culture of C. pasteurianum immediately stops nitrogenase synthesis (79). However pre-formed nitrogenase remains stable for about three generations, and N_2-fixation and ammonia assimilation will occur together.

Using chemostat cultures of A. chroococcum, Hill et al. (76) were able to obtain stable cultures in which states between fully induced and fully repressed populations existed, showing that N_2 fixation and ammonia assimilation were occurring at the same time. Electron micrographs showed a lowered internal membrane content in each organism, implying that each member of the population was partially repressed rather than the population being a mixture of totally repressed and fully induced organisms. Repressive concentrations of ammonia are proportional to cell density, an observation that may be of relevance to the natural environment where population densities are usually low and thus may be repressed by very low concentrations of fixed nitrogen. Ammonia below 10 μ$\underline{M}$ does not repress nitrogenase synthesis in $\underline{A}$. vinelandii (212).

The repression of nitrogenase activity in $\underline{A}$. vinelandii by ammonia is accelerated by endotoxins or cyclic AMP (213). Gordon and Brill (214) found that methionine sulfoximine and methionine sulfone blocked the repression by ammonia of nitrogenase formation in $\underline{A}$. vinelandii and $\underline{K}$. pneumoniae. The results suggest that ammonia alone is not the repressor for nitrogenase synthesis.

2.3.3.2 Nitrate and Nitrogen Oxides

Nitrate. Not all N_2-fixing organisms can utilize nitrate. Nitrate does not support growth of $\underline{A}$. agilis nor most strains of Beijerinckia. $\underline{B}$. polymyxa will grow on nitrate only under aerobic conditions and on dinitrogen only under anaerobic conditions. The assimilatory nitrate reductase of $\underline{A}$. chroococcum is inducible; its formation is prevented by tungstate (215). $\underline{A}$. chroococum will grow on nitrate in preference to dinitrogen; however in $\underline{A}$. vinelandii a high proportion of strains fix dinitrogen in the presence of nitrate. Azotobacter vinelandii utilized nitrate-nitrogen with little or no lag. $\underline{A}$. indicum used nitrate very slowly and nitrite accumulated. Other strains of Azotobacter rarely accumulated nitrite (114, 115).

Nitrous Oxide (N_2O). Nitrous oxide is a substrate for nitrogenase, being reduced to dinitrogen (see II.4). Being a substrate, N_2O competes with N_2 for the enzyme and it thus acts as a competitive inhibitor (K_i=0.08 atm) of growth on dinitrogen by $\underline{A}$. vinelandii (116), $\underline{C}$. butyricum (117), and $\underline{B}$. polymyxa (55). Growth on ammonia or nitrate was not affected by N_2O (118-120).

Nitrous oxide does not act as a source of nitrogen for growth. The very slight incorporation of label from $^{15}N_2O$ into $\underline{A}$. vinelandii (121) can be explained by the reduction of $^{15}N_2$ formed by the action of nitrogenase on $^{15}N_2O$.

Nitric Oxide (NO). Nitric oxide is a competitive inhibitor of nitrogenase (see II.5). Mozen (122) found that 0.001-0.01 atm NO completely blocked growth of $\underline{C}$. pasteurianum on N_2 and ammonia. The reactivity

Physiology					37

of NO has probably discouraged a more detailed study of its effects on whole cells.

Hyponitrous Acid ($H_2N_2O_2$). Hyponitrous acid, 10^{-3}M, was an irreversible inhibitor of respiration by A. vinelandii whether or not fixed nitrogen was supplied. It did not act as a nitrogen source (123).

2.3.3.3 Amines

Urea is an excellent source of reduced nitrogen and is often used instead of ammonia because it is more convenient to prepare nutrient media without changes in pH during autoclaving. Urease in A. vinelandii is induced in the presence of urea or thiourea (124). Ammonia and glutamate blocked urease synthesis in the presence of urea, both in growing cultures and washed cell suspensions.

Very few organic nitrogenous compounds are available as nitrogen sources for Azotobacter. Of the amino acids, only aspartic acid, asparagine, and glutamic acid support growth. Aspartic acid only partially represses nitrogenase in A. vinelandii; that bacterium can make simultaneous use of asparate and dinitrogen (216). The amino group of adenine can be used, but otherwise purine and pyrimidine derivatives are not utilized, nor are oximes and amides (125). Other N_2 fixers grow on amino acids.

Azotobacter grows poorly in peptone media, and this forms the basis of a convenient check for purity (126). Most contaminants will grow readily in peptone or milk; Azotobacter will grow slowly, and frequently display pleomorphic forms, apparently due to glycine in the peptone (127). Glutamic acid is not a favorable nitrogen source for Beijerinckia; N_2 fixation may continue in the presence of this amino acid (93).

Sorger (128) reported that two nonmetabolized amines, 2-methylalanine and methylamine, repressed the synthesis of nitrogenase by A. vinelandii. However St. John and Brill (129) found that methylalanine inhibited growth only when glucose or maltose was the carbon source. Their results indicate that methylamine is not a repressor of nitrogenase; the locus of its inhibition remains to be determined.

Hydroxylamine (NH_2OH). Hydroxylamine inhibited N_2 fixation by A. vinelandii and was inhibitory to its

growth at concentrations above 2 ppm (130). Below
this concentration, slow growth was attributed to the
abiological decomposition of NH_2OH to ammonia.

Nitramid (NO_2NH_2). Labelled nitramid was tested as a
possible intermediate in N_2 fixation (131), but there
was no evidence that nitrogen from this source could
be incorporated into growing A. vinelandii.

Azide (N_3^-). Azide was a powerful but nonspecific
inhibitor of N_2 fixation by C. pasteurianum (55),
while in A. vinelandii it was a specific reversible
inhibitor (132). Azide is a substrate for nitrogenase,
being reduced to dinitrogen and ammonia (see II.4).
C. pasteurianum grown on ammonia is incapable of reducing azide (133).

Carbamyl Phosphate. Carbamyl phosphate is not an
intermediate in nitrogen fixation (134). The addition
of 1 mM carbamyl phosphate to C. pasteurianum
inhibited acetylene reduction and repressed nitrogenase formation (135). However carbamyl phosphate
had no effect on nitrogenase-catalysed hydrogen evolution, or reductant dependent ATP hydrolysis. Carbamyl
phosphate binds to nitrogenase, and the dissociation
constant ($K=5\times10^{-5}$ M) is of the same order of magnitude as the endogenous carbamyl phosphate concentration in N_2-fixing cells. Whether it acts in
regulating nitrogenase in vivo remains to be proven.

2.3.3.4 Hydrazine and Diimide

If the fixation of dinitrogen is a totally reductive
process, then it must pass through valence stages
corresponding to diimide (HN=NH) and hydrazine
(H_2N-NH_2). Diimide is a very unstable compound, and
it is impossible to test it directly. Burris and co-workers (134) tested dipotassium azodicarboxylate and
anthracene-9,10-biimine which hydrolyse to diimide and
found inhibition of N_2 in extracts of C. pasteurianum.

Hydrazine was added to extracts of C. pasteurianum
fixing $^{15}N_2$, then isolated as benzalazine. There was
no excess ^{15}N in the azine, indicating that hydrazine
does not exist as a free intermediate in fixation
(134). The failure to observe reduction of added
hydrazine may be because hydrazine is protonated at
neutral pH, and may not bind to nitrogenase. Hydrazine

Physiology 39

is an inhibitor of N_2-fixation by A. vinelandii with growth as well as N_2 fixation being stopped until the organism converts hydrazine to acetyl hydrazine (136).

2.3.4 The Gas Phase

2.3.4.1 Nitrogen

Burk studied the relationship of $K_m(N_2)$ to N_2 fixation using his microrespirometric technique. Unaware that H_2 was an inhibitor he used that gas as a diluent, and his results were thus a mixture due to the effects of N_2 and H_2. The earliest correct assessment of the $K_m(N_2)$ was by Wyss et al. (137) who found a value of 0.01 atm for Azotobacter and saturation at 0.2 atm N_2. Higher values have been determined for other N_2-fixing bacteria (Table 2.3).

TABLE 2.3. Apparent Michaelis Constants of N_2 for N_2-Fixation

Bacterium	$K_M(N_2)$ (atm)	Reference
Azotobacter vinelandii	0.01	137
Rhodospirillum rubrum	0.07	77
Bacillus polymyxa	0.03	138
Clostridium pasteurianum	0.03	139
Klebsiella pneumoniae	0.10	140

The apparent Michaelis constant for N_2 with intact bacteria is less than has been observed with active cell-free preparations of nitrogenase. It may be that cells can concentrate dinitrogen at the active site so that there is a higher concentration near the enzyme than in the surrounding medium. Increased pressures of hydrogen elevate the apparent K_m of N_2, thus K_m values may be abnormally high where substantial hydrogen is evolved (141).

Low $K_m(N_2)$ values (Table 2.3) indicate that in a fermentor adequately purged with air or N_2, dinitrogen should never be a factor limiting N_2 fixation. This may not be the case, however, in a static culture.

Rice and Paul (83) using water-logged soil straw mixtures (III.2.3.4) found that N_2-fixation measured with $^{15}N_2$ decreased relative to nitrogenase activity measured by C_2H_2 reduction as the sample depth increased from 0.2 to 3.0 cm. Their conclusion was that N_2, which is poorly soluble in water, was limiting, while acetylene, which is more soluble, was not. Because of the cost of isotopic N_2 many researchers have used this gas at low partial pressures. In such cases N_2 may have been limiting in the experiments (see IV.12).

Zacharias (142) reported that growth of four species of Azotobacter in N-free medium was augmented if N_2 passed into the culture was first irradiated with ultraviolet light or ^{90}Sr. A reduced consumption of glucose/mg N fixed was associated with the presence of positive ions in the gas, but this effect could not be duplicated (106).

When a non-fixing culture depletes its supply of fixed nitrogen, there is a lag before dinitrogen utilization starts. The diauxic lag has been observed in A. vinelandii (143), K. pneumoniae (144), and C. pasteurianum (145). The lag can be eliminated if the cells are provided with amino acids, which presumably serve in the synthesis of nitrogenase.

Researchers have attempted to determine if N_2 is necessary for the induction of nitrogenase by measuring its formation in "nitrogen (N_2)-free" gas. The chemical inertness of dinitrogen precludes its easy removal from other gases. Strandberg and Wilson (143) studied A. vinelandii using gases with very low N_2 contamination, and concluded that if N_2 was an inducer for this organism, 0.0001 atm would suffice. Similarily, C. pasteurianum forms nitrogenase when the N_2 concentration is less than 30,000 molecules/cell (145). Under even more rigorous conditions, Parejko and Wilson (146) found that K. pneumoniae would form nitrogenase when the N_2 concentration was less than 100 molecules/microbial cell. It is, therefore, almost certain that nitrogenase is not substrate induced (see III.13.9).

In C. pasteurianum a partial repression occurred when a gas phase containing a small concentration of N_2 was replaced by N_2 (79). Under these conditions the cell probably produces a sudden excess of ammonia, which in turn stops synthesis of nitrogenase.

2.3.4.2 Oxygen

Nitrogenase in vitro is extremely sensitive to oxygen, being irreversibly denatured by it (see II.2 and II.3). N_2-fixation will occur then only when anaerobic conditions exist — either in the external environment, or via intracellular physiological or structural protective mechanisms.

To group N_2-fixing bacteria according to their oxygen sensitivity as aerobes or anaerobes and facultative bacteria is an oversimplification. There is almost a continuum of tolerance to O_2. N_2 fixation by the sulfate-reducing bacteria may be the most oxygen sensitive. Early attempts to demonstrate fixation may have been unsuccessful due to the presence of traces of oxygen. In the first demonstration of N_2 fixation (19) by Desulfovibrio vulgaris a medium containing 20-25 g N/ml in yeast extract was used. The added combined nitrogen apparently permitted enough growth to reduce the environment sufficiently for N_2 fixation to take place. N_2 fixation by the sulfate reducers was confirmed (18) in an N-free medium using Pankhurst tubes (147) in which traces of oxygen were absorbed by alkaline pyrogallol.

Such precautions are unnecessary with the clostridia, which can tolerate the slight leakage of air into typical laboratory apparatus. Oxygen may somehow repress nitrogenase synthesis in K. pneumoniae (219). Though most strains of Klebsiella only fix in anaerobic conditions, and are rapidly inactivated by air (74), a few have been isolated which fix in the presence of 0.03 atm O_2 (148).

Fixation by R. rubrum continued at 55% of the anaerobic level when the gas phase contained 0.018 atm O_2. Inhibition was complete with 0.04 atm O_2, but could be completely reversed by making the system anaerobic again (149).

Among the aerobes, Mycobacterium flavum (150) and D. gummosa (151) grow best at low pO_2 values, and are more oxygen sensitive than most azotobacters. A. paspali fixes N_2 optimally at only 0.04 atm O_2 while in association with Paspallum notatum (152). Beijerinckia indica reduces acetylene optimally at a pO_2 of 0.15 (220).

Oxygen concentrations above 0.3 atm are inhibitory to growth of A. vinelandii utilizing either NH_3 or N_2 (1). At lower concentrations O_2 acts as a competitive inhibitor of N_2, and Parker and Scutt (99) have suggested that O_2 and N_2 could be regarded as alternate respiratory acceptors (see III.1). If the reduction of dinitrogen were energy yielding, this would be a reasonable hypothesis. However the energy requirement for N_2 fixation makes this hypothesis unlikely.

The bacteria that fix N_2 aerobically obviously have some method for protecting nitrogenase from oxygen. Recent work indicates that two mechanisms are operating: (a) a respiratory protection in which the microbe increases its respiration if necessary to reduce the internal pO_2, and (b) a conformational protection (78).

Respiratory Protection. The respiratory activity of the Azotobacteraceae is among the highest recorded for microbes. The oxygen demand of A. vinelandii was studied in batch cultures by Phillips and Johnson (153). They noted that cells grown on ammonia or dinitrogen had an increased oxygen demand in the presence of excess oxygen, and proposed that an "oxygen-wasting system" might serve to reduce the intracellular oxidation potential. If respiration serves as a protective mechanism for nitrogenase, then the efficiency of N_2 fixation should decrease with increasing oxygen tensions, since more carbon would be diverted to respiration, and the ratio of N_2 fixed to C metabolized would be less.

A study of this was impracticable with batch cultures, for as the culture grows the dissolved oxygen tension decreases. Nagai et al. (154) grew chemostat cultures of N_2-fixing, glucose-limited A. vinelandii under different dissolved oxygen tensions. Glyceraldehyde-3-phosphate dehydrogenase, isocitric dehydrogenase, or isocitric lyase activities were the same under different oxygen tensions, but aldolase activity was proportional to the dissolved oxygen concentration. This would explain the increase in respiration and decrease in efficiency of fixation with higher pO_2, for an increase in aldolase directs more glucose to the pentose cycle, evolving more CO_2, and thus decreasing carbon available for the tricarboxylic acid cycle.

Oxygen tension may act at other sites. Increasing the pO_2 increases the cytochrome a content of A. chroococcum (155); at a pO_2 above 0.6 atm, the pyruvic

oxidase of A. vinelandii is apparently inactivated
(156). Oxygen may influence nitrogenase indirectly
by its effect on the ATP/ADP ratio or the ratio of
reduced to oxidized dinucleotides.

Conformational Protection. A. vinelandii cells grown
on dinitrogen contain a vast internal membrane network.
Cells grown on ammonia or amino acids contain smaller
amounts of membrane concentrated primarily near the
cell periphery, while nitrate-grown cells completely
lack the internal membrane (157). However, the oxygen
tension may be as important in controlling membrane
formation as the nitrogen source (73). Antibody
binding indicated that nitrogenase is bound to the
membranes in A. vinelandii (221), and nitrogenase containing vesicular membranes ("azotrophores") may be
isolated (222).

When A. vinelandii was disrupted in a French pressure cell the crude nitrogenase activity could be
sedimented by centrifugation and was much less sensitive to oxygen than nitrogenase extracted from C.
pasteurianum. When Azotobacter vinelandii was lysed
by osmotic breakage the nitrogenase so isolated was
much more sensitive to oxygen, and remained soluble
even when subjected to ultracentrifugation at
180,000 g for 3 hours (158).

It is, therefore, likely that nitrogenase is
associated with a membrane, but not tightly bound to
it. The internal membrane of Azotobacter, which
remains predominately within the osmotically lysed
cell, probably serves to protect nitrogenase and other
oxygen-sensitive enzymes from oxygen. However the
nitrogenase of Chromatium strain D is also apparently
associated with the membrane (223). Because this
organism is an anaerobe, the association of nitrogenase
with membrane may serve a function other than protection from oxygen.

When the aeration of batch-grown A. chroococcum
cultures was increased by shaking the flasks more
vigorously, there was a marked drop in nitrogenase
activity, as measured by acetylene reduction (159).
If the rapid shaking was not prolonged, a return to
normal agitation led to a return of normal nitrogenase
activity. A similar phenomenon may be demonstrated in
continuous cultures (78, 159). When A. chroococcum
was treated with EDTA and lysozyme, oxygen inhibition
was reversed without lag by reducing the oxygen

solution rate. This recovery by damaged cells indicates that the restoration of nitrogenase activity is not due to the synthesis of new enzyme.

Postgate and his co-workers have explained this "switch off" and "switch on" process by proposing a conformational change in the enzyme. When respiratory protection is inadequate to maintain a low intracellular oxygen tension, a change in the enzyme structure occurs that protects it from damage by oxygen. However, in its protective conformation the enzyme is unable to reduce acetylene or dinitrogen. Nucleotides are possibly involved; ATP increased the oxygen sensitivity of particulate nitrogenase. The internal membrane may provide a locus for this operation. It may be that a structural reorientation places the sensitive site in a phospholipid environment. Oxygen is less soluble in lipids than water, and N_2-fixing A. chroococcum contains more phospholipid than cells grown on ammonia. (160). The protective site for nitrogenase in A. vinelandii is sensitive to azide and 2,4-dinitrophenol (161).

The nitrogenase of M. flavum was particulate (150) but was as sensitive to inactivation by oxygen as extracts from C. pasteurianum or B. polymyxa. M. flavum is not capable of such vigorous respiratory activity as Azotobacter. The mechanisms for protecting nitrogenase from oxygen are obviously more primitive and this provides an explanation of why whole cells fix optimally at a pO_2 of only 0.05 atm.

Most Azotobacter produce copious amounts of slime, and it has been suggested (162) that the gummy capsule protects nitrogenase by slowing diffusion of oxygen to the bacterial cell. This would explain the results of Barooah and Sen (163), who found a positive correlation between slime production and N_2 fixation in sixteen strains of Beijerinckia indica growing on sucrose-agar. The thick capsular material of Azotobacter is closely associated with the cysts (164) and may protect these resting stages while they lack the power to lower the oxygen tension by respiration.

The argument for a protective role by slime is weakened by the observation that the anaerobe C. pasteurianum also produces a gummy capsule. The aquatic Azotobacter which lack a capsule and the non-gummy strains A. vinelandii OP (165), K. pneumoniae (A. aerogenes) M5AL (140) and A. beijerinckia O (97) fix as efficiently as other strains. Moreover, slime is produced by the N_2-fixing algae Anabaena, Plectonema, and Gloeocapsa. In these organisms, nitrogenase must be protected from O_2 generated within the organism. If the slime slowed O_2 diffusion, it would cause an increase in intracellular O_2 levels.

Not all polymers slow gas diffusion equally. It should be simple to prepare sufficient slime material to determine if those produced by the N_2-fixing microbes do indeed slow O_2 diffusion. The slime may simply be a reflection of the ready availability of carbohydrates in the common test media; the production and role of capsule would best be studied with cells grown on low concentrations of "natural" substrates such as organic acids and alcohols.

2.3.4.3 Hydrogen

Hydrogen acts as a weak competitive inhibitor of N_2 fixation by intact cells (see also II.5), and has no significant effect on cells growing on fixed nitrogen. In general, higher concentrations of hydrogen are required to suppress fixation in anaerobic bacteria which themselves produce H_2. For Azotobacter the $K_i(H_2)$ is approximately 0.08 atm; i.e., it is bound about one-eighth as well as N_2 (137). B. polymyxa can fix dinitrogen in 0.5 atm H_2 (56). The rate of N_2 fixation by C. pasteurianum was unaffected by 0.6 atm H_2, but the total N_2 fixed was reduced 10-20% (165). Westlake and Wilson (139) found that the $K_i(H_2)$ for C. pasteurianum was 0.5 atm.

Hydrogen inhibits fixation by K. pneumoniae ("Achromobacter sp.") (166) and R. rubrum (149), but 40% H_2 had no effect on fixation by D. desulfuricans (167). Hydrogen does not inhibit the formation of nitrogenase by A. vinelandii (143), even at a concentration of 0.8 atm which would completely inhibit N_2 fixation. Parker and Dilworth (168) have suggested that H_2 inhibition of N_2 fixation by A. vinelandii was due to a more prolonged lag resulting from a low pH_2, and that when cells were adapted to H_2, inhibition was not competitive.

2.3.4.4 Carbon Monoxide

Carbon monoxide is isoelectronic with N_2 and has the same molecular weight and similar physical properties. It is a potent inhibitor of nitrogenase (169) (see

II.5). Low concentrations of CO prevent the growth of N_2-fixing bacteria. Much higher levels are necessary to block growth on fixed nitrogen. Dinitrogen fixation by Azotobacter is inhibited by 0.002-0.006 atm CO; Clostridium is 80% inhibited by 0.003 atm and B. polymyxa, 50% inhibited by 0.03-0.07 atm.

The K_i for carbon monoxide inhibition of acetylene reduction by purified nitrogenase is 0.8 to 3×10^{-4}. The inhibition of acetylene reduction by carbon monoxide should be a confirmatory test to document the presence of nitrogenase, especially in novel putative N_2-fixing systems. Several anaerobic N_2-fixing bacteria can oxidize CO to CO_2 (224).

2.3.4.5 Acetylene

Acetylene is a substrate for nitrogenase, and inhibits N_2 fixation (170). If the concentration of C_2H_2 is sufficiently high, N_2 reduction will be slight, and the bacterium will be deprived of its nitrogen supply. Therefore C_2H_2 reduction assays conducted for long periods of time are being done on starved cells. Moreover, C_2H_2 may have other toxic effects; Clostridium exposed to C_2H_2 for long periods of time undergoes pleomorphic changes (171).

If the concentration of C_2H_2 is kept low (<0.001 atm), the inhibition of N_2 fixation becomes negligible. LaRue and Kurz (172) described a procedure for continually adding low levels of C_2H_2 to the gas aerating a fermentor. Assays of the effluent gas for C_2H_2 and C_2H_4 provided an estimate of nitrogenase. The growth rate, N increase, and morphology of A. vinelandii and C. pasteurianum were not effected by prolonged treatments with low levels of C_2H_2.

2.3.4.6 Inert Gases

Helium, ethane, neon, and argon have no effect on respiration or N_2 fixation by A. vinelandii (173), and may be used as inert diluents. Low levels of CO_2 are not inhibitory to most diazotrophs studied. Alkaline absorption of CO_2 from the gas phase over C. pasteurianum stimulated N_2 fixation, probably by minimizing the drop in pH (58). However, fixation by K. pneumoniae N4-B was increased by the reassimilation of CO_2 (43).

2.3.5 pH

Most bacteria fix dinitrogen optimally at a pH near neutrality. Though Azotobacter can grow on fixed nitrogen at pH 5.0 (93), it cannot fix dinitrogen below pH 6.0. C. pasteurianum will fix N_2 in the pH range 5.5-8.0. Beijerinckia can fix dinitrogen at pH 3.0-10.0; D. gummosa, over the range 5.0-9.0, and M. flavum at 4.0.

Beijerinckia is capable of fixing dinitrogen at alkaline pH (93), with inoculants into a medium with an initial pH of 10 able to grow and fix N_2. During growth acetic acid is produced, and the pH drops to as low as 6 (114). In contrast, Beijerinckia grown in an acid environment on glucose raises the pH. The alkaline substance(s) responsible has not yet been identified. Growth of photosynthetic bacteria has usually been found to be best at neutral or slightly alkaline reactions.

2.3.6 Temperature

As a group, the Azotobacter are mesophilic, and optimum growth is at approximately 30°C. Azotobacter from tropical soils may fix at 40°C, while A. beijerinckia may grow, though slowly, at 6°C.

The temperature growth range of Beijerinckia is more limited than that of Azotobacter. Both genera grow at 16°, Beijerinckia does not grow at or above 37°C whereas all Azotobacter strains tested grew up to 40 or 45° (93). In the Beijerinckia strains, storage at 3-4 months at 4°C did not affect the viability of the cells. After 7 or 8 months there was some decrease and after 12 months of storage most suspensions of slow growing Beijerinckia strains died.

2.3.7 Minerals

2.3.7.1 Molybdenum, Vanadium, and Tungsten

The observation that nitrogenase contains molybdenum came as no surprise, since Bortel found that molybdenum activated growth on N_2 by A. chroococcum (9). In every bacteria tested an increased need for that element during fixation has been found (174).

Molybdenum is also required for nitrate reduction, and activates the hydrogenase of C. pasteurianum (175). For most bacteria, the requirement is slight; concentrations of 0.1 ppm are adequate for near-optimal growth. Even if molybdate is omitted, the presence of Mo in other components of the medium, e.g., glucose (93) may satisfy the requirements. It is difficult to eliminate Mo entirely, and traces of Mo-containing nitrogenase have been isolated from Azotobacter vinelandii grown on "molybdenum-free" media (176). A vinelandii incorporates Mo in excess of that required for immediate growth (177), and several transfers through Mo-deficient medium may be necessary to demonstrate a requirement. The demand for Mo is generally greater for more rapidly growing species. The requirement will vary with physiological conditions. For example, the molybdenum requirement of B. indicum is higher at pH 3.7 than at 6.7. At both pH values N_2 fixation requires more molybdenum than does nitrate assimilation (93).

Dinitrogen-fixation and hydrogenase activity in C. pasteurianum are inhibited by trichloromethylsulfonyl benzoate, but the effect is reversed by sodium molybdate (178). That inhibitor, or a more stable analog, should prove useful in studying molybdenum metabolism.

Vanadium (V) may be able to replace Mo in N_2 fixation by A. vinelandii, A. chroococcum and M. flavum. Burns et al. (176) purified a V-containing nitrogenase and claimed it to be active. However, Benemann et al. (179) suggested that the activity was due to traces of Mo-nitrogenase, and that V had only a sparing effect, permitting Azotobacter to form effective nitrogenase from low levels of Mo in the medium.

Tungstate (W) inhibits the uptake of molybdate by A. vinelandii growing on N_2 (180), but does not compete with the metabolically functional Mo within the cells. Vanadate does not overcome this inhibitory action. Tungstate also inhibits growth of A. chroococcum on nitrate, but not on nitrite or ammonia (215).

Benemann et al. (225) reported the isolation of a W-nitrogenase from A. vinelandii. It was not clear if the slight activity observed was due to W-nitrogenase or the traces of normal Mo-nitrogenase present. Paschinger (226) concluded that W-nitrogenase in intact R. rubrum was active, but that there was an increased ATP requirement for N_2 fixation.

2.3.7.2 Iron

Iron is required for growth of all N_2-fixing bacteria. It is required for nitrogenase and ferredoxin, and is used in larger amounts when the bacteria is fixing N_2 than when ammonia is provided (174, 181). The requirement is large for a micronutrient (generally 2-10 ppm is used) and the availability of iron is often a problem in formulating nutrient media. Citrate or EDTA is usually added to maintain Fe in solution. The bacteria vary in their requirements and sensitivity for Fe. B. indica, from lateritic soils, will fix N_2 in a medium of 200 ppm Fe; fixation by A. vinelandii is severely impaired by such levels (93).

If the iron concentration is low, C. pasteurianum produces the non-metal electron carrier flavodoxin in place of ferredoxin (182). In iron-deficient media, A. vinelandii growing on nitrate or N_2 produces increased amounts of its characteristic yellow pigment. The pigment is composed of a peptide containing the rare amino acids, homoserine and β-hydroxyaspartic acid, and a fluorescent chromophore (183). In addition, the bacterium excretes large amounts of 2,3-dihydroxybenzoic acid and 2-N6-N-di-(2,3-dihydroxybenzoyl)-L-lysine (184). The relationship of these compounds to iron deficiency remains to be explained.

2.3.7.3 Other Elements

Cobalt. Optimum growth of C. pasteurianum occurs with 0.1-2 mg Co/l culture solution (185). A. vinelandii requires as little as 0.001 mg/l Co (186), and 1 mg/l is inhibitory to growth. Despite the most rigorous precautions it is difficult to remove such small amounts completely from the culture medium or vessels, and "0 Cobalt" media can support a growth rate of both species at over 50% of the optimum.

The Co requirement of A. vinelandii is higher when cells are grown on nitrate or N_2 than on ammonia or glutamate. There is no evidence that Co is directly involved in N_2 or nitrate reduction, and the low levels of B_{12} (25×10^{-9} g/g dry wt Azotobacter) make it doubtful that this cofactor functions directly in N fixation. The difference may be caused by an indirect effect, such as the capacity of a reduced nitrogen compound to leach Co from the culture vessel.

Calcium. Calcium is required for growth by A. vinelandii, A. chroococcum, and A. agilis, but not by B. indica (187, 203). In a detailed study Jakobsons et al. (188) demonstrated that A. vinelandii has a requirement for Ca whether growing on fixed nitrogen or dinitrogen. The requirement varied with the carbon source. If sucrose, mannitol, or ethanol was used as C source, somewhat more Ca (0.23 vs. 0.1 ppm) was required for 80% of maximum growth on N_2 than on urea. An excess of calcium in nutrient media may inhibit N_2 fixation or growth by precipitating trace elements, especially iron, during autoclaving (93, 187).

For A. vinelandii, Ca is necessary for the coordination of coat components into a rigid cyst resistant to dessication (189).

Phosphate. K. pneumoniae does not require more P when fixing N_2 than when grown on ammonia (227). The Azotobacter need at least 20 ppm phosphate for growth. The Beijerinckia can grow on the trace amounts present as contaminants in other constituents of the medium. Phosphate may protect against oxygen toxicity, since phosphate-limited cultures of A. chroococcum were much more sensitive to oxygen. Smith and Wyss (190) showed that A. vinelandii was susceptible to dilution in water and that phosphate was protective. Beijerinckia is tolerant to elevated phosphate levels, and one strain can grow and fix N_2 in 2% potassium phosphate (114), while Azotobacter is inhibited by one-tenth that amount, and the commonly used media are relatively low in P.

Glucose-1-phosphate, fructose-1,6-diphosphate, and β-glycerophosphate can serve as P source for A. chroococcum (191). Arsenate was a powerful inhibitor of fixation by washed cell suspensions of B. polymyxa (56), presumably by replacing phosphate in metabolic pathways.

Potassium phosphite (K_2HPO_3, $5 \times 10^{-3}M$) almost completely inhibited the growth of A. vinelandii on N_2, but did not alter the growth rate of cells in ammonia media (192). The inhibition may be related to a decrease in oxygen utilization.

Other Elements. Sodium stimulates short term ^{15}N incorporation into A. vinelandii (193). The effect is unexplained, and Na is required for growth on N_2 or ammonia. Sulfate, potassium, and magnesium are

Physiology

required for growth, but the demand is not generally increased by growth on dinitrogen. Potassium is intimately associated with metabolite transport in A. vinelandii (194). K. pneumoniae required 1.5 ppm Mg for optimum growth on N_2, while 1.0 ppm was adequate for growth on nitrate, ammonia, or glutamate (195). The magnesium requirement of Azotobacter varies with the phosphate source used in the nutrient (196). Copper was toxic to Azotobacter; iodine protected against copper toxicity. The sensitivity of a species to toxic levels of aluminum and titanium may depend on the natural origin of the strain (93). A. chroococcum can grow without borate, but both growth and fixation were better with 10 ppm B (197).

2.3.7.4 Osmotic Pressure and Ionic Strength

The addition of non-penetrating solutes to gram-negative bacteria usually results in the contraction of the cytoplasmic membrane away from the relatively rigid cell wall. In gram-positive organisms usually the whole cell contracts. The gram-negative A. vinelandii behaves like the gram-positive bacteria in that it does not plasmolyse (198), though the membrane may partially contract from the cell wall at one pole. Sucrose, and chloride salts of potassium, sodium, and magnesium at the same osmotic strength produce a similar decrease in packed cell volume and an increase in turbidity. Glycerol, which penetrates the cell, changes neither. There was little effect on respiration until the solute concentration exceeded 0.3 osM, above this it was increasingly inhibited.

Some strains of Azotobacter can tolerate high salt concentrations. Russian high-salinity soils have yielded A. chroococcum and A. galophilium [A. halophilum (?)] which will grow in remarkably high concentrations of $CaCl_2$ (2%), NaCl (3.5%), $MgCl_2$ (3.5%), and $MgSO_4$ (12%) (199).

The salt tolerance of bacteria will depend on the environment from which they are isolated. Fustec-Mathon et al. (200) isolated Azotobacter from seaside sands and muds. Strains from sand dunes would not grow in solutions containing more than 0.1% NaCl. Strains isolated from the upper beach were salt tolerant; they were able to grow at NaCl concentrations as high as 0.3%. Strains isolated from the lower beach, and closely influenced by seawater, were halophilic.

They grew optimally at concentrations from 0.2-0.4% NaCl, could tolerate 0.8%, and grew only slowly in concentrations below 0.2%. Similarly E. aerogenes from marine environments is salt tolerant (204).

Failure to recognize that many bacteria require an optimum ionic or osmotic strength for survival may lead to loss of viability if cells are transferred from one medium to another. Biologists frequently use either distilled water or "physiological saline" (0.9% NaCl) as diluents in counting bacteria. In both instances, Azotobacter suffers a dilution shock. This is partly due to a dilution to a concentration at which the bacteria cannot maintain the oxygen tension of the environment low enough to fix N_2 (201). In addition there is a loss of viability associated with the leaching of phospholipid from cells into the aqueous solution (190). Valid plate counts are possible only if the salts of the diluent mixture resemble those of the growth medium.

2.4 REFERENCES

1. Wilson, P. W., in Encyclopedia of Plant Physiology, W. Ruhland, Ed., Springer Verlag, Berlin, Vol. III, 9, 1958.
2. Wilson, P. W., in The Chemistry and Biochemistry of Nitrogen Fixation, J. R. Postgate, Ed., Plenum Press, London, 1, 1971.
3. Wingogradsky, S., Microbiologie du Sol: Problems et Methodes, Masson et Cie, Paris, 1949.
4. Waksman, S. A., Soil Microbiolgy, John Wiley and Sons, New York, 1952.
5. Burk, D., Ergeb Enzymforsch, 3, 23 (1934).
6. Winogradsky, S., Arch. Sci. Biol. St. Petersburg, 3, 297 (1895).
7. Beijerinck, M. W., Cent. f Bakt., II Abt., 7, 561 (1901); ibid, 9, 3 (1902).
8. Lipmann, J. G., N. J. St. Agr. Exp. Sta. Ann. Rep., 25, 237 (1904).
9. Bortels, H., Arch. Microbiol, 1, 333 (1930).
10. Bergey's Manual of Determinative Bacteriology, 7th ed., R. S. Breed, E. G. D. Murray, N. R. Smith, Eds., Williams and Wilkins, Baltimore, 1957.
11. Bergey's Manual of Determinative Bacteriology, 8th ed., R. E. Buchanan and N. E. Gibbons, Eds., Williams and Wilkins, Baltimore, 1974.

References

12. Winter, W. C., and D. I. Arnon, **Biochim. Biophys. Acta**, <u>197</u>, 170 (1970).
13. Lindstrom, E. S., R. H. Burris, and P. W. Wilson, **J. Bacteriol.**, <u>58</u>, 313 (1949).
14. Lindstrom, E. S., S. M. Lewis, and M. J. Pinsky, **J. Bacteriol.**, <u>61</u>, 481 (1951).
15. Lindstrom, E. S., S. R. Tove, and P. W. Wilson, **Science**, <u>112</u>, 197 (1950).
16. Evans, M. C. W. and R. V. Smith, **J. Gen. Microbiol.**, <u>65</u>, 95 (1971).
17. LeGall, J., J. C. Senez, and F. Pichinoty, **Ann. Inst. Pasteur.**, <u>96</u>, 223 (1959).
18. Postgate, J. R., **J. Gen. Microbiol.**, <u>63</u>, 137 (1970).
19. Reiderer-Henderson, M. A., and P. W. Wilson, **J. Gen. Microbiol.**, <u>61</u> 27 (1970).
20. Campbell, L. L., and J. R. Postgate, **Bacteriol. Revs.**, <u>29</u>, 359 (1965).
21. Jensen, H. L., **Bacteriol. Revs.** <u>18</u>, 165 (1954).
22. LeLey, J. and I. W. Park, **Antonie van Leeuwenhoek**, <u>32</u>, 6 (1966).
23. Norris, J. R., and H. M. Chapman, in **Identification Methods for Microbiologists**, Part B, B. M. Gibbs and D. A. Shapton, Eds., Academic Press, London, 19, 1968.
24. Burris, R. H., and C. E. Miller, **Science**, <u>93</u>, 114 (1941).
25. Tchan, Y. T., **Proc. Linn. Soc. N. S. Wales**, <u>78</u>, 85 (1953).
26. Baillie, A., W. Hodgkiss, and J. R. Norris, **J. Appl. Bacteriol.**, <u>25</u>, 116 (1962).
27. Jensen, H. L., **Acta Ag. Scand.**, <u>2</u>, 280 (1955).
28. Derx, H. G., **Proc. Acad. Sci. Amst.**, <u>54</u>, 342 (1951).
29. Pshenin, L. N., **Mikrobiologiya**, <u>33</u>, 615 (1964).
30. DeLey, J., **Antonie van Leeuwenhoek**, 34, 66 (1968).
31. Dobereiner, J., **Pesquisa Agropecuaria Brasileira**, 1, 357 (1966).
32. Hilger, F., Ann. **Institut Pasteur**, <u>109</u>, 406 (1965).
33. Starkey, R. L., and P. K. De, **Soil Sci.**, <u>47</u>, 329 (1939).
34. Dobereiner, J., and A. P. Ruschel, **Prev. Bras Port Biol.**, <u>1</u>, 261 (1958).
35. Derx, H. G., **Ann. Bogor.**, <u>1</u>, 1 (1950).
36. Tchan, Y. T., **Proc. Linn. Soc. N. S. Wales**, <u>82</u>, 314 (1957).

37. Roy, A. B., *Nature*, *182*, 120 (1958).
38. Jensen, H. L., E. J. Petersen, P. K. De, and R. Bhattacharya, *Arch. Microbiol.*, *36*, 182 (1960).
39. Hill, S., and J. R. Postgate, *J. Gen. Microbiol.*, *58*, 277 (1969).
40. Bergersen, F. J., and E. H. Hipsley, *J. Gen. Microbiol.*, *60*, 61 (1970).
41. Jensen, V., *Physiol. Plant.*, *9*, 130 (1956).
42. Mahl, M. C., P. W. Wilson, M. A. Fife, and W. H. Ewing, *J. Bacteriol.*, *89*, 1482 (1965).
43. Hamilton, I. R., R. H. Burris, and P. W. Wilson, *Proc. Nat'l. Acad. Sci. (US)*, *52*, 637 (1964).
44. Jensen, V., *Arch. Microbiol.*, *29*, 348 (1958).
45. Proctor, M. H., and P. W. Wilson, *Arch. Microbiol.*, *32*, 254 (1959).
46. Centifanto, Y. M., and W. S. Silver, *J. Bacteriol.*, *88*, 776 (1964).
47. Line, M. A., and M. W. Loutit, *J. Gen. Microbiol.*, *66*, 309 (1971).
48. Skerman, V. B. D., *A Guide to the Identification of the Genera of Bacteria*, 2nd ed., Williams and Wilkins, Baltimore, Md., 1967.
49. Bergersen, F. J., and E. H. Hipsley, *J. Gen. Microbiol.*, *60*, 61 (1970).
50. Raju, P. N., H. J. Evans, and R. J. Seidler, *Proc. Nat'l. Acad. Sci. (US)*, *69*, 3478 (1972).
51. Prevot, A., and V. Fredette (trans.), *Manual for the Classification and Determination of the Anaerobic Bacteria*, Lea and Febiger, Philadelphia 1965.
52. Wolf, I., and A. N. Barker, in *Identification Methods for Microbiologists, Part B*, Academic Press, London, 93, 1968.
53. Hino, S., and P. W. Wilson, *J. Bacteriol.*, *75*, 403 (1958).
54. Hino, S., *J. Biochem. (Tokyo)*, *47*, 482 (1960).
55. Hino, S., *J. Biochem. (Tokyo)*, *42*, 775 (1955).
56. Grau, F. H., and P. W. Wilson, *J. Bacteriol.*, *83*, 490 (1962), ibid., *85*, 446 (1963).
57. Witz, D. F., R. W. Detroy, and P. W. Wilson, *Arch. Mikrobiol.*, *55*, 369 (1967).
58. Rosenblum, E. D., and P. W. Wilson, *J. Bacteriol.*, *57*, 413 (1949); ibid., *59*, 83 (1950).
59. Whittenburg, R., K. C. Phillips, and J. F. Wilkinson, *J. Gen. Microbiol.*, *61*, 205 (1970).
60. Coty, V. F., *Biotechnol. Bioengineer.*, *9*, 25 (1967).

References

61. Fedorov, M. V., and T. A. Kalininskaya, Mikrobiologiya, 30, 9 (1961).
62. Hill, S., and J. R. Postgate, J. Gen. Microbiol., 58, 277 (1969).
63. Voets, J. P., and J. Debacker, Naturwiss., 43, 40 (1956).
64. Barker, H. A., Antonie van Leeuwenhoek, 6, 201 (1940).
65. Bryant, M. P., E. A. Wolin, and R. S. Wolfe, Arch. Mikrobiol., 59, 20 (1967).
66. Pine, M. J., and H. A. Barker, J. Bacteriol., 68, 589 (1959).
67. Seidler, R. J., P. E. Aho, P. N. Raju, and H. J. Evans, J. Gen. Microbiol., 73, 416 (1972).
68. Biggins, D. R., and J. R. Postgate, J. Gen. Microbiol., 65, 119 (1971).
69. Pfennig, N., Ann. Rev. Microbiol., 21, 285 (1967).
70. Burris, R. H., and P. W. Wilson, Botan. Gaz., 108, 254 (1946).
71. Kidman, A. D., R. Yanagihara, and R. N. Asato, Biochem. Biophys. Acta, 191, 170 (1969).
72. Dixon, R. O. D., Arch. Mikrobiol., 85, 193 (1972).
73. Pate, J. L., V. K. Shah, and W. J. Brill, J. Bacteriol., 114, 1346 (1973).
74. Bergersen, F. J., Aust. J. Biol. Sci., 23, 1015 (1970).
75. Wilson, P. W., and R. H. Burris, Ann. Rev. Microbiol., 7, 415 (1953).
76. Hill, S., J. W. Drozd, and J. R. Postgate, J. Appl. Chem. Biotechnol., 22, 541 (1972).
77. Munson, T. O., and R. H. Burris, J. Bacteriol., 97, 1093 (1969).
78. Dalton, H., and J. R. Postgate, J. Gen. Microbiol., 56, 307 (1969).
79. Daesch, G., and L. E. Mortenson, J. Bacteriol., 95, 346 (1968); ibid., 110, 103 (1972).
80. Kurz, W. G. W., and T. A. LaRue, Can. J. Microbiol., 21, 984 (1975).
81. van Niel, C. B., in Methods in Enzymology, A. San Pietro, Ed., Academic Press, New York, Vol. 23, 3, 1971.
82. Jensen, H. L., in Agronomy, 1965, 136 (1965).
83. Rice, W. A., and E. A. Paul, Can. J. Microbiol., 18, 715 (1972).
84. Monzon de Asconegui and P. Kaiser, Ann. Inst. Pasteur, 122, 1004 (1972).

85. Dommergues, Y., and S. Mutaftschiev, *Ann. Inst. Pasteur*, 109, 112 (1965).
86. Okuda, A., and M. Kobayashi, *Nature*, 192, 1207 (1961).
87. Okuda, A., T. Katayama, and M. Kobayashi, *Nippon Dojo-Hiryogaku Zasshi*, 36, 84 (1965); *C. A.*, 63, 18691 b (1965).
88. Line, M. A., and M. W. Loutit, *J. Gen. Microbiol.*, 74, 179 (1973).
89. Chan, E. C. S., P. Basavanand, and T. Liivak, *Can. J. Microbiol.*, 16, 9 (1969).
90. Hardisson, C., J. M. Sala-Trepat, and R. Y. Stanier, *J. Gen. Microbiol.*, 59, 1 (1969).
91. Kurz, W. G. W., and T. A. LaRue, *Can. J. Microbiol.*, 19, 321 (1973).
92. Jensen, V., *Nature*, 183, 1536 (1959).
93. Becking, J. H., *Plant Soil*, 14, 49 (1961).
94. Claus, D., and W. Hempel, *Arch. Mikrobiol.*, 73, 90 (1970).
95. Jensen, V., *Nature*, 190, 832, (1961).
96. Hitchins, V. M., and H. L. Sadoff, *J. Bacteriol.*, 113, 1273 (1973).
97. Senior, P. J., G. A. Beech, G. A. Ritchie, and E. A. Dawes, *Biochem. J.*, 128, 1193 (1972).
98. Stevenson, L. H., and M. D. Socolofsky, *Antonie van Leeuwenhoek*, 39, 341 (1973).
99. Parker, C. A., and P. B. Scutt, *Biochim. Biophys. Acta*, 38, 230 (1960).
100. Schmidt-Lorenz, W. and A. Rippel-Baldes, *Arch. Mikrobiol.*, 28, 45 (1957).
101. Winter, H. C., and R. H. Burris, *J. Biol. Chem.*, 243, 940 (1968).
102. Burk, D., *J. Gen. Physiol.*, 10, 559 (1927).
103. Wilson, P. W., and R. H. Burris, *Bact. Rev.*, 11, 41 (1947).
104. Bayliss, N. S., *Aust. J. Biol. Sci.*, 9, 364 (1956).
105. Senez, J. C., *Bact. Rev.*, 26, 95 (1962).
106. Kurz, W. G. W., and T. A. LaRue, *Can. J. Microbiol.*, 21, 738 (1975).
107. Stouthamer, A. H., and C. Bettenhaussen, *Biochim. Biophys. Acta*, 301, 52 (1973).
108. Nagai, S., Y. Nishizawa, and S. Aiba, *J. Gen. Microbiol.*, 59, 163 (1969).
109. van der Beek, E. G., and A. H. Stouthamer, *Arch. Mikrobiol.*, 89, 327 (1973).

References

110. Knowles, C. J., and L. Smith, Biochim. Biophys. Acta, 197, 152 (1970).
111. Baak, J. M., and P. W. Postma, FEBS Letters, 19, 189 (1971).
112. Shah, V. K., J. L. Pate, and W. J. Brill, J. Bacteriol., 115, 15 (1973).
113. Dilworth, M. J., Ann. Rev. Plant Physiol., 25, 81 (1974).
114. Becking, J. H., Plant Soil, 16, 171 (1962).
115. Green, M., and P. W. Wilson, J. Gen. Microbiol., 9, 89 (1953).
116. Wilson, T. G. G., and E. R. Roberts, Biochem. Biophys. Acta, 15, 508 (1954).
117. Virtanen, A. I., and S. Lundbom, Acta Chem. Scand., 7, 1223 (1953).
118. Molnar, D., R. H. Burris, and P. W. Wilson, J. Am. Chem. Soc., 70, 1713 (1948).
119. Mozen, M. M., R. H. Burris, S. Lundbom, and A. I. Virtanen, Acta Chem. Scand., 9, 1232 (1955).
120. Repaske, R., and Wilson, P. W., J. Am. Chem. Soc., 74, 3101 (1952).
121. Burris, R. H., in Inorganic Nitrogen Metabolism, W. D. McElroy and B. Glass, Eds., John Hopkins Press, Baltimore, Md., 316 (1956).
122. Mozen, M. M., Ph.D. Thesis, University of Wisconsin, 1955, cited by C. Bradbeer and P. W. Wilson in Metabolic Inhibitors, R. M. Hochster and J. H. Quastel, Eds., Academic Press, New York, 595, 1963.
123. Chaudharg, M. T., T. G. C. Wilson, and E. R. Roberts, Biochem. Biophys. Acta, 14, 507 (1954).
124. Mehta, S. L., M. S. Naik, and N. B. Das, Indian J. Biochem., 4, 194 (1967).
125. Horner, C. K., and F. E. Allison, J. Bacteriol., 47, 1 (1944).
126. Burk, D., and R. H. Burris, Ann. Rev. Biochem., 10, 587 (1941).
127. Vela, G. R., and R. S. Rosenthal, J. Bacteriol., 111, 260 (1972).
128. Sorger, G. T., J. Bacteriol., 95, 1721 (1968).
129. St. John, R. T., and W. J. Brill, Biochim. Biophys. Acta, 261, 63 (1972).
130. Segal, W., and P. W. Wilson, J. Bacteriol., 57, 55 (1949).
131. Mozen, M. M., and R. H. Burris, J. Bacteriol., 70, 127 (1955).

132. Rakestraw, J. A., and E. R. Roberts, Biochem. Biophys. Acta, 24, 555 (1957).
133. Schollhorn, R., and R. H. Burris, Proc. Nat. Acad. Sci. U. S., 57, 1317 (1967).
134. Burris, R. H., H. C. Winter, T. O. Munson, and J. Garcia-Rivera, in Non Heme Iron Proteins, E. San Pietro, Ed., Antioch Press, Yellow Springs, 315 (1965).
135. Seto, B., and L. E. Mortenson, J. Bacteriol., 117, 805 (1974).
136. Harris, P. L., A. A. Diamantis, and E. R. Roberts, Biochim. Biophys. Acta, 111, 11, 15 (1965).
137. Wyss, O., C. J. Lind, J. B. Wilson, and P. W. Wilson, Biochem. J., 35, 845 (1941).
138. Hiai, S., T. Mori, S. Hino, and T. Mori, J. Biochem. (Japan), 44, 839 (1957).
139. Westlake, D. W. S., and P. W. Wilson, Can. J. Microbiol., 5, 617 (1959).
140. Pengra, R. M., and P. W. Wilson, J. Bacteriol., 75, 21 (1958).
141. Dilworth, M. J., Biochim. Biophys. Acta, 99, 486 (1965).
142. Zacharias, B., Acta Chem. Scand., 17, 1599, 2055, 2221, 2225 (1963).
143. Strandberg, G. W., and P. W. Wilson, Can. J. Microbiol., 14, 25 (1967).
144. Mahl, M. C., and P. W. Wilson, Can. J. Microbiol., 14, 33 (1967).
145. Dalton, H., and L. E. Mortenson, Bact. Rev., 36, 231 (1972).
146. Parejko, R. A., and P. W. Wilson, Can. J. Microbiol., 16, 681 (1970).
147. Campbell, N. E. R., and H. J. Evans, Can. J. Microbiol., 15, 1342 (1969).
148. Klucas, R., Can. J. Microbiol., 18, 1845 (1972).
149. Pratt, D. C., and A. W. Frankel, Plant Physiol., 45, 333 (1959).
150. Biggins, D. R., and J. R. Postgate, Eur. J. Biochem., 19, 408 (1971).
151. Hill, S., J. Gen. Microbiol., 67, 77 (1971).
152. Dobereiner, J., J. M. Day, and P. J. Dart, J. Gen. Microbiol., 71, 103 (1972).
153. Phillips, D. H., and M. Johnson, J. Biochem. Microbiol. Technol. Eng., 3, 277 (1961).
154. Nagai, S., Y. Nichizawa, M. Onodera, and S. Aiba, J. Gen. Microbiol., 49, 513 (1971).

References

155. Drozd, J., and J. R. Postgate, *J. Gen. Microbiol.*, **63**, 63 (1970).
156. Dilworth, M. J., *Biochem. Biophys. Acta*, **56**, 127 (1962).
157. Oppenheim, J., and L. Marcus, *J. Bacteriol.*, **101**, 286 (1970).
158. Oppenheim, J., R. J. Fisher, P. W. Wilson, and L. Marcus, *J. Bacteriol.*, **101**, 292 (1970).
159. Dalton, H., and J. R. Postgate, *J. Gen. Microbiol.*, **54**, 463 (1968).
160. Drozyd, J. W., R. S. Tubb, and J. R. Postgate, *J. Gen. Microbiol.*, **73**, 221 (1972).
161. Shaw, V. K., J. L. Pate, and W. J. Brill, *J. Bacteriol.*, **115**, 15 (1973).
162. Postgate, J. R., in *The Chemistry and Biochemistry of Nitrogen Fixation*, J. R. Postgate, Ed., Plenum Press, London, 161, 1971.
163. Barooah, P. P., and A. Sen, *Arch. Mikrobiol.*, **48**, 381 (1964).
164. Vela, G. R., and G. D. Cagle, *Can. J. Microbiol.*, **18**, 371 (1972).
165. Bush, J. A., and P. W. Wilson, *Nature*, **184**, 381 (1959).
166. Goerz, R. D., and R. M. Pengra, *J. Bacteriol.*, **81**, 568 (1961).
167. LeGall, J., J. C. Senez, and F. Pechinoty, *Ann. Inst. Pasteur*, **96**, 223 (1959).
168. Parker, C. A., and M. J. Dilworth, *Biochem. Biophys. Acta*, **69**, 152 (1963).
169. Bradbeer, C., and P. W. Wilson, in *Metabolic Inhibitors*, R. M. Hoechster and J. H. Quastel, Eds., Academic Press, New York, 595, 1963.
170. Hardy, R. W. F., R. Holsten, and R. C. Burns, *Soil Biol. Biochem.*, **5**, 47 (1973).
171. Brouzes, R., and R. Knowles, *Can. J. Microbiol.*, **17**, 1483 (1971).
172. LaRue, T. A., and W. G. W. Kurz, *Can. J. Microbiol.*, **21**, 980 (1975).
173. Molnar, D. M., R. H. Burris, and P. W. Wilson, *J. Am. Chem. Soc.*, **70**, 1713 (1948).
174. Esposito, R. G., and P. W. Wilson, *Proc. Soc. Exp. Biol. Med.*, **93**, 564 (1957).
175. Kleiner, D., and R. H. Burris, *Biochim. Biophys. Acta*, **212**, 417 (1970).
176. Burns, R. C., W. H. Fuchsman, and R. W. F. Hardy, *Biochim. Biophys. Acta*, **42**, 354 (1971).

177. Keeler, R. F., and J. E. Varner, Arch. Biochem. Biophys., 70, 585 (1957).
178. Carnahan, J. E., L. E. Mortenson, and J. E. Castle, J. Bacteriol., 80, 311 (1960).
179. Benemann, J. R., C. E. McKenna, R. F. Lie, T. G. Traylor, and M. D. Kamen, Biochim. Biophys. Acta, 264, 25 (1972).
180. Bulen, W. A., J. Bacteriol., 82, 130 (1961).
181. Pengra, R. M., and P. W. Wilson, Proc. Soc. Exp. Biol. Med., 100, 436 (1959).
182. Knight, E., and R. W. F. Hardy, J. Biol. Chem., 241, 2752 (1966).
183. Fukasawa, K., and M. Goto, Biochim. Biophys. Acta, 320, 545 (1973).
184. Corbin, J. L., and W. A. Bulen, Biochemistry, 8, 757 (1969).
185. Nicholas, D. J. D., D. J. Fisher, W. J. Redmond, and M. Osborne, Nature, 201, 793 (1964).
186. Kleiwer, M., and H. J. Evans, Plant Physiol., 38, 99 (1962).
187. Tchan, Y. T., and C. J. Fernandes, Ann. Inst. Pasteur, 116, 799 (1969).
188. Jakobsons, A., E. A. Zell, and P. W. Wilson, Arch. Mikrobiol., 41, 1 (1962).
189. Stevenson, L. H., and M. D. Socolofsky, Antonie van Leeuwenhoek, 38, 605 (1972).
190. Smith, D. D., and O. Wyss, Antonie van Leeuwenhoek, 35, 84 (1969).
191. Overbeck, J., and H. Malke, Z. Allg. Mikrobiol., 7, 197 (1967).
192. Bulen, W. A., and D. S. Frear, Arch. Biochem. Biophys., 66, 502 (1957).
193. Bulen, W. A., and J. R. Lecomte, and H. E. Bleas, J. Bacteriol., 85, 666 (1963).
194. Postma, P. W., A. S. Visser, and K. van Dam, Biochim. Biophys. Acta, 298, 341 (1973).
195. Yoch, D. C., and R. M. Pengra, J. Bacteriol., 88, 808 (1964).
196. Jensen, H. L., Acta Agr. Scand., 4, 224 (1954).
197. Anderson, G. R., and J. V. Jordan, Soil Sci., 92, 113 (1961).
198. Knowles, C. J., and L. Smith, Biochim. Biophys. Acta, 234, 144 (1971).
199. Babak, N. M., Mikrobiologiya, 35, 162 (1966).
200. Fustec-Mathon, E., D. Neuville, and P. Daste., Ann. Inst. Pasteur, 119, 498 (1970).

References

201. Billson, S., K. Williams, and J. R. Postgate, J. Appl. Bacteriol., 33, 270 (1970).
202. Gogotov, I. N., N. A. Zorin, and L. V. Bogorov, Mikrobiologiya, 43, 5 (1974), trans. in Microbiology, 43, 1 (1974).
203. Becking, J. H., Soil Sci., 118, 196 (1974).
204. Werner, D., H. J. Evans, and R. Seidler, Can. J. Microbiol., 20, 59 (1974).
205. Aho, P. E., R. J. Seidler, H. J. Evans, and P. N. Raju, Phytopathology, 64, 1413 (1974).
206. Tunail, N., and H. G. Schlegel, Arch. Microbiol., 100, 341 (1974).
207. Gogotov, J. N., and H. G. Schlegel, Arch. Microbiol., 97, 359 (1974).
208. deBont, J. A. M., and E. H. Mulder, J. Gen. Microbiol., 83, 113 (1974).
209. Gray, B. H., C. C. Fowler, N. A. Nugent, N. Rigopoulos, and R. C. Fuller, Int. J. Syst. Bacteriol., 23, 256 (1973).
210. Carpenter, E. J., and J. L. Culliney, Science, 187, 551 (1975).
211. Chen, J. S., and L. E. Mortenson, Biochim. Biophys. Acta, 371, 283 (1974).
212. Kleiner, D., Arch. Microbiol., 101, 153 (1974).
213. Lepo, J. E., and O. Wyss, Biochem. Biophys. Res. Comm., 60, 76 (1974).
214. Gordon, J. K., and W. J. Brill, Biochem. Biophys. Res. Comm., 59, 967 (1974).
215. Guerrero, M. G., J. M. Vega, E. Leadbetter, and M. Losada, Arch. Michrobiol., 91, 287 (1973).
216. Gadkari, D., and H. Stolp, Arch. Microbiol., 96, 135 (1974).
217. Gordon, J. K., and W. J. Brill, Proc. Nat. Acad. Sci. U. S., 69, 3501 (1972).
218. Seto, B., and L. E. Mortenson, J. Bacteriol., 120, 822 (1974).
219. St. John, R. T., V. K. Shaw, and W. J. Brill, J. Bacteriol., 119, 266 (1974).
220. Spiff, E. D., and C. T. I. Odu, J. Gen. Microbiol., 78, 207 (1973).
221. Raveed, D., D. W. Reed and R. E. Toia, J. Cell Biol., 59, 282a (1973).
222. Reed, D. W., R. E. Toia, and D. Raveed, Biochem. Biophys. Res. Comm., 58, 20 (1974).
223. Winter, H. C., and J. A. Ober, Plant Cell Physiol., 14, 769 (1973).

224. Thauer, R. K., G. Fuchs, B. Kauffer, and V. Schnitker, Eur. J. Biochem., 45, 343 (1974).
225. Benemann, J. R., G. M. Smith, P. J. Kostel, and C. E. McKenna, FEBS Letters, 29, 219 (1973).
226. Paschinger, H., Arch. Microbiol., 101, 379 (1974).
227. Bergersen, F. J., J. Gen. Microbiol., 84, 412 (1974).

Related Publications, Added in Proof

Berndt, H., K.-P. Ostwal, J. Lalucat, C. Schumann, F. Mayer and H. G. Schlegel, Identification and physiological characterization of the nitrogen fixing bacterium Corynebacterium autotrophicum GZ29, Arch. Microbiol., 108, 17 (1976).

Day, J. M., and J. Dobereiner, Physiological aspects of N_2-fixation by Spirillum from Digitaria roots, Soil. Biol. Biochem., 8, 45 (1976).

Elliott, B. B., and L. E. Mortenson, Transport of molybdate by Clostridium pasteurianum, J. Bacteriol., 124, 1295 (1975).

Hill, S., Influence of atmospheric oxygen concentration on acetylene reduction and efficiency of nitrogen fixation in intact Klebsiella pneumoniae, J. Gen. Microbiol., 93, 335 (1976).

Hine, P. W. and H. Lees, The growth of nitrogen-fixing Azotobacter chroococcum in continuous culture under intense aeration, Can. J. Microbiol., 22, 611 (1976).

Kleiner, D., Ammonium uptake by nitrogen fixing bacteria, Arch. Microbiol., 104, 163 (1975).

Neilson, A. H., and S. Nordlund, Regulation of nitrogenase synthesis in intact cells of Rhodospirillum rubrum: Inactivation of nitrogen fixation by ammonia, L-glutamine and L-asparagine, J. Gen. Microbiol., 91, 53 (1975).

Wiezl, J., and H. G. Schlegel, Enrichment and isolation of nitrogen fixing hydrogen bacteria, Arch. Microbiol., 107, 139 (1976).

Yates, M. G., and C. W. Jones, Respiration and nitrogen fixation in Azotobacter, in Advances in Microbial Physiology, 11, 97 (1974).

CHAPTER 3

Blue-Green Algae

W. D. P. STEWART

Department of Biological Sciences
University of Dundee
Dundee, Scotland

3.1. Introduction, 64
3.2. Culturing and General Growth Requirements of N_2-fixing Algae, 67
    3.2.1. General, 67
    3.2.2. Pure Cultures, 68
    3.2.3. Growth Rates, 68
    3.2.4. Light, 69
    3.2.5. Heterotrophy and Photoheterotrophy, 71
    3.2.6. Temperature, 74
    3.2.7. pH, 76
    3.2.8. Desiccation, 77
    3.2.9. Oxygen, 77
    3.2.10. Various Ions, 81
        3.2.10.1. Iron, 81
        3.2.10.2. Molybdenum, 83
        3.2.10.3. Cobalt, 83
        3.2.10.4. Phosphorus, 84
        3.2.10.5. Combined Nitrogen, 86
3.3. Interrelations Between Nitrogenase Activity, Photosynthetic, and Nonphotosynthetic Processes, 90
    3.3.1. Photosynthesis, 90
    3.3.2. Photorespiration, 91
    3.3.3. Dark Processes, 92
3.4. The Heterocyst as a Site of N_2 Fixation, 93
3.5. Production of Extracellular Nitrogen by N_2-fixing Algae, 107
3.6. References, 110

3.1 INTRODUCTION

The blue-green algae (Cyanophyceae), once a neglected group of N_2-fixing microorganisms, are enjoying at present an unprecedented interest in their activities. This is due, perhaps, to two main reasons. First, they are the only group of N_2-fixing microorganisms that have a higher plant type of photosynthesis with two photosystems and an ability to use water as reductant, and thus evolve O_2. Secondly, there has been a resurgence of interest in the views of Mereschkowsky (1), and Famintzin (2) that the chloroplasts of higher plants may have originated from endophytic blue-green algae. If this is so, and if these organisms were able to fix N_2, the implications, in relation to obtaining N_2-fixing asymbiotic crop plants, may be enormous (3).

There is now evidence that the three main groups of blue-green algae (unicellular forms, filamentous nonheterocystous forms, and filamentous heterocystous forms) all have N_2-fixing representatives. It seems equally clear, despite the limited data yet available, that whereas virtually all heterocystous forms fix N_2, there is no good evidence that all nonheterocystous algae can do so. Furthermore, the groups differ in the degree to which O_2 inactivates nitrogenase activity in whole cells, with heterocystous and certain unicellular algae being able to fix N_2 aerobically, whereas nonheterocystous forms, in general, fix N_2 only under conditions of reduced O_2 tension. Reported N_2-fixing forms and the assays used to detect nitrogenase activity are listed in Table 3.1 (heterocystous algae) and Table 3.2 (nonheterocystous algae).

My remit in this chapter is to consider solely the more biological aspects of N_2 fixation by blue-green algae; biochemical aspects are dealt with in II.2, II.3, II.4, and II.5; ecological aspects in IV.3; and symbioses with lower plants, and with higher plants, in II.3 and III.4, respectively. Other recent reviews of N_2 fixation by blue-green algae include those of Fogg and Stewart (5, 6).

TABLE 3.1. Heterocystous Blue-Green Algae Reported to Fix N_2 in Pure Culture

Organism	Technique	Reference[a]
Anabaena ambigua	N_2	243
A. azollae	N_2	244
A. cycadeae	N_2	7
A. cylindrica	N_2	125, 245
	$^{15}N_2$	137
	C_2H_2	142
A. fertilissima	N_2	243
A. flos-aquae	N_2	246
	$^{15}N_2$	142
	C_2H_2	142
A. gelatinosa	N_2	247
A. humicola	N_2	125
A. levanderi	N_2	248
A. naviculoides	N_2	247
A. variabilis	N_2	249, 125
A. sp.	N_2	249
	C_2H_2	262
Anabaenopsis circularis	N_2	57, 250, 261
Anabaenopsis sp.	N_2	250, 261
Aulosira fertilissima	N_2	243
Calothrix brevissima	N_2	250, 261
C. elenkinii	N_2	251, 252
C. parietina	N_2	130
	$^{15}N_2$	223
C. scopulorum	N_2	25
	$^{15}N_2$	85
	C_2H_2	262
Chlorogloea fritschii	N_2	38
	N_2/Ar	38
	$^{15}N_2$	56
	C_2H_2	262

(Continued)

TABLE 3.1. (Continued)

Organism	Technique	Reference[a]
Cylindrospermum gorakhporense	N_2	243
C. licheniforme	N_2	125
C. maius	N_2	125
C. sphaerica	N_2	253
Fischerella major	N_2	254
F. muscicola	N_2	254, 255
Hapalosiphon fontinalis	N_2	251, 252
Mastigocladus laminosus	N_2	256
	$^{15}N_2$	257
Nostoc calcicola	N_2	125
N. commune	N_2	258
N. cycadae	N_2	259
	$^{15}N_2$	259
N. entophytum	N_2	25
	$^{15}N_2$	85
N. muscorum	C_2H_2	142
	$^{15}N_2$	142
	N_2	54, 55
N. paludosum	N_2	125
N. punctiforme	N_2	245
N. sphaericum	N_2	263
Scytonema arcangelii	N_2	248
S. hofmanni	N_2	248
Stigonema dendroideum	N_2	260
Tolypothrix tenuis	N_2	250, 259
	C_2H_2	142
Westiellopsis prolifica	N_2	39
	$^{15}N_2$	224
	C_2H_2	262

[a] Other references are given in these papers.

TABLE 3.2. Nonheterocystous Algae Reported to Fix N_2 in Pure Culture

Organism	Assay	Reference
Gloeocapsa 795	C_2H_2	106
Gloeocapsa 6501	C_2H_2	107
Lyngbya 6409	C_2H_2	264
Oscillatoria 6407	C_2H_2	264
Oscillatoria 6412	C_2H_2	264
Oscillatoria 6506	C_2H_2	264
Oscillatoria 6602	C_2H_2	264
Phormidium sp.	C_2H_2	262
Phormidium sp.	N_2	93
Plectonema boryanum 594	N_2	105
	$^{15}N_2$	105
	C_2H_2	105
Plectonema 6306	C_2H_2	264
Plectonema 6402	C_2H_2	264
Raphidiopsis indica	N_2	93
Trichodesmium	N_2	19*

*Impure

3.2 CULTURING AND GENERAL GROWTH REQUIREMENTS OF N_2-FIXING ALGAE

3.2.1 General

Blue-green algae as a group are photoautotrophic microorganisms that show little difference in their growth requirements irrespective of whether they are growing on N_2 or on combined nitrogen. They can grow on purely inorganic medium, and there is little new evidence to support some early reports (7) that certain strains of Nostoc or Anabaena which occur in symbiotic association with higher plants have lost their ability to grow photoautotrophically on inorganic media (8, 9). Inorganic culture media commonly used are those of Allen and Arnon (10) and Gorham et al. (11) for freshwater species while Van Baalen (12) grew a variety of marine blue-green algae on medium ASP-2 of Provasoli, McLachlan, and Droop (13). To date no N_2-fixing algae have been shown to have a vitamin requirement. Filamentous species in general grow equally well in liquid and solid (1.0-2.0% agar, wt/vol) media although unicellular forms such as Gloeocapsa grow best in liquid culture (14). The marine planktonic blue-green alga Trichodesmium, which fixes N_2 in the field (15-18), has not yet been grown in pure culture in the laboratory (see, however, ref. 19).

3.2.2 Pure Cultures

Pure cultures and genotypic uniformity are essential prerequisites to any critical study of N_2 fixation by blue-green algae. Genotypic uniformity can be achieved by standard dilution and plating techniques (20, 21). Pure cultures may be obtained by ultraviolet irradiation, the algae being more resistant to irradiation than bacteria (22). Gamma irradiation has also been used (23). The main disadvantage of such techniques is the possible mutagenic effect on the algal cells. Antibiotic treatment (24, 25), the use of various chemicals (3, 26-29), techniques that make use of the phototactic responses of blue-green algae (8, 30), and differential temperature treatments (21, 31) have all been used, and although the use of ultraviolet irradiation is often the most successful, this depends on the algal species under investigation as well as on the contaminating microorganisms. Fungal contaminants can usually be eliminated easily using cycloheximide (32). Further details of the methods used to purify blue-green algae are available (14, 33, 34).

3.2.3 Growth Rates

The general belief that blue-green algae grow only slowly is not borne out by the available data. Some algae such as the non-N_2-fixing Synechococcus sp. (Anacystis nidulans) can double every 2 hours (35). The maximum growth rate of N_2-fixing algae, however, is usually slower and in continuous culture is about 20-24 hours for Anabaena cylindrica and Nostoc muscorum. Growth rates are comparable under N_2 and nitrate-nitrogen if the conditions are optimized (10, 36).

3.2.4 Light

When N_2-fixing algae are growing on purely inorganic media in the light the necessary reductant and ATP come ultimately from photosynthetic processes. Thus there is a close correlation in long-term experiments (more than a few hours) between the effect of light intensity on photosynthesis and its effect on N_2 fixation [see Fogg and Than Tun, (37) for *Anabaena cylindrica*, Fay and Fogg (38) for *Chlorogloea fritschii*, and Pattnaik (39) for *Westiellopsis prolifica*]. The optimum illumination varies with such factors as culture density (compare, e.g., refs. 36 and 37), CO_2 concentration, the design of the culture vessel and light intensity. As a general guide, white light intensities of 3,000-6,000 lumens/m^2 supplied at the surface of the vessel by fluorescent tubes are usually optimum at culture $O.D._{750}$ levels of 0.2-0.3.

Light quality can exert an important effect on N_2 fixation. In short-term studies nitrogenase activity is supported by photosystem I light, i.e., light absorbed primarily by chlorophyll *a*. Evidence for this has come from three independent, but complementary, studies. Fay (40) compared the action spectrum of C_2H_2 reduction with that of photosynthetic O_2 evolution. He showed that per unit energy, maximum nitrogenase activity occurs at 675 nm (absorbed by chlorophyll *a*) whereas maximum O_2 evolution occurs at 625 nm, where phycocyanin, which is associated mainly with photosystem II, shows maximum absorption. Second, Bothe and Loos (41) found that far-red light (above 695 nm), which is mainly absorbed by photosystem I and which allows photophosphorylation to proceed, did not support CO_2 fixation in the presence of the photosystem II inhibitor DCMU, although C_2H_2 reduction was scarcely affected by this treatment. Third, Lyne and Stewart (42) studied possible Emerson enhancement (43) of nitrogenase activity in *Anabaena* using as their light sources a sodium light with emission lines at 589 and 589.6 nm (which is absorbed mainly by photosystem II), and a red light (>660 nm) absorbed mainly by photosystem I. They found (Table 3.3) that there was an enhancement of CO_2 fixation when both light sources were supplied simultaneously but there was no such enhancement of C_2H_2 reduction under aerobic or microaerobic conditions. This finding of enhancement of carbon fixation agrees with other studies on

TABLE 3.3. Enhancement of C Fixation but not of Acetylene Reduction by <u>Anabaena</u> incubated under Aerobic and Microaerobic Conditions (after Lyne and Stewart, ref. 42)

Conditions[a]	Air	Enhancement	Argon/CO_2	Enhancement
Carbon fixation (μmoles C fixed·mg chl a·h)				
Dark	2.9	-	0.5	-
Phot. I	6.0	-	6.7	-
Phot. II	43.7	-	64.6	-
Phot. I + Phot. II	69.8	1.52	92.7	1.31
Acetylene reduction (μmoles·mg chl a·h)				
Dark	16.8	-	1.0	-
Phot. I	23.0	-	10.3	-
Phot. II	23.4	-	13.1	-
Phot. I + Phot. II	26.1	0.73	21.5	0.96

[a]Photosystem I light = red light (5.6 mW/cm^2) absorbing above 660 nm; photosystem II light = sodium light (0.75 mW/cm^2) absorbing at 589 and 589.6 nm. The light intensities used were limiting, and the incubation period was 15 min.

Anacystis nidulans (44), and the lack of enhancement of C_2H_2 reduction indicates a prime involvement of photosystem I in N_2 fixation. Light quality may also affect the morphogenesis of N_2-fixing blue-green algae (45-48). In long-term experiments the situation may change because some algae show chromatic adaptations. This has been noted in the N_2-fixing <u>Tolypothrix tenuis</u> where the ratio of phycocyanin/phycoerythrin, for example, varies depending on light quality (49, 50).
   The importance of light period has been studied by various workers. Allen (36) concluded that N_2-fixing <u>Anabaena cylindrica</u> grows as well in a 13:11, light (16,000 lux):dark regime, as do cultures illuminated continuously and this accords with some recent findings which show that nitrogenase activity in the dark can continue at high rates under suitable conditions. There is evidence that the rate at which N_2 fixation proceeds is governed largely by a pool of substrate made available by photosynthesis (51, 52), and that

this subsequently provides reductant for nitrogenase either via dark reactions alone, or perhaps via dark reactions and photosystem I (138). These substrates presumably supply not only the reductant, but also the ATP that becomes available through oxidative phosphorylation in the dark. The length of time during which photoautotrophic blue-green algae can continue to fix N_2 in the dark in inorganic medium usually varies between 1 and 24 hours.

3.2.5 Heterotrophy and Photoheterotrophy

The ability of certain blue-green algae to grow and fix N_2 heterotrophically has been the subject of some controversy but, despite some doubt in the past, there is good evidence to support the idea of N_2 fixation by blue-green algae growing under heterotrophic conditions. In fact it has been suggested (5) that nitrogenase activity in blue-green algae may have been geared initially to heterotrophic rather than to autotrophic mechanisms. As early as 1917, Harder (53) showed that a strain of Nostoc punctiforme isolated from the angiosperm Gunnera grew heterotrophically, but it was Allison and Morris (54) and Allison, Hoover, and Morris (55) who first demonstrated N_2 fixation under heterotrophic conditions. They found that a strain of Nostoc muscorum could grow on N_2 in the dark in the presence of glucose or sucrose. Later Fay (56) reported that Chlorogloea fritschii could be adapted with time to fix N_2 when supplied in the dark with sucrose, and to a lesser extent with glucose, and that there was some growth on glycine, glutamine, and maltose. Watanabe and Yamamoto (57) also showed heterotrophic N fixation by Anabaenopsis circularis, which fixes N_2 best under those conditions on glucose, and to a lesser extent on fructose, sucrose, and maltose. More recently Khoja and Whitton (58) listed strains of blue-green algae that grew heterotrophically on N_2. These belonged to the genera Anabaenopsis, Calothrix, Chlorogloea, Nostoc, and Scytonema.

Despite the fact that certain blue-green algae fix N_2 heterotrophically, heterotrophic growth is slow compared with photoautotrophic growth. This may be due to a limited uptake of organic compounds (see 59-64) or to one or more of the following reasons.

First, there is evidence that certain blue-green algae have an incomplete tricarboxylic acid cycle (59-61, 63) due to the absence of α-ketoglutarate dehydrogenase and succinyl CoA synthetase. This could lead in the dark to a decreased availability of reductant and ATP compared with algae with a complete cycle. There are reports, however, that the enzymes of the glyoxylate shunt are present in blue-green algae (59) although Hoare et al. (64) question whether the reported levels of some of the enzymes are significant and note that those reported for isocitrate lyase are lower than those considered to be negative in photosynthetic bacteria (65). Smith, London, and Stanier (61) found that only a small proportion of the carbon assimilated in unicellular blue-green algae was metabolized via the tricarboxylic acid cycle, and, more recently, the importance of the oxidative pentose phosphate pathway as the major route of hexose oxidation in the dark has been emphasized (175, 176) (see also below).

An indication that shortage of ATP may limit nitrogenase activity in the dark comes from studies which show that certain algae that cannot grow heterotrophically in the dark can do so photoheterotrophically. Lazaroff and Vishniac (67) found that although a strain of the N_2-fixing alga Nostoc muscorum grew heterotrophically on glucose in the dark, it grew better on glucose at light intensities that were too low to support photoautotrophic growth. Similar results have since been obtained by Van Baalen, Hoare, and Brandt (68) for non-N_2-fixing species and by Hoare, Ingram, Thurston, and Walkup (9) for a N_2-fixing Nostoc sp. In addition, Fay (56) observed maximum N_2 uptake by Chlorogloea fritschii under simultaneous CO_2 fixation and sucrose uptake, and we have obtained similar effects on C_2H_2 reduction by Westiellopsis prolifica (Fig. 3.1). Rippka (287) has obtained good growth of Chlorogloea on 0.5% glucose in the presence of $1 \times 10^{-5} \underline{M}$ DCMU.

Another view on the limited heterotrophic growth of blue-green algae is that these organisms are unable to exercise control at the transcription level over synthesis of certain of their enzymes which thus do not increase in the presence of substrate. This results in a failure of these algae to increase their growth rate in response to substrates which are assimilated and metabolized (59, 66, 288).

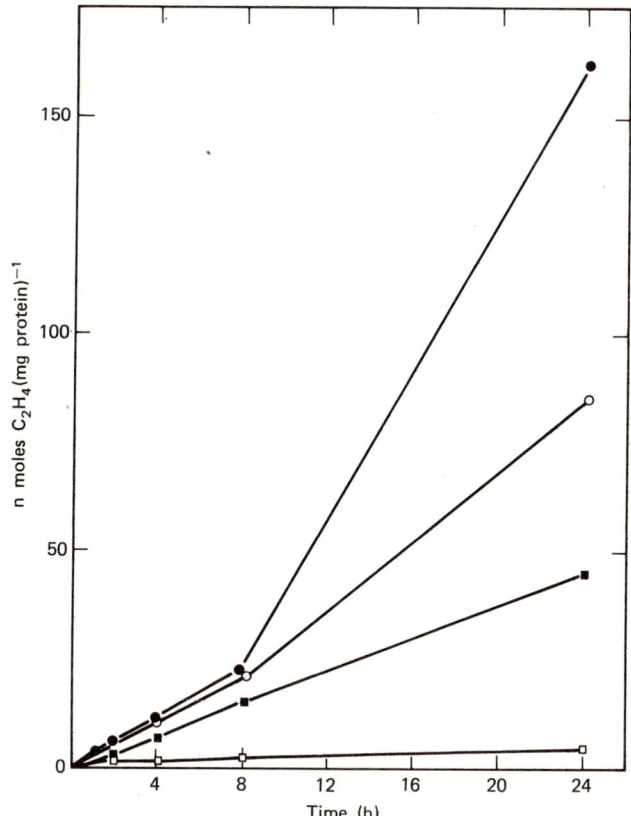

Fig. 3.1. Nitrogenase activity by <u>Westiellopsis prolifica</u> incubated in the presence and absence of sucrose (0.1% wt/vol) in the light (3,500 lux) and in the dark at 30°C. ●—●, light + sucrose; ○—○, light - sucrose; ■—■, dark + sucrose; □—□, dark - sucrose.

A suggestion, put forward by Pan (62), is that obligate autotrophy may result from the production in the dark of toxic compounds that are liberated by the alga and that if these are removed, limited heterotrophic growth can occur even though biochemical lesions exist. So far, there has been no additional experimental evidence that supports this concept in blue-green algae.

3.2.6. Temperature

Blue-green algae, including N_2-fixing species, usually have a temperature optimum around 32.5-35°C. This is appreciably higher than that of most other algae, and Allen and Stanier (21) and Wieringa (31) have used high-temperature treatments to purify Cyanophyceae from mixed cultures of algae. N_2 fixation by Nostoc commune and by the Nostoc-containing lichen Collema from Antarctica (69, 70) has been recorded at temperatures approaching 0°C, and in the Arctic, Kallio, Suhonen, and Kallio (71) found that C_2H_2 reduction by the Nostoc-containing lichens Nephroma arcticum and Solorina crocea is maximum at 15°C with a lower limit near 0° and -5°C for Nephroma and Solorina, respectively. Hitch and Stewart (72) observed nitrogenase activity of the temperate Lichina pygmaea at -1°C. Warmling (268) has obtained high rates of C_2H_2 reduction by Calothrix scopulorum from the supra-littoral zone of the Oslofjord at temperatures between 0° and 11°C.

In general there is a rapid stimulation of nitrogenase activity whenever the temperature increases and this may be particularly important in arctic and antarctic regions where, in summer, high temperatures occur for only a short time each day. Henriksson (73) concludes, however, from studies in Sweden, that temperature variations have little effect on nitrogenase activity by algae, and we have confirmed that, in the field, temperature effects are usually less marked than are the effects of desiccation and light intensity (Fig. 3.2).

Various studies that show an active nitrogenase at temperatures near freezing suggest that unlike cell-free extracts (74) the enzyme in vivo is not cold labile, because it is unlikely that nitrogenase synthesis could be responsible for the rapid increase in nitrogenase activity which occurs when these organisms are transferred from low to high temperatures. Reported Q_{10} values for nitrogenase activity are: Anabaena cylindrica, 6 (37); Nostoc commune, 4-6 (69); and Calothrix symbiotic in Lichina confinis, 3-4 (79). Arrhenius plots for C_2H_2 reduction and N_2-fixation have not yet been carried out, but they are biphasic in other organisms with activation energy values of approximately 14 kcal/mole above 20°C and 35 kcal/mole below 20°C (75).

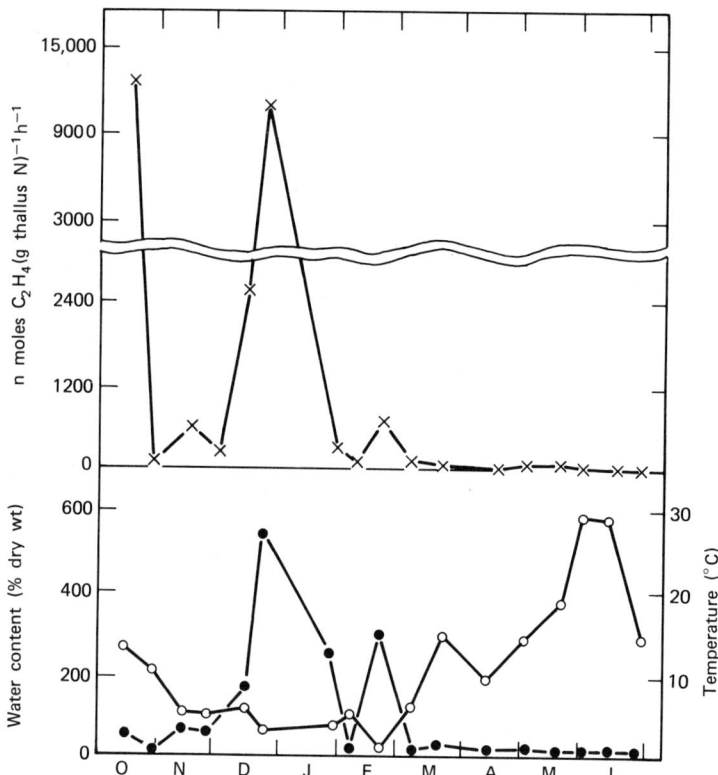

Fig. 3.2. Seasonal variation in in situ rates of acetylene reduction by Lichina confinis (X-X) during the period October 1968 to June 1969 and corresponding values for percentage water content of the thalli (●-●) and the temperature (o-o) during the sampling period. (after Hitch and Stewart, ref. 72).

Nitrogenase activity at high temperatures has been studied using blue-green algae from hot spring regions. Stewart (76) recorded N_2 fixation, probably by Mastigocladus species, at 46°C. Nitrogenase activity at even higher temperatures (60°C), again by Mastigocladus, has been noted by Stewart et al. (77). They found that the algae at these high temperatures were adapted to the temperature at which they occurred, and that although algae growing at 57°C continued to reduce acetylene when placed at 60°C, algae growing at 45°C ceased to reduce acetylene when placed at 54°C. It is possible that different strains, or even species, are characteristic of habitats of different temperatures (78). Mastigocladus laminosus can adapt, in long term studies, to temperatures that are considerably different to those from which it was originally isolated. Fogg (79) has shown, for example, that a strain isolated from a hot spring region in New Zealand grew well in laboratory culture at 32°C. In cell-free systems nitrogenase extracted from Anabaena cylindrica is inactivated irreversibly on heating at 60° for 10 min (80). Jutono (81) reports that, in rice paddy soils in Indonesia, the water temperatures where blue-green algae grow may reach 40-45°C.

3.2.7 pH

In general blue-green algae grow best under neutral or slightly alkaline conditions, with growth being very poor below pH 6. The data for Nostoc muscorum are fairly typical. This alga grows within the pH range 5.7-9.0, has an optimum of 7.0-8.5, and there is a marked decrease in activity below pH 6.5 (55). It will be appreciated, of course, that although these observations are typical, the results may vary depending on the culture conditions and on the strains used. The optimum pH for nitrogenase activity in cell-free extracts is broad within the pH range of 7.0-7.5 (82, 83).

The pH range for growth on N_2 and on combined nitrogen is fairly similar. Walp and Schopbach (84), however, found that Nostoc muscorum grew at pH extremes on combined nitrogen at which it could not grow on N_2, but the techniques that they used to measure growth (cell counts) are not very reliable. It has also been shown, for example, that at high pH levels growth may be poorer on NH_4^+ than on N_2, due to increased NH_4^+ toxicity with increase in pH (85). In natural ecosystems, abundant growths of algae are seldon seen in acid areas. In a survey of various soils in Scotland, Stewart and Harbott confirmed the finding of John (86) and Lund (87) that blue-green algae are rare in acid soils. However certain localized growths of N_2-fixing

Growth Requirements of N_2-fixing Algae

algae do occur in acid areas. For example, Stewart (88) reported on N_2-fixing filaments of Hapalosiphon in acid boglands, and Granhall and Selander (89) reported similarly that certain N_2-fixing algae occurred there associated with Sphagnum. Dooley and Houghton (90, 91) found no, or little, C_2H_2 reduction in various Irish soils of pH 4.2-6.9. They concluded that N_2 fixation by algae was unlikely to be important in these acidic habitats. In thermal hot spring regions, N_2-fixing algae occur particularly in alkaline streams, are rare in neutral streams, and are absent from acid streams where the anomalous eukaryote, Cyanidium caldarium, predominates.

3.2.8 Desiccation

A characteristic of blue-green algae is their capacity to withstand desiccation and this is probably one of the most important factors governing nitrogenase activity by these organisms in the field. The inhibitory effect of desiccation on N_2 fixation has been reported from antarctic (69, 70), tropical (92, 93), and temperature regions (72, 73, 94). According to Shtina (94) the optimum soil moisture content for the development of blue-green algae is 80-100% of the soil dry weight. Most N_2-fixing algae have thick mucilaginous sheaths, and this, together with their prokaryotic cell structure, enables them to absorb moisture very quickly and to lose it slowly. So long as the algae are metabolically active, they recover activity rapidly on being rewetted, as the data for the marine Lichina pygmaea show (Fig. 3.3) (95). Hitch and Stewart (72) showed that Lichina confinis and Peltigera canina thalli with water contents as low as 6.7 and 3.8% of their dry weights restarted reducing acetylene within 1-2 hours of being moistened.

3.2.9 Oxygen

Interrelations between O_2 sensitivity and N_2 fixation in blue-green algae are considered again in III.3.4. Here it is sufficient to note that a pO_2 above 0.2 atm may be inhibitory to nitrogenase activity and, to a lesser extent, to $^{14}CO_2$ fixation and to dark respiration (96). At high light intensities higher rates of

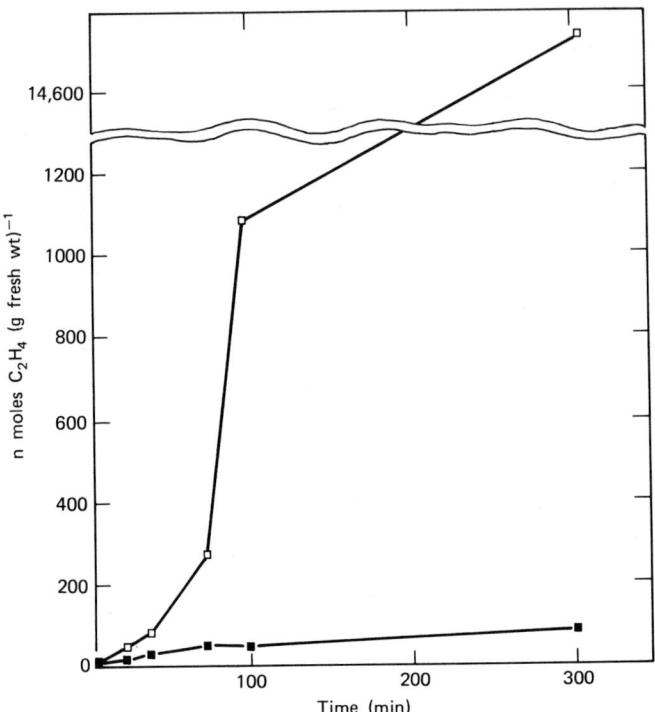

Fig. 3.3. Time course of recovery of nitrogenase activity by
Lichina pygmaea after 4 hours in a desiccated condition. □-□,
seawater added; ■-■, no water added. Temperature during experiment 20°C; light intensity, 4000 lux (after Stewart, ref. 95).

nitrogenase activity may be obtained at O_2 levels below
0.2 atm due presumably to O_2 evolution in photosynthesis satisfying some of the intracellular O_2 requirement, although in the dark, or at low light intensities, nitrogenase activity may increase with an
increase in pO_2 up to 0.2 atm (52, 97) (Fig. 3.4).
Because of their O_2 sensitivity a helpful technique in
culturing blue-green algae is to add low levels of
sodium sulphide to the medium (29). In the field
Cyanophyceae are particularly common in areas where
high O_2 levels do not accumulate and persist, e.g., in
slow-moving eutrophic waters rich in H_2S (98).
Sirenko et al. (99) have reported on how, in tank

experiments, Aphanizomenon flos-aquae filaments move from water with high levels of dissolved O_2 to waters with lower O_2 levels.

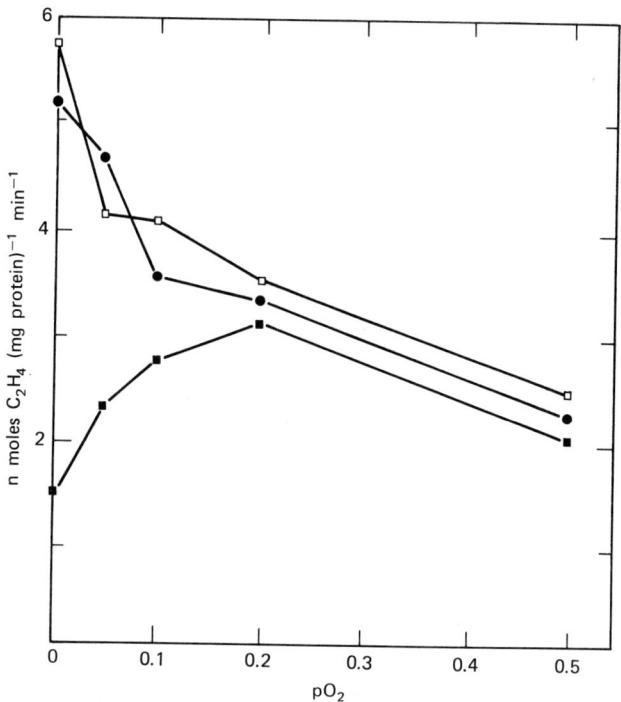

Fig. 3.4. Effect of pO_2 on acetylene reduction by Anabaena cylindrica at 6400 lux (□–□), 2000 lux (●–●), and 600 lux (■–■) (after Lex, Silvester, and Stewart, ref. 52).

The inhibitory effect of O_2 may differ in vitro and in vivo. Thus, in cell-free extracts, activity is irreversibly inhibited on exposure to O_2 for short periods (74, 100) (e.g., 10% O_2 for 5 min causes almost complete inactivation, whereas in whole cells the O_2 inhibition is reversible (96, 101). There are several possible reasons for this reversibility. One is a "switch-on, switch-off" mechanism of the type proposed for Azotobacter by Dalton and Postgate (104) and which envisages a reversible conformational change of the enzyme to an O_2-protected non-N_2-fixing form on exposure to high pO_2. Another possibility, which has been studied in some detail by Bone (102) and which he considers more likely, is that high pO_2 inactivates the enzyme irreversibly and that the return of nitrogenase activity is due to resynthesis of the enzyme. He found that, when inhibitors of protein synthesis (e.g., chloramphenicol and puromycin) were added to cells containing O_2-inactivated nitrogenase, there was no recovery of activity when the cells were returned to low O_2. Similarly, ammonia prevented the return of nitrogenase activity in O_2-inactivated cells. Bone (102) concluded that the maintenance of steady state levels of nitrogenase in blue-green algae is due to a balance between enzyme inhibition caused by O_2 and NH_3 and the synthesis of new nitrogenase. Weare and Benemann (103) studied the return of nitrogenase when cells that had lost their activity through incubation in the dark were returned to the light. They found that some recovery took place even in the presence of chloramphenicol, which suggested that biosynthesis of new nitrogenase alone was insufficient to account for the recovery. They considered that in their experiments the lag in recovery was probably due in part to the time required for the movement of photosynthate from the vegetative cells into the heterocysts. They also obtained evidence that the O_2 sensitivity of nitrogenase varies depending on the age of the culture and that younger cultures are less sensitive to O_2 than are older ones.

Certain nonheterocystous, filamentous algae fix N_2 only under microaerobic conditions (see Table 3.2) and here the most useful technique is to grow the algae in air on low levels of combined nitrogen and then to place the cultures under a gas phase without O_2 and without combined nitrogen in the medium, to induce an active nitrogenase (105). When this is done activity is usually noted after 24-36 hours presumably with nitrogenase synthesis occurring once the intracellular nitrogen is depleted, and when there is no build up of O_2 to inhibit the enzyme.

Unicellular algae that fix N_2 in air (106, 107) also show better nitrogenase activity at, or below, 0.2 atm than at higher O_2 levels (101), and although Wyatt and Silvey (106) carried out all their experiments in air, Rippka et al. (107) incubated their isolate of <u>Gloeocapsa</u> under $N_2:CO_2$ (99.5:0.5, vol/vol) in tests for nitrogenase activity. Gallon, LaRue, and

Kurz (108), however, failed to grow Gloeocapsa LB795 under $N_2:CO_2$ (95:5, vol/vol) and found that gassing with $A:CO_2$ (95:5, vol/vol) before harvesting also resulted in a loss of activity. It appears that 5% CO_2 is inhibitory to this alga.

3.2.10 Various Ions

The inorganic nutrients required for the growth of N_2-fixing blue-green algae are similar to those of blue-green algae in general [see (3)]. Several, however, are of particular importance to N_2-fixing algae.

3.2.10.1 Iron

Iron is an essential component of both the Mo-Fe protein and Fe protein of nitrogenase, and in blue-green algae is involved in the photosynthetic electron transport chain that includes cytochromes and ferredoxin. Cytochromes of the respiratory chain are also involved in oxidative phosphorylation and in electron transport. The optimum levels of iron for growth of Nostoc muscorum on N_2 and on nitrate nitrogen are, respectively, 10 and 1 ppm (109). Because of its low solubility iron is usually supplied in a chelated form, often as an EDTA complex.

Anabaena has a typical nitrogenase (see II.2) with reduced thiol groups and iron at the active site (83). Smith, Telfer, and Evans (110) partially purified the enzyme into two components which probably corresponded to the Mo-Fe protein and Fe protein of nitrogenase. Neither component showed nitrogenase activity alone, but they did so in combination. The Mo-Fe protein fraction complemented the Fe protein fraction of Chloropseudomonas, but the reciprocal cross was inactive.

Cytochromes of the c-type are most common in blue-green algae. Cytochrome c_{554} replaces the cytochrome f of higher plants in the photosynthetic electron transport chain of Anacystis (111) and Fujita and Myers (112) also isolated a cytochrome c_{552} which was rather similar to mammalian cytochrome c, and a cytochrome c_{549} which resembles that found by Webster and Hackett (113, 114) in the colorless blue-green alga Vitreoscilla. The latter workers also obtained evidence for the presence of b-type cytochromes and

Biggins (115), who demonstrated the presence of cytochrome b_6, obtained evidence that, at least in Phormidium, it was closely bound to the photosynthetic lamellae. Recently, the presence of cytochrome c_{554} and cytochrome b_{559} has been demonstrated in isolated heterocysts of Anabaena cylindrica (289). Although Webster and Hackett were unable to find any a-type cytochromes, Biggins and Dietrich (116) reported their presence in Saprospira grandis.

Ferredoxin has been characterized from a variety of blue-green algae including Anacystis nidulans (117), Nostoc sp. (118) and Microcystis flos-aquae (119). Purified ferredoxin from the N -fixing alga Nostoc forms brownish-red needle-like crystals with absorption spectra characteristic of chloroplast-type ferredoxin with absorption maxima at 276, 331, 423, and 470 nm. There is also a high ferredoxin:chlorophyll ratio that is rather similar to the ratio of bound ferredoxins to chlorophyll found in chloroplasts and photosynthetic bacteria. These ferredoxins contain two iron atoms per mole and one (117) or two (119) labile sulphur atoms per mole. As in chloroplasts, all are one-electron carriers, rather than two-electron carriers, and in this respect are dissimilar to the ferredoxins of photosynthetic and heterotrophic bacteria. Bound and soluble ferredoxins have both been demonstrated in isolated heterocysts of Anabaena cylindrica (289). Under conditions of iron deficiency, the flavoprotein phytoflavin may develop and replace ferredoxin, as Smillie (120) found in Anacystis nidulans. Phytoflavin resembles flavodoxin, which acts as an electron carrier in Clostridium pasteurianum (121).

Bothe (82, 122) has shown that Anabaena ferredoxin can mediate in the transfer of electrons from spinach chloroplasts to Anabaena nitrogenase and that it can be substituted by spinach ferredoxin, or phytoflavin. He also found that DSPD (disalicylidene proanediamine), which inhibits ferredoxin-dependent reactions, also inhibited CO_2 fixation and acetylene reduction to a similar extent. Smith, Noy, and Evans (123) and Smith and Evans (124) obtained evidence for a role of ferredoxin as an electron carrier to nitrogenase. They also showed that spinach chloroplasts, or photosynthetic lamellar fractions from Anabaena could, when supplied with ascorbate/dichlorophenol-indophenol and an ATP-generating system, transfer electrons to nitrogenase via ferredoxin. Ferredoxin also mediated in a

transfer of electrons from an isocitrate/NADP system
and from pyruvate to nitrogenase, but they considered
that the latter dark reactions were of limited
physiological significance in Anabaena. Haystead and
Stewart (83) have shown that Anabaena ferredoxin can
mediate in electron transport from H_2 via Clostridium
hydrogenase to nitrogenase.

3.2.10.2 Molybdenum

The Mo-Fe protein of nitrogenase contains molybdenum,
and its necessity for the growth of blue-green algae
on N_2 was realized by Bortels (125), confirmed by
Fogg (126), and investigated in detail by Wolfe (127).
The latter worker found that Anabaena cylindrica had
no molybdenum requirement when grown on ammonium-
nitrogen, but that 0.1 and 0.2 ppm, respectively, were
required for optimum growth on nitrate and on N_2. A
more recent report by Eyster (109) suggests that the
optimum molybdenum concentration for growth on N_2 is
1×10^{-7}M and that for nitrate utilization it is
1×10^{-10}M. Minimum concentrations necessary for N_2
fixation to occur are put at 1×10^{-10}M.

In his early studies Bortels (125) reported that
vanadium could substitute for molybdenum in blue-green
algae, and although this was disputed subsequently
(10) the validity of Bortels' work has been verified
by the finding (128, 129) that vanadium can substitute for molybdenum to give an active nitrogenase
(see II.4). In cultures grown on N_2, molybdenum
deficiency leads to a loss of phycocyanin, chlorosis
of the alga, and increased numbers of heterocysts
(126). Fay and Vasconcelos (269) have observed that,
as heterocyst numbers increase under these circumstances, there is a decrease in the nitrogenase
activity of the culture.

3.2.10.3 Cobalt

A cobalt requirement in plants was first demonstrated
by Holm-Hansen, Gerloff and Skoog (130) who found that
blue-green algae responded to the addition of cobalt
and that the response was greatest in N_2-fixing
species. They showed further that vitamin B_{12} could
substitute for cobalt. Buddhari (131) confirmed the
cobalt requirement for blue-green algae, and more
recently Johnson, Mayeux and Evans (132), studying the

Anabaena-Azolla association, found that although cobalt was not required for growth of the fern on combined nitrogen it was required for growth on N_2. In the absence of added cobalt, there was severe chlorosis of the alga, but this was relieved by the addition of as little as 0.01 µg/l of the element. This resulted in very large increases in yield, chlorophyll content, and N_2 fixation by the symbiosis. The role of cobalt in N_2-fixing blue-green algae is unknown, but its effect on nitrogenase activity appears to be indirect. Various views on its role in N_2-fixing symbioses have been summarized by Evans and Russell (133).

3.2.10.4 Phosphorus

Best growth of N_2-fixing blue-green algae occurs on inorganic phosphorus, particularly K_2HPO_4, which also serves as a good buffer for growth. Monobasic phosphate is usually inhibitory because of the resultant low pH. Optimum levels of dibasic phosphate are near 0.05% with higher levels being toxic, perhaps due to an intracellular accumulation of ADP. Organic sources of phosphorus, including some detergents, also support growth, and these are broken down to inorganic phosphate as a result of the action of phosphatases. Bone (134), who used phosphate limitation to slow down growth in chemostat cultures of the N_2-fixing Anabaena flos-aquae, found highest levels of alkaline phosphatase in phosphorus-starved cells, particularly in the presence of high levels of KNO_3 (15 mM). Phosphatases also break down storage polyphosphate that accumulates within the Cyanophyceae under conditions of phosphorus sufficiency. The phosphorus released in this way supports subsequent growth and N_2 fixation for a period, even in phosphorus-depleted waters (135, 136).

The availability of phosphorus markedly affects nitrogenase activity in blue-green algae. Phosphorus-limited algae, for example, show a marked stimulation in nitrogenase activity (within 15 min) when phosphorus is provided (136). Such a stimulation in acetylene reduction has been noted in phosphorus-deficient Anabaena cylindrica, A. flos-aquae, Anabaenopsis circularis, and Chlorogloea fritschii with a response being obtained to as little as 3-5 µg/l of phosphorus. The addition of phosphorus also stimulates dark respiration, but the response in the rate of $^{14}CO_2$ fixation is slower (136). Phosphorus-sufficient cells, which assimilate phosphorus more slowly, show little response in activity on the addition of phosphorus. This technique of measuring levels of available phosphorus on the basis of the response in C_2H_2-reducing activity, which is obtained when phosphorus is supplied to phosphorus-depleted cells, may, because of its sensitivity and rapidity, prove to be a particularly useful bioassay (135). Phosphorus deficiency probably depletes the cellular levels of ATP required for nitrogenase activity. The correlation between the intracellular ATP pool and growth on N_2 has been studied in <u>Anabaena cylindrica</u> where Bottomley and Stewart (270) conclude that cyclic photophosphorylation can supply all the necessary ATP for nitrogenase activity. In N_2-fixing blue-green algae the nitrogenase may also show ATP-dependent H_2 evolution that leads to a "wastage" of reductant under non-N_2-fixing conditions (74).

The sources of ATP for nitrogenase activity in blue-green algae are largely photophosphorylation and oxidative phosphorylation. Evidence for photophosphorylation as an important source of ATP comes from several lines of research, all of which center around the apparent key role of photosystem I in nitrogenase activity (40-42, 137, 138). In particular, Cox and Fay (137) found that, in carbon-sufficient cells where photosystem II was inactivated by CMU, there was an increase in C_2H_2 reduction with increase in light intensity. Because C_2H_2 reduction has no requirement for carbon skeletons and as dark-generated reductant was available in quantity, the most logical explanation is that the increased nitrogenase activity is due to an increased availability of ATP from cyclic photophosphorylation. In these experiments, which were carried out in an anaerobic gas phase in the presence of CMU, the importance of oxidative phosphorylation in the light could not be gauged and, thus, the relative contribution of photophosphorylation and oxidative phosphorylation is uncertain. Cox and Fay (137) believe, however, that photophosphorylation is the sole source of ATP for nitrogenase activity in the light. In the dark, oxidative phosphorylation is sufficient to drive nitrogenase, although activity is usually less than in the light. Photophosphorylation and oxidative phosphorylation both occur in heterocysts (139, 286) and the possible significance of this is considered later. The reported levels of ATP required for nitrogenase activity vary markedly but, according to Dalton and Mortenson (140), studies using purified nitrogenase indicate ATP/2e$^-$ ratios of 4.3 for <u>Azotobacter</u> and 3 for <u>Clostridium</u>. No information is available for blue-green algae.

3.2.10.5 Combined Nitrogen

The inhibitory effect of NH_4^+ on N_2 fixation by algae was noted by Fogg (126), and its addition results in a repression of nitrogenase synthesis. In <u>Anabaena flos-aquae</u>, urea, which is broken down to ammonia by a constitutive urease, has a similar effect (141). Inhibition of nitrogenase by nitrate-nitrogen is seldom complete. For example, in <u>Nostoc muscorum</u> activity is reduced to about 16 per cent of the original level by 7.1 mM KNO_3 (142). In <u>Anabaena flos-aquae</u> a similar inhibition is obtained with 1.5 mM KNO_3 (102).

A critical factor regulating nitrogenase synthesis appears to be the free intracellular pool of NH_3 within the alga, or some product of it, rather than the level in the external medium. This is suggested by the finding that under a N_2-free gas phase, 1.5 mM nitrate or 0.5 mM urea have little effect on nitrogenase activity although they inhibit nitrogenase activity by cells incubated under N_2 (102, 143). Ammonia, or its product, appears to regulate nitrogenase sythesis by acting in a repressor-depressor role, and N_2 does not act as an inducer because increased nitrogenase activity occurs in nitrogen-depleted N_2-free cultures (100, 143, 144). Recent evidence (271-273) suggests that the factor regulating nitrogenase synthesis in <u>Anabaena cylindrica</u> is not ammonia <u>per se</u>. It has been found that when L-methionine DL-sulphoximine, which inhibits glutamine synthetase, is supplied exogenously to N_2-fixing <u>Anabaena cylindrica</u>, NH_3 accumulates in the medium and nitrogenase activity continues. That is, when NH_3 assimilation via glutamine synthetase is blocked, nitrogenase continues to function in the presence of NH_3. Stewart and Rowell (272) showed further that even in NH_3-grown cultures, which lacked nitrogenase activity (and heterocysts), the addition of analogue resulted in the formation of new heterocysts, followed by nitrogenase resynthesis. There are no reports of NH_3 inhibition of existing nitrogenase in blue-green algae [see however (141) for <u>Anabaena</u> and (145) for <u>Azotobacter</u>], and in <u>Nostoc muscorum</u> (142) nitrogenase activity continues at a linear rate for at least 60 min in the presence of NH_3. In longer term studies there is a drop-off in nitrogenase activity as enzyme synthesis ceases in the presence of NH_4^+ and enzyme breakdown occurs (141). Long periods of incubation in the presence of C_2H_2 lead to increased nitrogenase activity, as this results in nitrogen-starvation which depletes the intracellular NH_3 pool. The papers by Bone (102, 134, 141) and Stewart and Rowell (271, 272) should be consulted for further information on factors regulating nitrogenase synthesis and inactivation.

In view of the above findings, it is important that N_2-fixing algae have an efficient mechanism for removing newly fixed NH_3 from around the site of nitrogenase synthesis. This can be achieved at the biochemical level by the NH_3 being incorporated rapidly into organic compounds, either near the site of nitrogenase and then being transported away from it, or else being transported away from nitrogenase as NH_3 and then being incorporated. There is evidence that: (i) certain organo-nitrogen compounds such as arginine and glycine (146) do not inhibit nitrogenase activity, (ii) the intracellular glutamine levels may be important in N_2-fixing algae (291), and (iii) various species can grow on L-alanine, L-glutamate, and L-asparagine as sole nitrogen sources (148).

The primary organic products into which newly fixed NH_3 is incorporated in blue-green algae have been studied by a variety of workers. <u>In vitro</u> enzymatic data show that glutamic <u>dehydrogenase</u> (GDH) which is the primary NH_3-assimilating enzyme in many plant species, although not measurable in early studies on blue-green algae (149), is present in low but significant amounts in a variety of blue-green algae (59, 139, 148, 105). The enzyme is $NADP^+$ specific with activity being barely detectable when $NADP^+$ is replaced by NAD^+. In view of the low detectable activities of GDH, other NH_3 assimilating enzymes were searched for in N_2-fixing blue-green algae and the most significant of these appears to be glutamine synthetase (GS). This enzyme, first detected by the Dundee group (147, 150) is present in quantity in both the vegetative cells and heterocysts of Anabaena cylindrica. With its low K_m^{app} for NH_4^+ (1 - 2 mM), this enzyme could scavenge NH_4^+ from around the site of nitrogenase (150), and if the results of Tubb (265) and Shanmugam et al. (266) for Klebsiella are applicable to blue-green algae, then GS may also play a role in the regulation of the nif operon. It is of particular interest that L-methionine DL-sulphoximine, which inhibits the activity of GS in Anabaena cylindrica in vivo and in vitro (272) releases the inhibition of heterocyst production and nitrogenase synthesis which occurs in the presence of NH_4^+ in the medium. The report that glutamate synthase [glutamine amide-2-ketoglutarate amino transferase (oxido-reductase)], i.e. GOGAT, occurred in blue-green algae (147) appeared equivocal at one stage when Haystead, Dharmawardene and Stewart (150) found that glutamine- and 2-ketoglutarate-dependent NADPH oxidation could be due to glutamic dehydrogenase acting on ammonia produced on the hydroysis of glutamine by glutaminase (150). However the presence of GOGAT in blue-green algae, using a ferredoxin linked assay, has been clearly demonstrated (274). Good activity of alanine dehydrogenase occurs in blue-green algae. It may be NAD^+ (148) or $NADP^+$ (139) specific, and the former type is present in both vegetative cells and heterocysts (275, 276). It has a molecular weight of 270,000, has six identical subunits and in this, and in its enzymic properties, is very similar to the enzyme of Bacillus subtilis. In addition to these enzymes, aspartate dehydrogenase (150) and carbamyl phosphate synthetase (150, 151), as well as the aminotransferases, glutamate:oxaloacetate amino transferase (150) and glutamate:pyruvate amino transferase (150) have been found. Aspartase and asparagine synthetase have not been detected (150). Qualitative interrelations are shown in Fig. 3.5. Quantitative data based on enzymic data (above), kinetic studies that show that newly fixed NH_3 is incorporated initially into the amide group of glutamine, glutamate and alanine (275), ^{14}C-labelling kinetics following the addition of N_2 and NH_3 to cultures fixing $^{14}CO_2$ photosynthetically (277), and the inhibition of assimilation of newly fixed NH_3 by methionine sulphoximine all support the view that the glutamine synthetase — glutamate synthase pathway is probably the main NH_3 assimilatory pathway in N_2-fixing

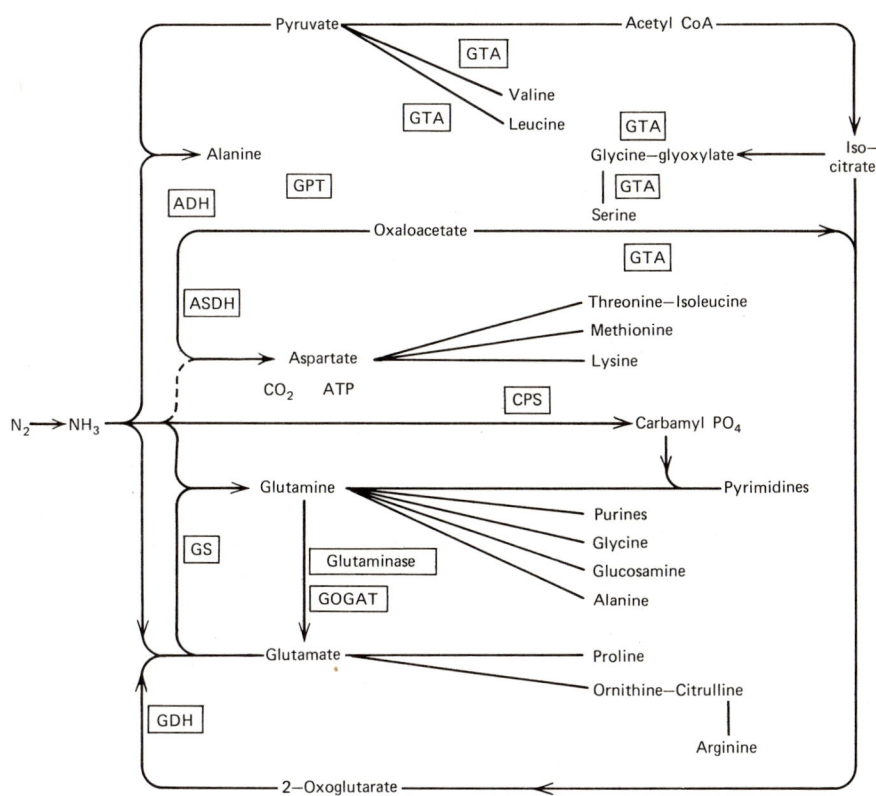

Fig. 3.5. Ammonia incorporation in the N_2-fixing alga <u>Anabaena cylindrica</u>. GDH, glutamic dehydrogenase; GS, glutamine synthetase; GOGAT, glutamate synthase; ASDH, aspartic dehydrogenase; ADH, alanine dehydrogenase; CPS, carbamyl phosphate synthetase; GPT, glutamate:pyruvate amino transferase; GTA, glutamate dependent transaminases (modified from Haystead, Dharmawardene, and Stewart, ref. 150).

algae and that alanine dehydrogenase may also be important. The organo-nitrogen compounds so formed thus provide nitrogen for general algal metabolism. Excess nitrogen is stored as phycocyanin (105, 143, 152), a rapidly mobilized nitrogen reserve and as structured granules (153-156, 271) which are co-polymers of aspartic acid and arginine (153, 277, 278).

3.3 INTERRELATIONS BETWEEN NITROGENASE ACTIVITY, PHOTOSYNTHETIC, AND NONPHOTOSYNTHETIC PROCESSES

In vitro studies on algal nitrogenase show that this enzyme complex, like that of other N_2-fixing microorganisms, has an absolute requirement for ATP and reductant (74, 82, 100, 108). ATP is supplied usually as an ATP-generating system and the optimal ATP level is near 2 mM (83). Reductant is routinely supplied as $Na_2S_2O_4$ (157) and the optimal level, which is not sharp, is near 2.5-5.0 mM (83). In vivo, photosynthesis provides the reductant and ATP in photoautotrophically grown cells, but the relative importance of a direct contribution by photosynthesis and a contribution via other routes such as respiration is not entirely settled.

3.3.1 Photosynthesis

Various studies have shown that there is a close relationship between N_2 fixation and photosynthesis in whole cells of blue-green algae, and early stoichiometric data (158-160) suggested that the photolysis of water directly supplied electrons for N_2 reduction in these organisms. More recent studies, using cell-free extracts, have shown that in vitro it is possible to provide reductant for nitrogenase via the photosynthetic electron transport chain (see above), and Smith, Noy, and Evans (123) conclude that this is the main in vivo source of reductant for nitrogenase. Many workers doubt, however, that the reductant is supplied by electrons generated directly on the photolysis of water. Some of the reasons for this doubt are as follows: First, CMU (p-chlorophenol-1,1-dimethylurea) and DCMU (3,4-dichlorophenyl-1,1-dimethylurea), inhibitors of the photolysis of water, do not inhibit nitrogenase activity immediately (82, 96, 137, 138, 161) although they have an immediate effect on carbon fixation (137, 159, 161). Second, photoautotrophically grown N_2-fixing blue-green algae continue to fix N_2 for a period when placed in the dark. Third, there is evidence that certain blue-green algae fix N_2 indefinitely when grown heterotrophically in the dark (see above), although the rates of N_2 fixation under such conditions are lower than in the light. Fourth, evidence that photosystem II is not involved directly in N_2 fixation comes from work on the effect of light quality on nitrogenase activity (see III,3.2.4). Fifth, it has been shown that the stoichiometric relationship between O_2 evolution and N_2-fixation can be varied depending on the environmental conditions (161). It is thus difficult to accept that the photolysis of water directly provides the main source of reductant for nitrogenase. This does not mean, of course, that photosystem I is not involved in electron transport to nitrogenase. Indeed there is evidence, detailed earlier, that photosynthetic ferredoxin is a component of the electron transport chain to nitrogenase in the light and in the dark, and there is preliminary evidence that photosystem I may be involved in the transport of electrons from dark reactions to nitrogenase. There is good evidence that cyclic photophosphorylation is an important source of ATP for photoautotrophically grown cells (see III,3.2.4).

3.3.2 Photorespiration

It appears that the ultimate source of reductant for nitrogenase is photosynthesis which provides a pool of substrates which in subsequent processes, some, if not all of which, are dark reactions, supply electrons for nitrogenase. Any factor which regulates the size of this pool may thus affect N_2 fixation.

One such factor may be photorespiration, a light-enhanced O_2 uptake, which occurs in various photosynthetic systems, including N_2-fixing blue-green algae [see (162, 163)]. The exact mechanism of O_2 uptake is uncertain, but there are several possibilities, e.g., the oxygenation of ribulose 1,5-diphosphate (164), the oxidation of glycollate (165), the oxidation of a reduced product of photosystem 1 (166), etc. The end result, in any case, is that photorespiration and nitrogenase may compete for reductant, and that factors such as low CO_2, high O_2, and high light intensities, all of which stimulate photorespiration, reduce the rate of nitrogenase activity (52). The question of how much of the O_2 uptake actually competes with nitrogenase for the reductant pool is unanswered as yet, but in any case photorespiration is an additional parameter that has to be considered in elucidating the regulation of nitrogenase activity in intact blue-green algae.

3.3.3. Dark Processes

Hexoses are the end product of carbon fixation in blue-green algae, all of which show a typical Calvin-cycle-type pathway rather than that of Hatch and Slack (3). These hexoses may be used in metabolic processes or may be stored as glycogen-type polyglucan granules. The latter which have been isolated and purified [see (67, 168)], are largely composed of highly branched polyglucosyl units, and develop particularly in cultures grown under heterotrophic conditions. They provide materials for metabolic reactions including N_2 fixation under conditions of carbon starvation. In addition to polyglucoside granules, poly-β-hydroxybutyrate granules are also present as storage products in blue-green algae (169, 170). It has been supposed (see ref. 133) that these may play some role in bacteroids as a source of reductant for nitrogenase, but certain species of nodulated legumes fix N_2 rapidly and do not accumulate poly-β-hydroxybutyrate.

The specific sources of reductant for nitrogenase in the dark may vary depending on the growth conditions and the physiological state of the alga. Carbohydrate breakdown via glycolysis to pyruvate has been suggested as an important metabolic route, and there is some evidence that pyruvate may be the main source of reductant to nitrogenase (137) as happens in <u>Clostridium</u>. Under some conditions (123, 137, 161) it may stimulate nitrogenase activity in whole cells, and the stoichiometry of N_2 uptake:CO_2 evolution from the decarboxylation of pyruvate suggested to Cox and Fay (137) that it could be the sole source of reductant to nitrogenase. Leach and Carr (171) subsequently showed the presence of a pyruvate:ferredoxin oxidoreductase (but not pyruvate dehydrogenase) in a non-N_2-fixing strain of <u>Anabaena variabilis</u>, and the enzyme of <u>Anabaena cylindrica</u> has since been reported on and partially characterized by Bothe and his co-workers (280, 281). In <u>Anabaena cylindrica</u> the activity of the pyruvate:ferredoxin oxidoreductase in NH_4 grown cells is only one-fifth of that occurring in cells grown on N_2 or nitrate nitrogen, and these workers conclude that a physiological role of the reaction is to generate reduced ferredoxin for the production of NH_3 from N_2. Recently, Codd, Rowell and Stewart (290) have shown pyruvate-dependent C_2H_2 reduction in extracts of <u>Anabaena cylindrica</u>, and were unable to detect pyruvate dehydrogenase activity in such extracts, although pyruvate:ferredoxin oxidoreductase was present at levels of activity sufficient to support the in vitro rate of pyruvate-supported C_2H_2 reduction. Smith, Noy, and Evans (123), on the basis of the poor stimulation noted on adding pyruvate to an in vitro assay system, conclude however that pyruvate may be of limited significance as a physiological electron donor in Anabaena.

The oxidative pentose phosphate pathway which operates in a variety of blue-green algae (172-176) could also provide a source of reductant for nitrogenase. The pathway has been studied in more detail in the unicellular Cyanophyceae, Aphanocapsa 6308, Aphanocapsa 6714, Synechococcus 6301, and Synechococcus 6307. However, in N_2-fixing Anabaena cylindrica extracts, it has been shown that a glucose-6-phosphate, glucose-6-phosphate dehydrogenase, NADP, ferredoxin, and ferredoxin-NADP oxidoreductase system supports good rates of nitrogenase activity in the light. More recently Winkenbach and Wolk (282) showed that heterocysts of Anabaena cylindrica had 7- to 8-fold higher activities of glucose-6-phosphate dehydrogenase and 6-phosphogluconate dehydrogenase, and two-fold more hexokinase activity, than did whole filaments per milligram of cell-free extract soluble protein. They showed further that about 74-80% of the total activity of glucose-6-phosphate dehydrogenase present in the filaments was located in the heterocysts. Supporting data have since been obtained by Lex and Carr (283). Thus, the oxidative pentose phosphate pathway may be an important source of reductant for algal nitrogenase.

It seems possible that the most important role of the tricarboxylic acid cycle/glyoxylate shunt is to provide carbon-skeletons for newly fixed NH_3, and evidence supporting this latter role has been obtained (277).

3.4 THE HETEROCYST AS A SITE OF N_2 FIXATION

Cell-free extract studies have established clearly that in vitro the nitrogenase of blue-green algae is as sensitive to O_2 as is that of the strict anaerobe Clostridium (74, 100). In addition, blue-green algae evolve O_2 in photosynthesis, their rate of dark respiration is low (115, 177), and even in the light, where there is a photostimulation of O_2 uptake, this rate seems to be insufficient to afford a protective mechanism of the type which may occur in Azotobacter (see 104, 178). Indeed, light-enhanced O_2 uptake may actually compete with nitrogenase for a pool of reductant.

In blue-green algae there appears to be a compartmentalization of the O_2-evolving sites and the N_2-fixing sites. Such a compartmentalization, which could involve both morphological and physiological mechanisms, is obviously not very efficient in N_2-fixing nonheterocystous filamentous types like Plectonema which fix N_2 only under microaerobic conditions, or in the unicellular algæ, Gloeocapsa which, although fixing N_2 in air, is highly sensitive to O_2. It is only in the heterocystous algae that a robust protective mechanism against O_2 inactivation of nitrogenase has been evolved. The nature of this protective mechanism has been the subject of much experimentation and some controversy.

The general consensus of opinion now is that in heterocystous algae the non-O_2-evolving cells, the heterocysts, are the sole sites of N_2 fixation under aerobic conditions and that the vegetative cells are the sites of carbon fixation and O_2 evolution. I subscribe to this view which is supported by most, but not all, workers on N_2 fixation. The divergent views are discussed more fully by Fogg, Stewart, Fay, and Walsby (3), Fay (179), and Stewart (6). Various other roles have been ascribed to the heterocyst. These include: a breakage point for filament fragmentation (180); reproductive units (181), organs that help to attach the filaments to the substratum (182, 183), and regulators of spore formation (184-186). The fact that the heterocysts appear to function as N_2-fixing sites is no good reason to reject these additional roles. However, if they are important (and I question whether all of these are) it does seem that they are probably of secondary importance compared with N_2 fixation.

Morphologically, heterocysts are cells which appear empty under the light microscope. They are usually, but not always (e.g. in Chlorogloea fritschii), larger than the vegetative cells from which they develop (Plates 3.1 and 3.2), but are typically smaller than the spores which develop adjacent to the heterocysts in some species (Plate 1a). Heterocysts

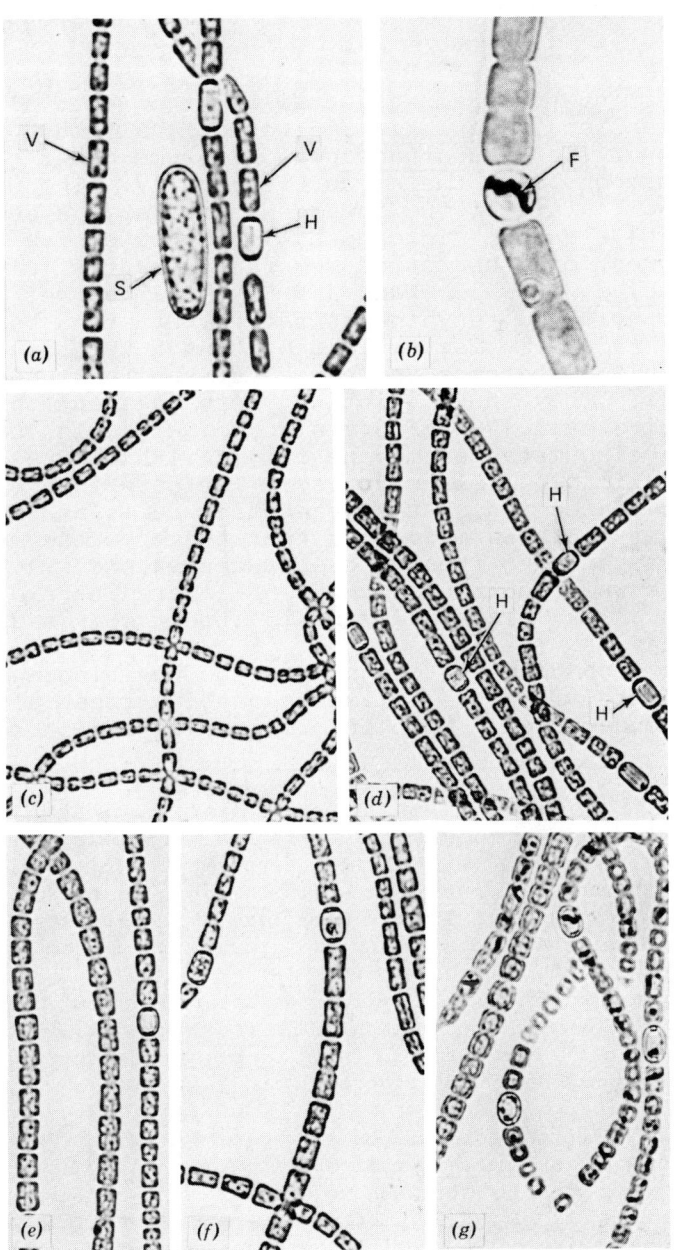

Plate 3.1. (a) Vegetative cells (V), heterocysts (H), and spores (S) of the N_2-fixing Anabaena cylindrica (x 1050); (b) filament

(Continued)

occur at intervals along the filament in most algae (e.g. Anabaena, Nodularia and Nostoc), although in members of the Rivulariaceae and in some other genera, e.g. Nostochopsis, they occur terminally. The outstanding feature of the mature heterocyst, as seen under the electron microscope, is the extensive development of a contorted membranous system (Plate 3.2.a, b, d, e) which Lang and Fay (187) suggest is due to thylakoid resynthesis rather than to a rearrangement of existing thylakoids. Such heterocysts are rich in ribosomes and have lost their structured granules. They show a distinct pore adjacent to each vegetative cell (Plates 3.2.a, c, d, e), which under certain circumstances may be filled with a dense osmiophilic plug (Plate 3.2.c). The pore channels allow a connection of the vegetative cells and heterocysts via microplasmadesmata that traverse the membranes separating the two cell types (3, 187). The envelope which surrounds the four-layered heterocyst wall is thick (Plate 3.2) and is itself composed of three layers.

The mechanism that regulates the positioning of heterocysts along the filaments of blue-green algae is poorly understood. Most information is available for

---

Plate 3.1. (Continued). of Anabaena cylindrica treated with 0.1% triphenyl-tetrazolium chloride for 15 min showing the deposition of formazan (F) within the heterocyst (x 1750); (c) filaments of Anabaena cylindrica grown in the presence of NH_4^+-N (50 mg N/l) in continuous culture. Note lack of heterocysts (x 840); (d) filaments of Anabaena cylindrica grown on N_2 in continuous culture. Note abundant heterocysts (H) (x 840); (e, f, g) air-grown filaments of Anabaena cylindrica after treatment with 0.1% (wt/vol) triphenyltetrazolium chloride for 0 min (e), 15 min (f), and 30 min (g). Note formazan in heterocysts only in f and in both heterocysts and vegetative cells in g (x 1200).

Plate 3.2. Electron micrographs of vegetative cells (V) and heterocysts (H) of Anabaena cylindrica. Note thick heterocyst envelope (E), pores (P) between vegetative cells and heterocyst; dense osmiophilic plug (PP); contorted lamellae (CL) and protoplasmic continuity between heterocysts and vegetative cells; polyphosphate bodies (PB) in vegetative cells but not in heterocysts. (a x 3960; b x 5280; c x 3960; d x 9900; e x 9900).

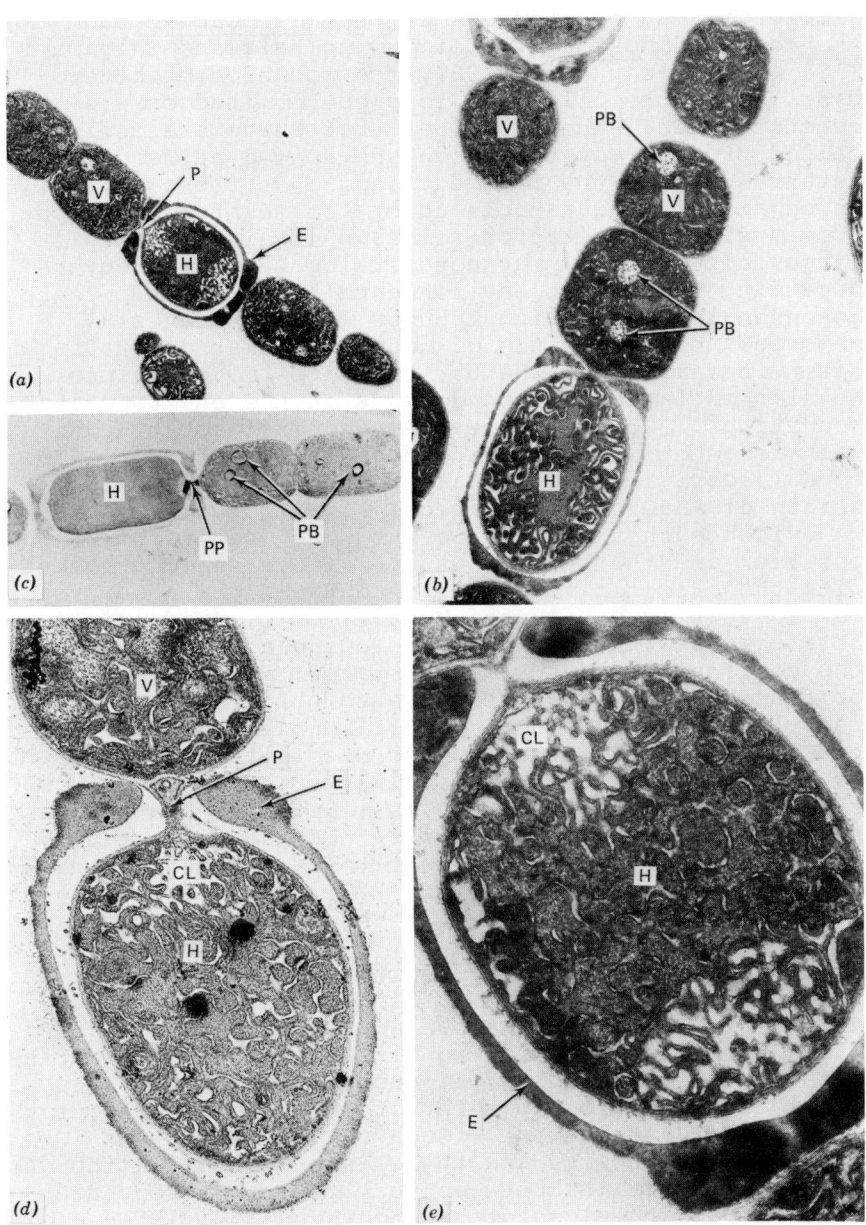

simple filamentous members of the Nostocaceae such as
Anabaena and Nostoc. According to Kale and Talpasayi
(188) and Wilcox (189, 190), the position of heterocysts along the filaments is predetermined on a
genetic basis. They consider that there are cells,
called proheterocysts, which are spaced at regular
intervals along the filaments and that these cells,
which are distinguishable under the light microscope
from ordinary vegetative cells, by their slightly
larger size, may or may not develop into heterocysts
depending on various environmental parameters. There
is an analogy here with legumes where Nutman (191)
reported that the sites of nodule initiation are predetermined and that whether the sites develop into
nodules or lateral roots depends on the environmental
conditions. However, the view that heterocyst position is genetically predetermined has been challenged
by Kulasooriya, Lang, and Fay (192). Using the same
strain of Anabaena cylindrica as Wilcox, they obtained
no evidence that proheterocysts were predetermined in
position and their findings were in general agreement
with the original view (126), that the position of new
heterocysts is determined by the position of existing
heterocysts with new ones developing at a point where
the concentration of some substance, perhaps ammonia,
produced by existing heterocysts is lowest (see also
ref. 186). We have no data for Anabaena cylindrica,
but do have a mutant of Nostoc muscorum which is unable
to fix N_2, but which has regularly positioned heterocysts along the filament. Thus, in this alga at least,
newly fixed ammonia produced by nitrogenase activity
in heterocysts does not appear to be the factor which
regulates heterocyst spacing (273).

   The effect of combined nitrogen on heterocyst
formation is certainly marked. There is ample evidence
to show that, as with legume root nodules, heterocyst
development is inhibited by high levels of combined
nitrogen, particularly ammonia (126, 135, 189, 190,
193-195). Thus, by growing heterocystous algae in the
presence of ammonia in continuous culture, it is
possible to obtain a population of nonheterocystous
filaments (Plate 3.1.c). Such filaments show no
nitrogenase activity and there is a general, although
not exact, parallel between the loss of nitrogenase
activity and the loss of heterocysts (135, 195).
Stewart and Rowell (272) have shown that in Anabaena
cylindrica, the addition of L-methionine

DL-sulphoximine relieves the inhibitory effect of NH_4^+ on heterocyst formation and that the return of heterocysts is accompanied by renewed nitrogenase activity.

The correlation between the rate of nitrogenase activity and heterocyst development can be followed readily by monitoring the development of each on transfer of heterocyst-free filaments from medium containing ammonium-nitrogen to medium free of combined nitrogen. The development of proheterocysts is noted first, and at that stage there is no nitrogenase activity. However, in aerobic cultures, when proheterocysts develop into young heterocysts (characterized by the absence of polyphosphate bodies and structured granules and the development of a cell envelope), nitrogenase activity becomes measurable, and a few hours after nitrogenase induction there is a rapid increase in activity which can be attributed to rapid nitrogenase synthesis during a period when the algae are nitrogen-starved prior to a build-up of nitrogenous products. Results obtained with Anabaena flos-aquae are shown in Fig. 3.6. The kinetics of nitrogenase induction in this alga are similar to those reported for Anabaena strain 6411 by Neilson et al. (143) and by Stewart and Lex (105) for Plectonema boryanum.

In heterocysts of algae that are rapidly fixing N_2, the pigments involved in photosystem II activity and O_2 evolution in blue-green algae:phycocyanin (196-200), allophycocyanin, and chlorophyll a 670 (199) cannot be detected. This may be due either to a destruction of these pigments as the heterocysts develop, or to a bleaching of the pigments. Heterocysts isolated from N_2-fixing filaments show a gross uptake of O_2 rather than an evolution of O_2 (201). These heterocysts also show strongest reducing activity, particularly near the poles, evidenced by using the photographic emulsion technique of Stewart, Haystead, and Pearson (200, 202). The latter findings thus confirm earlier results using tetrazolium salts (203, 204) (see Plate 3.1.b, e, f, g) and solutions of silver salts (205). N_2-fixing heterocysts have a C/N ratio of about 8:1, and according to Kulasooriya, Lang, and Fay (192) N_2 fixation is not observed until the C/N ratio of the filament has increased to this level. As heterocysts subsequently age they show a decrease in acetylene-reducing activity as well as in general

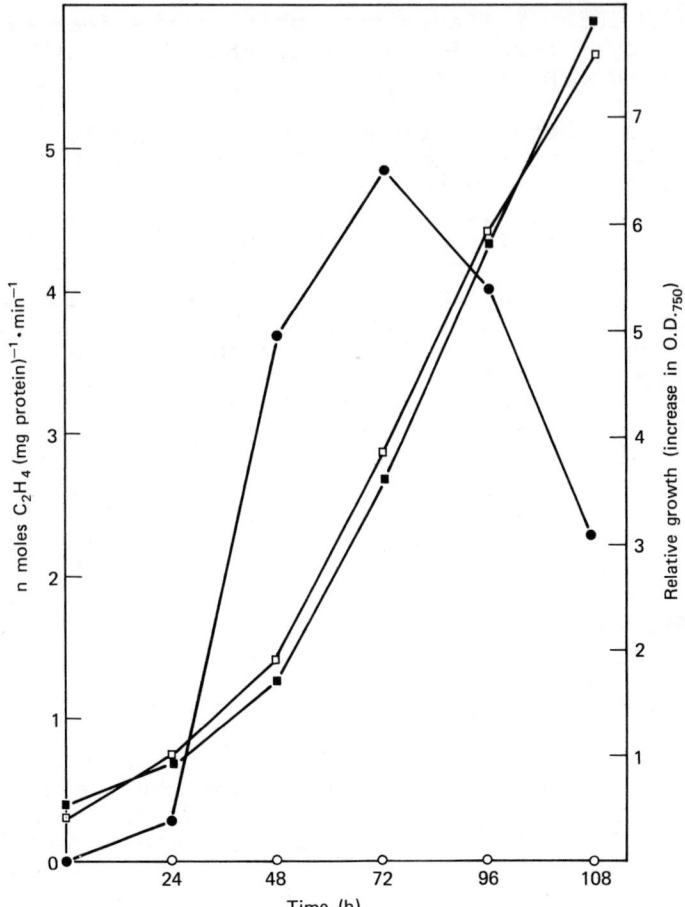

Fig. 3.6. Growth and acetylene reduction by <u>Anabaena flos-aquae</u> in batch culture. The algae had been grown on NH_4^+-N (50 mgN/l) in continuous culture and at 0 time one series was transferred to NH_4^+-free medium; the other was incubated into fresh medium containing 50 mg/l of NH_4^+-N. ■—■, growth on N_2; □—□, growth on NH_4^+-N; ○—○, C_2H_2 reduction by the N_2 grown culture; ○—○, C_2H_2 reduction by the NH_4^+-N grown culture.

reducing activity, and in old heterocysts the photosystem II pigments reappear and both the nitrogenase activity and the general reducing activity of the cells cease (202). Thus it is clear that metabolic activity varies with age of the heterocyst and it is unwise to consider that the heterocyst cannot be the site of N_2 fixation because the total number of heterocysts does not always tally with the rate of N_2 fixation (206).

It has, in fact, been demonstrated directly that heterocysts do fix N_2. Stewart, Haystead, and Pearson (200) found that isolated heterocysts, prepared from aerobic cultures, supplied with a source of ATP and reductant ($Na_2S_2O_4$) reduced acetylene to ethylene. They also found a loss of nitrogenase activity as the heterocysts became detached from the vegetative cells in the absence of added reductant, which suggested: One, that the vegetative cells, which remained in short filaments, did not fix N_2 in air and two, that if the heterocysts were the N_2-fixing sites, the substrates required for nitrogenase activity came from the vegetative cells. Such a loss of activity on detachment of heterocysts from vegetative cells has been confirmed by Wolk and Wojciuch (207), Fay and Kulasooriya (208) and Weare and Benemann (103) although the period during which nitrogenase activity continues after detachment varies in different studies. It has also been shown that isolated heterocysts do not reduce triphenyltetrazolium chloride (208). Direct evidence of a transfer of carbon compounds from vegetative cells into heterocysts has also been noted by Wolk (209). He showed that ^{14}C fixed in the light by vegetative cells of Anabaena cylindrica moved into the heterocysts, which themselves did not fix CO_2. Stewart, Haystead, and Pearson (200) noted further that there was some fixation of ^{14}C in developing heterocysts but that the ability to fix $^{14}CO_2$ was lost in mature heterocysts. Wolk and Wojciuch (210) later obtained evidence from cavitation studies that isolated heterocysts contained up to 30-40% of the activity of intact filaments. Tel-Or and Stewart (286) have recently been able to separate heterocyst and vegetative cell fractions of aerobically grown Anabaena cylindrica and demonstrated that when cell-free extracts of each fraction were assayed for nitrogenase activity over 90% of the total detectable activity was located in the heterocysts. Wolk (210) also obtained evidence that in the absence of added ATP, or reductant, heterocysts reduce acetylene in the light, which indicates indirectly that these cells can both photoreduce and photophosphorylate.

There is also a variety of indirect evidence which suggests that the heterocysts are the sole, N_2-fixing sites in heterocystous algae in air. Earlier findings have been detailed elsewhere (3, 212, 213) and only the recent results of Fleming and Haselkorn (284, 285) need be mentioned here. These workers carried out a gel electrophoretic study of the proteins of vegetative cells and heterocysts of aerobically grown Nostoc muscorum and found that as nitrogenase activity developed on transfer of algae grown on combined nitrogen to nitrogen-free medium several new proteins were differentiated in heterocysts but not in the vegetative cells. The electrophoretic pattern of certain of these proteins corresponded with those of Mo-Fe subunits of Azotobacter nitrogenase and, thus, may be nitrogenase components.

The sources of reductant and ATP for nitrogenase in heterocysts are known in general terms at least. First, a direct photoreduction of N_2 by electrons from the photolysis of water cannot occur as the photosystem II pigments are absent from the heterocysts of N_2-fixing algae. However, as we have discussed, carbon fixed in the vegetative cells is transferred into the heterocysts, which themselves do not fix CO_2. This transfer of carbon appears to be across the membrane separating the vegetative cells and heterocysts and is not due to carbon liberated extracellularly by the vegetative cells being reassimilated by the heterocyst. Wolk (209) noted that the appearance of ^{14}C-labeled compounds occurred in the heterocysts 20 min after the start of the exposure period to $^{14}CO_2$ and Weare and Benemann (103) consider that the rate of movement of substrate from the vegetative cells to the heterocysts may regulate the rate of development of nitrogenase activity under certain conditions. The nature of the compounds transferred is not known, but it has been suggested that it may be an oligosaccharide, or a sugar other than glucose, and that this is converted into glucose-6-phosphate in the heterocyst (282).

The carbon compound(s) entering the heterocyst is thus the ultimate source of reductant for nitrogenase activity occurring there and as mentioned earlier it could be metabolized further in several ways, e.g., in the tricarboxylic acid cycle/glyoxylate shunt, as pyruvate, or in the oxidative pentose phosphate pathway. The idea that the oxidative pentose phosphate pathway may be important is attractive. Although this pathway is likely to be inoperative in vegetative cells in the light, which fix CO_2 and accumulate ribulose-1,5-diphosphate thereby causing an allosteric inhibition of the pentose phosphate pathway, such an inhibition would not occur in the non-CO_2-fixing heterocysts.

An alternative route of reductant to nitrogenase within the heterocyst may be via photosystem I. Smith, Noy and Evans (123) have obtained *in vitro* evidence that photosystem I may be implicated in electron transfer to nitrogenase, Wolk and Wojciuch (207) have produced evidence that isolated heterocysts may photoreduce, and Lex and Stewart (138) have preliminary evidence that one route of reductant to nitrogenase could be from the Krebs cycle via photosystem I. Ferredoxin has been demonstrated in isolated heterocysts (286, 289).

Cyclic photophosphorylation in heterocysts could supply the necessary ATP, and, indeed, photosystem I pigments and photophosphorylation have been demonstrated. Indirect evidence for the latter comes from the finding (207) that nitrogenase activity in heterocysts is stimulated in the light and from the direct demonstration of photoproduction of ATP by isolated heterocysts (139, 286). These findings support the view from work on intact filaments that cyclic photophosphorylation is the main source of ATP for N_2 reduction in the light (137, 270). The importance of dark generated ATP should not be ruled out, however, because, as we have seen, various algae fix N_2 in the dark, though at lower rates than in the light. Under these circumstances nitrogenase activity is presumably dependent on oxidative phosphorylation with activity falling off rapidly when filaments are incubated in the dark in an O_2-free gas phase (Table 3.4). Oxygen uptake by isolated heterocysts has been observed. If, as suggested, oxidative phosphorylation in heterocysts can support nitrogenase activity, it is likely that

TABLE 3.4. Acetylene Reduction by Anabaena cylindrica when a Photoautotrophically Grown Culture is Placed in the Dark under Aerobic and Anaerobic Conditions

	C_2H_4 Production nmoles/1.0-ml sample/min Time (min)		
Treatment	0	30	60
Light, air	0.13	0.13	0.16
Light, N_2/CO_2 (99.96/0.04)	0.14	0.13	0.13
Dark, air	0.14	0.09	0.07
Dark, N_2/CO_2 (99.96/0.04)	0.14	0.02	0.01

The inoculum was obtained from a continuous culture grown aerobically at 3000 lux and 26°C. Each C_2H_2 reduction assay was for 15 min starting at the times shown. Dark bottles (7.0-ml capacity serum bottles containing 1.0 ml of algae) were wrapped in aluminum foil. Each value is the mean of triplicates.

there is some compartmentation within the cell of the sites of nitrogenase activity and those of oxidative phosphorylation. It is also possible, of course, that ATP may be generated by oxidative phosphorylation in the vegetative cells and transferred to the heterocyst.

Although it is generally accepted that heterocysts fix N_2, there is less certainty about whether vegetative cells of heterocystous algae develop an active nitrogenase. The hypothesis that the vegetative cells of heterocystous blue-green algae possess nitrogenase was suggested by Wolk (211), who concluded that its activity there, presumably under aerobic conditions, depended on some substance produced by the heterocysts. Smith and Evans (100) subsequently obtained evidence that the nitrogenase which they extracted from Anabaena cylindrica, incubated previously in the absence of added O_2, came from the vegetative cells. More recently Kurz and LaRue (206) considered that, because higher rates of nitrogenase activity were obtained in Anabaena cultures with few heterocysts than in those with many, the vegetative cells and not the heterocysts were the sites of N_2 fixation. Ohmori and Hattori (214) have since reported that short filaments of Anabaena cylindrica without heterocysts fixed $^{15}N_2$ in the presence of O_2 and concluded that heterocysts are not essential for N_2 fixation. We have been unable to confirm the latter finding. Wolk et al. (235) also have ^{13}N data which imply that there may be some vegetative cell nitrogenase in aerobically grown cultures of Anabaena cylindrica. Previous $^{13}N_2$ fixation tests on other organisms, however, have been far from satisfactory, and more conclusive data for an active vegetative cell nitrogenase in aerobic cultures of Anabaena cylindrica are required.

The possibility that the vegetative cells fix N_2 implies a compartmentalization between the N_2-fixing and O_2-evolving sites within these cells, and, indeed, there is ample evidence of compartmentalization. For example, Lyne and Stewart (42) found no evidence of competition between carbon fixation and nitrogenase activity in short-term experiments under both aerobic and microaerobic conditions (Table 3.5). Whether this could be due to compartmentalization within the vegetative cells or between vegetative cells and heterocysts is uncertain although the evidence of

TABLE 3.5. Lack of Competition between Carbon Fixation and Acetylene Reduction of Aerobically and Microaerobically Grown Anabaena cylindrica (after Lyne and Stewart, ref. 42)

Conditions	Carbon fixation (µmoles/mg chl a·h)		Acetylene reduction (µmoles/mg chl a·h)	
	$-C_2H_2$	$+C_2H_2$	$-NaHCO_3$	$+2mM$ $NaHCO_3$
Aerobic				
Dark	1.4	0.6	0.2	0.2
Light	58.8	58.7	19.8	19.8
Microaerobic				
Light	54.9	55.7	31.5	33.1

Algae were taken from continuous cultures and experimental treatment $\pm C_2H_2$ and $\pm NaHCO_3$ was 15 min. Each value is the mean of triplicates.

Ohmori and Hattori (214) certainly implies compartmentalization within the vegetative cell. An alternative possibility is that the vegetative cells fix N_2, but only under microaerobic conditions, when O_2 is not available to inactivate the enzyme or prevent nitrogenase synthesis. Such a system operates in certain nonheterocystous filamentous species (see Table 3.2). It does not, of course, explain the more O_2 stable vegetative cell nitrogenase of Gloeocapsa (106, 107) and possibly Trichodesmium (15-19), although even in the former alga the enzyme is more O_2 sensitive than that of heterocystous algae. It could, however, account for the results of Smith and Evans (100), who extracted their nitrogenase from microaerobically grown filaments. In addition, Van Gorkom and Donze (215) followed the appearance and disappearance of phycocyanin (a readily mobilized nitrogen reserve, as well as a photosynthetic pigment) by fluorescence microscopy and observed that when nitrogen-starved Anabaena filaments were supplied with N_2 under aerobic conditions there was a decreasing gradient of phycocyanin development away from the heterocysts. Under microaerobic conditions there was no phycocyanin gradient, the pigment developing equally in all the vegetative cells, which suggests that under these conditions the vegetative cells also fix N_2. However, there are also objections to this idea. For example, under microaerobic conditions the increase in nitrogenase activity is paralleled by an increase in heterocysts (192); there is a compartmentalization of nitrogenase activity and carbon fixation activity both under microaerobic and aerobic conditions (42); the lack of phycocyanin gradient noted by Van Gorkom and Donze could possibly be due to increased nitrogenase activity in the heterocysts under microaerobic conditions, although this would not explain easily the highly O_2 sensitive nitrogenase that develops under microaerobic conditions only (215). Also, in the experiments of Smith and Evans (100) the nitrogenase may have come from broken heterocysts. Thus the question of whether the vegetative cells of heterocystous algae ever possess an active nitrogenase is not resolved unequivocally.

3.5 PRODUCTION OF EXTRACELLULAR NITROGEN BY N_2-FIXING ALGAE

Blue-green algae characteristically liberate into the medium a proportion of the N_2 that they fix, and although there are many reports of these extracellular products, which appear to be mainly ammonia, amides, peptides and polypeptides [see (6)], most studies have been observational only and few attempts have been made, apart from those of Fogg (216) and Hood and Carr (217), to relate these to the physiology and biochemistry of the organisms.

The pattern of production of total extracellular products, first noted by Fogg (218), is characteristic of N_2-fixing blue-green algae grown in batch culture. During log phase and post-log phase the percentage of extracellular nitrogen is highest, and it is lowest during lag phase. This has been confirmed in a variety of species (192, 219-222). In general 20-30% of the total nitrogen fixed may be liberated extracellularly, but values range from a high of 60-80% (222) to a low of less than 5% (223).

Using ^{15}N as tracer, Fogg and Pattnaik (224) showed that newly fixed NH_3 was an extracellular product during the early stage of growth of the N_2-fixing blue-green alga Westiellopsis prolifica and that subsequently this NH_3 was reassimilated by the alga (224). A release and subsequent reassimilation of NH_3 has also been reported for Anabaena (225). Evidence that extracellular NH_3 is produced when the pathway of NH_3 assimilation via glutamine synthetase and glutamate synthase is blocked has been obtained by Stewart and Rowell (272).

The production of extracellular amides by Anabaena cylindrica (218) and by Westiellopsis prolifica (224) has also been noted. In Westiellopsis, amides (possibly glutamine and asparagine) were liberated in greatest quantity during the latter stages of growth, and these showed a lower ^{15}N labeling than did the intact alga, indicating that they are not associated specifically with the N_2 fixation process. However, this is by no means certain because appreciable intracellular levels of glutamine (147) are found in blue-green algae under N_2-fixing conditions.

Peptides and polypeptides are perhaps the most abundant extracellular products found in filtrates taken from batch cultures of blue-green algae. They have been reported for Anabaena cylindrica (218, 225), Lyngbya (227), Calothrix scopulorum (220-222), Nostoc entophytum (220), Nostoc commune (238), Anabaena azollae (229), Anabaena variabilis, Anabaena sp., Calothrix sp., Chlorogloea fritschii, Cylindrospermum sp., Nostoc muscorum, and Nostoc sp. (230). Their origin and role are poorly understood, but they are produced from healthy cultures and do not have the same composition as that of the alga from which they are produced (218). Walsby (226, 231) reported on at least 12 different peptides in filtrates from Anabaena cylindrica and found that a proportion of these occurred as serine-threonine complexes that were not produced in NH_4^+ grown cultures, suggesting, perhaps, an involvement in N_2 fixation. Jakob (232, 233) reported that cultures of Nostoc muscorum with heterocysts produced an extracellular substance that was inhibitory to other microorganisms, but that this compound was apparently absent from nonheterocystous cultures of the same strain. Fogg (218) also noted that the extracellular products of Anabaena cylindrica are produced in reduced amounts when molybdenum, a component of nitrogenase, is limiting, but iron limitation had no effect on extracellular nitrogen production.

A relationship between heterocysts and the production of extracellular ammonia has been noted in Anabaena by Thomas and David (225). They obtained evidence that very high levels of NH_4^+ (60-80% of the total nitrogen fixed) may be liberated extracellularly. They concluded that under conditions of nitrogen limitation heterocysts are induced, and that when these begin to fix N_2 there is an accumulation of NH_3 that prevents the development of additional heterocysts, until the ammonia levels decrease once more to a critical level which allows the induction of fresh heterocysts. It has been suggested that liberation into the medium may be an important route of transfer of newly fixed NH_3 from the heterocysts to the vegetative cells in N_2-fixing algae, but this seems rather unlikely in natural ecosystems. It is possible, however that NH_3 may be an important extracellular product in the transfer of nitrogen in symbiotic systems from one organism to another. Experiments in my laboratory have shown that when the cephalodia of Peltigera aphthosa are incubated under N_2 extracellular NH_3 is liberated, but on incubation under argon there is little extracellular ammonia in the filtrates.

There is now good evidence that some product of NH_4^+ assimilation inhibits the synthesis of new nitrogenase either directly, or by affecting a regulator of nitrogenase such as glutamine synthetase (see 272). It is thus possible that intracellular nitrogenase activity is dependent on the extent to which extracellular nitrogen production, as well as intracellular protein synthesis reduces the intracellular pool of inhibitor. This could explain, in part at least, the following findings. First, there is evidence of more rapid N_2-fixation when free-living algae are grown together with bacteria than when they are grown axenically (240). The usually accepted explanation for this is that the heterotrophic bacteria stimulate algal growth by reducing the oxygen tension and increasing, through respiration, the availability of carbon dioxide for algal photosynthesis. It seems possible that the bacteria could also play some part by increasing the size of the sink for extracellular nitrogenous compounds produced by the algae and in this way reduce the intracellular concentrations of the inhibitor of nitrogenase. The rapid fixation noted when blue-green algae occur in the rhizosphere of plants such as rice (92), Suaeda maritima, Agrostic stolonifera, Bryum pendulum (241) and Artemesia sp. (242) could also be explained partially in the same way, as could the finding that blue-green algae growing in continuous culture often show higher rates of nitrogenase activity than do those growing in batch culture.

In the case of symbiotic blue-green algae a slightly different situation may apply. It is well known that in symbiotic associations blue-green algae may show much higher rates of acetylene reduction than do free-living blue-green algae growing under identical conditions (234). There is also evidence of a rapid transfer of fixed nitrogen from the alga to the host. Millbank and Kershaw (236) showed a significant increase in ^{15}N labeling of the lichen thallus minus blue-green alga after about 3 hours, but I suspect that evidence of an even more rapid transfer of nitrogen could be obtained. In other such symbioses the host also provides a sink for extracellular nitrogen. In cycad root nodules that have an active nitrogenase (142, 230, 237), labeling can be detected almost throughout the host within 48 hours of exposure of the alga to $^{15}N_2$ (238). In Gunnera, Silvester and Smith (239) showed that, within 1·5 hours of exposure of plants to $^{15}N_2$, there was evidence that nitrogen fixed by the alga had been transported to the leaves. Likewise we have found in Anthoceros (267) that there is a rapid transfer of fixed nitrogen from the phycobiont (a Nostoc sp.) to the host gametophyte and sporophyte. We (291) have evidence that in Peltigera canina the fungal hyphae modify the ammonia-assimilating mechanism of the alga so that much of the newly fixed nitrogen is released extracellularly and that this results in less inhibition of nitrogenase synthesis by the alga so that in the symbiotic state high levels of activity can be sustained. Detailed studies on the role of the environment in regulating nitrogenase synthesis and on the inter-relations of nitrogen-fixing algae and their symbionts are urgently required.

3.6 REFERENCES

1. Mereschkovsky, C., Biol. Zbl., 25, 593 (1905).
2. Famintzin, A., Biol. Zbl., 27, 353 (1907).
3. Fogg, G. E., W. D. P. Stewart, P. Fay, and A. E. Walsby, The Blue-green Algae, Academic Press, London, 1973.
4. Fogg, G. E., in Algal Physiology and Biochemistry, W. D. P. Stewart, Ed., Blackwell, Oxford, pp. 560-582, 1974.

5. Stewart, W. D. P., in *The Biology of the Blue-Green Algae*, N. G. Carr and B. A. Whitton, Eds., Blackwell, Oxford, 1973.
6. Stewart, W. D. P., *Ann. Rev. Microbiol.*, 27, 283, (1973).
7. Winter, G., *Beitr. Biol. Pflanz.*, 23, 295 (1935).
8. Bowyer, J. W., and V. B. D. Skerman, *J. Gen. Microbiol.*, 54, 299 (1968).
9. Hoare, D. S., L. O. Ingram, E. L. Thurston, and R. Walkup, *Arch. Mikrobiol.*, 78, 310 (1971).
10. Allen, M. B., and D. O. Arnon, *Plant Physiol.*, 30, 366 (1955).
11. Gorham, P. R., J. S. McLachlan, U. T. Hammer, and W. K. Kim, *Verh. Int. Ver. Limnol.*, 15, 596 (1964).
12. Van Baalen, C., *Bot. Mar.*, 4, 129 (1962).
13. Provasoli, L., J. J. A. McLaughlin, and M. R. Droop, *Arch. Mikrobiol.*, 25, 392 (1956).
14. Stanier, R. Y., R. Kunisawa, M. Mandel, and G. Cohen-Bazire, *Bacteriol. Rev.*, 35, 171 (1971).
15. Dugdale, R. C., D. W. Menzel, and J. H. Ryther, *Deep Sea Res.*, 7, 298 (1961).
16. Dugdale, R. C., J. J. Goering, and J. H. Ryther, *Limnol. Oceanogr.*, 9, 507 (1964).
17. Bunt, J. S., K. E. Cooksey, M. A. Heeb, C. C. Lee, and B. F. Taylor, *Nature*, 227, 1163 (1970).
18. Taylor, B. F., C. C. Lee, and J. S. Bunt, *Arch. Mikrobiol.*, 88, 205 (1973).
19. Ramamurthy, V. D., and S. Krishnamurthy, *Curr. Sci.*, 37, 21 (1968).
20. Van Baalen, C., *J. Phycol.*, 3, 154 (1967).
21. Allen, M. M., and R. Y. Stanier, *J. Gen Microbiol.*, 51, 203 (1968).
22. Gerloff, G. C., G. P. Fitzgerald, and F. Skoog, *Am. J. Bot.*, 37, 216 (1950).
23. Kraus, M. P., *Nature (London)*, 211, 310 (1966).
24. Pintner, I. J., and L. Provasoli, *J. Gen. Microbiol.*, 18, 190 (1958).
25. Stewart, W. D. P., *Ann. Bot. Lond.*, N. S., 26, 439 (1962).
26. Fogg, G. E., *J. Exp. Biol.*, 19, 78 (1942).
27. Galloway, R. A., and R. W. Krauss, *Am. J. Bot.*, 46, 40 (1959).

28. Middlebrook, J. B., and R. O. Bowman, Appl. Microbiol., 12, 44 (1964).
29. Allen, M. B., Arch. Mikrobiol., 17, 34 (1952).
30. Bunt, J. S., Nature, 192, 1275 (1961).
31. Wieringa, K. T., Anton. van Leeuwenhoek, 32, 183 (1966).
32. Zehnder, A., and P. R. Gorham, Can. J. Microbiol., 6, 645 (1960).
33. Cyanophyceenkulturen. Anreicherungs- und Isolierverfahren. Zentbl. Bakt. Parasitenk., 1, Supplement Heft 1, 415 (1964).
34. Stewart, W. D. P., Plant Soil, 32, 555 (1970).
35. Kratz, W. A., and J. Myers, Am. J. Bot., 42, 282 (1955).
36. Allen, M. B., Sci. Mo., 83, 100 (1956).
37. Fogg, G. E., and Than-Tun, Proc. Roy. Soc. Lond. B., 153, 111 (1960).
38. Fay, P., and G. E. Fogg, Arch. Mikrobiol., 42, 310 (1962).
39. Pattnaik, H., Ann. Bot., 30, 231 (1965).
40. Fay, P., Biochim. Biophys. Acta, 216, 353 (1970).
41. Bothe, H., and E. Loos, Arch. Mikrobiol., 86, 241 (1972).
42. R. L. Lyne, and W. D. P. Stewart, Planta (Ber.), 109, 27 (1973).
43. Myers, J., Ann. Rev. Plant Physiol., 22, 289 (1971).
44. Gibbs, M., C. A. Fewson, and M. D. Schulman, Yb. Carnegie Inst., Wash., 62, 352 (1963).
45. Lazaroff, N., J. Phycol., 2, 7 (1966).
46. Lazaroff, N., in Taxonomy and Biology of Blue-green Algae, T. V. Desikarchary, Ed., University of Madras, Madras, pp. 521-544, 1972.
47. Kaushik, M., and H. D. Kumar, Arch. Mikrobiol., 74, 52 (1970).
48. Peat, A., and B. A. Whitton, Arch. Mikrobiol., 57, 155 (1967).
49. Hattori, A., and Y. Fujita, J. Biochem., 46, 521 (1959).
50. Hattori, A., and Y. Fujita, J. Biochem., 46, 1259 (1959).
51. Donze, M., H. J. Van Gorkom, and A. J. P. Raat, J. Gen. Microbiol., 69, xiv (1971).

References

52. Lex, M., W. B. Silvester, and W. D. P. Stewart, Proc. Roy. Soc. London. B., 180, 85 (1972).
53. Harder, R., Zeit. Bot., 9, 145 (1917).
54. Allison, F. E., and H. J. Morris, Sci. N. Y., 71, 221 (1930).
55. Allison, F. E., S. R. Hoover, and H. J. Morris, Bot. Gaz., 98, 433 (1937).
56. Fay, P., J. Gen. Microbiol., 39, 11 (1965).
57. Watanabe, A., and Y. Yamamoto, Nature (London), 214, 738 (1967).
58. Khoja, T., and B. A. Whitton, Arch. Mikrobiol., 79, 280 (1972).
59. Pearce, J., C. K. Leach, and N. G. Carr, J. Gen. Microbiol., 55, 371 (1969).
60. Hoare, D. S., and R. B. Moore, Biochim, Biophys. Acta, 109, 622 (1965).
61. Smith, A. J., J. London, and R. Y. Stanier, J. Bacteriol., 94, 972 (1967).
62. Pan, P., Can. J. Microbiol., 18, 275 (1972).
63. Pearce, J., and N. G. Carr, Biochem. J., 105, 45p. (1967).
64. Hoare, D. S., S. L. Hoare, and A. J. Smith, in Taxonomy and Biology of Blue-green Algae, T. V. Desikachary, Ed., University of Madras, Madras, pp. 501-507, 1972.
65. Kornberg, H. L., and J. Lascelles, J. Gen. Microbiol., 23, 511 (1960).
66. Carr, N. G., W. Hood, and C. K. Leach, in Taxonomy and Biology of Blue-green Algae, T. V. Desikachary, Ed., University of Madras, Madras, pp. 494-500, 1972.
67. Lazaroff, N., and W. Vishniac, J. Gen. Microbiol., 25, 365 (1961).
68. Van Baalen, C., D. S. Hoare, and E. Brandt, J. Bacteriol., 105, 685 (1971).
69. Fogg, G. E., and W. D. P. Stewart, Brit. Antarct. Surv. Bull., 15, 39 (1968).
70. Horne, A. J., Brit. Antarct. Surv. Bull., 27, 1 (1972).
71. Kallio, P., S. Suhonen, and H. Kallio, Rep. Kevo Subarctic Res. Stat., 9, 7 (1972).
72. Hitch, C. J. B., and W. D. P. Stewart, New Phytol., 72, 509 (1973).
73. Henriksson, E., Plant Soil, Special Vol., 415 (1971).

74. Haystead, A., R. Robinson, and W. D. P. Stewart, Arch. Mikrobiol., 74, 235 (1970).
75. Hardy, R. W. F., R. C. Burns, and R. D. Holsten, Soil Biol. Biochem., 5, 47 (1973).
76. Stewart, W. D. P., Phycologia, 9, 261 (1970).
77. Stewart, W. D. P., T. Mague, T. Brock, and R. H. Burris, (unpublished).
78. Peary, J. A., and R. W. Castenholz, Nature, 202, 720 (1964).
79. Fogg, G. E., J. Exp. Bot., 2, 117 (1956).
80. Robinson, R., and W. D. P. Stewart, (unpublished).
81. Jutono, Soil Biol. Biochem., 5, 91 (1973).
82. Bothe, H., Ber. Dtsch. Bot. Ges., 83, 421 (1970).
83. Haystead, A., and W. D. P. Stewart, Arch. Mikrobiol., 82, 325 (1972).
84. Walp, L., and R. Schobach, Growth, 6, 33 (1942).
85. Stewart, W. D. P., J. Gen. Microbiol., 36, 415 (1964).
86. John, R. P., Ann. Bot. Lond., 6, 323 (1942).
87. Lund, J. W. G., New Phytol., 46, 35 (1947).
88. Stewart, W. D. P., Nitrogen Fixation in Plants, Athlone Press of the University of London, London, 168 pp. (1966).
89. Granhall, U., and H. Selander, I. B. P. Swedish Tundra Biome Project Tech. Rep., 11 (1972).
90. Dooley, F., and J. A. Houghton, Brit. Phycol. J., 8, 289 (1973).
91. Dooley, F., and J. A. Houghton, Brit. Phycol. J., 8, 295 (1973).
92. Singh, R. N., The Role of Blue-green Algae in Nitrogen Economy of Indian Agriculture, Indian Council Agr. Res., 1961.
93. Singh, R. N., Physiology and Biochemistry of Nitrogen Fixation by Blue-green Algae (Final technical report 1967-1972), Dept. Bot., Banaras Hindu University, Varanasi-5, India, 66 pp.
94. Shtina, E. A., in Taxonomy and Biology of Blue-green Algae, T. V. Desikachary, Ed., University of Madras, Madras, pp. 294-295, 1972.
95. Stewart, W. D. P., J. Mar. Biol. Assoc., (in press).
96. Stewart, W. D. P., and H. W. Pearson, Proc. Roy. Soc. Lond. B., 175, 293 (1970).
97. Wolk, C. P., Ann. N. Y. Acad. Sci., 175, 641 (1970).
98. Bozniak, E. G., and L. L. Kennedy, Can. J. Bot., 46, 1259 (1968).

99. Sirenko, L. A., N. M. Stetsenko, V. V. Arendarchuk, and M. I. Kuz'menko, Microbiology, 37, 199 (1968).
100. Smith, R. V., and M. C. W. Evans, J. Bacteriol., 105, 913 (1971).
101. Stewart, W. D. P., Plant Soil, Special Vol., 377 (1971).
102. Bone, D. H., Arch. Mikrobiol., 80, 234, 242 (1971).
103. Weare, N. M., and J. F. Benemann, Arch. Mikrobiol., 90, 323 (1973).
104. Dalton, H., and J. R. Postgate, J. Gen. Microbiol., 54, 463 (1969).
105. Stewart, W. D. P., and M. Lex, Arch. Mikrobiol., 73, 250 (1970).
106. Wyatt, J. T., and J. K. G. Silvey, Sci. N. Y., 165, 908 (1969).
107. Rippka, R., A. Neilson, R. Kunisawa, and G. Cohen-Bazire, Arch. Mikrobiol., 76, 341 (1971).
108. Gallon, J. R., T. A. LaRue, and W. G. W. Kurz, Can. J. Microbiol., 18, 327 (1972).
109. Eyster, C., in Taxonomy and Biology of Blue-green Algae, T. V. Desikachary, Ed., University of Madras, Madras, pp. 508-520, 1972.
110. Smith, R. V., A. Telfer, and M. C. W. Evans, J. Bacteriol., 107, 574 (1971).
111. Holton, R. W., and J. Myers, Biochim. Biophys. Acta, 131, 362, 375 (1967).
112. Fujita, Y., and J. Myers, Arch. Biochem. Biophys., 112, 519 (1965).
113. Webster, D. A., and D. P. Hackett, Plant Physiol., 39, lix (1964).
114. Webster, D. A., and D. P. Hackett, Plant Physiol., 41, 599 (1965).
115. Biggins, J., J. Bacteriol., 99, 570 (1969).
116. Biggins, J., and W. E. Dietrich, Arch. Biochem. Biophys., 128, 40 (1969).
117. Yamanaka, T., S. Takenami, K. Wada, and K. Okunuki, Biochim. Biophys. Acta, 180, 196 (1969).
118. Mitsui, A., and D. I. Arnon, Physiol. Plant., 25, 135 (1971).
119. Rao, K. K., R. V. Smith, R. Cammack, M. C. W. Evans, and D. O. Hall, Biochem. J., 120, 1159 (1972).
120. Smillie, R. M., Biochem. Biophys. Res. Commun., 20, 621 (1965).

121. Knight, Jr., E., and R. W. F. Hardy, J. Biol. Chem., 241, 2752 (1966).
122. Bothe, H., Proc. Int. Congr. Phot. Stresa, 2nd, pp. 2169-2178 (1972).
123. Smith, R. V., R. J. Noy, and M. C. W. Evans, Biochim. Biophys. Acta, 253, 104 (1971).
124. Smith, R. V., and M. C. W. Evans, Proc. Int. Congr. Phot. Stresa, 2nd (1972).
125. Bortels, H., Arch. Mikrobiol., 11, 155 (1940).
126. Fogg, G. E., Ann. Bot. Lond. N. S., 13, 241 (1949).
127. Wolfe, M., Ann. Bot. Lond. N. S., 18, 299 (1954).
128. McKenna, C. E., J. R. Benemann, and T. G. Traylor, Biochem. Biophys. Res. Commun., 41, 1501 (1971).
129. Burns, R. C., W. H. Fuchsman, and R. W. F. Hardy, Biochem. Biophys. Res. Commun., 42, 853 (1971).
130. Holm-Hansen, O., G. C. Gerloff, and F. Skoog, Physiol. Plant., 7, 665 (1954).
131. Buddhari, W., Ph. D. Thesis, University of California, Berkeley; quoted in Ref. 133.
132. Johnson, G. V., P. A. Mayeux, and H. J. Evans, Plant Physiol., 41, 852 (1966).
133. Evans, H. J., and S. A. Russell, in The Chemistry and Biochemistry of Nitrogen Fixation, J. R. Postgate, Ed., Plenum Press, London/New York, pp. 191-244, 1971.
134. Bone, D. H., Arch. Mikrobiol., 80, 147 (1971).
135. Stewart, W. D. P., G. P. Fitzgerald, and R. H. Burris, Proc. Nat. Acad. Sci. U. S., 66, 1104
136. Stewart, W. D. P., and G. Alexander, Freshwat. Biol., 1, 389 (1971).
137. Cox, R. M., and P. Fay, Proc. Roy. Soc. London. B, 172, 357 (1969).
138. Lex, M., and W. D. P. Stewart, Biochim. Biophys. Acta, 291, 64 (1973).
139. Scott, W. E., and P. Fay, Brit. Phycol. J., 7, 283 (1972).
140. Dalton, H., and L. E. Mortenson, Bacteriol. Rev., 36, 231 (1972).
141. Bone, D. H., Arch. Mikrobiol., 86, 13 (1972).
142. Stewart, W. D. P., G. P. Fitzgerald, and R. H. Burris, Arch. Mikrobiol., 62, 336 (1968).
143. Neilson, A., R. Rippka, and R. Kunisawa, Arch. Mikrobiol., 76, 139 (1971).
144. Ohmori, M., and A. Hattori, Arch. Mikrobiol., 13, 589 (1972).

145. Shah, V. K., L. C. Davis, and W. J. Brill, Biochim. Biophys. Acta, 256, 498 (1972).
146. Wyatt, J. T., G. G. Lawley, and R. D. Barnes, Naturwissenschaften, 58, 570 (1971).
147. Dharmawardene, M. W. N., W. D. P. Stewart, and S. O. Stanley, Planta (Ber.), 108, 133 (1972).
148. Neilson, A., and M. Doudoroff, Arch. Mikrobiol., 89, 15 (1973).
149. Hoare, D. S., S. L. Hoare, and R. B. Moore, J. Gen. Microbiol., 49, 351 (1967).
150. Haystead, A., M. W. N. Dharmawardene, and W. D. P. Stewart, Plant Sci. Lett., 1, 439 (1973).
151. Hood, W., and N. G. Carr, J. Gen. Microbiol., 73, 417 (1972).
152. Allen, M. M., and A. J. Smith, Arch. Mikrobiol., 69, 114 (1969).
153. Simon, R. D., Proc. Nat. Acad. Sci. U. S., 68, 265 (1971).
154. Lang, N. J., and K. A. Fisher, Arch. Mikrobiol., 67, 173 (1969).
155. Stewart, W. D. P., Proc. Roy. Soc. Edinb. B, 71, 209 (1972).
156. Lang, N. J., in Taxonomy and Biology of Blue-green Algae, T. V. Desikachary, Ed., University of Madras, Madras, pp. 6-12, 1972.
157. Bulen, W. A., R. C. Burns, and J. R. LeComte, Proc. Nat. Acad. Sci. U. S., 53, 532 (1965).
158. Fogg, G. E., and Than-Tun, Biochim. Biophys. Acta, 30, 209 (1958).
159. Cobb, H. D., and J. Myers, Am. J. Bot., 51, 753 (1964).
160. Davis, E. B., and R. G. Tischer, Nature (London), 212, 302 (1966).
161. Cox, R. M., Arch. Mikrobiol., 53, 263 (1966).
162. Jackson, W. A., and R. J. Volk, Ann. Rev. Plant Physiol., 21, 385 (1970).
163. Tolbert, N. E., in Algal Physiology and Biochemistry, W. D. P. Stewart, Ed., Blackwell, Oxford, pp. 474-504, 1974.
164. Anderson, L. E., G. B. Price, and R. C. Fuller, Science, 161, 482 (1968).
165. Grodzinski, B., and B. Colman, Plant Physiol., 45, 735 (1970).
166. Heber, U., and C. S. French, Planta (Ber.), 79, 99 (1968).
167. Giesy, R. M., Am. J. Bot., 51, 388 (1964).

168. Chao, L., and C. C. Bowen, in Taxonomy and Biology of Blue-green Algae, T. V. Desikachary, Ed., University of Madras, Madras, pp. 12-17, 1972.
169. Carr, N. G., Biochim. Biophys. Acta, 120, 308 (1966).
170. Jensen, T. E., and L. M. Sicko, J. Bacteriol., 106, 683 (1971).
171. Leach, C. K., and N. G. Carr, Biochim. Biophys. Acta, 245, 165 (1971).
172. Cheung, W. Y., and M. Gibbs, Plant Physiol., 41, 351 (1965).
173. Wildon, D. C., and T. Rees, Plant Physiol., 40, 332 (1965).
174. Pearce, J., and N. G. Carr, J. Gen Microbiol., 54, 451 (1969).
175. Pelroy, R. A., and J. A. Bassham, Arch. Mikrobiol., 86, 25 (1972).
176. Pelroy, R. A., R. Rippka, and R. Y. Stanier, Arch. Mikrobiol., 87, 302 (1972).
177. Kratz, W. A., and J. Myers, Plant Physiol., 30, 275 (1955).
178. Postgate, J. R., in The Chemistry and Biochemistry of Nitrogen Fixation, J. R. Postgate, Ed., Plenum Press, London, pp. 161-190, 1971.
179. Fay, P., in The Biology of Blue-green Algae, N. G. Carr and B. A. Whitton, Ed., Blackwell, Oxford, pp. 238-259, 1973.
180. Borzi, A., Nuovo Giorn. Bot. Ital., 10, 236 (1878).
181. Singh, R. N., and D. N. Tiwari, Nature (London), 221, 62 (1969).
182. de Puymaly, A., Botaniste, 41, 209 (1957).
183. Allsopp, A., Nature (London), 220, 810 (1968).
184. Carter, H. J., Ann. Mag. Nat. Hist. Ser. II, 18, 115, 221 (1856).
185. Fritsch, F. E., Proc. Linn. Soc. London., 162, 194 (1951).
186. Wolk, C. P., Am. J. Bot., 53, 260 (1966).
187. Lang, N. J., and P. Fay, Proc. Roy. Soc. Lond. B, 178, 193 (1971).
188. Kale, S. R., and E. R. S. Talpasayi, Ind. Biol., 1, 19 (1969).
189. Wilcox, M., Nature (London), 228, 686 (1970).
190. Wilcox, M., Nature (London), 239, 110 (1972).
191. Nutman, P. S., Ann. Bot. Lond. N. S., 12, 81 (1948).

References

192. Kulasooriya, S. A., N. J. Lang, and P. Fay, Proc. Roy. Soc. Lond. B, 181, 199 (1972).
193. Fay, P., H. D. Kumar, and G. E. Fogg, J. Gen. Microbiol., 35, 351 (1964).
194. Ogawa, R. A., and J. F. Carr, Limnol. Oceanogr., 14, 342 (1969).
195. Jewell, W. J., and S. A. Kulasooriya, J. Exp. Bot., 21, 874 (1970).
196. Donze, M., J. Haveman, and P. Schiereck, Biochim. Biophys. Acta, 256, 157 (1972).
197. Fay, P., Arch. Mikrobiol., 67, 62 (1969).
198. Wolk, C. P., and R. D. Simon, Planta (Ber.), 96, 92 (1969).
199. Thomas, J., Nature (London), 228, 181 (1969).
200. Stewart, W. D. P., A. Haystead, and H. W. Pearson, Nature (London), 224, 226 (1969).
201. Bradley, S., and N. G. Carr, J. Gen. Microbiol., 68, xiii (1971).
202. Thomas, J., and K. A. V. David, Nature, New Biol., 238, 219 (1972).
203. Drawert, H., and I. Tischer, Naturwissenschaften, 43, 132 (1956).
204. Tischer, I., Arch. Mikrobiol., 27, 400 (1957).
205. Talpasayi, E. R. S., Curr. Sci., 7, 190 (1967).
206. Kurz, W. G. W., and T. A. LaRue, Naturwissenschaften, 58, 417 (1971).
207. Wolk, C. P., and E. Wojciuch, Planta (Ber.), 97, 126 (1971).
208. Fay, P., and S. A. Kulasooriya, Arch. Mikrobiol., 87, 341 (1972).
209. Wolk, C. P., J. Bacteriol., 96, 2139 (1968).
210. Wolk, C. P., and E. Wojciuch, J. Phycol., 7, 339 (1971).
211. Wolk, C. P., Ann. N. Y. Acad. Sci., 175, 641 (1970).
212. Fay, P., W. D. P. Stewart, A. E. Walsby, and G. E. Fogg, Nature (London), 220, 810 (1968).
213. Stewart, W. D. P., in Taxonomy and Biology of Blue-green Algae, T. V. Desikachary, Ed., University of Madras, Madras, pp. 227-235, 1972.
214. Ohmori, M., and A. Hattori, Pl. Cell Physiol., 13, 589 (1972).
215. Van Gorkom, H. J., and M. Donze, Nature (London), 234, 231 (1972).
216. Fogg, G. E., Oceanog. Mar. Biol. Ann. Rev., 4, 195 (1966).

217. Hood, W., A. G. Leaver, and N. G. Carr, Biochem. J., 114, 12 (1969).
218. Fogg, G. E., Proc. Roy. Soc. Lond. B, 139, 372 (1952).
219. Fogg, G. E., Arch. Hydrobiol. Beih. Ergebn. Limnol., 5, 1 (1971).
220. Stewart, W. D. P., Nature, 200, 1020 (1963).
221. Jones, K., and W. D. P. Stewart, J. Mar. Biol. Ass. U. K., 49, 475 (1969).
222. Jones, K., and W. D. P. Stewart, J. Mar. Biol. Ass. U. K., 49, 701 (1969).
223. Williams, A. E., and R. H. Burris, Am. J. Bot., 39, 340 (1952).
224. Fogg, G. E., and H. Pattnaik, Phykos, 5, 58 (1966).
225. Thomas, J. and K. A. V. David, in Proc. Symp. Cellular Processes in Growth, Development and Differentiation, Bhabha Atomic Res. Centre, Nov. 22-24, 1971, pp. 401-411.
226. Walsby, A. E., Ph. D Thesis, University of London (1971).
227. Goryunova, S. V., and G. N. Rhzanova, in Biology of the Cyanophyta, V. D. Federov and M. M. Telitchenko, Eds., Moscow University Press, Moscow, pp. 111-118, 1964.
228. Taha, E. E. M., and A. E. M. H. Elrefai, Arch. Mikrobiol., 43, 67 (1962).
229. Venkataraman, G. S., and H. K. Saxena, Ind. J. Agr. Sci., 33, 21 (1963).
230. Whitton, B. A., J. Gen. Microbiol., 40, 1 (1965).
231. Walsby, A. E., Brit. Phycol. Bull., 2, 514 (1965).
232. Jakob, H., C. R. Hebd. Seanc. Acad. Sci., Paris, 238, 2018 (1954).
233. Jakob, H., C. R. Hebd. Seanc. Acad. Sci., Paris, 244, 19680 (1957).
234. Millbank, J. W., New Phytol., 71, 1 (1972).
235. Wolk, C. P., S. M. Austin, J. Bortins, and A. Galonsky, J. Cell. Biol., 61, 440 (1974).
236. Millbank, J. W., and K. A. Kershaw, New Phytol., 68, 721 (1969).
237. Bergersen, F. J., G. S. Kennedy, and W. Whittman, Aust. J. Biol. Sci., 18, 1135 (1965).
238. Boulware, M. A., and J. A. Edmisten, A. S. B. Bull., 16, 45 (1969).
239. Silvester, W. B., and D. R. Smith, Nature, 224, 1231 (1969).

References

240. Bjälfve, G., Physiol. Plant., 15, 122 (1962).
241. Stewart, W. D. P., Nature (London), 214, 603 (1967).
242. Mayland, H. F., and T. H. McIntosh, Nature (London), 209, 421 (1966).
243. Singh, R. N., Ind. J. Agr. Sci., 12, 743 (1942).
244. Venkataraman, G. S., Ind. J. Agr. Sci., 32, 22 (1962).
245. Fogg, G. E., J. Expt. Biol., 19, 78 (1942).
246. Davis, E. B., R. G. Tischer, and L. R. Brown, Physiol. Plant., 19, 823 (1966).
247. De, P. K., Proc. Roy. Soc. Lond. B, 127, 121 (1939).
248. Cameron, R. E., and W. H. Fuller, Soil Sci. Soc. Am. Proc., 24, 353 (1960).
249. Drewes, K., Zentralbl. Bakteriol. Parasitenk. II, 76, 84 (1928).
250. Watanabe, A., Arch. Biochem. Biophys., 34, 50 (1951).
251. Taha, M. S., Microbiology, 32, 421 (1963).
252. Taha, M. S., Microbiology, 33, 352 (1964).
253. Venkataraman, G. S., Proc. Nat. Acad. Sci., India, 31, 100 (1961).
254. Pankow, H., Naturwissenschaften, 51, 274 (1964).
255. Mitra, A. J., Proc. Nat. Acad. Sci., India, Sec. A, 31, 98 (1961).
256. Fogg, G. E., J. Exp. Bot., 2, 117 (1951).
257. Schneider, K. C., C. Bradbeer, R. N. Singh, L. C. Wang, P. W. Wilson, and R. H. Burris, Proc. Nat. Acad. Sci. U. S., 46, 726 (1960).
258. Herisset, A., Bull. Soc. Chim. Biol. Paris, 34, 532 (1952).
259. Watanabe, A., and T. Kiyohara, in Microalgae and Photosynthetic Bacteria, Jap. Soc. Plant Physiol., Tokyo, pp. 189-196, 1963.
260. Venkataraman, G. S., Ind. J. Agr. Sci., 31, 213 (1961).
261. Watanabe, A., J. Gen. Appl. Microbiol. Tokyo, 5, 21 (1959).
262. Davies, M., and W. D. P. Stewart, (unpublished).
263. Pankow, H., and B. Martens, Arch. Mikrobiol., 48, 203 (1964).
264. Kenyon, C. N., R. Rippka, and R. Y. Stnaier, Arch. Mikrobiol., 83, 216 (1972).

265. Tubb, R. S. *Nature (London)*, *251*, 481 (1974).
266. Shanmugam, K. T., S. L. Streicher, C. Morandi, F. Ausubel, R. B. Goldberg and R. C. Valentine, *Proc. 1st Int. Symp. on Nitrogen Fixation*, W. E. Newton and C. J. Nyman, Eds., *2* (1976).
267. Rodgers, G. A., and W. D. P. Stewart, *Brit. Phycol. J.*, *9*, 223 (1974).
268. Warmling, P., *Bot. Mar.*, *16*, 237 (1973).
269. Fay, P., and L. Vasconcelos, *Arch. Microbiol.*, *99*, 221 (1974).
270. Stewart, W. D. P., and P. J. Bottomley in *International Symposium on Nitrogen Fixation*, W. Newton and C. Nyman, Eds., Washington State University Press, (in press, 1975).
271. Stewart, W. D. P., *Proc. 50th Soc. Expt. Biol. Meeting*, Pergamon, New York, pp. 235-246, 1976.
272. Stewart, W. D. P., and P. Rowell, *Arch. Microbiol., Biochem. Biophys. Res. Comm.*, *65*, 846 (1975).
273. Stewart, W. D. P., P. Rowell, and E. El-Tor, *Biochem. Soc. Trans.*, *3*, 357 (1975).
274. Lea, P., and B. J. Miflin, *Biochem. Soc. Trans.*, *3*, 381 (1975).
275. Stewart, W. D. P., A. Haystead, and M. W. N. Dharmawardene, in *Nitrogen Fixation by Free-Living Micro-Organisms*, W. D. P. Stewart, Ed., Cambridge University Press, pp. 129-158, 1975.
276. Rowell, P., and W. D. P. Stewart, *Arch. Microbiol.*, *107*, 115 (1976).
277. Lawrie, A. C., G. A. Codd, and W. D. P. Stewart, (unpublished).
278. Simon, R. D., *Arch. Microbiol.*, *92*, 115 (1973).
279. Simon, R. D., *J. Bacteriol.*, *114*, 1213 (1973).
280. Bothe, H., and B. Falkenberg, *Plant Sci. Letts.*, *1*, 151 (1973).
281. Bothe, H., B. Falkenberg, and U. Nolteernsting, *Arch. Microbiol.*, *96*, 291 (1975).
282. Winkenbach, F., and C. P. Wolk, *Plant Physiol.*, *52*, 480 (1973).
283. Lex, M., and N. G. Carr, *Arch. Microbiol.*, *101*, 161 (1974).
284. Fleming, H., and R. Haselkorn, *Proc. Nat. Acad. Sci. U. S.*, *70*, 2727 (1973).
285. Fleming, H., and R. Haselkorn, *Cell*, *3*, 159 (1974).

References

286. Tel-Or, E., and W. D. P. Stewart, Biochim. Biophys. Acta, 423, 189 (1976).
287. Rippka, R., Arch. Microbiol., 87, 93 (1972).
288. Carr, N. G. and S. Bradley, Symp. Soc. Gen. Microbiol., 23, 161 (1973).
289. Tel-Or, E., and W. D. P. Stewart, (unpublished).
290. Codd, G. A., P. Rowell, and W. D. P. Stewart, Biochem. Biophys. Res. Comm., 61, 374 (1974).
291. Rowell, P., Enticott, S., and W. D. P. Stewart (in preparation).

CHAPTER 4

Lower Plant Associations

J. W. MILLBANK

Botany Department
Imperial College
London, England

4.1. Introduction, 126
4.2. Associations with Fungi; Lichens, 127
    4.2.1. Culture Methodology; Growth, 127
    4.2.2. Physiology, 129
        4.2.2.1. Morphology of the Thallus, 129
        4.2.2.2. Dinitrogen-fixing Ability, 129
        4.2.2.3. The Role of N_2-fixing Bacteria, 133
        4.2.2.4. The Rate of N_2 Fixation, 134
        4.2.2.5. Release of Fixed Nitrogen by the Phycobiont, 136
        4.2.2.6. Translocation of Nitrogen Compounds between the Symbionts, 139
        4.2.2.7. The Effect of Environmental Factors: Moisture, Light and Temperature, 139
4.3. Associations with Green Plants, 141
    4.3.1. Liverworts, 141
        4.3.1.1. Anthoceros, 141
        4.3.1.2. Blasia, 142
        4.3.1.3. Cavicularia, 143
    4.3.2. Mosses, 144
    4.3.3. Pteridophytes, 144
        4.3.3.1. Azolla, 144
4.4. Other Associations, 147
    4.4.1. Endocyanoses, 147

4.5. The Relationship of Lichens to Other N_2-fixing Symbiotic Systems, 147
4.6. References, 148

4.1 INTRODUCTION

The Cyanophyceae, or blue-green algae, in addition to being widespread in distribution in the free-living state, are able to form associations with a range of other organisms. The algae are frequently motile, an attribute that is believed to contribute to their ability to invade spaces and enter tissues in their host plants. Dinitrogen fixation is not related here to the symbiotic mode of life, all the blue-green algae found living in associations being capable of fixation when isolated and cultured independently. These N_2-fixing symbioses are, therefore, fundamentally different from the legume-Rhizobium association, in which fixation is strictly dependent on the symbiotic association except under very specific experimental conditions.

The range of variation in these relationships between blue-green algae and lower plants is immense, from Glaucocystis, an intracellular association with a colorless form of the green alga Oocystis, through the liverworts, e.g., Anthoceros in which the alga may inhabit cavities on the lower surface of the thallus, to the very much defined and consistent associations with fungi in lichens. Higher up the scale in the plant kingdom, the pteridophyte Azolla is noted for its association with Anabaena azollae in cavities on the leaves. Finally, there are the root nodules of the cycads, e. g., Ceratozamia and Encephalatos, which have Nostoc as the endophyte, and the swellings at the bases of the petioles of the angiosperm Gunnera, in which Nostoc is also found, growing intracellularly and heterotrophically. These last, being classified as higher plants are outside the scope of this account and will be dealt with in III.6.

It is very evident that there is little or no uniformity in the associations with the various plant groups, and so the association with each will be dealt with separately. Of the groups, the lichens have received by far the most attention, their attraction for biologists exceeding all the other groups combined.

4.2 ASSOCIATIONS WITH FUNGI; LICHENS

4.2.1 Culture Methodology; Growth

Due to the slow rate of growth of lichens, experimentation on the intact thallus has had to be carried out using material collected from the field. The growth rate is quite closely correlated with climatic conditions, periods of high humidity and cool temperatures being optimal. However, growth rates vary considerably among individual thalli of the same species subjected to differing microclimates, and even between adjacent lobes of the same thallus. Annual growth increments of many lichens can be as little as 1 mm or less, the crustose forms being slowest. The thallose forms are much more rapid growers, and of these Peltigera species are among the fastest, annual radial increments of thallus growth of 2-3 cm being not uncommon (1).

Although the slow growth rates and sensitivity to environmental change present considerable obstacles to experimental work, a number of techniques for studying the intact thallus have been developed. It is usually most advantageous to collect material in spring and autumn in temperate climates, when it is moist, and to transport it to the laboratory as rapidly as possible in closed polyethylene bags. Subsequent storage should be on moist filter paper in petri dishes at 5-10°C in the light [250 candela (cd)], and if no specialized incubator is available the material should be used immediately or as soon as possible within 48 hours. Dry material, and material arriving by mail should always be washed clean with distilled water and stored moist in the light at 5-10°C for 24 hours; in this way much metabolic variation can be avoided. For periods of storage of up to 3 months, a specialized incubator is necessary, providing cool (8-10°C), moist conditions with a regular light cycle and air free from SO_2, CO, and organic gases and vapors (2).

Before experimentation, thorough washing of the moist thallus with sterile distilled water (analogous to rain) and careful removal of vegetable material enmeshed in the rhizinae are necessary. Surface sterilization to minimize epiphytic bacteria and algae is of very doubtful value, the damage far exceeding any physiological contribution presumed to arise from the epiphytes.

The thallus is then sampled, discs of known area being widely employed. In dinitrogen fixation studies individual 1 cm^2 discs are of ample size for very short-term estimates using the acetylene reduction technique (IV.12).

The discs are shaken at 25°C in 0.5 ml of mineral salts medium (e.g., Allen and Arnon, ref. 3) in rubber serum capped glass vials of 6-7 ml capacity with an appropriate gas phase and illuminated at 600-700 cd. The shaking rate must be adequate to permit rapid gas equilibration; 120 cycles/min typical of Warburg respirometer practice is satisfactory. This technique can demonstrate the variability of material within a single thallus specimen. For many purposes, especially those requiring estimates of internal metabolites, much larger numbers of discs may be used to reduce the variation between replicate batches.

Since any N_2-fixing activity is due to the blue-green alga symbiont, meaningful comparisons can only be made when fixation is related to the algal content of a thallus specimen, especially as this is known to vary with habitat, climate, and season. Estimates of cell numbers are often useful, and Hill and Woolhouse (4) describe a technique in which the thallus is homogenized after preliminary treatment with chromium trioxide (5% wt/vol). Extracted pigments can also provide a useful means of estimating algal populations (4, 5). For many studies, e.g., the fate of fixed nitrogen, growth in pure culture, and as standards for pigment content, separated symbionts are required. Quantitative separation of the fungal symbiont is not at present practicable, but dissection techniques can be useful and are available for some lichens, e.g., Peltigera polydactyla (6). This also gives a preparation in which the alga is very substantially freed of fungus. More complete separations of the alga require differential centrifugation procedures (7, 8). In the latter reference sucrose density gradient techniques gave a quantitative separation of the green alga, Coccomyxa, from Peltigera aphthosa; quantitative recovery of the blue-green algal component of any lichen has not yet been achieved, due to the fragility of the cells; very large amounts are lost during the initial homogenization (5, 71).

Associations with Fungi; Lichens   129

   For isolation of the symbionts in pure culture, a
combination of differential centrifugation, micro-
manipulation and selective media have been successful,
in conjunction with the more routine microbiological
techniques of dilution, streaking, ultraviolet irradiation, and the use of antibiotics either for preliminary washing or in the growth media. The algae are
relatively more straightforward to isolate and grow
than the fungi, which often have exacting nutritional
requirements or ascospores of extremely low viability.
   Detailed accounts of isolation and pure culture
methods and techniques for studying the whole thallus
are available (9, 10).

4.2.2   Physiology

4.2.2.1   Morphology of the Thallus

The lichen symbiosis has evolved into very many
morphological forms and a description of these is not
included, but two major forms of internal organization
are diagrammatically illustrated in Fig. 4.1A and B.
The heteromerous or layered type (B) is the more
advanced, with the phycobiont or algal symbiont confined to a specific region of the mycobiont or fungal
tissue, usually just below the upper cortex. Also
shown (C) are structures known as cephalodia. These
occur on certain lichens either internally or on the
upper surface of the thallus. They contain blue-green
algae enclosed within fungal mycelia. Such lichens
are thus composed of three components: a fungus and
two algae, one of the latter being blue-green. In the
homoiomerous type (A) the phycobiont is uniformly
distributed. Dinitrogen-fixing lichens of both structural types occur. The electron micrograph section of
a heteromerous lichen, Peltigera canina (Fig. 4.2)
shows that the algal cells are very closely apposed to
the fungal hyphae and surrounded by an electron-
transparent mucilagenous sheath. The possible significance of these features in connection with N_2-fixing
activity will be referred to later.

4.2.2.2   Dinitrogen-fixing Ability

Only lichens with a phycobiont from the Cyanophyceae
have been shown to fix N_2. Eight genera of blue-green

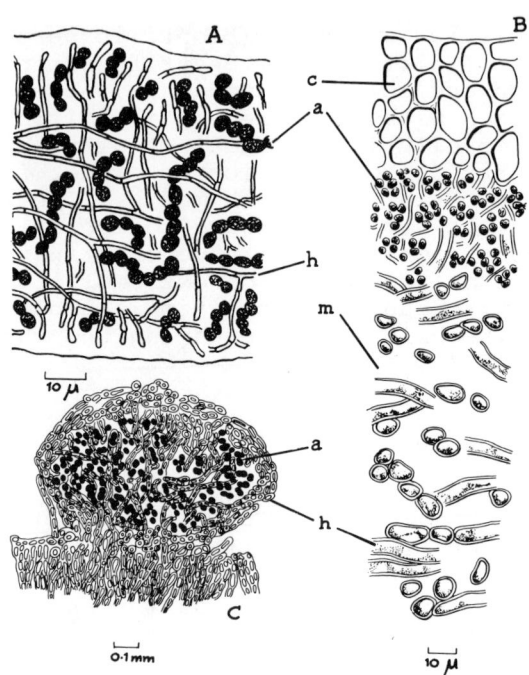

Fig. 4.1. Lichens. Structure of the thallus. (A) Homoiomerous type, Collema sp. (B) Heteromerous type, Peltigera horizontalis. (C) A cephalodium of P. aphthosa. Abbreviations: a, algal cells; h, fungal hyphae; c, cortex; m, medulla.

algae have been listed as lichen phycobionts (11, 12). The phycobionts of Collema tenax (13) and Peltigera virescens (14) fix N_2 when isolated, and it can be assumed that fixation takes place in the lichen as well. In the first definitive demonstration of N_2 fixation in an intact lichen (15), $^{15}N_2$ was used on Collema granosum and Leptogium lichenoides. Those lichens and isolated phycobionts that have been definitively shown to fix are listed in Table 4.1.

Fig. 4.2. Electron micrograph of a section of the thallus of the lichen, Peltigera canina. Note the close juxtaposition of algal cells with their electron-transparent sheath and fungal hyphae. Photograph by courtesy of A. D. Greenwood and H. B. Griffiths, using 1000 Kv EM7 microscope, Imperial College, London.

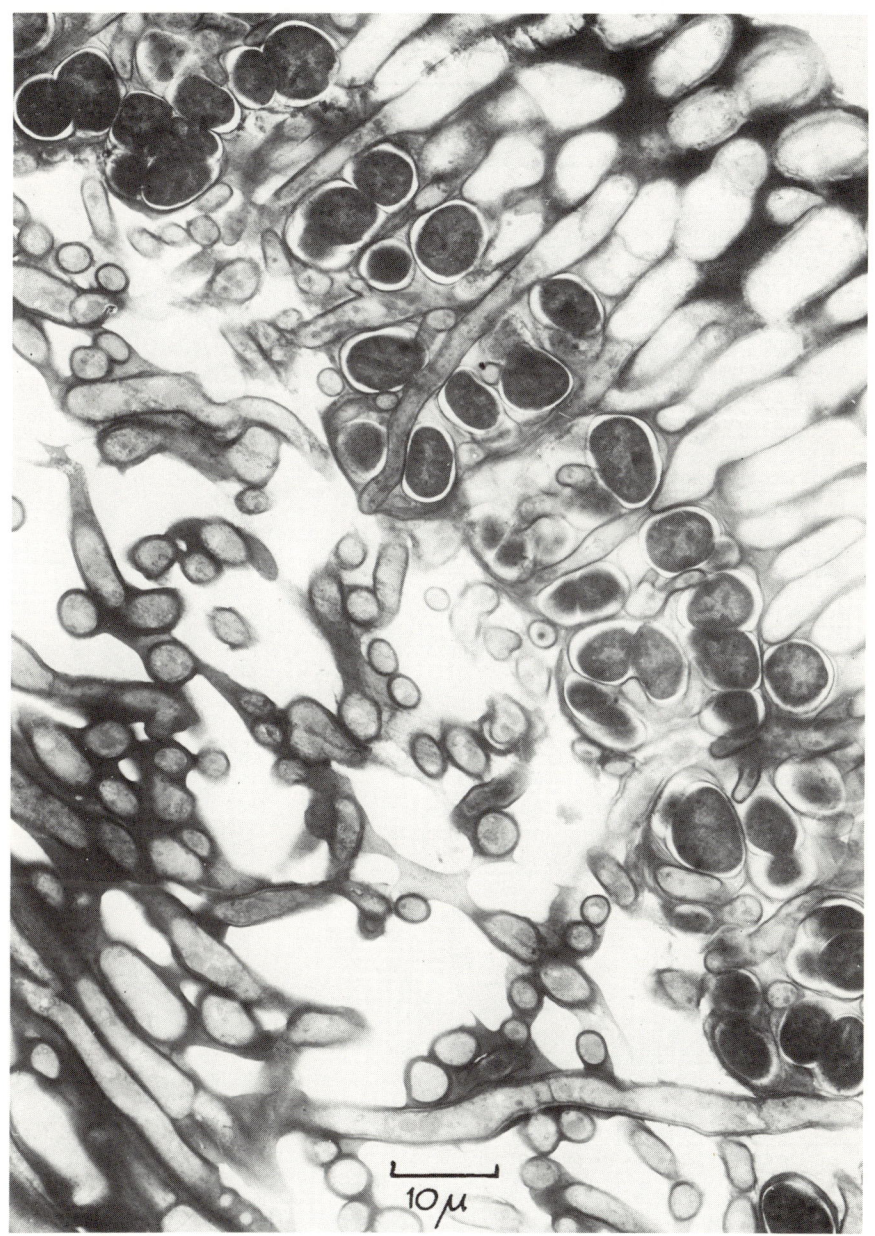

TABLE 4.1. Established Dinitrogen-fixing Lichen Species

Lichen	Phycobiont genus	Technique used	Reference
Collema auriculatum	Nostoc	C_2H_2	72
C. coccophorus	Nostoc	$^{15}N_2$	16
C. crispum	Nostoc	C_2H_2	17
C. fluviatile	Nostoc	C_2H_2	72
C. furfuraceum	Nostoc	C_2H_2	72
C. granosum	Nostoc	$^{15}N_2$	15
C. pulposum	Nostoc	$^{15}N_2$	18
C. subfervum	Nostoc	C_2H_2	72
C. tuniforme	Nostoc	C_2H_2	19
Dendriscocaulon umhausense	Scytonema	C_2H_2	73
Ephebe lanata	Stigonema	C_2H_2	72
Leptogium burgessii	Nostoc	C_2H_2	72
L. lichenoides	Nostoc	$^{15}N_2$	15
L. sinuatum	Nostoc	C_2H_2	72
L. teretiusculum	Nostoc	C_2H_2	72
Lichina confinis	Calothrix	$^{15}N_2$	20
L. pygmaea	Calothrix	$^{15}N_2$	20
Lobaria pulmonaria	Nostoc (cephalodia)	C_2H_2	72
L. scrobiculata	Nostoc	C_2H_2	72
Massalongia carnosa	Scytonema	C_2H_2	72
Nephroma arcticum	Nostoc	C_2H_2	75
N. laevigatum	Nostoc	C_2H_2	72
Pannaria pezizoides	Nostoc	C_2H_2	72
P. rubiginosa	Nostoc	C_2H_2	72
Parmeliella atlantica	Nostoc	C_2H_2	72
P. plumbea	Nostoc	C_2H_2	72

(Continued)

TABLE 4.1. (Continued)

Lichen	Phycobiont genus	Technique used	Reference
Peltigera aphthosa var. variolosa	Nostoc (cephalodia)	$^{15}N_2$	21
P. canina	Nostoc	C_2H_2	21
P. evansiana	Nostoc	C_2H_2	74
P. polydactyla	Nostoc	$^{15}N_2$	14
P. praetextata	Nostoc	$^{15}N_2$	22
P. pruinosa	Nostoc	$^{15}N_2$	14
P. rufescens	Nostoc	C_2H_2	19
Placopsis gelida	n. d. (cephalodia)	C_2H_2	17
Placynthium nigrum	Dichothrix	C_2H_2	72
P. pannariellum	Dichothrix	C_2H_2	72
Polychidium muscicola	Scytonema	C_2H_2	72
Pseudocyphellaria thouarsii	Nostoc	C_2H_2	72
Solorina crocea	Nostoc (intl. layer)	C_2H_2	75
S. saccata	Nostoc (intl. layer)	C_2H_2	72
S. spongiosa	Nostoc (cephalodia)	C_2H_2	72
Stereocaulon sp.	Nostoc (cephalodia)	$^{15}N_2$	18
S. paschale	Nostoc (cephalodia)	C_2H_2	76
Sticta limbata	Nostoc	C_2H_2	72
S. fuliginosa	Nostoc	C_2H_2	72

4.2.2.3 The Role of N_2-fixing Bacteria

Reports on the association of N_2-fixing bacteria with lichens have appeared from time to time. Their presence in the cephalodia of Peltigera aphthosa was reported by Cengia Sambo (23), who put forward a theory of polysymbiosis. Both Cengia Sambo (24) and Goebel (25) considered cephalodia to be analogous to leguminous root nodules. Goebel, however, decided that the Nostoc in the cephalodia was the N_2-fixing organism. Azotobacter was reported as associated with epiphytic lichens (26), and these findings were later extended (27, 28) to other lichens, including Cladonia, which had originally been reported as not possessing Azotobacter. Scott (22) refers to the unpublished findings of Metcalfe on the isolation of Azotobacter from numerous British lichens, including Cladonia impexa. However, other findings (29, 30) were wholly negative, with no Azotobacter detected in any of the 40 samples of lichens examined. Rather more recently Panosyan and Nikogosyan (31) did not find Azotobacter in any of the Armenian lichens they examined and attributed this to the absence of any respirable substrate. In their investigation of various symbiotic systems for N_2 fixation, using $^{15}N_2$, Bond and Scott (15) concluded that although Azotobacter could have been present epiphytically on the thallus samples that they studied, they made no significant contribution to the N_2 fixation observed. This conclusion was based on the lack of the necessary respirable substrate, and the absence of anything like the number of bacterial cells needed. Scott (22) extended these studies to Peltigera praetextata where he demonstrated, again with $^{15}N_2$, that dinitrogen fixation was entirely due to the Nostoc symbiont. Any contribution by epiphytic Azotobacter was regarded as insignificant. Furthermore, he reported that his trials with Cladonia impexa were uniformly negative for Azotobacter despite frequent earlier reports of its presence.

Finally, electron micrography of Peltigera canina failed to reveal the presence of a third symbiont (32), and it must be concluded that Azotobacter can only be a casual epiphyte of lichens, making an insignificant contribution to their nitrogen economy.

4.2.2.4 The Rate of N_2 Fixation

The majority of the studies referred to in III.4.2.2.2 have dealt with N_2 fixation either qualitatively or on a "whole thallus" basis. Millbank and Kershaw (8) attempted to relate N_2 fixation in lichen thalli to the population of the N_2-fixing symbiont. They found that the total nitrogen content of Peltigera aphthosa was approximately 3% of the dry weight, and that the algal cells contained 7.5% of the total nitrogen in the thallus and 2.4% in the cephalodia. Later work (5, 71) showed that the figure of 2.4% was much too low due to separation losses. More reliable figures are 35% for the cephalopodia of P. aphthosa and 8% for P. canina. Thus the phycobiont was less abundant than reported for P. praetextata (22), where Scott estimated the algal content by visual observation of sections, a simple and straightforward technique, but liable to rather large errors and uncertainties. Millbank and Kershaw (8) further showed that the total nitrogen in the cephalodia was 2-4% of that in the entire lichen, so clearly the Nostoc nitrogen is only a very small fraction (35% of 3%) of the total thallus nitrogen. Thus, although the total amount of N_2 fixed per unit of thallus was small, the rate of fixation by the Nostoc was large, and, if used entirely for growth, would have permitted a generation of 18 hours at 25° and 24 hours at 12°, a representative noon temperature for the natural habitat in spring. This is similar to a reported generation time of 19.5 hours for Nostoc muscorum in shake culture at 25° on nitrogen-free medium (33).

Later studies (5, 71) on Peltigera canina, using the acetylene reduction technique, confirmed that fixation was rapid, at least in Peltigera. The Nostoc cells were again found to be a small proportion of the total thallus material with the algal nitrogen representing 8% of the total thallus nitrogen (8.8 x 10 cells/cm^2 thallus). The algal cells were very much larger than when grown in the free-living state, with average dimensions of 8.3 x 6.7 µm, compared with 4.0 x 3.7 µm.

The rate of acetylene reduction was examined (71) in a large number of specimens of thallus and was found to be rather variable. However, the most important finding was that the rate was rapid, with a mean value of 2.7 nmoles C_2H_4 evolved/min·mg Nostoc protein. This rate was about as fast as the usual rate in free-living alga. Furthermore, heterocysts, although present, were rather infrequent, representing about 3.3% of the total algal cells (32, 72). This confirms Peat's observations (34), and probably accounts for the inability of Drew and Smith (7) to observe any heterocysts in homogenized thalli under the light microscope.

If N_2 fixation only takes place in heterocysts as has been supposed (35), their nitrogenase activity becomes equivalent to 90 nmoles C_2H_4 evolved/min·mg heterocyst protein. Data on the specific activity of nitrogenase (36) reveals a representative figure of 900 nmoles/min·mg. On this basis, one must conclude that about 10% of the total heterocyst protein is nitrogenase; thus fixation may possibly occur in the vegatitive cells. This could be feasible as a consequence of possible low internal pO_2 and high pCO_2 produced by the respiratory activity of the fungi and emphasized by their close association with encapsulated algal cells (Fig. 4.2). Such conditions would also tend to depress photorespiration in the algal cells. This latter process has been shown (37) to be antagonistic to nitrogenase activity in Anabaena. Thus the findings are in general accord with the hypothesis concerning the site of N_2 fixation in algae proposed by Stewart (38) and supported by the results of van Gorkom and Donze (39). Recent findings (71, 72) have shown that the proportion of heterocysts in lichens with one phycobiont is circa 4%; in lichens with two phycobionts (one blue-green) it is circa 20-30%, and the N_2 fixing activity is reflected in these figures. They thus provide more support for the heterocyst theory. See also III.3.

Henriksson and Simu (19) have also expressed their results for nitrogenase activity in Collema tuneforme and Peltigera rufescens on a quantitative basis, but based on the dry weight of the entire thallus. If a number of assumptions are made, the nitrogenase activity per milligram Nostoc protein is of the same order as found by Hitch and Millbank (71). It may well be that high rates of fixation are a characteristic of lichen Nostoc under appropriate conditions of moisture, light, and temperature.

Utilizing the sensitivity of the acetylene technique to analyze small thallus samples, a study was made (5) of the variation of nitrogenase activity over the thallus area. The highest rate of fixation, on both an area and a cell number basis, occurred in the more mature part of the thallus, about 3 cm behind the growing edge.

4.2.2.5 Release of Fixed Nitrogen by the Phycobiont

Henriksson (13) reported that between 19 and 28% of the nitrogen fixed by the Nostoc phycobiont of Collema tenax was released into the growth medium. The release of organic nitrogen is a normal feature of the metabolism of blue-green algae (see, e.g., ref. 40 and III.3). Henriksson inferred that the alga was behaving similarly within the lichen thallus, and this was confirmed by Scott (22). Later work on the isolated phycobiont (41-43) has shown that a Nostoc from Collema released about 5% of the nitrogen that it fixed into the growth medium, together with biotin, nicotinic, and pantothenic acids, riboflavin, and thiamine. It also released unknown compounds that inhibited the growth of its cultured mycobiont. Henriksson also reported (44) that an isolated strain of Nostoc released substances that supported the growth of the mycobiont, although their nature was not established. Many lichen fungi are deficient in biotin and thiamine, and the release of these compounds by the phycobiont could be vital to the maintenance of healthy growth of the combination.

Millbank and Kershaw (8), working with Peltigera aphthosa and using $^{15}N_2$, showed that fixation in this lichen was confined to the cephalodia, which contained the Nostoc phycobiont. There was no incorporation of $^{15}N_2$ into thallus from which the cephalodia had been removed by dissection. Trials extending over periods of 25 days showed that incorporation of N into the thallus was linear (Fig. 4.3) and that the amount of $^{15}N_2$ labeling in the cephalodia became constant after 10 days. It was concluded that there was a continuous flow of nitrogen via the Nostoc in the cephalodia and that virtually all the nitrogen fixed by the Nostoc was released to the mycobiont.

The authors, although considering the mycobiont to control the metabolism of the N_2-fixing phycobiont, were unable to supply any evidence as to the form of the control. They favored some form of chemical control of the alga, affecting either its nitrogen utilization, or its retentive ability for organic nitrogen. As yet no substances or extracts from lichen thalli have been demonstrated to influence the rate of N_2 fixation or nitrogen loss from blue-green algal cells. That the closely proximate mycobiont mycelium is acting as a nitrogen sink and affecting the rate of loss is worth considering; the possibility that the control is connected with the oxygen tension in the thallus is at least equally likely and at present tends to be favored as an explanation.

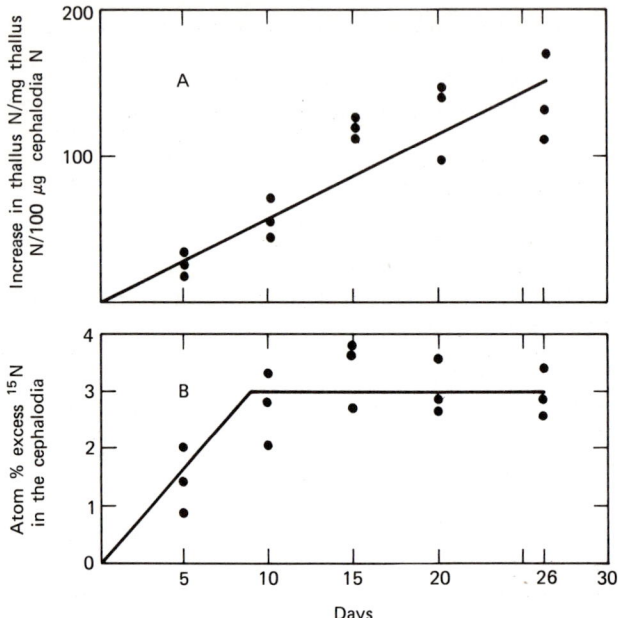

Fig. 4.3. Uptake of $^{15}N_2$ into the thallus and cephalodia of <u>Peltigera</u> <u>aphthosa</u> at 12°C. (<u>A</u>) Nitrogen uptake by the thallus. (<u>B</u>) $^{15}N_2$ enrichment of the cephalodia exposed to N_2 enriched to 30% ^{15}N. (From Millbank and Kershaw, ref. 8.)

There are few reports on the form of nitrogen released by the alga in the symbiotic state, but studies (73) in the author's laboratory indicate that they are at least in part composed of low molecular weight peptides. These may be eluted from specimens of <u>Peltigera</u> <u>polydactyla</u> from which the medulla has been removed by dissection (6) leaving the alga-enriched cortical zone. Other, simpler, dissection techniques are used to enable the cephalodia of <u>Peltigera</u> <u>aphthosa</u> and <u>Placopsis</u> <u>gelida</u> to be investigated. High voltage paper electrophoresis of the eluates reveals a range of substances with mobilities similar to the peptides released into the growth medium by free-living algae (45). Gel filtration techniques indicated their molecular weight to be of the order of 1100.

4.2.2.6 Translocation of Nitrogen Compounds between the Symbionts

Kershaw and Millbank (46) investigated the movement of $^{15}N_2$ in <u>Peltigera aphthosa</u> over periods of up to 55 days. This is a three-component lichen containing two algae and a fungus. One of the algae is a blue-green, <u>Nostoc</u>. The labeled nitrogen, which has been fixed and then released from the <u>Nostoc</u> in the cephalodia, predominated in the fungus of the thallus. After considering the proportions of fungus and green alga in the thallus, only about 3% of the expected amount of nitrogen was actually found in the green algal (<u>Coccomyxa</u>) cells. The fungus had evidently captured virtually all the fixed nitrogen released by the <u>Nostoc</u>. When this lichen was maintained in a suitable enclosure (2) it benefited greatly by the regular addition of combined nitrogen and it seemed that the cephalodia were of only slight value as a source of nitrogenous nutrient to the green algal component (46). In the case of <u>Peltigera canina</u>, with a tenfold proportion of blue-green alga per unit of thallus (71), the <u>Nostoc</u> may provide a more adequate supply of nitrogen, but the observation (47, p. 68) that the majority of lichens contain non-N_2-fixing Chlorophyceae as phycobionts and do as well on inhospitable substrates as the N_2-fixing ones is, nevertheless, true.

4.2.2.7 The Effect of Environmental Factors: Moisture, Light and Temperature

There are few studies of temperature effects on N_2 fixation by lichens. Fogg and Stewart (18) reported that N_2 fixation by lichens under arctic conditions "showed some relation to temperature." They suggested that "although appreciable fixation occurs in the vicinity of 0°C, the rates increased rapidly with rise in temperature and it is possible that the bulk of fixation is accomplished in brief periods when the micro environment reaches a temperature of 10° or more." Horne (48) has extended these studies and has confirmed that N_2 fixation by <u>Collema</u> was severely limited by low temperatures; ground (algal) temperatures above +1° were necessary for fixation. Desiccation, however, was found to be a much more critical

factor. Millbank and Kershaw (8) found that the
fixation rate of Peltigera aphthosa under laboratory
conditions exhibited a Q_{10} of approximately 2 over the
temperature range 12-25°C. Hitch and Stewart (17)
have shown that under constant conditions of light
intensity (700 cd) and saturated moisture content
nitrogenase activity in Lichina confinis was maximal
at 20°C with lower and upper limits of -3 and 35°C.
Peltigera rufescens had a similar temperature range,
-3° to 46°, with maximum rate at 31°C. The diurnal
variation in fixation rate of Peltigera polydactyla
and Lichina confinis was correlated directly with
light intensity and temperature when the thallus was
adequately moist. More detailed studies of photosynthesis and N_2 fixation (75, 76), using the lichens
Nephroma, Solorina, and Stereocaulon, have shown that
N_2 fixation ceased at +2°, 0° and +5° respectively,
and was optimal at about 15-20°C. Net carbon assimilation, however, was rapid at +5°, and the discrepancy
in the optima between the two processes may account in
part for the N deficiency in the arctic.

Thallus moisture content is a critical factor,
especially as the uptake and loss of water by lichen
thalli is largely a physical process. Henriksson and
Simu (19) reported the ability of Collema tuneforme
and Peltigera rufescens to recover N_2-fixing activity
after long periods of desiccation. Storage was for
periods of up to 30 weeks in plastic containers in the
dark at 12°C. After storage for periods longer than
4 weeks, the N_2-fixing activity upon rewetting was
about 50% of that observed with material stored for 1
day only; but these rates were achieved after a recovery period of only 24 hours, and in the case of
Collema a recovery period of only 1 hour restored 50%
of the original activity after 22 weeks' storage, and
25% after 30 weeks'. These storage conditions were
very exacting and show the lichen's remarkable
capacity to withstand severe environmental stress.

Hitch and Stewart have reported (17) the effects
of moisture content in Collema, Lichina, and Peltigera.
When samples collected from the field in a desiccated
state were moistened, N_2 fixation commenced after lag
periods of 60, 20, and 35 min, respectively. The
thallus moisture contents, however, changed from the
original level of 20% of the oven dry weight to 200%
within 5 min. In these lichens when the moisture

content became less than 80-90% of the oven dry weight (equivalent to "60% of saturated"), acetylene reduction stopped. If these results are applicable to other species, then under field conditions and with fast drying rates of the thallus, very little N_2 fixation may occur at exposed sites. Kershaw (74) has shown that N_2 fixation ceases in Canadian specimens of <u>Peltigera canina</u>, <u>P</u>. <u>evansiana</u>, <u>P</u>. <u>polydactyla</u> and <u>P</u>. <u>praetextata</u> at thallus moisture contents of 20%, 15%, 30% and 25% of saturation respectively; these results give a tolerable correlation with the observed distribution of the lichens in habitats of varied moisture content.

Concerning the effect of light, Hitch and Stewart (17) found that N_2-fixing activity was greatest at intensities above about 400 cd in <u>Peltigera</u> and <u>Lichina</u>. In the dark, there was a progressive loss of activity, the rate of loss probably being a reflection of the level of carbon reserve in the thallus.

4.3 ASSOCIATIONS WITH GREEN PLANTS

4.3.1 Liverworts

4.3.1.1 Anthoceros

Mucus-filled cavities on the lower surface of the thallus become infected with a <u>Nostoc</u> species (49), which apparently then stimulates the formation of papillae. The <u>Nostoc</u> filaments ramify among these structures, forming colonies. The formation of papillae is evidently characteristic of <u>Nostoc</u>, other algal invaders being ineffective. The effectiveness of the endophyte in fixing N_2 and its value to the host plant was debated for some time, with Peirce (50) averring that specimens without <u>Nostoc</u> colonies were more healthy in appearance, and Lhotsky (51) concluding that the host did not derive any nutritional advantage from the endophyte whose ability to fix N_2 was doubtful. Lhotsky's findings were contradicted by Ridgway (52), who concluded that the <u>Nostoc</u> associate was of considerable benefit to the liverwort when grown on nitrogen-free liquid medium, and Stewart and Rodgers (private communication) support this finding. Using $^{15}N_2$ they showed that N_2 fixation occurred and was

algal in origin. They have also obtained gametophytes
with and without blue-green algal symbionts and cultured them on media with and without combined nitrogen.
In the absence of both combined nitrogen and algae the
thalli died; in the presence of either or both growth
occurred, with growth on combined nitrogen better than
in the presence of algae alone. Thus the Nostoc can
evidently provide a valuable source of nitrogen under
exacting or subsistence conditions, and the early
observations of Prantl (53) on the ability of Nostoc
in Anthoceros to fix N_2 and pass it on to the host
have been ultimately shown to be the most perceptive.
There are at present no data on the rates of fixation,
and the effects of climatic and seasonal factors.

4.3.1.2 Blasia

The morphology of the symbiotic system in which
colonies of Nostoc are found in cavities within small
growths on the lower surface of the thallus near the
edges has been described in detail (54) and is illustrated in Fig. 4.4. Gargeanne (55) cultured the
liverwort on sand irrigated with nitrogen-deficient
mineral medium. He found little or no effect of
inoculation with Nostoc, and inferred that the Nostoc
conferred only a marginal benefit, if any, on its
host. When combined nitrogen was supplied there was
again little or no difference between specimens with
and without Nostoc. However these results were not
entirely definitive, and Gargeanne considered that
sand culture was not wholly suitable for study, and
also that under field conditions, where the material
would be expected to be more vigorous, his results
might well not be applicable. Pankow and Martens (56)
and Watanabe and Kiyohara (14) have identified the
endophyte as Nostoc sphaericum.

The definitive demonstration by Bond and Scott
(15) that the alga fixes N_2 used an atmosphere containing N_2 enriched with ^{15}N. After 7 days' incubation at
19°C replicate analyses showed an enrichment of the
order of 0.2%, representing an increase in total thallus nitrogen of approximately 1%. The authors deci-
sively ruled out the possibility of bacterial contamination being responsible for their results and
reported strong presumptive evidence that the fixed
nitrogen had moved from the alga to the liverwort

Associations with Green Plants 143

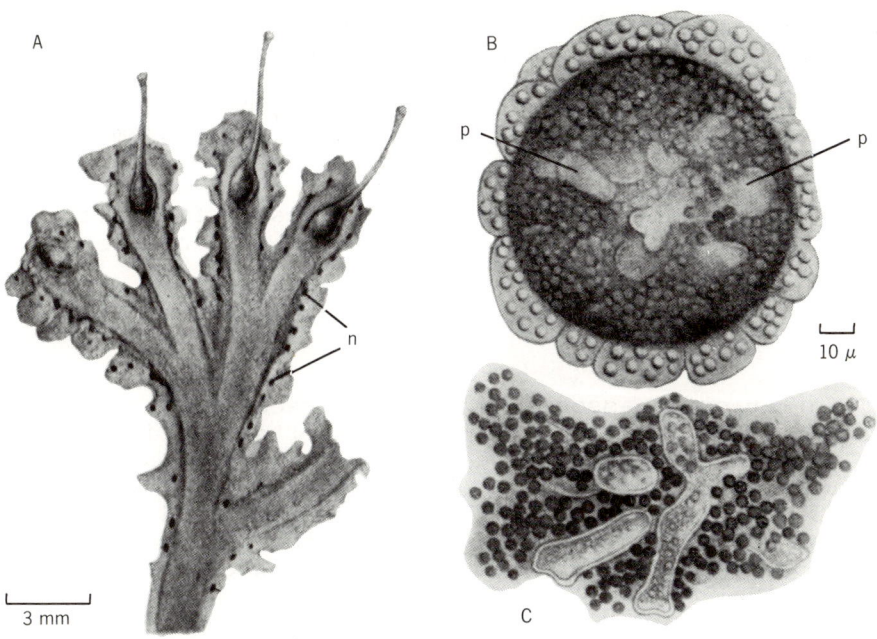

Fig. 4.4. Blasia. (A) Thallus, with Nostoc colonies (n). (B) Nostoc colony, showing papillae (p). (C) Squashed preparation of Nostoc colony showing papillae. From Molisch, 1926 (54).

thallus. As far as the content of algal cells per unit of thallus is concerned, no data are available and so the rates of fixation of the symbiotic alga are not known. Watanabe and Kiyohara (14), also using Blasia pusilla, have confirmed these findings, and have also shown that the algal symbiont fixed N_2 vigorously when grown in pure culture.

4.3.1.3 Cavicularia

Molisch (54) has also dealt with the morphology of this liverwort, which closely resembles Blasia pusilla; Takesige (57) isolated the algae from both liverwort hosts. He was further able to show that the alga from either host could successfully infect the other.

Watanabe and Kiyohara (14) and Pankow and Martens (56)
identified the alga as Nostoc sphaericum, and the
former authors showed that ^{15}N was incorporated into
the thallus after exposure to N_2 enriched in ^{15}N for
5 weeks. The very low levels of incorporation may
have been due to the use of low partial pressures of
N_2 (5%) in the experimental setup.

No data are available on the rates of fixation by
the Nostoc symbiont, its abundance in the cavities,
and the effects of environmental factors.

4.3.2 Mosses

Information on associations with mosses is extremely
scanty, but Stewart (47, p. 68) reports that in
Britain a species of Hapalosiphon can occasionally be
found in association with Sphagnum, where it apparently becomes trapped within the hyaline cells of the
moss. Preliminary $^{15}N_2$ data indicate that the association fixes nitrogen, but no further information is
available.

4.3.3 Pteridophytes

4.3.3.1 Azolla

The small freshwater fern, Azolla, has four species
and is of worldwide distribution, although occurring
most abundantly in the tropics. It floats on the
water surface and can grow very rapidly, giving rise
to a thick carpet of vegetation, capable of overwhelming the existing plant growth (58). It is extensively
used as a green manure in rice cultivation (59). The
blue-green alga consistently associated with the fern
is Anabaena azollae. It occurs in a mucilage-filled
cavity or pore in the ventral surface of the dorsal
lobe of each leaf (Fig. 4.5). It has not proved
possible to isolate the alga and maintain it in axenic
culture (77, 80).

Oes (60) reported that the association fixed N_2,
but in 1933 Huneke (61) disputed this, reporting that
the fern could not develop on a nitrogen-deficient
substrate. Her investigations, and those of Wildemann
(62), also showed that the fern could exist successfully in the absence of the alga. Later, Nickell (63)

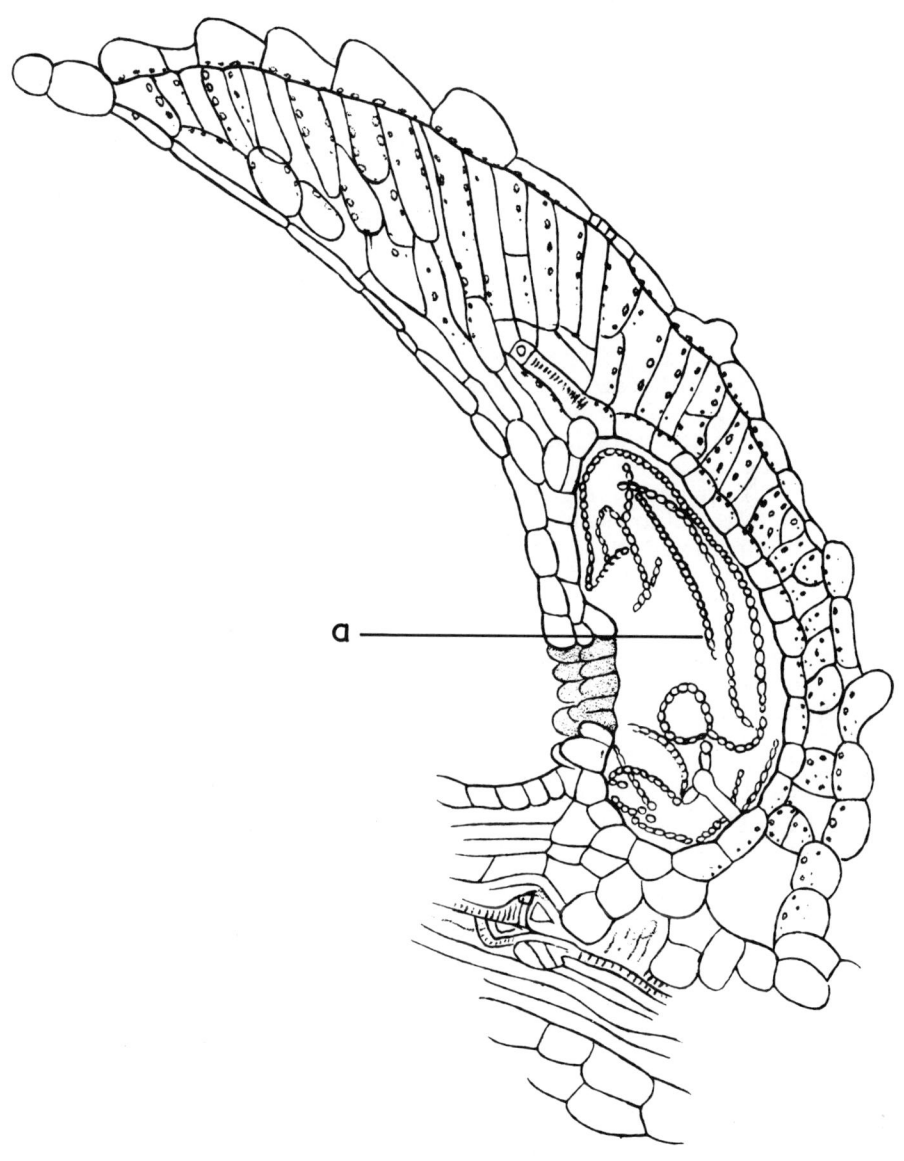

Fig. 4.5. Azolla. Section through the upper lobe of a leaf. a, Anabaena filaments. (From Strasburger.)

has described a technique for freeing Azolla from its Anabaena associate, and gives details of a medium (containing nitrate) on which the fern may be successfully grown in isolation.

Bortels (64) reexamined the question and showed that fixation did occur in the combination; and that the process was enhanced by the addition of molybdenum, now known to be an essential constituent of nitrogenase. For a discussion of the requirement for cobalt, see III.3.

A careful but necessarily limited study by Saubert (59) showed that the fern-alga combination can grow successfully in nitrogen-deficient medium of appropriate pH, considerable trouble being taken to ensure the availability of Fe and Ca ions. In field studies the nitrogen fixed was of great benefit to the associated rice crop.

Still relying on classical total nitrogen techniques, Venkataraman (65) concluded that the isolated endophyte fixed N_2, and liberated combined nitrogen into the growth medium (66).

All these observations have been extended by Olsen (58), who reported on laboratory and field studies using several Danish lakes. He confirmed that the fern grew luxuriantly at a pH of 6-7 and brought about a major improvement in nitrogen status, up to 95 kg fixed N_2/ha during one summer season. The availability of Fe and Ca ions and their balance was especially important. Reducing conditions on the lake bottom giving rise to ferrous ions are essential at neutral pH; ferric ions are only available at or below pH 5 and in these conditions a level of Ca^{++} above about 10 μg/ml seems necessary to avoid a toxic level of Fe^{+++}.

More recently, Peters and Mayne have made a detailed study (78, 79) of the symbiosis, and have provided data on the rates of $N_2(C_2H_2)$ fixation under aerobic and microaerobic conditions in the light and dark. Microaerobic dark fixation was negligible. Confirmation was obtained that the alga-free fern can grow only in media containing combined nitrogen, and also that the symbiotic system can thrive in media free from combined nitrogen. A considerable transport of carbon compounds from the fern to the alga was demonstrated. The alga releases combined nitrogen in the form of ammonium ions and as ninhydrin-positive compounds (81).

4.4 OTHER ASSOCIATIONS

4.4.1 Endocyanoses

These are associations between a member of the Cyanophyceae and another microorganism that is lacking chlorophyll; the alga is intracellular. There is little information on the physiology of these associations, and none on their N_2-fixing activities. Glaucocystis nostochinearum is an association between an achlorophyllous form of Oocystis and a species of Nostoc (67, 68). Geosiphon pyriforme is a curious association between Nostoc sphaericum (which can fix N_2 in pure culture) and a phycomycete, the latter becoming vesicular in the symbiotic state (69). The Nostoc cells are engulfed if contacted by a fungal hypha, the hyphal contents being released to surround the alga and a new wall formed around the whole. The symbiosis has been shown to be autotrophic, the Nostoc functioning as a chromatophore; it is possible that N_2 fixation may also occur.

4.5 THE RELATIONSHIP OF LICHENS TO OTHER N_2-FIXING SYMBIOTIC SYSTEMS

Scott (70) arrived at a definition of symbiosis as a "state of equilibrated physiological interdependence of two or more organisms involving no permanent stimulation of defensive reaction mechanisms." This is a synthesis of six criteria, the possession of any four of which he considers adequate to constitute a symbiotic association. These criteria are that the association should (a) be a permanent feature of the organisms' life cycles; (b) involve physical contact between the participants; (c) involve unilateral or bilateral movement of metabolites; (d) ameliorate environmental status, thus giving rise to an extension of ecological range; (e) give rise to morphogenetic effects; and (f) provide opportunity for the production of metabolites not formed by either of the organisms separately.

How do lichens, which conform to all these criteria except perhaps the first, relate to the other N_2-fixing symbioses? These last comprise the legume-Rhizobium association; the non-leguminous angiosperm-root nodule systems (endophytes unidentified); the leaf nodule association of, e.g., Klebsiella-Psychotria; and the relationship between blue-green algae with liverworts, ferns, Gymnosperms, and Angiosperms.

Since the endophyte(s) in the non-leguminous nodule system have not yet been identified it is not possible to compare their physiology and attributes to the other systems. Ignoring them, therefore, it seems on the one hand that as far as N_2 fixation is concerned, the N_2-fixing lichens are much more closely analogous to the Blasia, Azolla, Cycas, and Gunnera associations, which have criteria (b), (c), (d), (f), and sometimes (e), than to the strict N_2-fixing symbiosis, the legume-Rhizobium system. This association not only possesses criterion (a) but is the only strict N_2-fixing symbiosis; in all the other systems the process is carried on by an organism able to fix N_2 in pure culture or free-living in nature. Furthermore, in the associations considered here the faculty of release of combined nitrogen by the fixer is also normal, and is enhanced rather than initiated by the specialized environment. On the other hand, the morphological distinctiveness of lichens, in which there is little or no resemblance between the product of the combination and any of the constituent partners is one of the most outstanding features of the symbiosis, and distinguishes it completely from the other associations considered. It is evidently a much more intimate relationship than the others, giving rise to profound morphological and biochemical changes but retaining at least some of the fundamental biochemical attributes of the individual partners. If it is accepted also that mutual advantages are features of a symbiotic relationship, it seems most appropriate to consider that the major benefit to the alga(e) in the lichen association is a considerable extension of ecological range, at the expense of a greatly reduced rate of growth and what can be thought of as a state of controlled parasitism by the fungus.

4.6   REFERENCES

1. Smith, D. C., Lichenologist, 1, 209 (1961).
2. Kershaw, K. A., and J. W. Millbank, Lichenologist, 4, 83 (1969).

3. Allen, M. B., and D. I. Arnon, Plant Physiol., 30, 366 (1955).
4. Hill, D. J., and H. W. Woolhouse, Lichenologist, 3, 207 (1966).
5. Millbank, J. W., New Phytol., 71,1 (1972).
6. Smith, D. C., Symp. Soc. Gen. Microbiol., 13, 31 (1963).
7. Drew, E. A., and D. C. Smith, New Phytol., 66, 379 (1967).
8. Millbank, J. W., and K. A. Kershaw, New Phytol., 68, 721 (1969).
9. Richardson, D. H. S., Methods in Microbiology, C. Booth, Ed., Academic Press, New York, Chap. 10, 4, 1971.
10. Ahmadjian, V., and M. E. Hale, Eds., The Lichens, Academic Press, New York, 1973.
11. Ahmadjian, V., Phycologia, 6, 127 (1969).
12. Duncan, U., assisted by P. W. James, Introduction to British Lichens, T. Buncle and Co., Arbroath, Scotland, 1970.
13. Henriksson, E., Physiol. Plantarum, 4, 542 (1951).
14. Watanabe, A., and T. Kiyohara, Plant Cell Physiol. (Tokyo), Studies on Microalgae and Photosynthetic Bacteria, 189 (1963).
15. Bond, G., and G. D. Scott, Ann. Botany (London) N. S., 19, 67 (1955).
16. Rogers, R. W., R. T. Lange, and D. J. D. Nicholas, Nature (London), 209, 96 (1966).
17. Hitch, C. J. B., and W. D. P. Stewart, New Phytol., 72, 509 (1973).
18. Fogg, G. E., and W. D. P. Stewart, Brit. Antarct. Survey Bull., 15, 39 (1968).
19. Henriksson, E., and B. Simu, Oikos, 22, 119 (1971).
20. Stewart, W. D. P., Plant Soil, 32, 555 (1970).
21. Millbank, J. W., and K. A. Kershaw, New Phytol., 69, 595 (1970).
22. Scott, G. D., New Phytol., 55, 111 (1956).
23. Cengia-Sambo, M., Atti Soc. Ital. Sci. Nat., Museo Civico Storia Nat. Milano, 62, 226 (1923).
24. Cengia-Sambo, M., Atti Soc. Ital. Sci. Nat., Museo Civico Storia Nat. Milano, 64, 191 (1926).
25. Goebel, K., Ann. Roy. Bot. Gdn. Buitenz., 36, 1 (1926).
26. Henckel, P. A., and L. A. Yuzhakova, Bull. Inst. Rech. Biol. Perm. (Molotov), 10, 315 (1936).
27. Henckel, P. A., Bull. Soc. Nat. Moscow (Biol.), N. S., 47, 13 (1938).

28. Iskina, R. Y., Bull. Inst. Rech. Biol. Perm. (Molotov), 11, 133 (1938).
29. Quispel, A., Rec. Trav. Bot. Neerl., 40, 413 (1945).
30. Krasilnikov, N. A., Mikrobiologiya, Moskow, 18, 3 (1949).
31. Panosyan, A. K., and K. Nikogosyan, Biol. Zh. Arm., 19, 3 (1966).
32. Griffiths, H. B., A. D. Greenwood, and J. W. Millbank, New Phytol., 71, 11 (1972).
33. Kratz, W. A., and J. Myers, Am. J. Bot., 42, 282 (1955).
34. Peat, A., Arch. Mikrobiol., 61, 212 (1968).
35. Fay, P., W. D. P. Stewart, A. E. Walsby, and G. E. Fogg, Nature (London), 220, 810 (1968).
36. Hardy, R. W. F., R. C. Burns, R. R. Hebert, R. D. Holsten, and E. K. Jackson, Plant Soil, Special Vol., 561 (1971).
37. Lex, M., W. B. Silvester, and W. D. P. Stewart, Proc. Roy. Soc. Ser. B, 180, 87 (1972).
38. Stewart, W. D. P., Plant Soil, Special Vol., 377 (1971).
39. van Gorkom, H. J., and M. Donze, Nature (London), 234, 231 (1971).
40. Fogg, G. E., Physiology and Biochemistry of Algae, R. A. Lewin, Ed., Academic Press, New York, 475, 1962.
41. Henriksson, E., Physiol. Plant., 10, 943 (1957).
42. Henriksson, E., Physiol. Plant., 13, 751 (1960).
43. Henriksson, E., Physiol. Plant., 14, 813 (1961).
44. Henriksson, E., Svensk. Botanisk Tidskrift, 52, 391 (1958).
45. Walsby, A. E., Ph. D. Thesis, University of London (1971).
46. Kershaw, K. A., and J. W. Millbank, New Phytol., 69, 75 (1970).
47. Stewart, W. D. P., Nitrogen Fixation in Plants, Athlone Press, London, 1966.
48. Horne, A. J., Brit. Antarct. Survey Bull., 27, 1 (1972).
49. Leitgeb, H., Sitzungsberichte Akad. Wiss. Wein., 77, 411 (1878).
50. Peirce, G. J., Bot. Gaz., 42, 55 (1906).
51. Lhotský, S., Stud. Bot. Cechoslovaca, 7, 20 (1946).
52. Ridgway, J. E., Ann. Mo. Bot. Gdn., 54, 95 (1967).
53. Prantl, K., Hedwigia, 28, 135 (1889).
54. Molisch, H., Sci. Rep. Tohoku Univ., 1, 169 (1926).
55. Garjeanne, A. J. M., Ann. Bryologicii, 3, 97 (1930).

References

56. Pankow, H., and B. Martens, Arch. Mikrobiol., 48, 203 (1964).
57. Takesige, K., Bot. Mag. (Tokyo), 51, 332 (alternatively, 514), (1937).
58. Olsen, C., C. R. Trav. Lab. Carlsb., 37, 269
59. Saubert, G. G. P., Ann. Roy. Bot. Gdn. Buitenz., 51, 177 (1949).
60. Oes, A., Z. Bot., 5, 145 (1913).
61. Huneke, A., Beitr. Biol. Pflanz., 20, 315 (1933).
62. Wildemann, L., Dissertation, Munster University (1934).
63. Nickell, L. G., Am. Fern J., 48, 103 (1958).
64. Bortels, H., Arch. Mikrobiol., 11, 155 (1940).
65. Venkataraman, G. S., Ind. J. Agr. Sci., 32, 22 (1962).
66. Venkataraman, G. S., and H. K. Saxana, Ind. J. Agr. Sci., 33, 21 (1963).
67. Geitler, L., Arch. Protistenkunde, 47, 1 (1923).
68. Bourrelly, P., Compt. Rend., 251, 416 (1960).
69. Knapp, E., Ber. Deut. Bot. Ges., 51, 210 (1933).
70. Scott, G. D., Plant Symbiosis, Edward Arnold and Co., London, 1969.
71. Hitch, C. J. B., and J. W. Millbank, New Phytol., 74, 473 (1975).
72. Hitch, C. J. B., and J. W. Millbank, New Phytol., 75, 239 (1975).
73. Millbank, J. W., New Phytol., 73, 1171 (1974).
74. Kershaw, K. A., Can. J. Bot., 52, 1423 (1974).
75. Kallio, P., S. Suhonen and H. Kallio, Rep. Kevo Subarctic Res. Sta., 9, 7 (1972).
76. Kallio, S., Rep. Kevo Subarctic Res. Sta., 10, 34 (1973).
77. Hill, D. J., Planta, (Berl.), 122, 179 (1975).
78. Peters, G. A., and B. C. Mayne, Pl. Physiol., 53, 813 (1974).
79. Peters, G. A., and B. C. Mayne, Pl. Physiol., 53, 820 (1974).
80. Moore, A. W., Bot. Rev., 35, 17 (1969).
81. Peters, G. A., Proc. Symp. Dinitrogen Fixation, Pullman, Wash. (1974).

CHAPTER 5

Foliar Associations in Higher Plants

W. S. SILVER

Department of Biology
University of South Florida
Tampa, Florida, U.S.A.

5.1. The Foliar Habitat, 154
    5.1.1. The Leaf as a Microenvironment, 154
    5.1.2. Resident Microorganisms, 156
        5.1.2.1. Procaryotes, 156
        5.1.2.2. Eucaryotes, 159
5.2. Dinitrogen Fixation in the Phyllosphere, 159
    5.2.1. Identifiable Microbes, 159
    5.2.2. Potential of the Habitat, 160
5.3. Leaf Nodule Plants, 161
    5.3.1. Systematics and Geographical Distribution, 161
    5.3.2. Biology of the Cyclic Symbiosis, 164
        5.3.2.1. <u>Rubiaceae</u>, 165
        5.3.2.2. <u>Myrsinaceae</u>, 169
        5.3.2.3. <u>Discoreaceae</u>, 171
5.4. Leaf Nodule Bacteria, 171
    5.4.1. Isolation Methodology, 171
    5.4.2. Bacteria Isolated from Nodulated Leaves, 172
5.5. Does N_2 Fixation Occur in the Symbiotic State?, 178
    5.5.1. Evidence from $^{15}N_2$ and C_2H_2 Reduction Studies, 179
    5.5.2. Evidence from Habitat Studies, 180
5.6. References, 180

5.1 THE FOLIAR HABITAT

5.1.1 The Leaf as a Microenvironment

Generally, microbiological studies of the microflora of plants have been concerned with those organisms associated with the root, probably as a result of interest in nodulation of the agriculturally important legumes. It was not until the studies of Ruinen (1) that microbiologists began to consider the nature of, and importance of, foliar epiphytes. The realization that the foliar environment (phyllosphere) supported N_2 fixation (2), coupled with the International Biological Program (IBP) focus on N_2 fixation in tropical regions (3), has stimulated interest in the foliar habitat and its microbes. Furthermore, an understanding of the normal leaf microflora may provide better insight of the events leading to the diseased state, another aspect of plant biology important to food production.

Despite the obvious fact that world food production depends upon photosynthesizing leaves, the nature of the leaf surface is poorly understood. This prompted the recent appearance of a symposium volume devoted to the ecology of foliar microorganisms (4). Some of the points discussed in that volume will be summarized here when the matter has bearing upon the main theme of this treatise — dinitrogen fixation.

From the viewpoint of the microbe the leaf surface is a very heterogeneous surface both in a physical and chemical sense. Those factors most important to the microbe inhabiting the terrestrial plant are the availability of water and the amount of leached nutrients available for growth, as well as resistance to deleterious radiation and heat. First, let us consider the factors other than atmospheric conditions that regulate the amount of moisture on the leaf surface.

Considerable water may be released from within the plant during the night by guttation from special openings located at the leaf margin. The natural exudates may contain substances stimulatory (5), as well as inhibitory (6), to microbial growth, and the availability of one may influence the effect of the other. Since the leaf surface leachings are low in nitrogen but contain carbon sources the habitat favors colonization initially by oligonitrophilic and N_2-fixing organisms. In addition to water released from within the plant by guttation and the transpiration stream, there is an intermittent input from the external environment in the form of either dew or rain. Not all of the rainfall is lost rapidly to the soil, and the rate of water disappearance is a function of a host of factors — surface geometry, degree and arrangement of the waxy layer, relative humidity, and temperature. Accurate measurements of the surface humidity are difficult to obtain. From the microbiological point of view, it is the surface humidity of the microenvironment which is important. Under most conditions the relative humidity above the plant never falls below 40% during the day, the highest humidities being noted in the densest region of the stand (7).

The boundary layers (1-10 mm thick), which vary greatly depending upon the plant type, can have a marked influence on the success of colonization by microorganisms, either in a direct manner (via the microclimate or solute movement in the surface moisture) or indirectly — the thicker the layer, the more resistant the plant to intrusion by possible pathogens (8). In addition to the effect of mechanical abrasion on the imperviousness of the cuticle, the microbial action may disturb the layer so that a lipolytic strain might pave the way for growth of an organism lacking this property. Indeed the common leaf epiphytes, Cryptococcus laurentii and Rhodotorula glutinis, secrete lipases which are active against the cuticular waxes of Aloe sp. and Sanseviera sp. (9).

It should be borne in mind that the route carrying the chemical traffic on the leaf surface is a two-way street, i.e., both the plant and microbe can exchange substances. Plant leachates may contain in addition to free sugars, inorganic ions, amino acids, and growth regulators (10). It should not be assumed that quantitatively this represents only trace amounts. For example, apple foliage can lose 20-30 kg K, 10.5 kg Ca, 9 kg Na, and 800 kg carbohydrate/ha·yr [cited in (4)] and the materials leached become available to the plant as well as the microbe (11). Some generalizations can be made: nutrient losses will be greater from (a) nutrient deficient plants, (b) damaged plants, (c) dark-grown plants, and (d) old leaves, the latter perhaps reflecting the loss of surface waxes (4).

The microbe may effect the plant surface by means unrelated to direct enzymatic attack. Leakage of low molecular weight substances from bacterial and fungal cells is known to occur and, most important, the microbial production of plant growth regulators may evoke a plant response. That microbial plant growth regulator production can be quantitatively significant was shown in the studies of Libbert and co-workers, who noted that IAA producing epiphytic bacteria contributed significantly to the conversion of typtophan to IAA by homogenates of septic plant tissues (12, 13). Although most of their studies were done with roots and shoot tissue, leaves also possessed this property. Furthermore IAA destruction by bacterial epiphytes could also be important (14). If these effects apply to a variety of situations, the fact that other microbes produce cytokinins (15) and ethylene (16), coupled with the variety of microbes found on leaf tissues, suggests that this type of plant-microbe interaction may be important.

Competition for nutrients between epiphytes can also be important (17). The interaction between microbes, as well as its effect on the plant response to pathogens, while of great interest from the viewpoint of plant pathology (18-21), will not be discussed here unless it relates to N_2 fixation.

5.1.2 Resident Microorganisms

5.1.2.1 Procaryotes

In order to fulfill the purpose of this review the discussion of the foliar flora will be restricted to those facets that deal either directly or indirectly with N_2-fixing organisms and their function in the foliar region. It is a well-known principle of microbial ecology, implicit in the pioneering studies of Winogradsky and Beijerinck, and known as "Beijerinck's Rules," that (a) everything is everywhere, and (b) the milieu selects (22). The foliar environment, like other regions, is rich in microbial propagules; however, unlike the subterranean part of a plant, the air interface is a much more restrictive environment. Though the air may provide a continuing source of inocula, only a restricted proportion become established as residents. Residents, in contrast to "casuals," are those epiphytes which multiply on (or in) the plant without apparently harming the host — i.e., they are not pathogens and they find the environment hospitable. Comments will be restricted here to the microscopic epiphytes, not the epiphylleae.

Any qualitative or quantitative assessment of microbial epiphytes is greatly influenced by the methodology employed. Sampling becomes a problem only if one is attempting to quantitate the population on a per surface area basis. The method employed in washing the leaf surface can be important. Nonrigorous water washing undoubtedly removes more casuals than residents, the latter being more tightly adhered by the mucigel. The use of surfactive additives (like the Tweens) must be done with caution for some organisms are lysed in the process. Leaf maceration is usually unsatisfactory, for many toxic plant substances are released from the tissue.

Direct observations either with the scanning electron microscope or indirectly with the long-used leaf-impression method are of value mostly with fungi since many are identifiable by their morphology. The former method particularly has amply demonstrated that epiphytes appear as clones and that the surface of the leaf is very heterogeneous.

Definition of the surface flora by culturing is a function of the enrichment media used. Frequently the media used are excessively rich and, therefore, are greatly unlike the relatively nutrient-poor state of the leaf surface under most conditions. Common media employed include N-free, cellulose, lipid, and pectin agar, each being highly selective (23).

In one thorough study of the bacterial microflora of common agricultural plants, it was noted that pigmented bacteria made up a large proportion of the flora (24), perhaps a reflection of the known radiation resistance of carotenoid containing bacteria (25). In the study cited above the following types were predominant: Pseudomonas, Mycobacterium, Chromobacterium, micrococci, and some actinomycetes. Other studies have revealed the presence of Corynebacterium, Flavobacterium, Lactobacillus, and Aerobacter (Enterobacter) [cited in (8)]. The latter is of considerable interest from the viewpoint of N_2 fixation in view of the widespread occurrence of the N_2-fixing Klebsiella and its common confusion with some aerobacters (see III.2).

Jensen conducted a study of the microflora of beech leaves with the aim of assessing the productivity of a beech forest (26). Total counts were greatly influenced by weather conditions, and, for the temperate region, phyllosphere microbes represented only a small part of the total microflora of the ecosystem. Four hundred bacterial isolates of the early summer flora were classified both morphologically and nutritionally, and 94% of them belonged to the fluorescent pseudomonads, yellow-pigmented rods, and lactic acid bacteria groups. Diversity increased later in the growing season. It is of interest that classification of isolates on the nutritional basis, originally employed by Lockhead and Chase (27) to characterize rhizosphere bacteria, indicated that the leaf flora was nutritionally less exacting, a not unexpected reflection of the nutritional state of the foliar environment.

One cannot ignore the age of the leaf for it e exists as a continuum in time, each plant species and each new leaflet being populated by successive waves of saprophytes. As Ruinen has so thoroughly documented for tropical plants, the pioneering microbial community is dominated by oligonitrophilic and N_2-fixing species, a reflection of the low availability of N and the usually nonexacting requirement for energy source of these types (1). Once the diazotrophs have increased the combined N content of the phyllosphere to permit the growth of more restrictive types, the nature of the succeeding populations will be determined largely by the physiological state of the foliage.

Some generalizations are in order despite the seemingly hopeless complexity of the interaction of a multitude of factors. Gram negative species are more common than gram positive species, especially in the earlier stages of development, and diazotrophs are readily demonstrated (see III.5.2). The outgrowth of succeeding types during leaf senescence ultimately results in a beneficial effect on the plant, for the nutrients, including any combined N sequestered by N_2 fixation, can be returned to the plant when the leaf litter decomposes.

5.1.2.2 Eucaryotes

The lack of the ability of any known eucaryote to carry out N_2 fixation in the absence of a procaryotic symbiont dictates that only a few comments are appropriate here. Fungi, including yeasts, make up a major portion of the leaf microflora established later during leaf maturation (1). Although recent reports claim that Aspergillus terreus (28) and Lipomyces starkeyii (29) fix dinitrogen, the validity of this seems very unlikely since the latter has not been confirmed (30), and all earlier reports of a similar nature have not held up to more critical examination (31). More likely, at least in case of the phyllosphere yeasts, their presence reflects the fact that they are oligonitrophilic and mucoid, two attributes favorable to growth in the phyllosphere (32).

Fungi, due to their O_2-sequestering capacity, can affect N_2 fixation in the phyllosphere, much as any aerobe can protect facultative N_2-fixing bacteria from O_2 inhibition. This has been shown in vitro for a culture of Derxia, originally isolated from leaves, which had greater $N_2(C_2H_2)$-fixing activity in mixed culture with Pencillium than in pure culture (33), a result reminiscent of a similar, earlier report of stimulation of N_2 fixation by Lipomyces (109).

5.2 DINITROGEN FIXATION IN THE PHYLLOSPHERE

5.2.1 Identifiable Microbes

One of the first to note the occurrence of N_2-fixing microbes in the phyllosphere was Ruinen (1) who pointed out that in the tropics where leaf fall is not seasonal, the possibility of considerable increases in nitrogen via such organisms was great (2). Following her reports there have been many similar examples of isolations of N_2-fixing bacteria from a wide array of plant leaves, stems, and aerial roots (34-45). The most common N_2 fixers isolated were the enterics, presumably Klebsiella (see, for example, ref. 42), although Azotobacter and Beijerinkia are common on tropical plants. In many investigations the specific type of bacterium was not revealed. The important point from the view of a consideration of N_2 fixation is that there are diazotrophs present and that the high sugar content of leaf surface moisture makes it a favorable site for their proliferation.

Aquatic plants represent a special case, for the availability of foliar nutrients may depend upon whether they are submerged, emergent, or floating. The prolific growth rate of the water hyacinth (<u>Eichornia crassipes</u>), a floating vascular aquatic plant, prompted Iswaran et al. to determine whether diazotrophs might provide some of the plants N (43). <u>Azotobacter chroococcum</u> was readily isolated from the surface of and from within the leaves (43). How specific this may be is not clear since a search was not made for blue-green bacteria (algae), which are also very likely to be present. It has been noted in this author's laboratory that the roots of this plant also harbor diazotrophs (45). In the case of marine grasses the foliar diazotrophic epiphytes appear to be predominantly blue-greens, for dark bottle samples had lower C_2H_2-reducing activity than corresponding light bottle samples (110).

5.2.2 Potential of the Habitat

Earlier in a consideration of the leaf surface of terrestrial plants as a microenvironment (III.5.1.1) it was noted that foliage in the moist condition can lose vast amounts of carbohydrate (800 kg/ha·yr for apple). In addition to the large population of a variety of common bacteria on the foliar surface (1, 46) the populations of diazotrophs can be considerable (1, 2). In assessing the potential of the foliar habitat for N_2 fixation several questions may be raised: (a) Specifically, what N_2-fixing organisms are present and how extensive are the populations? (b) Does the plant leachate support N_2 fixation by such bacteria? (c) Is the pattern similar in temperate and tropical plants? Insofar as the author is aware, only one investigation was conducted with sufficient depth to attempt to answer these questions (42).

In this study the principal plant species used were corn (<u>Zea mays</u>) and guatemala grass (<u>Tripsacum laxum Nash.</u>). In the former, a temperate plant, plate counts of sheath water on azotobacter agar ranged from 10^4 to 10^9/ml. However, most isolates tested, being oligonitrophilic, failed to fix dinitrogen. The C/N ratios of sheath water was relatively low and would not favor N_2 fixation. In contrast, in the tropical guatemala grass sheath water, populations of apparent diazotrophs were higher, the C/N ratio was very high (654 in one case) and most of the isolates belonged to the Enterobacteriaceae and a few to the Achromobacteriaceae. It is of interest that Azotobacter and Beijerinckia were never isolated. Further studies by growth independent methods revealed that all C_2H_2-reducing strains were klebsiellas, a not unexpected result considering their distribution on plant tissues (35). When entire plants were studied only 8×10^{-3} mg N/plant (based on C_2H_2 reduction assays) was gained in 5 days. Further experiments with $^{15}N_2$ on guatemala grass growing in a field in Surinam gave an estimate of 263 g N/ha based on calculations from a 4-day assay. A considerable amount of the N_2 fixed was found in the blades and stems of the plant.

If one may generalize from this specific study, the potential for N_2 fixation in the phyllosphere is considerable in climates where the C/N ratio of sheath water is high and is probably insignificant in temperate regions where evaporation and periodic drying must place a great stress on plant surface microbes.

5.3 LEAF NODULE PLANTS

5.3.1 Systematics and Geographical Distribution

It is now well known that many plant species of the Rubiaceae possess foliar nodules that usually contain bacterial symbionts. The presence of these protuberances, or "warts," was first noted by Trimen (47), but it was Zimmermann who associated the leaf nodules with the constant presence of bacteria (48). The detailed accounts of von Faber clearly established the cyclic ("heredity") nature of the symbiosis and suggested the possible function of the microsymbiont in N_2 fixation (49, 50). While it is not appropriate to discuss details of the systematic botany of these plants here, some comment on the genera involved and their distribution is warranted, for only a few species have been investigated from the viewpoint of dinitrogen fixation and it may not be reasonable to extrapolate these results to others.

Among the works on the systematics of the leaf nodule plants that of Bremenkamp, spanning over 30 years, is noteworthy (51-54), as well the recent monographs of Petit on the African species (55, 56). Nodulation is widespread among some of the genera — in Bremenkamp's monographs, 339 of 427 species of Pavetta are described as nodulated. Nodulation, though not always obvious and having a different appearance in various species (Fig. 5.1), can be a taxonomically useful trait (56). It will be noted in Table 5.1 that in contrast to the wide distribution of root nodulation in the non-legumes (see III.6.2.2), foliar nodulation is limited to three families, the Rubiaceae, Myrsinaceae, and Dioscoreaceae. Nodulated species are restricted to tropical and subtropical regions, a situation perhaps related to the establishment of large populations of leaf epiphytes, a likely prerequisite for the establishment of symbiosis. The leaf nodule plants may be found in many nonarid locations between 30°N and 30°S latitude, particularly in the more equatorial regions. For example, Petit lists 41 species of Psychotria in the Democratic Republic of the Congo which lies in the 5°N-12°S region while in the more temperate South Africa only two species were noted (55). This genus is also found in Oceania on large islands like Java and Madagascar and on isolated, minor coral islands like Aldabra (88).

Since only few of the many botanically described species were investigated experimentally, Table 5.1 lists only those species in which some attempt was made to assess the importance or identification of the microsymbiont. Among these, three species stand out — Pavetta zimmermanniana, Psychotria punctata (bacteriophila), and Ardisia crispa. The interesting biology of these symbioses is discussed below.

Fig. 5.1. Leaf nodules of various species of African Psychotria [from Petit (56)]. (A) P. Schliebenii var. parvipaniculata; (B) P. calva; (C) P. punctata var. punctata; (D) P. faucicola; (E) P. cryptogrammata; (F) P. heterosticta var. heterosticta; (G) P. Guerkeana; (H) P. Molleri; (I) P. Kikwitensis. Note how easily nodules may be seen in P. punctata (C) as compared to P. Schliebenii (A). This author has seen a specimen of hairy P. punctata in which nodules were obscured by the hirsute leaf surface.

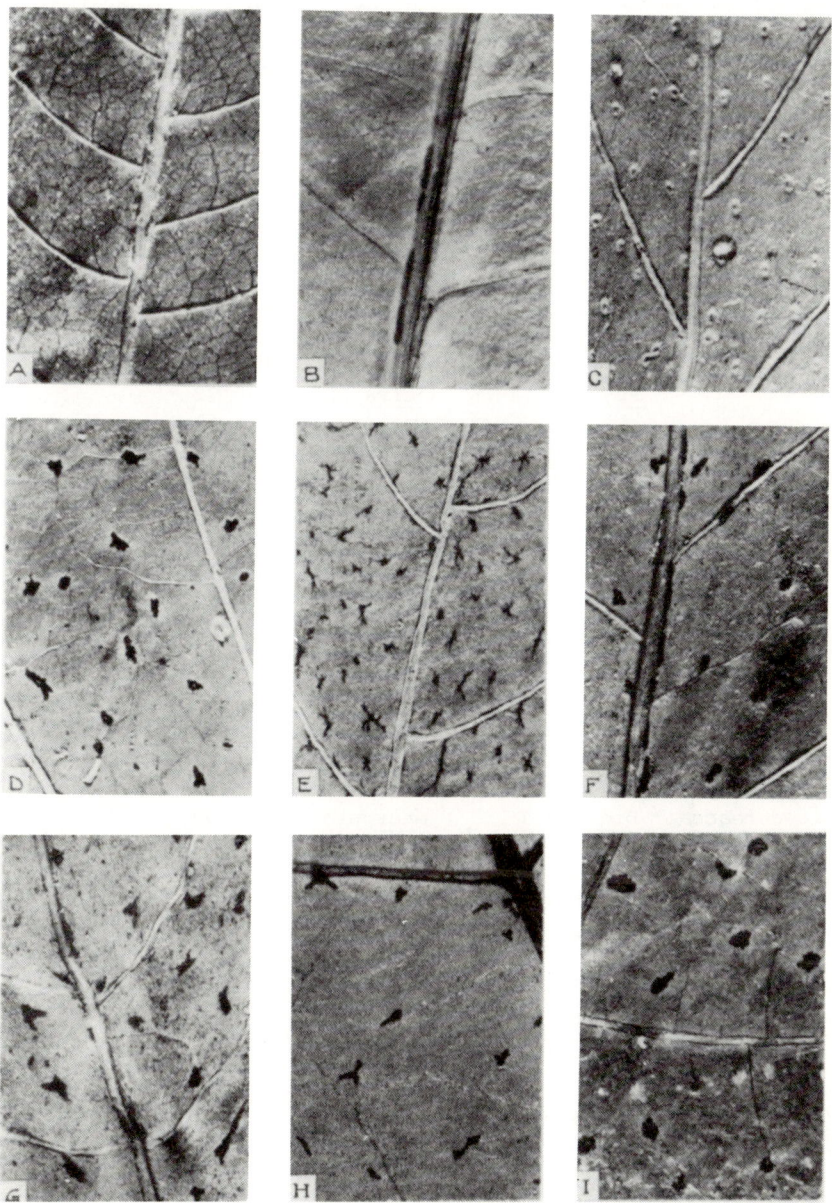

TABLE 5.1. Distribution of Foliar Nodulation in Angiosperms

Plant family	Genus	Nodulated species in which the microsymbiont has been experimentally investigated	Reference
Rubiaceae	Pavetta	grandiflora	57
		indica	58
		reinwardtii	68
		zimmermanniana	48, 49, 68, 71
	Psychotria	calva	72
		emetica	57
		mucronata	61
		nairobiensis	57
		punctata (bacteriophila)	59, 60, 68, 76-78
	Neorosea	androgensis	75
Myrsinaceae	Ardisia	crenata	64, 94
		crispa	57, 62-71, 94
		hortorum	70
	Amblyanthopsis	*a	
	Amblyanthus	*b	
Dioscoreaceae	Dioscorea	macroura	73, 74

*Botanical descriptions of 2 species (a) and 3 species (b) have been reported but no microbiological investigations were made (79).

5.3.2 Biology of the Cyclic Symbiosis

Although the cyclic symbiosis in the Rubiaceae and Myrsinaceae are similar in many respects, they are sufficiently different to warrant separate discussions.

5.3.2.1 Rubiaceae

Although Zimmermann was the first to recognize the microbial nature of the symbiosis (48), it was the detailed work of von Faber which provided an understanding of the intricate relationship existing between the host plant and the microsymbiont (49, 50). According to him the bacterial leaf nodules are derived from rather primitive secretory structures whose formation is not dependent upon the presence of the microsymbiont. During leaf development some of the stomata located in the region of the secretory glands, which exude a gummy secretion and contain bacteria, become infected (Fig. 5.2D). This eventually becomes walled off in the leaf and is readily discernable as a macroscopic nodule. Since this may occur in floral as well as leaf primordia, the seed usually becomes infected, thereby establishing a cyclic process as shown below.

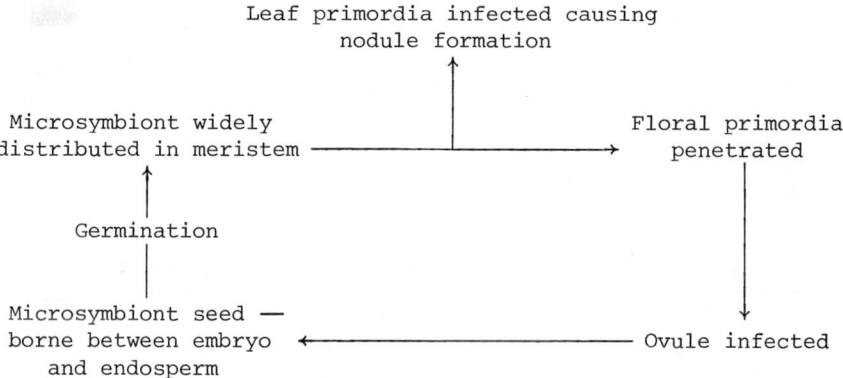

Von Faber was able to "rid" seeds of the microsymbiont by heat treatment (25 min at 50°C for Pavetta). Plants grown from such seeds did not develop normally and these dwarf plants were termed "cripples." Rarely did a non-nodulated specimen survive more than about 1-1½ years. However, some dwarf plants did revert to the normal growth habit — this was invariably accompanied by the appearance of nodules on the new leaves. These observations were remarkably accurate and subsequent investigations have supported most of the data (57, 59, 60). Von Faber succeeded in isolating the

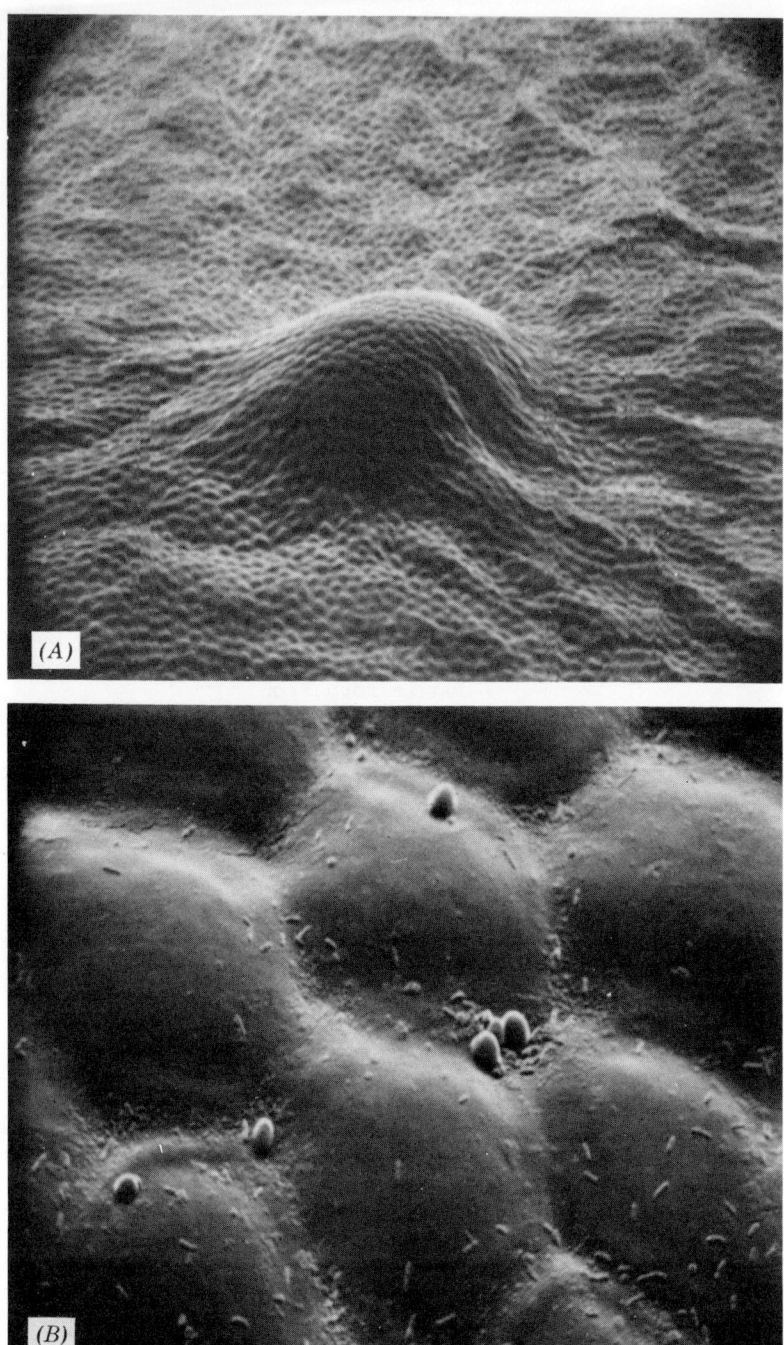

Fig. 5.2. Leaf nodules of <u>Psychotria</u> <u>punctata</u>. (<u>A</u>) Scanning electron micrograph (SEM) of adaxial leaf surface at a nodule (x50). (<u>B</u>) Same area as (A) but at x900 magnification. Note that leaf surface is heavily populated with microbial epiphytes, an important point to consider when nodular N_2 fixation is being

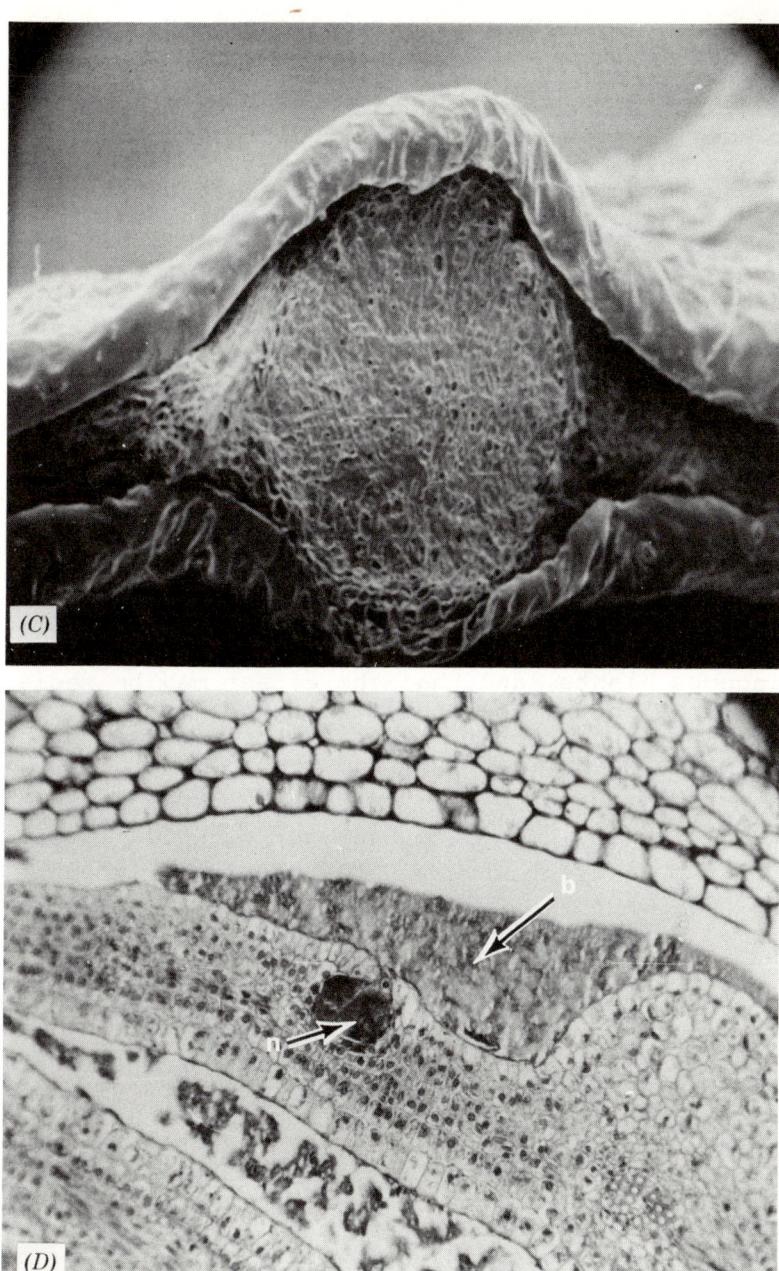

assessed (see 5.5). (C) SEM view of section through a leaf nodule (x120). (D) Light microscope photomicrograph of section through developing leaflets. Note bacteria (b) in gummy matrix and in nodule (n) which has not yet been walled off (x120). Photographs (A)-(C) by Ms. Patricia Carlyle using a Cambridge Stereoscan Model IIA SEM, 10 KC accelerating voltage. Preparations were mounted with silver-base paint, evaporated on a Denton VD-502 and coated with 30 nm of gold.

microsymbiont from three nodulated plant species, and concluded that the endophytes of different species were basically similar.

More recently the development and structure of the leaf nodules of Psychotria punctata have been subjected to the greater resolution of the electron and fluorescent light microscopes (71, 76, 77). These studies confirm that the bacteria in the mucilaginous material surrounding the shoot apex enter via stomates in leaf primordia into the substomatal chamber, where the enlarging chambers develop into leaf nodules when the young leaves leave the apical region. The fine structure of the nodule bacteria resembles that of gram negative bacteria, and they are usually encapsulated. The bacteria are intercellular and an extensive membrane system occurs between them (76). It has been suggested that the thick-walled mesophyll cells surrounding the bacteria supply substrate for the microsymbiont which eventually degenerates (76). Recently Lersten has suggested that there is a relationship between colleter morphology and nodulation in rubiaceous plants, and that the effect may be mediated by a change in the chemical secretion of the colleter (111, 112), but another investigation revealed no correlation between gland type and symbiotic ability (114). Since no one has unequivocally proven that non-nodulated plants are entirely free of symbiotic bacteria it has been difficult to establish whether purported endophytes reinfect non-nodulated plants, or whether endogenous bacteria are responsible (see 5.4). Unfortunately attempts to obtain bacteria-free plants from tissue culture have not been successful (78).

Although von Faber had analytical data suggesting that the microsymbiont was responsible for N gains presumably from N_2 fixation, evidence has gradually accumulated which suggests that plant hormones, perhaps cytokinins (61), are involved (59, 60, 79). This is in agreement with the deformed growth habit characteristic of non-nodulated plants (57, 59, 61, 79). Evidently the microsymbiont supplies the host plant with hormonal substances or precursors the plant cannot synthesize, or it removes some natural inhibitor that interferes with normal plant growth. The great similarity in the nodular structure of many species of Pavetta and Psychotria, and the distribution of the microsymbiont in the plant, suggest that the symbiosis is very similar throughout the group (57). Indeed, in the lesser studied Neorosea andongensis (Hiern) N. Halle (Tricalysia andongensis Hiern), where mature nodules appear as long, dark protuberances along both sides of the midrib and petiole, similarity to the Pavetta and Psychotria species was noted (75).

5.3.2.2 Myrsinaceae

The symbiosis in this group is different from that of the Rubiaceae in certain aspects although the involvement of plant hormones and N_2 fixation has been suggested. The scheme given above for the cyclic symbiosis in the Rubiaceae generally also applies to the situation in Ardisia crispa, the species that has been studied the most extensively (for reviews see refs. 80-82). Again the detailed studies of the early investigators, in this case H. Miehe (83-86), have proven to be essentially correct. In Ardisia the microsymbiont is located in the mature leaf in a regular fashion along the crenated leaf margin, where it is most readily observed from the abaxial surface due its darker appearance than the remainder of the leaf. That the bacteria are distributed throughout the meristem except for roots and anthers is well established (57, 66) (see Fig. 5.3). De Jongh has distinguished several morphological forms of the endophyte: Stage I — a short, motile rod (viable); a larger, nonmotile rod (viable, Stage II; or nonviable, Stage III); and a "bacteroid" form which was nonviable, Stage IV (66). The microsymbiont has been isolated in pure culture (1, 62, 66, 84, 87). In Ardisia host plants lacking the microsymbiont fail to form leaves from leaf primordia, thereby giving rise to a warty mass of undifferentiated tissue (66). Such plants eventually die. As in the case of the nodulated species of the Rubiaceae, plants may be freed of bacteria by heat treatment (7 min at 52°C). Similar deformity also occurs spontaneously in about 50% of the seedlings in mass plantings (86). Although Miehe ruled out N_2 fixation as being a likely factor in the symbiosis, it was not until the work of de Jongh that possible hormonal effects were tested experimentally (66). He found elevated peroxidase levels in bud tips of "cripples" only but β-IAA treatment was without effect. More recently direct evidence for the

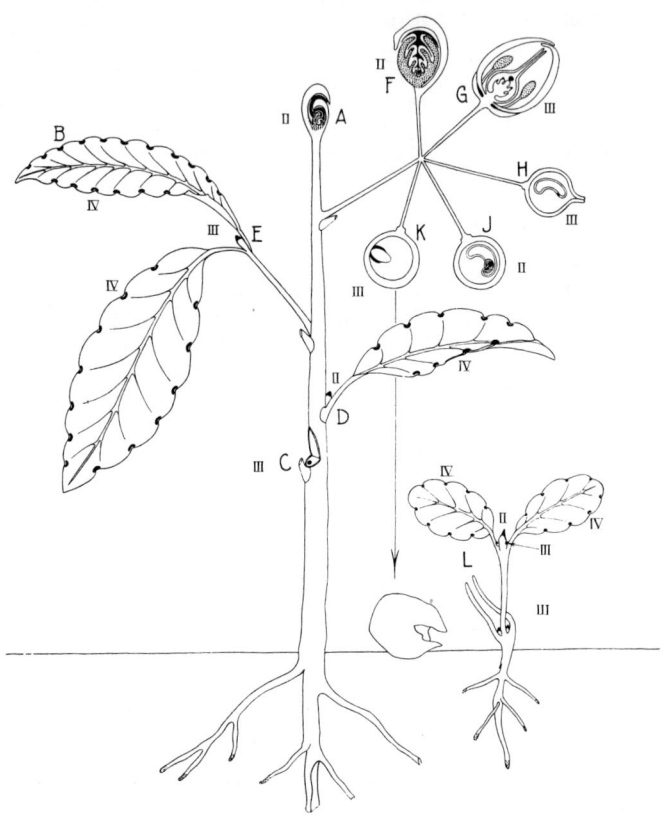

Fig. 5.3. Schematic representation of the cyclic symbiosis in
<u>Ardisia</u> (De Jongh, ref. 66). The bacteria are shown in black,
meristem stippled. The Roman numerals refer to the bacterial
stages in 5.3.2.2. It is noteworthy that all meristem except
that of roots and anthers have associated bacteria. (<u>A</u>) Terminal
bud; (<u>B</u>) foliar nodules; (<u>C</u>) dormant bud; (<u>D</u>) axillary bud; (<u>E</u>)
inflorescence bud; (<u>F</u>) young developing flower; (<u>G</u>) closed; (<u>H</u>)
small fruit; (<u>J</u>) large fruit; (<u>K</u>) mature fruit; (<u>L</u>) seedling.

Leaf Nodule Bacteria

production of cytokinins by the isolated strains of endophyte has been presented, as well as the inference that this is the function of the microsymbiont in the association (64).

5.3.2.3 Discoreaceae

In this case there is little evidence that any obligate type of association exists between the host plant (Dioscorea macoura) and the microorganisms that are found massed in closed furrows along the upper leaf surface. Endophytic N_2-fixing bacteria were isolated from the fleshy tips of leaf primordia (89). Schaede, however, failed to find bacteria in vegetative cones of rudimentary leaves except at tips of primordia where a variety of "casuals" were noted microscopically in sections (74). Although not definitive, the evidence points toward a nonessential, nonspecific relationship, for the removal of all the leaf tips had no apparent effect upon the plant (81). This appears to be an example of the loose type of association which may have preceded the obligate variety during the course of their evolution (92).

Recent observations (89) have failed to confirm an earlier claim that foliar bacterial associations were common in Coprosma and other New Zealand plants (90, 91). At this time one must conclude that foliar symbioses occur in only two plant families, but in almost 400 species.

5.4 LEAF NODULE BACTERIA

5.4.1 Isolation Methodology

Most workers agree that mature nodulated leaves are not good sources for the attempted isolation of the microsymbiont (except in Ardisia) for the nodular form is "bacteroidal" in nature, i.e., it is not viable and colony formation on appropriate media does not occur. In addition, the complete removal of surface epiphytes from leaves is more difficult than from seeds, which, being impervious, can withstand rigorous sterilizing treatment. Only two persons have reported successful isolations from nodulated leaves (71, 94).

The tissue of choice is the germinating seedling for the seed can be readily sterilized, grown aseptically until germination has proceeded sufficiently to crack the seed coat, and then homogenized and plated after the embryo and endosperm have been removed from the seed coat (49, 59). In one study in which this method was used only one plate out of hundreds had colonies of a single type of contaminant (59). Isolations may also be made from buds (48, 71). It is likely that any growing point contains viable cells which may be cultured; however, many of these tissues would be more difficult to render contaminant-free than the seed.

5.4.2 Bacteria Isolated from Nodulated Leaves

A tabulation of all reported cases of the isolation or alleged isolation of the microsymbionts from various leaf nodule symbioses is given in Table 5.2. In the case of Ardisia there is apparent agreement that the endophyte is some type of gram negative, motile rod; however, only some of the strains are reported to fix dinitrogen. In one case two reports on the same culture are not in agreement with regard to this trait (62, 67). The bulk of the evidence suggests that the culture originally isolated by Miehe (which he named Bacillus foliicola) was indeed the endophyte, but it has been suggested that the legitimate name is Phyllobacterium follicola (Miehe) comb. nov. Horner and Lersten (95). Although some of the isolated strains have been shown to fix dinitrogen in pure culture (57, 70), there is no good evidence that this occurs in the plant. Indeed, there is good reason to believe that a plant growth regulator is involved, presumably a cytokinin (64). The symbiosis in Ardisia is sufficiently interesting to warrant more work, but certainly not from the viewpoint of N_2 fixation.

In Psychotria and Pavetta there is less agreement that purported isolants were the microsymbiont. The detailed reports of von Faber (49, 50) leave little doubt that he had succeeded in isolating the microsymbiont for he was able to infect very young seedlings from heat-treated, "bacteria-free" seeds in which the seed coat had been artificially opened so that the inoculated vegetative cone could become infected early in the plant's growth. Within 2 months 18 of 29 inoculated plants were indistinguishable from the nodulated control plants (49). Although one might raise the criticism that this might be the expression of reversion from dwarf to normal habit, it has been this author's experience that reversion does not occur with sufficient regularity (in Psychotria punctata) to explain the observed changes in so many of the inoculated seedlings. Von Faber presented a detailed description of his culture and appeared convinced that the microsymbionts from the different plant species he used were very similar (49, 50). The only point which seems to have been in error was his designation that the bacterium was acid fast for his staining method was not the Ziehl-Neelsen type and known nonacid fast bacteria stain positively by it (57, 60). Otherwise his culture appears very similar to the Klebsiella isolated from P. punctata by Centifanto and Silver (60). In the later case the isolated organism conformed in all respects, including % GC (96, 97), to the genus Klebsiella, and was therefore named K. rubiacearum in accord with the rules of nomenclature (60). This strain, which has been referred to by others as K. pneumoniae (98), was agglutinated by klebsiella type 24 antiserum (60). The widespread occurrence of klebsiellas in nature (99) has led to some doubt that the isolant was the microsymbiont especially since Becking was unable to nodulate "cripples" with cultural suspensions (61). However, he used a different plant species as host, and the source of inoculum, while authentic, was prepared from cultures that had been maintained in vitro for 8 years after the initial isolation (61). These discrepancies notwithstanding, the fact remains that the original report provided three types of data in support of the claim: (a) the same bacterium was repeatedly isolated from seedlings germinated under aseptic conditions, (b) nodules indistinguishable from normally occurring ones were induced on the leaves of "cripples" by abrasion with cultured cells, and (c) bacteria taken directly from nodules of normal plants reacted with antiserum prepared from the isolated culture (60).

In another investigation a variety of bacteria related to the Rhizobiaceae were isolated from Ardisia, Pavetta, and Psychotria (57) (see Table 5.2). The traits of a dozen cultures were very similar except that the ability to fix dinitrogen was limited to half of the strains isolated (57). Further investigation identified two cultures isolated from Ardisia crispa

TABLE 5.2. Bacteria Isolated from Leaf Nodule Plants

Host plant	Isolant	Morphological characteristics			
		Shape	Endo-spore	Gram stain	Acid-fast stain
Ardisia crispa	Bacillus foliicola	rod	-	neg	-
	Bacillus foliicola	rod	-	neg	-
	Bacterium foliicola[a]	rod	-	neg	-
	Bacterium foliicola	rod	-	0	0
	C$_2$	rod	-	neg	0
	Mycoplana rubra	rod	-	neg	0
	Ac. 1	rod	-	neg	-
	Ac. 2	rod	-	neg	-
	Ac. 3	rod	-	neg	-
	Phyllobacterium myrsinaceacum[d]	rod	0	neg	-
Ardisia hortorum	Xanthomonas hortoricola	rod	0	0	0
Ardisia humilis	L-forms	rod	0	0	0
Ardisia punctata	P$_3$	rod	-	neg	0
Ardisia crenata	Phyllobacterium myrsinaceacum[d]	rod	0	neg	-
Pavetta indicum	Mycobacterium rubiacearum	rod	-	neg	+
Pavetta grandiflora	Pg. 1	rod	-	neg	-
	Pg. 2	rod	-	neg	-

Morph. char.	Nitrogen source for growth			Evidence that isolant is the microsymbiont	Reference
Motility	N_2	Inorg. N	Org. N		
+	−	−	+	Infection of cripples	84
+	−	+	+	Infection of cripples	66
+	prob −[b]	+	+	0	62
+	0	0	0	0	71
+	−	+	+	Infection of cripples	87
+	−	0	+	None[c]	68
+	−	−	+	Immunological	57
+	+	+	+	Immunological	57
−	+	−	+	Immunological	57
0	−	+	+	0	94
+	+	0	+	Infection of cripples	70
0	0	0	0	0	93
+	−	+	+	Infection of cripples	67
0	−	+	+	0	94
−	+	+	+	0	48
−	±	−	+	Immunological	57
−	−	−	+	Immunological	57

(Continued)

TABLE 5.2. (Continued)

			Morphological characteristics		
Host plant	Isolant	Shape	Endo-spore	Gram stain	Acid-fast stain
Pavetta zimmermanniana	Bacterium rubiacearum	rod	0	0	-
	Mycobacterium rubiacearum	rod	0	0	+
Psychotria calva	Flavobacterium sp.	rod	-	pos	-
Psychotria emetica	Pe. 1	rod	-	neg	-
	Pe. 2	rod	-	neg	-
	Pe. 3	rod	-	neg	-
Psychotria nairobiensis	Chromobacterium lividum	rod	-	neg	-
	Pn. 1	rod	-	neg	-
	Pn. 2	rod	-	neg	-
	Pn. 3	rod	-	neg	-
	Pn. 4	rod	-	neg	-
Psychotria punctata	Mycobacterium rubiacearum	rod	0	0	+
	Klebsiella rubiacearum	rod	-	neg	-

[a] Two strains, K and V were isolated.
[b] Glaubitz reported slight N_2 fixation (62), but Yamada reported no N_2 fixation using the same culture (67).
[c] The cultures isolated were leaf surface epiphytes.
[d] An invalid name = Phyllobacterium foliicola.
[e] FA prepared from cultures isolated from Psychotria nairobiensis gave specific fluorescence with bacteria present in crushed

Morph. char.	Nitrogen source for growth				
Motility	N_2	Inorg. N	Org. N	Evidence that isolant is the microsymbiont	Reference
−	0	0	0	0	71
−	+	+	+	Infection of cripples	48, 49
−	0	0	0	0	72
+	+	−	+	Immunological	57
+	−	+	+	Immunological	57
+	−	−	+	Immunological	57
+	+	+	+	Immunological	65
+	−	+	+	Immunological	57
+	−	+	+	Immunological	57
+	−	+	+	Immunological	57
+	+	−	+	Immunological	57
−	+	+	+	0	48
+[f]	+	+	+	Infection of cripples, immunological	60

tissue of P. emetica, P. hirtelli, P. kirkii, P. capensis, Pavetta grandiflora, P. gardenifolia, P. revoluta, P. lanceotota, and Ardisia crispa, but not with the non-nodulated A. ellipita (65).
[f] Although the culture originally isolated was motile, this trait was lost upon subculture.
0 indicates that the test was not reported.

and two from Psychotria nairobiensis as Chromobacterium lividum (65). Fluorescent antibody prepared from the isolated culture cross reacted with bacteria directly from crushed nodules, seed embryo, and flowers, but not with the tissue from a non-nodulated species of Ardisia (65). This is the first investigation that infers that the microsymbionts in the Myrsinaceae and Rubiaceae are identical, but the validity of the earlier isolations of Bacillus foliicola (84) by Miehe, Glaubitz (62), and Yamada (87) remains unanswered. Furthermore no mention was made about the other cultures Gordon isolated (57) which could not fix dinitrogen. It is of interest that Klebsiella rubiacearum as well as Klebsiella sp. serotype 24 were not serologically related to Chromobacterium lividum (65).

One may conclude from this discussion that a variety of bacteria may be isolated from tissue of leaf-nodulated plants, and that many of the isolants fix dinitrogen in pure culture. However, several points remain unanswered: the host range specificity of leaf nodule bacteria, a rigid demonstration of Koch's postulates using purported endophytes, and whether N_2 fixation occurs in the nodulated plant when a qualitatively and quantitatively sufficient population of the microsymbiont is present. The latter point is considered below (5.5).

5.5 DOES N_2 FIXATION OCCUR IN THE SYMBIOTIC STATE?

There is sufficient evidence from the reports of several investigations to conclude that N_2-fixing bacteria may be isolated from a variety of nodulated plant species (48, 57, 60, 65) and in one case N_2 fixation by the isolated culture was studied in detail (59, 60, 100-102). Although the designation of some of these cultures as true endophytes has been disputed (61), the important point from the viewpoint of dinitrogen fixation is whether this process occurs in the nodulated plant, and if so, whether the amount is ecologically significant.

5.5.1 Evidence from $^{15}N_2$ and C_2H_2 Reduction Studies

Except for several brief reports (103, 104) which the original authors, as well as others, could not subsequently confirm (61, 108), there has been no conclusive evidence supporting the contention that N_2 fixation occurs in nodulated leaves (113). Moreover, analysis of the N content of leaf segments of _Psychotria calva_ by the micro-Kjeldahl method did not indicate that such tissues had any greater N content than did the leaves from surrounding trees (105). Such type of evidence is, however, merely suggestive, for the N content of leaves of different plant species may not necessarily be comparable. It is not surprising that the use of $^{15}N_2$ failed to indicate any uptake of the isotope for, if the rate of N_2 fixation were low as is likely in slow growing plants like _Psychotria_, the technique might very well lack sufficient sensitivity, especially since the exposure times in most investigations were short. In one attempt to demonstrate N_2 fixation, intact plants were used in which the foliar region was isolated in a glass vessel that was evacuated and refilled with an artificial atmosphere containing $^{15}N_2$ and 5% CO_2. No enrichment was found even after 3 days, at which time the experiment was terminated because the leaves began to appear abnormal (W. S. Silver, unpublished experiments).

Failure to obtain positive evidence by the acetylene reduction method is, however, much more indicative of low or negligible rates of N_2 fixation because of the superior sensitivity of the technique. At this time there have been a number of such investigations all with negative results (61, 106, 108), even with incubation periods as long as 18 hours (107). One must conclude that (a) the populations of microsymbionts that are in a physiological state to reduce dinitrogen or acetylene are too small to show up during the test period, or (b) the process is inhibited by O_2 produced in the leaves during photosynthesis, or (c) isolated N_2-fixing bacteria are not the plant endophytes, or (d) the process does not occur in these symbiotic associations, whether or not the microsymbiont fixes N_2 in pure culture. At this time there is insufficient information to make the choice with any finality.

5.5.2 Evidence from Habitat Studies

In the case of the root-nodulated legumes and non-legumes it is often possible to infer that N_2 fixation is associated with the plant (see IV.4.2.1) from the observation that a plant is a successful pioneer in low N soils and that it shows no signs of N deficiency in such soils, whereas non-nodulated plants are N deficient. It has become more apparent recently that this type of phenomenon may also be due to plant epiphytes, the association being of a looser nature (see IV.2.3.2). A similar situation could exist in plants with leaf nodules. Although there is really no good evidence that leaf-nodule-bearing plants are predominant in N-deficient niches, in tropical regions the phyllosphere is very likely an area of intense microbial activity, favoring the proliferation of N_2-fixing and oligonitrophilic varieties. It is likely that this type of N_2 fixation may be of some importance and not that which might occur in leaf glands (see IV.2.6). Further study of this habitat and the biology of the organisms involved is required in order to properly assess this possibility.

5.6 REFERENCES

1. Ruinen, J., Plant Soil, 15, 81 (1961).
2. Ruinen, J., Plant Soil, 22, 375 (1965).
3. Lie, T. A., and E. G. Mulder, Plant Soil, Special Volume (1971).
4. Preece, T. E., and C. H. Dickinson, Eds., Ecology of Leaf Surface Micro-Organisms, Academic Press, New York, 1971.
5. Dunn, C. L., K. I. Beynon, K. F. Brown, and J. T. W. Montagne, in Ecology of Leaf Surface Micro-Organisms, T. E. Preece and C. H. Dickinson, Eds., Academic Press, New York, 505, 1971.
6. DiMenna, M. E., in Ecology of Leaf Surface Micro-Organisms, T. E. Preece and C. H. Dickinson, Eds., Academic Press, New York, 170, 1971.
7. Burrage, S. W., in Ecology of Leaf Surface Micro-Organisms, T. E. Preece and C. H. Dickinson, Eds., Academic Press, New York, 99, 1971.
8. Last, F. T., and R. C. Warren, Endeavor, 31, 143 (1973).
9. Ruinen, J., Ann. Inst. Pasteur (Paris), 3, 343 (1966).

References

10. Tukey, H. B., Jr., Ann. Rev. Plant Physiol., 21, 305 (1970).
11. Ruinen, J., Plant Soil, 33, 661 (1970).
12. Wichner, S., and E. Libbert, Physiol. Plant., 21, 227 (1968).
13. Wichner, S., and E. Libbert, Physiol. Plant., 21, 500 (1968).
14. Libbert, E., and H. Risch, Physiol. Plant., 22, 51 (1969).
15. Phillips, D. A., and J. G. Torrey, Plant Physiol., 49, 11 (1972).
16. Chou, T. W., and S. F. Yang, Arch. Biochem. Biophys., 157, 73 (1973).
17. Sztynberg, A., and J. P. Blakeman, J. Gen. Microbiol., 78, 15 (1973).
18. Leben, C., Ann. Rev. Plant Pathol., 3, 209 (1965).
19. Ellingboe, A. H., Ann. Rev. Plant Pathol., 7, 317 (1969).
20. Wood, R. K. S., Proc. Roy. Soc. Lond. Ser. B, 181, 213 (1972).
21. Kelman, A., and L. Sequeiria, Proc. Roy. Soc. Lond. Ser. B, 181, 247 (1972).
22. Wiebe, W. J., in Fundamentals of Ecology, 3rd ed., E. P. Odum, Ed., Saunders, Philadelphia, 485, 1971.
23. Dickinson, C. H., in Ecology of Leaf Surface Micro-Organisms, Academic Press, New York, 136, 1971.
24. Klincare, A. A., D. J. Kreslina, and I. V. Mishke, in Ecology of Leaf Surface Micro-Organisms, Academic Press, New York, 136, 1971.
25. Singer, C. E., and B. N. Ames, Science, 170, 822 (1970).
26. Jensen, V., in Ecology of Leaf Surface Micro-Organisms, Academic Press, New York, 464, 1971.
27. Lockhead, A. G., and F. E. Chase, Soil Sci., 55, 185 (1943).
28. Nowakowska-Waszezuk, A., Acta Microbiol. Polonica B, 4, 75 (1972).
29. Anderson, J. R., and E. A. Drew, J. Gen. Microbiol., 70, 43 (1972).
30. Benjamin, M., and W. S. Silver, Abstr. Ann. Meet. Am. Soc. Microbiol., 168 (1973).
31. Millbank, J. W., Arch. Mikrobiol., 68, 32 (1969).
32. Ruinen, J., Ant. van Leeuwenhoek J. Microbiol. Serol., 29, 425 (1963).
33. Jensen, V., and E. Holm, in Nitrogen Fixation and the Biosphere, W. D. P. Stewart, Ed., Cambridge University Press, 101 (1975).

34. Ruinen, J., in Ecology of Leaf Surface Micro-Organisms, Academic Press, New York, 567, 1971.
35. Duncan, C. L., and A. R. Colmer, Appl. Microbiol., 12, 173 (1964).
36. Balasundaram, V. R. and A. Sen, Indian J. Microbiol., 8, 33 (1968).
37. Sen, A., R. B. Rewari, N. B. Paul, and A. N. Sen, Proc. Nat. Acad. Sci., India, Sect. A, 24, 5 (1955).
38. Bond, G., Advan. Sci., 15, 382 (1959).
39. Bhurat, M. C., and A. Sen, Indian J. Agr. Sci., 38, 319 (1968).
40. Jones, K., Ann. Bot., 34, 239 (1970).
41. Jones, K., E. King, and M. E. Eastlick, Ann. Bot., 38, 222 (1974).
42. Bessems, E. P. M., Agr. Res. Rep., 786, Wageningen, 1973.
43. Iswaran, V., A. Sen, and R. Apte, Plant Soil, 39, 461 (1973).
44. Vasantharajan, V., and J. Bhat, Plant Soil, 28, 258 (1968).
45. Silver, W. S., A. Jump, and D. Pukatzki, Abstr. Ann. Meet. Am. Soc. Microbiol., 327 (1974).
46. Dunleavy, J. M., Am. J. Bot., 59, 665 (1972).
47. Trimen, P., Handbook of the Flora of Ceylon, Dulau, London, 1894.
48. Zimmermann, A., Jahr. Wiss. Bot., 37, 1 (1902).
49. von Faber, F. C., Jahr. Wiss. Bot., 51, 285 (1912).
50. von Faber, F. C., Jahr. Wiss. Bot., 54, 243 (1914).
51. Bremenkamp, C. E. B., J. Bot. (London), 71, 271 (1933).
52. Bremenkamp, C. E. B., Rep. Spe. Nov. Regn. Vegetabilis, 37, 1 (1934).
53. Bremenkamp, C. E. B., Rep. Spe. Nov. Regn. Vegetabilis, 47, pp. 12, 81 (1939).
54. Bremenkamp, C. E. B., Mus. Nat. Hist. Nat. Notulae Syst., 16, 41 (1960).
55. Petit, E., Bull. Jard. Bot. d'Etat, Bruxelles, 34, 1 (1964).
56. Petit, E., Bull. Jard. Bot. d'Etat, Bruxelles, 36, 65 (1966).
57. Gordon, J. F., Ph. D. Thesis, Imperial College, University of London (1963).
58. Rao, A. K., Agr. J. India, 18, 132 (1923).
59. Centifano, Y. M., Ph. D. Dissertation, University of Florida, Gainesville (1964).
60. Centifano, Y. M., and W. S. Silver, J. Bacteriol., 88, 776 (1964).

References

61. Becking, J. H., Plant Soil, Special Vol., 361 (1971).
62. Glaubitz, W., Diplom-Arbeit, Hamburg (1957).
63. Hofstra, J. J., and T. Koch-Bosma, Acta Bot. Neerl., 19, 665 (1970).
64. Rodrigues-Pereira, A. S., P. J. W. Houwen, H. W. J. Deurenberg-Vos, and E. B. F. Pey, Zeit. Pflanzenphysiol., 68, 170 (1972).
65. Bettleheim, K. A., J. F. Gordon, and J. Taylor, J. Gen. Microbiol., 54, 177 (1968).
66. de Jongh, P., Nederl. Akad. Wetensch. Afd. Nat. II, 37, 1 (1938).
67. Yamada, T., Bull. Fac. Ed. Chiba Univ., 9, 1 (1960).
68. de Vries, J. T., and H. G. Derx, Ann. Bogor., 1, 53 (1950).
69. Miehe, H., Jahrb. Wiss. Bot., 53, 1 (1914).
70. Hanada, K., Jap. J. Bot., 14, 235 (1954).
71. Ziegler, H., Z. Naturforsch., B, 13, 297 (1958).
72. Adjanohoun, E., C. R. Acad. Sci., 245, 576 (1957).
73. Orr, M. Y., Notes Roy. Bot. Gard. Edinburgh, 14, 57 (1923).
74. Schaede, R., Jahrb. Wiss. Bot., 88, 1 (1939).
75. van Hove, C., Ann. Bot., 36, 259 (1972).
76. Lersten, N. R., and H. T. Horner, Jr., J. Bacteriol., 94, 2027 (1967).
77. Whitmoyer, R. E., and H. T. Horner, Jr., Bot. Gaz., 131, 193 (1970).
78. LaMotte, C. E., and N. R. Lersten, Am. J. Bot., 59, 89 (1972).
79. Humm, H. J., J. N. Y. Bot. Gard., 45, 193 (1944).
80. Schwartz, W., in Handbuch der Pflanzenphysiologie, Springer Verlag, New York, Vol. 11, pp. 566-572, 1959.
81. Schaede, R., Die pflanzenlichen Symbiosen, 3rd ed., Fischer, Stuttgart, 1962.
82. Lange, R. T., in Symbiosis, S, M, Harvey, Ed., Academic Press, New York, Vol. I, pp. 99-170, 1966.
83. Miehe, H., Abh. Math.-Phys. Kl. Sachs Acad. Wiss. (Leipzig), 4, 399 (1911).
84. Miehe, H., Jahrb. Wiss. Bot., 53, 1 (1914).
85. Miehe, H., Ber. Deut. Bot. Ges., 34, 576 (1916).
86. Miehe, H., Jahrb. Wiss. Bot., 58, 29 (1919).
87. Yamada, T., Bull. Fac. Ed. Chiba Univ., 10, 75 (1961).
88. Renvoize, S. A., Phil. Trans. Roy Soc. Lond., Ser. B, 260, 227 (1971).
89. van Hove, C., and A. S. Craig, Ann. Bot., 37, 1013 (1973).

90. Stevenson, G., *Ann. Bot.*, 17, 343 (1953).
91. Stevenson, G., *Ann. Bot.*, 23, 622 (1959).
92. Silver, W. S., in *Chemistry and Biochemistry of Nitrogen Fixation*, J. R. Postgate, Ed., Plenum, London, 245, 1971.
93. Bose, S. R., *Nature (London)*, 175, 395 (1955).
94. Knosel, D., *Zentralbl. Bakteriol. Parasitenk. Infectionskr. Hyg.*, Abt. 2, 116, 79 (1962).
95. Horner, H. J., Jr., and N. R. Lersten, *Int. J. Sys. Bacteriol.*, 22, 117 (1972).
96. DeLey, J., *Anton. van Leeuwenhoek J. Microbiol. Serol.*, 31, 203 (1965).
97. Ouellete, C. A., R. H. Burris, and P. W. Wilson, *Anton. van Leeuwenhoek J. Microbiol. Serol.*, 35, 275 (1969).
98. Mahl, M. C., P. W. Wilson, M. A. Fife, and W. H. Ewing, *J. Bacteriol.*, 89, 1482 (1965).
99. Evans, H. J., N. E. R. Campbell, and S. Hill, *Can. J. Microbiol.*, 18, 13 (1972).
100. Heeb, M. J., M. S. Thesis, University of Florida, Gainesville, 1968.
101. Neelands, M. L., M. S. Thesis, University of Florida, Gainesville, 1967.
102. Hammer, S. M., M. S. Thesis, University of Florida, Gainesville, 1968.
103. Silver, W. S., Y. M. Centifanto, and D. J. D. Nicholas, *Nature (London)*, 199, 396 (1963).
104. Grobbelaar, N., J. M. Strauss, and E. G. Groenewald, *Plant Soil, Special Vol.*, 325 (1971).
105. Lohr, E., *Physiol. Plant.*, 21, 1156 (1968).
106. Silvester, W. B., and S. Astridge, *Plant Soil*, 35, 647 (1971).
107. Silver, W. S., *Proc. Roy. Soc. Lond., Ser. B*, 172, 389 (1969).
108. Grobbelaar, N., and E. G. Groenewald, *Z. Pflanzenphysiol*, 73, 103 (1974).
109. Dommergues, Y., and S. Mutaftschiev, *Ann. Inst. Pasteur (Paris)*, 109, 112 (1965).
110. McRoy, C. P., J. J. Goering, and B. Chaney, *Limnol. Oceanog.*, 18, 998 (1973).
111. Lersten, N. R., *Amer. J. Bot.*, 61, 973 (1974).
112. Lersten, N. R., *Bot. J. Linn. Soc.*, 69, 125 (1974).
113. Becking, J. H., in *Symbiotic Nitrogen Fixation in Plants*, P. S. Nutman, Ed., Cambridge University Press, 1975.
114. van Hove, C., and K. Kagoyre, *Ann. Bot.*, 38, 989 (1974).

CHAPTER 6

Dinitrogen-fixing Associations in Higher Plants other than Legumes

J. H. BECKING

Institute for Atomic Sciences in Agriculture
Wageningen, The Netherlands

6.1. Introduction, 187
6.2. Nodule-bearing Nonleguminous N_2-fixing Systems, 188
    6.2.1. Taxonomic Aspects, 188
6.3. Stem Symbiosis in Higher Plants, 198
    6.3.1. Taxonomic Aspects, 199
6.4. Present Distribution of Nonleguminous N_2-fixing Systems According to Main Floral Regions, 199
    6.4.1. Eurasia, 200
    6.4.2. North America, 201
    6.4.3. South America, 204
    6.4.4. Subtropical and Tropical Africa, 204
    6.4.5. Southeast Asia, 205
    6.4.6. Australia and New Zealand, 206
6.5. Fossil Distribution, 208
6.6. Evidence of N_2 Fixation by Root-Nodule Symbioses, 210
    6.6.1. Growth Experiments with Whole Plants, 210
    6.6.2. Experiments with Excised Root Nodules, 215
        6.6.2.1. $^{15}N_2$ Tests, 215
        6.6.2.2. Acetylene Reduction Tests, 218

6.6.3. Experiments with Nodule Breis, 223
 6.6.3.1. $^{15}N_2$ Tests, 223
 6.6.3.2. Acetylene Reduction Experiments, 224
6.7. Evidence of N_2 Fixation in the Gunnera Symbiosis, 224
 6.7.1. Growth Experiments with Whole Plants, 224
 6.7.2. $^{15}N_2$ Tests, 225
 6.7.3. Acetylene Reduction Tests, 225
6.8. Physiological Aspects of N_2 Fixation by Root-Nodule Systems, 226
 6.8.1. Isolation of Nitrogenous Compounds from Nodules, 226
 6.8.2. Influence of Photosynthate, 227
 6.8.3. Transfer of Nitrogenous Compounds from Nodule to Shoot, 228
 6.8.4. Mineral Requirements of the Dinitrogen-fixing Symbiosis, 228
  6.8.4.1. Molybdenum, 229
  6.8.4.2. Cobalt, 232
 6.8.5. Effect of Gaseous Environment on Dinitrogen Fixation, 232
  6.8.5.1. Dinitrogen, 233
  6.8.5.2. Oxygen, 233
  6.8.5.3. Hydrogen, 233
  6.8.5.4. Carbon Monoxide, 234
 6.8.6. Presence of Hemoglobin, Poly-β-Hydroxybutyrate and Growth Substances in Root Nodules, 235
  6.8.6.1. Hemoglobin, 235
  6.8.6.2. Poly-β-hydroxybutyrate, 236
  6.8.6.3. Growth Stimulating and Inhibiting Substances, 236
6.9. Root Nodules of Nonlegumes, 237
 6.9.1. Structure, 237
  6.9.1.1. Alnus Type, 238
  6.9.1.2. Myrica/Casuarina Type, 239
 6.9.2. Formation, 239
 6.9.3. Cytology, 244
6.10. Endophyte of the Root Nodule System, 247
 6.10.1. Morphology and Cytology, 247
 6.10.2. Classification, 252
 6.10.3. Isolation, 254
6.11. Endophyte of the Gunnera Symbiosis, 258
6.12. Discussion and Conclusions, 259
6.13. References, 264

6.1 INTRODUCTION

Like leguminous plants a number of nonleguminous plants in association with microorganisms have the capacity to fix N_2 and, thus, to contribute to the nitrogen status of soils. These plants often play an important role in natural ecosystems by the colonization of bare soils deficient in combined nitrogen. Because of this capacity they frequently are the basis of the chronosequence of vegetation in plant successions, e.g., in the afforestation of eroded areas, in the plant occupation of bare areas by the recent withdrawal of ice by glaciers, or in the restoration of plant cover destroyed by volcanic activity.

All nonleguminous plants having the capacity to fix dinitrogen belong to the Dicotyledones. Their association with microorganisms is either in the root nodules or in the stem like in Gunnera. In most root-nodule producing nonleguminous plant species the causative organism is not a true bacterium like in leguminous plants (order Eubacteriales, genus Rhizobium), but the symbiont is an actinomycete (order Actinomycetales, genus Frankia).

There is only one well checked exception, i.e., the recent discovery of root nodulation in Trema cannabina var. scabra (not Trema aspera, pers. comm. M. J. Trinick) of the family Ulmaceae, Order Urticales. These root nodules proved to be inhabited by a Rhizobium species, which was cross-inoculable with certain legumes (218). This observation urges the reexamination of reported nodulation in the family Zygophyllaceae in the species Zygophyllum album, Z. coccineum, Z. decumbens, Z. simplex, Fagonia arabica, and Tribulus alatus, which root nodules are mentioned to contain Rhizobium and to be associated with nitrogen fixation (219, 220). These claims were refuted by Allen and Allen (221), who in a study of the anatomy of root nodules of a green-house plant of Tribulus cistoides, came to the conclusion that these were storage organs. Moreover, the rhizobial nature of the endophyte and their ability to cross-inoculate legumes is doubted. In the light of the recent finding of the association of Trema with Rhizobium, a reinvestigation of nodulation in this family is highly desirable. Finally, Farnworth and Clawson (222) and Clawson et al. (223) enumerated a number of plant species growing on rangeland and forest soils of Northern Utah, U. S., possessing root nodulation and nitrogen fixation capacity. These plants were mainly <u>Artemisia ludoviciana</u> and <u>A. michauxiana</u> of the Asteraceae (Compositae), but also phylogenetically widely unrelated plant species such as representatives of the Violales, Cactales, Umbellales, and Boraginales. Before these claims can be accepted they should be confirmed as the evidence so far presented is rather scanty.

In the N_2-fixing species of the genus <u>Gunnera</u> (order Myrtales, family Gunneraceae) no root nodules are formed, but the symbiont (a blue-green alga of the Nostocaceae) is an inhabitant of special glands in the stem.

Most nodulated nonleguminous plants with the capacity of fixing dinitrogen are woody. Therefore, their economic application will be mostly in forestry whereas the legumes have this position in agriculture. Currently the significance of these nonleguminous plants for afforestation of eroded areas and their use as understory plantings of commercial wood species is not fully appreciated and profitably used. Also <u>Gunnera</u> although herbaceous can be used for the green cover of landslides and eroded soils in wet habitats. It is anticipated that in the near future the use of nonleguminous nitrogen-fixing plants will be further explored and practiced.

The following sections will give information on the types of nonleguminous plants fixing dinitrogen, aspects of the symbiosis and the N_2-fixing capacity, the nature of the endophyte, and the infection process, especially with regard to the root-nodule associations. Only a brief account of the role of the plants in natural ecosystems and their application for soil fertility will be given for this is treated elsewhere in this treatise (IV.4).

6.2 NODULE-BEARING NONLEGUMINOUS N_2-FIXING SYSTEMS

6.2.1 Taxonomic Aspects

All plants in this group that fix dinitrogen belong to the class Dicotyledones of the Angiospermae. As is evident from Table 6.1 these N_2-fixing nonlegumes comprise diverse and phylogenetically unrelated plant taxons covering in total 8 orders, 9 families, and

Nodule-bearing Nonleguminous N_2-fixing Systems

18 genera. The genus <u>Arctostaphylos</u>, about 70 species of the Ericaceae, is not included in the table for although nodulation has been reported in <u>Arctostaphylos uva-ursi</u> (1), the nature of the endophyte is unknown, and there is no record of N_2 fixation by this species. It is likely that the latter symbiosis is mycorrhizal, as in other Ericaceous plants such as <u>Calluna</u> <u>vulgaris</u> (2), and that no N_2 fixation occurs.

TABLE 6.1. Classification of Nonleguminous Dinitrogen-fixing Angiosperms

Order	Family	Genus	Number of Symbiotic Species. In parenthesis total number of species.[a]
Root-nodule symbiosis with <u>Actinomycetes</u> species			
Casuarinales	Casuarinaceae	Casuarina	18 (45)
Myricales	Myricaceae	Myrica	20 (35)
		Comptonia	1 (1)
Fagales	Betulaceae	Alnus	33 (35)
Rhamnales	Elaeagnaceae	Elaeagnus	14 (45)
		Hippophaë	1 (3)
		Shepherdia	2 (3)
	Rhamnaceae	Ceanothus	31 (55)
		Discaria	2 (10)
		Colletia	2 (17)
Coriariales	Coriariaceae	Coriaria	13 (15)
Rosales	Rosaceae	Dryas	3 (4)
		Purshia	2 (2)
		Cercocarpus	3 (20)
Root-nodule symbiosis with <u>Rhizobium</u> species			
Urticales	Ulmaceae	Trema	1 (55)
Malpighiales	Zygophyllaceae[b]	Zygophyllum	4 (100)
		Fagonia	1 (40)
		Tribulus	2 (20)

(Continued)

TABLE 6.1. (Continued)

Order	Family	Genus	Number of Symbiotic Species. In parenthesis total number of species.[a]
	Stem symbiosis with blue-green algae		
Myrtales	Gunneraceae (or Haloragaceae)	Gunnera	22 (54)

[a] Total number of species according to Willis (13), only with Dryas, Trema and Gunnera another species classification has been followed. Generic names according to Hutchinson (224).
[b] Root nodulation of the Zygophyllaceae (219, 220) is provisionally included in this survey, although the experimental evidence needs confirmation. The reported root nodulation and nitrogen fixation in Asterales, Violales, Cactales, Umbellales and Boraginales with indicated Rhizobia symbioses (222, 223) are not included, since the experimental evidence is rather weak. Also the reported root nodulation and nitrogen fixation by Calamagrostis arundinacea (L.) Roth. containing Rhizobium-like endophytes (272) have been left out, because this evidence needs better description and documentation. Moreover, the root nodulation in some Ericaceae like Arctostaphylos (1) and Calluna (2) have been omitted, because there is evidence that no nitrogen fixation is involved and that these symbioses are mycorrhizal (225).

Most phylogenetic schemes of the Angiospermae start with the Casuarinales comprising a single family with a single genus, in order to show its relationship with the more primitive Gymnospermae (3). All Casuarina species examined in their natural habitat appeared to be nodulated, although over half of the species in this genus have not been examined for this property. The Myricales are also regarded by most plant taxonomists to be a rather primitive group. By placing the orders Casuarinales and Myricales in close proximity, the similarity in the type of nodulation of both groups is emphasized in contrast to all other orders of nodulated nonleguminous plants. In Casuarina and Myrica the apex of each nodule lobe gives rise to a normal, but negatively geotropic root, which is endophyte free, the endophyte occurring only in the basal part of the root nodule. Because of this peculiar root growth the root nodules of Casuarina and Myrica appear to be covered with upward growing rootlets (Plate 6.1A). This type which I will call here the Myrica/Casuarina nodulation-type differs from the Alnus nodulation-type found in all other nodule-bearing nonleguminous species. In the latter the apical meristems shows only a very restricted growth making these root nodules similar to those of leguminous plants (Plate 6.1B, C). Comptonia asplenifolia of the Myricales has the Alnus-type of root nodules, since root nodules of this species depicted by Ziegler (4) did not show the apical growth of the meristem resulting in the formation of negative geotropic roots. Moreover, under certain conditions some Alnus species may produce root nodules with negative geotropic roots, which occur only under aberrant conditions, as when Alnus rubra plants from western North America are inoculated with the endophyte of European Alnus glutinosa. In this case the nodules were rather ineffective in N_2 fixation and showed the peculiarity that each nodule lobe gave rise to a negative geotropic root (5-7) (Plate 6.1D). In addition to Alnus of the Fagales, all other nodulated species present in the Rhamnales, Coriariales, and Rosales have the Alnus-type of root nodules with a superficial resemblance to legume root nodules except for frequent dichotomous branching (Plate 6.1D), a feature unfrequently found in leguminous plants.

Alnus in the family Betulaceae occupies a rather isolated position, for of the six other genera in this family, e.g., Betula, Corylus, Carpinus, etc., none produce root nodules, whereas all examined representatives of the genus Alnus are nodulated.

While there is controversy regarding the taxonomic position of the Elaeagnaceae, a close relationship exists between the Elaeagnaceae and the Rhamnaceae, for members of both families show the same type of coralloid root nodules. Therefore in the present classification (Table 6.1), Hutchinson's proposal is followed in which the Elaeagnaceae are placed in the Rhamnales (8). Since the Coriariaceae obviously have an isolated position as nearly all of them are nodulated in contrast to some other similar families, Hutchinson's taxonomy is followed here also.

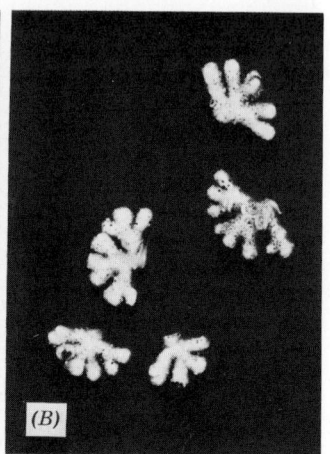

Nodule-bearing Nonleguminous N_2-fixing Systems

In the Rosaceae of the Rosales, three genera, i.e., Dryas, Purshia, and Cercocarpus, of the more than 100 genera within this family are nodulated. Surprisingly Dryas octopetala is nodulated in Alaska, but not in Eurasia (Europe to Japan), where it also is native. Moreover only three species of Cerocarpus are nodulated, although twenty are known.

As may be noted in Table 6.1, the largest number of nodulated species are present in the genera Alnus and Ceanothus with 33 and 31 nodulated species, respectively, which is followed by Myrica, Casuarina, Elaeagnus, and Coriaria with 12-20 nodulated species. Of the 14 genera with this type of symbiosis 139 of 290 known species are recorded as nodulated.

The most convincing evidence that Rhizobium may infect nonleguminous Dicotyledones is the report of root nodulation in Trema cannabina var. scabra (218). In Hutchinson's scheme (8) of phylogenetic relationships the Ulmaceae, to which Trema belongs, has family serial number 68. The closely related Casuarinaceae is assigned serial number 67, and the order serial numbers are also adjacent. Thus both orders are in the same evolutionary line as most other nodule-bearing taxa.

Table 6.2 gives a complete list of all root-nodulated nonleguminous species now known with the author who first reported nodulation cited.

Plate 6.1. Root-nodule types of nonleguminous plants. (A) Root nodule of Casuarina equisetifolia showing that the apex of each nodule lobe gives rise to a negative geotropic root (x 1.0), (B) Coralloid root nodules of Alnus glutinosa. Detached and divided root nodules showing the dichotomous branching of the nodule lobes (x 1.5), (C) Coralloid root nodules of Alnus glutinosa on the root (x 1.0), (D) Root nodules of Alnus rubra produced by an Alnus glutinosa inoculum. Some nodular lobes produced negatively geotropic rootlets like those in Casuarina and Myrica species. (x 1.0)

TABLE 6.2. List of All Nonleguminous Angiosperms Reported to Have a Root Nodule or Stem Symbiosis Nomenclature

Root Nodule Symbioses with Actinomycetes Species

CASUARINA

Reported symbiotic species 18, total number of species 45. C. cristata* Miq. (according to I. K. syn. C. stricta) (Becking, unpublished), C. cunninghamiana Miq. (126), C. equisetifolia L. (169), C. fraseriana Miq. (109), C. glauca Sieber (170), C. huegeliana Miq. (171), C. junghunniana Miq. (7), C. lepidophloia F. Muell. (109), C. montana Leschen. (160), C. muricata* DC ex. Miq. or Roxb. (according to I. K. syn. C. montana or C. equisetifolia (172), C. nodiflora Forst. f. (173), C. quadrivalvis Labill. (172), C. rumphiana Miq. (7), C. stricta Ait. (174), C. sumatrana Jungh. (109), C. tenuissima Sieber ex. Spreng. (109), C. torulosa Ait. (pers. comm. of Mackintosh to Bond, ref. 20), C. triangularis (not listed in I. K.) (109).

MYRICA

Reported symbiotic species 20, total number of species 35. M. adenophora Hance (170), M. californica Cham. and Schlectdl. (125), M. brevifolia E. Mey. ex C. DC (35), M. carolinensis Mill (175), M. cerifera L. (176), M. cordifolia L. (35), M. diversifolia Adamson (35), M. faya Ait. (226), M. gale L. (177), M. gale L. var. tomentosa C. DC. (22), M. humilis Cham. and Schlectdl. (35), M. integra (A. Chev.) Killick (35), M. javanica Blume (5), M. kraussiana Buching. ex Meisn. (35), M. pensylvanica Loisel. (176), M. pilulifera Rendle (pers. comm. of Dale to Bond, ref. 20, 35), M. pubescens Wilk (pers. comm. of Smit to Bond, ref. 20), M. quercifolia L. (35), M. ruba Sieb. et Zucc. (178), M. sapida Wall. var. longiflora (Teysm. and Binned.) (176), M. serrata Lam. (pers. comm. of Scott to Bond, ref. 20, 35).

COMPTONIA*

Monotypic genus. Symbiotic species 1, total number of species 1. C. asplenifolia* (L.) Banks (179) [according to I. K. syn. Myrica asplenifolia L. and Comptonia peregrina* (L.) Coult.].

(Continued)

TABLE 6.2. (Continued)

Root Nodule Symbioses with Actinomycetes Species

ALNUS

Reported symbiotic species 33, total number of species 35.
A. cordata (Lois.) Desf. (180), A. crispa (Ait.) Pursh (181),
A. fauriei Lev. et Vnt. (22), A. firma Sieb. et Zucc. (170),
A. formosana Mak. (182), A. fruticosa Rupr. (183), A. glutinosa
(L.) Gaertn. (184), A. glutinosa (L.) Gaertn. var. subbarbata
(185), A. hirsuta Turcz. (186), A. hirsuta Turcz. var.
tinctoria Kudo (22), A. incana (L.) Moench (185), A. incana
(L.) Moench var. glauca Ait. (178), A. inokumai Murai et Kusaka
(pers. comm. of Mackintosh to Bond, ref. 20), A. japonica*
(Thunb.) Stendel (170), A. japonica* (Thunb.) Stendel var.
arguta (Regel) Callier (22), A. jorullensis Kunth (pers. comm.
of Smit to Bond, ref. 20), A. jorullensis Kunth var. spachii
(Regel) Kunth (187), A. maritima Nutt. (136), A. matsumurae
Callier (22), A. maximowiczii Callier (22), A. mayrii Callier
(22), A. mollis Fernald (188), A. multinervis Matsumura (136),
A. nepalensis D. Don (171), A. nitida Endl. (160), A. oregona*
Nutt. (91) (according to I. K. syn A. rubra), A. orientalis
Done (171), A. pendula Matsumura (22), A. rubra Bongard (syn.
A. oregona*) (26), A. rugosa (Du Roi) Spreng. (24),
A. serrulata (Ait.) Willd. (pers. comm. of Silver to Bond,
ref. 20), A. serrulatoides Callier (22), A. sieboldiana
Matsumura (136), A. sinuata Rydb. (129), A. tenuifolia Nutt.
(27), A. tinctoria Sarg. var. glabra Call. (136), A.
trabeculosa Handel-Mazzett (22), A. undulata Willd. (177),
A. viridis Regel (180).

ELAEAGNUS

Reported symbiotic species 14, total number of species 45.
E. angustifolia L. (189), E. argentea Pursh (181), E. commutata
Bernh. (1), E. edulis Siebold ex. E. May (190), E. grabra
Thunb. (22), E. longipes A. Gray (191), E. macrophylla Thunb.
(128), E. matsuoana Makino (22), E. multiflora Thunb. (192),
E. multiflora Thunb. var. angustifolia Makino et Nemoto (22),
E. multiflora Thunb. var. edulis (Carr.) Schneider (193),
E. multiflora Thunb. var. hortensis Maxim. (22), E. murakamiana
Makino (22), E. pungens Thunb. (177), E. pungens Thunb. var.
simonii (Carr.) Nicholson (173), E. rhamnoides (L.) A. Nelson
(190), E. umbellata Thunb. (193), E. yoshinoi Makino (22).

TABLE 6.2. (Continued)

Root Nodule Symbioses with Actinomycetes Species

HIPPOPHAË

   Reported symbiotic species 1, total number of species 3.
   H. rhamnoides L. (194).

SHEPHERDIA

   Reported symbiotic species 2, total number of species 3.
   S. argentea (Pursh) Nutt. (195), S. canadensis Nutt. (196),

CEANOTHUS

   Reported symbiotic species 31, total number of species 55.
   C. americanus L. (197), C. azureus Desf. (179), C. corduiatus
   Kell. (198), C. crassifolius Torr. (199), C. cuneatus (Hook.)
   Nutt. (61), C. delilianus Spach. (179), C. divaricatus Nutt.
   (61), C. diversifolius Kell. (198), C. fendleri A. Gray (179),
   C. foliosus Parry (200), C. freshensis Dudley (198), C. glaber
   Spach (201), C. gloriosus Howell var. exaltatus Howell (61),
   C. greggii A. Gray var. vestitus Greene (199), C. griseus
   (Trel.) McMinn (61), C. impressus Trel. (198), C. incanus Torr.
   et Gray (61), C. integerrimus Hook. et Arn. (198), C.
   intermedius* Hook. (202) (according to I. K. syn. C. ovatus or
   C. americanus depending on the sense using "intermedius"),
   C. jepsonii Greene (61), C. leucodermis Greene (203),
   C. microphyllus Michx. (179), C. oliganthus Nutt. (199),
   C. ovatus Desf. (179), C. parvifolius Trel. (198), C.
   prostratus Benth. (198), C. rigidus Nutt. (200), C. sanguineus
   Pursh (129), C. sorediatus Hook. et Arn. (61),
   C. thyrsiflorus Esch. (200), C. velutinus Dougl. (81).

DISCARIA

   Reported symbiotic species 2, total number of species 10.
   D. toumatou Raoul (204), Discaria sp from South America (227).

COLLETIA

   Reported symbiotic species 2, total number of species 17.
   Colletia paradoxa (Spreng.) Escalante and C. armata Miers.?
   (227).

(Continued)

TABLE 6.2. (Continued)

Root Nodule Symbioses with Actinomycetes Species

CORIARIA

Reported symbiotic species 12, total number of species 15.
C. angustissima Hook. f. (44), C. arborea Lindsay (62), C. intermedia Matsum. (170), C. japonica A. Gray (170), C. kingiana Col. (44), C. lurida T. Kirk (44), C. myrtifolia L. (18), C. nepalensis Wall. (227), C. plumosa W. R. B. Oliv. (205), C. pottsiana W. R. B. Oliv. (44), C. pteridoides W. R. B. Oliv. (205), C. sarmentosa Forst. f. (44), C. thymifolia Humb. et Bonpl. ex Willd. (pers. comm. of Smit to Bond, ref. 20), and the hybrids of the New Zealand species (44).

DRYAS

Reported symbiotic species 3, total number of species 4.
D. drummondii Richardson (31), D. drummondii Richardson var. eglandulosa Porsild (19), D. integrifolia Vahl. (19), D. octopetala L. (19).

PURSHIA

All species known are nodulated.
P. glandulosa Curran (206), P. tridentata (Pursh) DC. (206).

CERCOCARPUS

Reported symbiotic species 3, total number of species 20.
C. betuloides Nutt. (207), C. montanus Raf. (132), C. paucidentatus Britt. (132).

Root Nodule Symbiosis with Rhizobium Species

TREMA

Reported symbiotic species 1 in 1 variety, total number of species 55. T. cannabina Lour. var. scabra (Bl) de Wit (T. cannabina Lour. var. glabrescens de Wit is non-nodulated small!) (218).

ZYGOPHYLLUM

Reported symbiotic species 4, total number of species 100.
Z. album L.f., Z. coccineum L., Z. decumbens Del., Z. simplex L. (219, 220).

(Continued)

TABLE 6.2. (Continued)

Root Nodule Symbioses with Rhizobium Species

FAGONIA
  Reported symbiotic species 1, total number of species 40.
  F. arabica L. (219, 220).

TRIBULUS
  Reported symbiotic species 1, total number of species 20.
  T. alatus Del. (219, 220); nodular development but no endophyte
  in T. cistoides L. (221).

Stem Symbiosis with Nostoc Species

GUNNERA
  Reported symbiotic species 22, total number of species 54.
  G. albocarpa Cockayne (14), G. arenaria Cheesem ex. T. Kirk (208),
  G. bracteata Steud. (according to I. K. syn. G. scabra*) (9), G.
  chilensis Lam. (according to I. K. syn. G. scabra*) (209), G.
  commutata Blume (9), G. cordifolia Hook. f. (9), G. densiflora
  Hook f. (208), G. dentata T. Kirk (208), G. hamiltonii T. Kirk
  ex W. Ham (208), G. herteri Osten. (210), G. insignis Oerst (9),
  G. integrifolia Blume (9), G. lobata Hook. f. (9), G. macrophylla
  Blume (72), G. magellanica Lam. (209), G. manicata Linden (Kanitz,
  pers. comm. to Reinke (9), G. microcarpa T. Kirk (208), G. monoica
  Raoul (209), G. peltata Phil. (9), G. perpensa L. (209), G.
  petaloides Gaudich (9), G. prorepens Hook. f. (9), G. scabra*
  Ruiz. et Pav. (according to I. K. syn. G. chilensis) (209),
  G. strigosa Colenso (14).

---

*Nomenclature according to the Index Kewensis (168) and their
Supplements. Literature citation refers to author reporting the
symbiosis. In text Index Kewensis will be abbreviated I. K.
Names of species marked with an asterisk are synonyms according
to the Index Kewensis.

6.3 STEM SYMBIOSIS IN HIGHER PLANTS

Another type of symbiosis markedly different from the
nodule-bearing symbiosis of certain nonlegumes is
found in the genus Gunnera (Dicotyledones of the
Angiospermae). Here the symbiont is present in special
glands occurring in pairs at the junction of each leaf

to the stem. The endophyte is a N_2-fixing blue-green alga of the family Nostocaceae and has been isolated in pure culture (9-12). The endophyte proved to be Nostoc punctiforme. Growth of the alga starts in the mucous-producing glands first intercellularly, but later on intracellularly (12, IV.4).

6.3.1 Taxonomic Aspects

The Gunnera symbiosis is the sole blue-green alga symbiosis in higher plants. The only other blue-green algal symbiosis in plants are found in Pteridophyta (Azolla species with an Anabaena symbiosis) (275, 276) Gymnospermae (Cycadinae: Cycas, Dioon, Zamina, etc., also with a Nostoc symbiosis) (277), in liverwort species (Hepaticae: Blasia and Anthoceratales), and in lichens (277). Also some flagelates appear to harbor endophytic blue-green algae.
   Gunnera species range from small herbs to stout plants up to 1.0-1.5 m in height (G. hamiltonii and G. macrophylla). They are present only in wet habitats. In South America 40 Gunnera species are listed (13); in New Zealand, 10 (14, 15), and in Southeast Asia (Indonesia, G. macrophylla) and South Africa (G. capense and G. perpensa), several. In total there are over 50 Gunnera species of which 22 are recorded as forming a Nostoc symbiosis (Tables 6.1 and 6.2). So far all Gunnera species examined in their natural habitat live symbiotically and it is still to be proved that natural nonendophytic Gunnera exists or that it is possible to grow Gunnera without endophyte. Although it cannot be assumed that in Gunnera the symbiosis is compulsory for growth, experiments with endophyte-free plants have not yet been reported.

6.4 PRESENT DISTRIBUTION OF NONLEGUMINOUS N_2-FIXING SYSTEMS ACCORDING TO MAIN FLORAL REGIONS

Nonleguminous N_2-fixing associations have a worldwide distribution for they are present in arctic, temperate, subtropical, and tropical regions and are absent only in polar regions completely devoid of vascular plants. In subarctic regions they play an important role in plant succession by the colonization of bare soil recently exposed by the regression of glaciers (16).

Similarly they are important in tropical regions on eroded soils or in the restoration of soil cover in regions where the vegetation is destroyed by volcanic activity such as lava streams, e.g., Myrica javanica (17).

In the following sections a brief outline will be given of the main nonleguminous, dinitrogen-fixing species in each of the various main floral regions.

6.4.1 Eurasia

Undoubtedly the most important nonleguminous plant in this region is Alnus. In Europe four species are indigenous, i.e., Alnus glutinosa (Black Alder), Alnus incana (Gray Alder), Alnus cordata (Italian Alder), and Alnus viridis (Green Alder). Unquestionably the dominant species is Alnus glutinosa, which is widely distributed and occupies various ecological niches such as marshy land and reforested gravel sands or anthracite mining wastes. The next most common species, Alnus incana, is particularly prominent in middle Europe and Asia. Alnus cordata occurs in south and southeastern Europe, whereas Alnus viridis is a typical montane species occurring only at the higher altitudes in the Alps. In Eurasian moorlands and bogs Myrica gale is common, reaching as far north as the Arctic Circle. Near the sea coast Hippophaë rhamnoides is prominent, occupying bare dune sand and, thus, acting as pioneer vegetation at these sites. In the Mediterranean region, especially in southern France and Spain, Coriaria myrtifolia (Coriariaceae) establishes a complete ground cover frequently under evergreen oak (Quercus ilex) (18). In central and south Portugal the nodulated Myrica faya occurs, but this species was probably introduced because its main distribution is in the Azores, Madeira, and the Canary Islands (226).

In northern regions near the Arctic Circle a small herb, Dryas octopetala (Rosaceae), forms dense mats in the tundra vegetation. This and other Dryas species are nodulated in North America (19). However European Dryas as examined in Scotland and in the Alps (7, 19-21) was nonnodulated, the same being true for Dryas octopetala var. asiatica in Japan (22).

In the eastern Eurasion continent the dominant position of Alnus as the most common nonleguminous, dinitrogen-fixing plant is not appreciably changed. Uemura (22) listed 12 Japanese species with root nodulation and some of them were of economic importance for forestry (Alnus firma, A. hirsuta, A. japonica, and A. sieboldiana). In addition, he mentioned 8 Japanese species of Elaeagnus with root nodulation, i.e., Elaeagnus macrophylla, E. multiflora, E. umbellata, E. grabra, E. matsuoana, E. murakamiana, E. pungens, and E. yoshinoi (22). In this region Myrica gale var. tomentosa and M. rubra are nodulated, together with a Coriaria japonica, giving a total picture more or less the same as in Europe including the southern regions. Also in more southeastern regions of this continent, Alnus appears to be plentiful. For instance, Wilson (23) reported that A. formosana is the most common deciduous tree on Taiwan. It was found from sea level to 2500-m altitude, and it was the first tree to appear on bare slopes after landslides. The native people use this tree to restore soil fertility in exhausted millet fields. In the Himalayas (Sikkim) and China a number of Coriaria species occur such as Coriaria terminalis, C. nepalensis, and C. sinica, but nothing is known about their nodulation.

6.4.2 North America

The northern part of the North-American floral region has some similarities to the Eurasian floral region where Alnus species play a dominant role. Alnus crispa in the Arctic region has an important function in the occupation of recently deglaciated areas. This plant is also found on sandhills, eroded slopes, and mountain slopes swept by snow slides (24). Further south, A. rugosa, a small tree, is common along water courses throughout Canada. Near the Pacific Northwest coast A. rubra is very abundant on poor soils and because of this capacity this tree plays an important role in the forestry of that region (25, 26). Still more to the south A. tenuifolia may occur along lakes and streams, and, according to Goldman (27), it adds substantially to the nitrogen status of that environment. As in Eurasia, Myrica gale is found in the far northern bogs and moorlands and there are at least three Elaeagnaceae species native to Canada: Shepherdia canadensis (Canadian buffaloberry), S. argentea (Buffaloberry), and Elaeagnus commutata (Wolfwillow, Silverberry). These plants are usually found on sandy soil, eroded slopes, or gullies. On the eastern coast of North America Comptonia asplenifolia (= Myrica asplenifolia) is common on the bare soil of landslides, roadcuts, or eroded soil; it also plays an important function in the recolonization of mining wastes (28). In southern parts of North America (Florida) Myrica cerifera is a significant constituent of the native flora, and is found at similar but generally drier sites (29). For Florida, Silver and Mague (30) reported nodulation and N_2 (C_2H_2) fixation by the Asia/Australia-imported Casuarina cunninghamiana (Casuarinaceae).

A very important contribution to soil fertility in this floral region is made by Ceanothus species (Rhamnaceae), a large group of plants indigenous to northwest North America. There are about 55 Ceanothus species of which 31 species have been reported as bearing nodules (Tables 6.1 and 6.2; Plate 6.2). Ceanothus is a major component of the semiarid fire-type scrub association known as chaparral covering extensive areas of low-fertility soils. Together with two other endemic genera and species belonging to the Rosaceae (Purshia glandulosa, P. tridentata, and Cercocarpus betuloides), they contribute substantially to the nitrogen status of soil of these drier habitats.

As in Eurasia, Dryas occurs in North America in the northern regions, and also in the Rocky Mountains at high elevations (comparable to Dryas in the Alps in Europe). Nodulation has been reported for Dryas drummondii, D. integrifolia, and D. octopetala in Alaska (1, 16, 19, 31, 32) and dinitrogen fixation has been proven with $^{15}N_2$ tests (19). In Alaska Dryas plays an important role in plant succession and vegetation development of recently exposed areas after ice regression, or on very infertile rock soils or soils exposed to extensive erosion or drainage. Dryas can withstand these severe conditions because of its deep rooting system. As already stated, it is noteworthy

---

Plate 6.2. Ceanothus velutinus Dougl. (A) Plant growing in a pot culture in perlite with a nitrogen-free nutrient solution. (x 0.46), (B) Field material of root nodules collected at 1500 m on the Rogue River National Forest, southwestern Oregon, U. S. (x 0.44), (C) Close-up of root nodule showing the dichotomous branching of the nodular lobes. (x. 2.5)

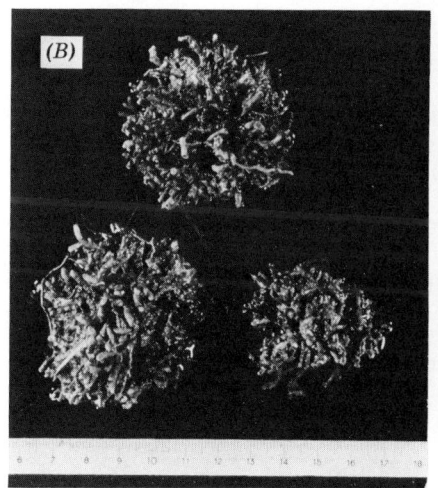
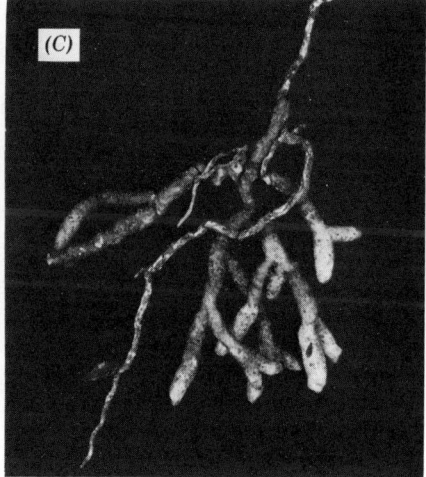

that one of the Dryas species, Dryas octopetala, was
not nodulated in the Eurasion floral region, in spite
of their extensive native distribution in this region
as shown by fossil remains.

6.4.3 South America

The most important nonlegume with dinitrogen-fixing
capacity is undoubtedly Alnus jorullensis, which is an
active colonizer of bare soil along roads and on landslides (33). Two Coriaria species, C. ruscifolia and
C. thymifolia, rather widely distributed on this
continent from the Andes of Chile and Peru to the up-
lands of Central America (North Mexico) and a slight
eastward extension along the mountains of northern
Venezuela. However, nothing about the nodulation of
these species is known. In South America there are
40 species of Gunnera (13), which grow particularly on
wet sites such as drainage-induced landslides. Many
of these species have been reported to bear Nostoc
glands as observed either in the field (or from
herbarium material) or from living specimens present
in botanical gardens (Tables 6.1 and 6.2).

6.4.4 Subtropical and Tropical Africa

In the Mediterranean and subtropical part of Africa no
dinitrogen-fixing nonlegumes are known, except
Casuarina, which has been introduced from Asia.
Casuarina trees examined by the author in Tunis were
not nodulated, but Dommergues (34) found extensive
nodulation in Casuarina (C. equisetifolia) at the Cape
Verde Islands, off the coast of Dakar.
 In the tropical part of Africa certainly some
native Myrica species are present, but nothing is reported about nodulation. In South Africa, the occurrence of nine Myrica species has been reported, and
all are nodulated (35, 38). These nine species are
M. brevifolia, M. cordifolia, M. diversifolia, M.
humilis, M. integra, M. kraussiana, M. pilulifera, M.
quercifolia, and M. serrata. Bond (36, 37) has shown
nodulation and dinitrogen fixation with $^{15}N_2$ tests in
two of these species, M. cordifolia and M. pilulifera.
Also using $^{15}N_2$, Jansen van Ryssen and Grobbelaar (35)
and Grobbelaar et al. (38) proved N_2 fixation in the

other six native Myrica species of South Africa. Myrica cordifolia is of particular ecological interest, because it is a low shrub that flourishes on and stabilizes sand dunes on the South African coast, much like Hippophaë rhamnoides does in the Eurasian floral region. Myrica pilulifera is a tree and occurs in tree-shrub associations at rather high altitudes in South Africa, Rhodesia, Malawi, and Swaziland (39). Myrica brevifolia is a shrub often growing in dense macchia associations on steep mountain slopes in sandy acid soil. Myrica humilis is also a shrub or a small tree preferring steep mountain slopes. Myrica quercifolia grows more on the lower mountain slopes (35, 38).

6.4.5 Southeast Asia

Compared to the Eurasian and North American floral regions Southeast Asia is relatively poor in nodulated nonlegumes. A characteristic species of wide distribution on the sandy beaches of the many islands is Casuarina equisetifolia. At these sites it is often the dominant species because it is salt tolerant. In the drier mountainous parts, e.g., eastern Java, Indonesia, the nodulated Casuarina junghuhniana forms extensive, nearly pure stands over savannah-like grass like at the Hiang and Idjen Plateau, Mt. Semeru and Mt. Baluran. No nodulated nonleguminous species have been reported for the wet tropical rain forest. The only N_2-fixing nonleguminous plants of this region occur in the tropical mountain forest reaching to the elfin forest of tropical alpine regions. At this site Myrica javanica is a common tree along streams and brooks and, moreover, it is an active colonizer of recently erupted lava streams or forest destructed by volcanic activity (17). Gunnera macrophylla prefers the same habitat, but at wetter sites. This plant favors reed marshes, areas near waterfalls or places of seepage, and along brooks or swiftly running mountain streams. Its distribution is not restricted to Java, as this plant is also found on Sumatra, North Borneo (Mt. Kinabalu), the Philippines, Sangihe, Celebes, and New Guinea (up to 3300 m) (40). Becking (41, 42) showed with in situ experiments using the acetylene reduction test that the Nostoc symbiosis present in the stem of the plant fixes N_2. Up to now the two Elaeagnus species, E. conferta and E. latifolia, of the montane flora of Java have not been examined for root nodulation.

Abundant root nodulation was observed in Trema cannabina var. scabra growing as a weed between rows of tea in the Pangia District of New Guinea (218) (Plate 6.3). This plant which is a 12 to 15-m-high tree when full grown, ranges in New Guinea from the lowlands to highland altitudes of 2,700-3,000 m. It is noteworthy that other varieties and species of Trema (T. cannabina var. glabrescens, T. orientalis, and T. aspera), which grow regularly at the same sites, proved to be nonnodulated (personal communication, M. J. Trinick). Of about 55 Trema species in tropical and subtropical regions (224) nodulation has been reported in only one variety of one species, indicating that the relationship of the endophytic Rhizobium to Trema must be highly specific.

6.4.6 Australia and New Zealand

To date no indigenous N_2-fixing nonlegumes have been reported in Australia. Native conifers such as Podocarpus, Dacrydium, Libocedrus, Phyllocladius, or Agathis spp. sometimes reported to fix N_2 are excluded in the present survey (Tables 6.1 and 6.2), because these conifers actually do not fix dinitrogen and the measured fixation is due to associative organisms (IV.4). Native Casuarina species occur in Australia, but little about their nodulation has been reported. Many Coriaria species (Coriariaceae) indigenous to New Zealand are able to fix N_2 by a root-nodule symbiosis, but apparently these Coriaria species do not occur in Australia (43).

All eight Coriaria species reported for New Zealand, i.e., C. arborea, C. sarmentosa, C. kingiana, C. lurida, C. pteridoides, C. angustissima, C. plumosa, and C. pottsiana are nodulated as well as their mutual hybrids (44). Moreover, Discaria toumatou (Rhamnaceae) is nodulated on New Zealand, and N_2 fixation by these nodules has been proven with $^{15}N_2$ (45). Discaria toumatou is a characteristic plant on loess, alluvial fans, and gravel screes and is, moreover, found as a common component of sparse vegetation of the arid regions of Central Otago (204, 205). Coriaria species occur on gravel screes, sandy soils, rock faces, river terraces, and in tussock grasslands of this region (46).

Plate 6.3. Nodulated plant of <u>Trema cannabina</u> var. <u>scabra</u> de Wit. The picture gives the general habitus of the plant and the position of the <u>Rhizobium</u> root nodules. (Photograph by Dr. M. J. Trinick). (x 0.56)

6.5 FOSSIL DISTRIBUTION

Fossil evidence shows that a number of the nonleguminous plants mentioned in the previous sections had in the past a much broader distribution than at present. Their wide distribution in the past was due mainly to climatic conditions at that time. For instance during the Pleistocene Dryas octopetala was a major component of the tundra or semiarctic vegetation covering large areas of north-central Europe and Asia, since its typical crenated tomentose leaves have been found abundantly in clay deposits of fossil glacier lake beds together with remnants of Salix polaris, S. herbaceae, and Betula nana. In Central Europe, Gross (47) recognized three Dryas phases in late glacial and postglacial time: "Oldest Dryas," "Older Dryas," and "Younger Dryas," all named after Dryas octopetala as index plant. In this context, it is remarkable that Dryas octopetala of Eurasia is nonnodulated, in contrast to the nodulation of this species and other Dryas species in the north of the American continent (Plate 6.4). It is unlikely that nodulation, which is surely an advantage for such a pioneer plant, has been lost by the "survivors" of the Dryas octopetala stock still occurring in Europe and Asia in similar niches such as the Alps and some other mountains (Scotland and Japan) and in the tundra vegetation of the extreme North. Though rather speculative, it may suggest that the symbiotic relation of Dryas in the north of North America is in geological terms a rather recent achievement and that it has developed completely independently of the Eurasian Dryas octopetala stock.

From Pleistocene glacial deposits fossil remains of Myrica gale (Myricaceae), which in recent time is a well-known nodulated species, are also described. In addition fossil remains of Hippophaë rhamnoides (Elaeagnaceae) were found. The presence of this species is an indicator of temporary warm conditions in the Pleistocene Interstadials.

Fossil evidence of nonleguminous N_2-fixing plants in floral associations of other geographical regions are scare or lacking, with the exception of North America, where Dryas remains belonging to the species D. integrifolia are described for the silty sand glacial sediments of Minnesota (48, 49).

Plate 6.4. <u>Dryas drummondii</u> Richards. (<u>A</u>) Plant growing in nitrogen-free sand culture mixed with some peat (x 2.3), (<u>B</u>) Root nodules showing their attachment to very thin lateral roots. Root nodules are thicker and dichotomously branched. (x 2.2)

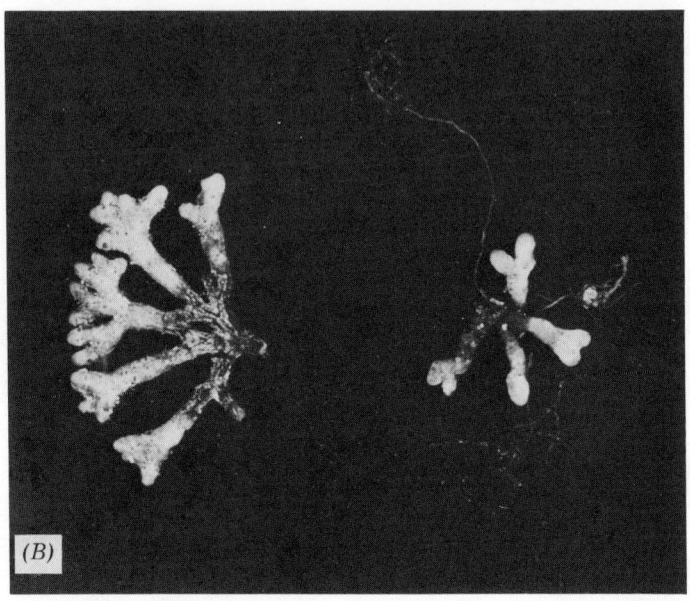

(B)

6.6 EVIDENCE OF N_2 FIXATION BY ROOT-NODULE SYMBIOSES

6.6.1 Growth Experiments with Whole Plants

Dinitrogen fixation by nodulated nonleguminous plants can be demonstrated by growing these plants in nodulated condition on nitrogen-free or nitrogen-deficient substrates, e.g., soil, vermiculite, or water culture. Under these conditions non-N_2-fixing plants or non-nodulated plants will show a very restricted growth and severe nitrogen deficiency symptoms in the leaves.

This method has been applied by Bond (50) using a nitrogen-free nutrient solution for <u>Alnus glutinosa</u>, <u>Myrica gale</u>, <u>Hippophaë rhamnoides</u>, and <u>Casuarina cunninghamiana</u>, by Rodriguez-Barrueco (20) for <u>Alnus jorullensis</u>, and by Becking (5-7) for <u>A. glutinosa</u>, <u>A. cordata</u>, A. incana, <u>A. rubra</u>, and some <u>Casuarina</u> species such as <u>C. cristata</u>, <u>C. equisetifolia</u>, <u>C. junghuhniana</u>, <u>C. rumphiana</u>, and <u>C. sumatrana</u>. In recent unpublished experiments, this author established

N$_2$ fixation in some Japanese Alnus species such as
A. firma, A. firma var. sieboldiana, A. japonica var.
genuiina, A. inokumai, A. pendula, A. tinctoria var.
obtusiloba, and in Japanese Myrica rubra, and in
Javanese M. javanica (Plate 6.4).

Growth of these nonleguminous plants was best on a
nitrogen-free Van der Crone's solution (Table 30A in
ref. 51) of pH 6.2, with added trace elements as in
Hoagland and Snyder's solution (52), or the trace-
element prescription of Hewitt (51, p. 238). Young
seedlings received ¼- or ½-strength Crone's solution,
but at later stages full-strength Crone's solution.
The growth of Alnus glutinosa on nitrogen-free culture
solution is shown in Figure 6.1; after inoculation and
nodule formation both dry matter per plant and milli-
grams of N per plant increased concomittantly producing
in 48 weeks (one season) a biomass of 30 g dry wt and

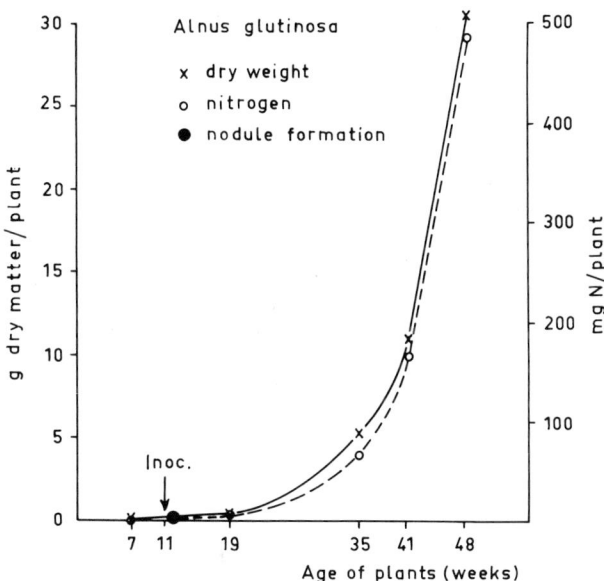

Fig. 6.1. Dry matter and nitrogen increase of Alnus glutinosa
plants grown in introgen-free nutrient solution. At point ● root
nodules appeared.

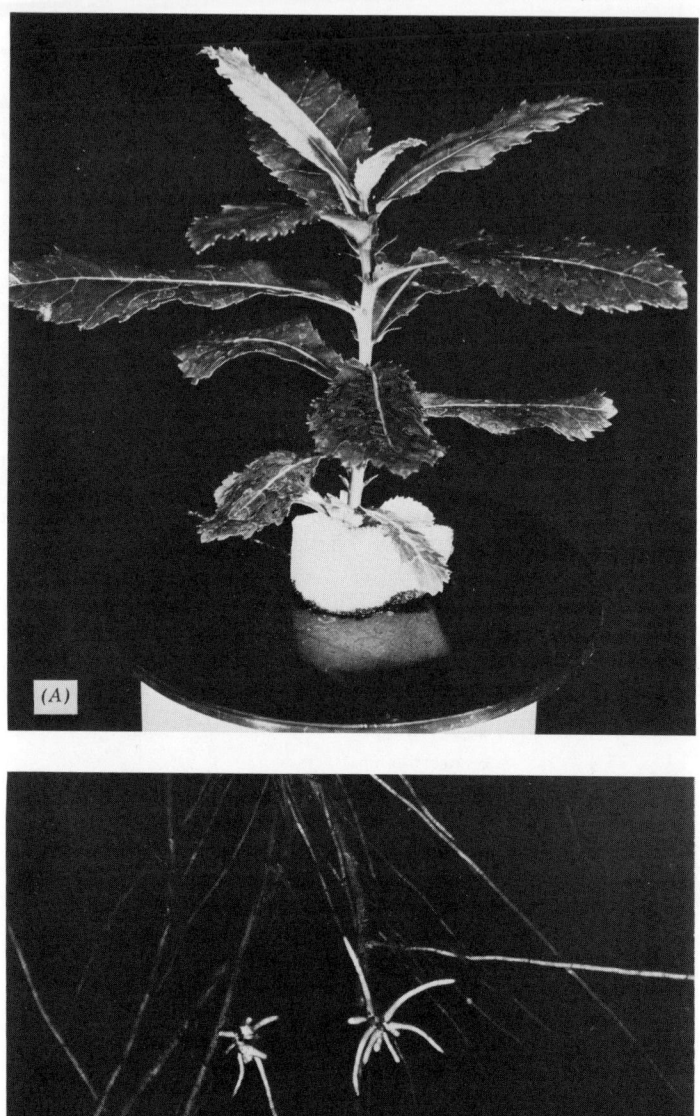

Plate 6.5. <u>Myrica</u> <u>javanica</u> Blume. (<u>A</u>) Plant growing in nitrogen-free nutrients solution (x 0.54), (<u>B</u>) Root nodules on the root system. (x 0.9)

500 mg N per plant. Figure 6.2 presents plant N at various ages subdivided into leaves, stems, and roots; one-half of the nitrogen is present in the leaves. Plate 6.6 depicts A. glutinosa plants at the age of 48, 41, and 35 weeks from such an experiment.

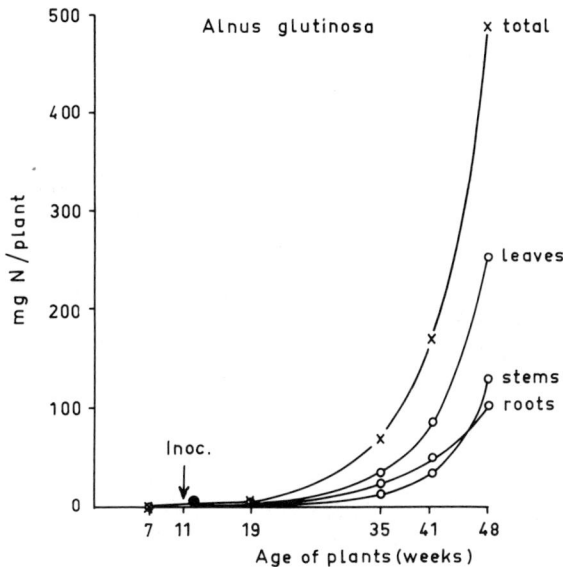

Fig. 6.2. Nitrogen contents of total plants, stems, leaves, and roots of Alnus glutinosa plants grown in nitrogen-free nutrient solution at point ● root nodules were formed.

Other nonleguminous plants such as Hippophaë rhamnoides, Myrica gale, and Casuarina spp. show about the same yields or somewhat less as was obtained with A. glutinosa (Becking, unpublished). Russell and Evans (53) showed that nodulated Ceanothus velutinus var. laevigatus plants in a nitrogen-free medium fixed between 2.6 and 25.8 mg N/day·g dry wt nodule. An average N_2 fixation of 286 mg N per plant was measured for an 18-week period, or a calculated rate of 762 mg N per plant for 48 weeks which is similar to that observed by us in A. glutinosa.

The method of growing whole plants in a nitrogen-free medium in water culture or vermiculite, and application of crushed root nodules often fails to

Plate 6.6. General view of <u>Alnus glutinosa</u> plants of various ages growing in a nitrogen-free nutrient solution. These are the same plants as those for which data are presented in Figures 6.1 and 6.2. From the left to the right the pairs of plants are 48, 41, and 35 weeks of age, respectively. (x 0.09)

produce nodulation in certain plant species. In such
a case (36), it is better to sow the plants first in
habitat soil and transfer them to a nitrogen-free
rooting medium such as water culture, perlite, or
vermiculite when root nodules have developed (36).

6.6.2 Experiments with Excised Root Nodules

6.6.2.1 $^{15}N_2$ Tests

One of the characteristics of the legume root-nodule
symbiosis is that the N_2-fixing capacity is rapidly
lost after detachment of the nodule from the root (54-
56). Bond (57) used $^{15}N_2$ to show that detached root
nodules of Alnus glutinosa had a marked retention of
N_2-fixing ability after separation from the plant. In
later experiments the time course of N_2 fixation by
detached root nodules of Alnus glutinosa and Hippophaë
rhamnoides was determined with $^{15}N_2$ (58). In Alnus
an appreciable N_2 fixation was found until 12 hours
from the time of detachment, but thereafter there was
little further fixation. The N_2 fixation in Hippophaë
rhamnoides was more active and, moreover, it was much
longer-lived as there was a substantial increase in
atom percent ^{15}N in the sample up to 24 hours. Using
a gas mixture containing 20% (vol/vol) N_2 labeled with
36% ^{15}N, the appreciable concentration of 0.645 atom %
excess ^{15}N was found in the root nodules of Hippophaë.
Similarly, Becking (5, 7) found in Alnus glutinosa
root nodules an increase of ^{15}N up to 12 hours
(Fig. 6.3).

The capacity of nonleguminous root nodules for
long-termed N_2 fixation makes them a good subject for
N_2-fixation studies of intact root nodules. The reason
for the longevity of the N_2-fixing capacity is not
known; it may be related to the more woody structure
of the nonleguminous root nodules compared to the
leguminous ones.

The $^{15}N_2$ method has been used mostly as a qualitative test for N_2 fixation in nonlegumes. Definite N_2
fixation has been proven with $^{15}N_2$ in excised root
nodules of a large number of nonleguminous species
including Alnus glutinosa and Hippophaë rhamnoides as
mentioned above, and in addition Casuarina cunninghamiana (58), Myrica gale (57), M. cerifera (36, 59),
South African Myrica species (35), M. cordifolia,

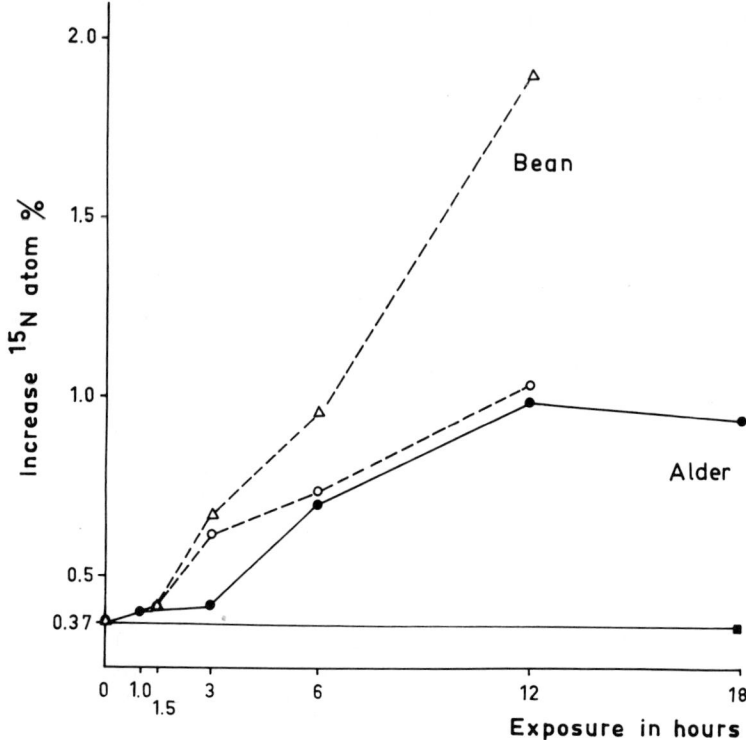

Fig. 6.3. Atom % ^{15}N in excised root nodules of alder (Alnus glutinosa) and in nodulated roots of bean (Phaseolus vulgaris) exposed to an atmosphere containing 19% (vol/vol) N_2 (labeled with 35.0 atom % ^{15}N) 21% (vol/vol) O_2, 3% (vol/vol) CO_2, and 57% (vol/vol) Ar. Bean: roots, △; root nodules, O. Alder: mature root nodules, ●; very young root nodules, ■.

M. pilulifera (37), Comptonia asplenifolia (60), Shepherdia canadensis (58), Ceanothus species (61), C. velutinus (36), Discaria toumatou (45), Coriaria arborea (62, 63), C. myrtifolia (36), Dryas drummondii (19), and Purshia tridentata (64).

The $^{15}N_2$ method has seldom been used to assess the quantity of N_2 fixed, there being no quantitative figures with ^{15}N published for nonlegumes. This lack arises from the complexity of the calculation, which requires

N_2 Fixation by Root-Nodule Symbioses

more parameters than merely ^{15}N enrichment of the nodules and dry weight or N content of nodules. Consequently, the $^{15}N_2$ method has been used mainly for qualitative evidence of N_2 fixation, and the acetylene reduction method has been applied to estimate the quantity of N_2 fixed. However, the $^{15}N_2$ method is suitable for this purpose and is certainly a more direct proof of N_2 fixation than the acetylene reduction test.

For the experiment presented in Figure 6.1, it is possible to calculate the amount of N_2 fixed for each point, based on the composition of the gas atmosphere and the ^{15}N enrichment. This value should be determined by mass spectrometry rather than based on the manufacturer's indicated enrichment of the chemical used for the preparation of $^{15}N_2$. In our experiment an artificial gas atmosphere containing 2% (vol/vol) N_2, 20% (vol/vol) O_2, 66% (vol/vol) Ar, and 2% (vol/vol) CO_2 was used. At the end of the incubation period both the gas composition of the atmosphere as well as its ^{15}N-enrichment were determined by mass spectrometry. The values obtained were corrected for a small air leakage due to sampling by the formula

$$\frac{[^{14}N^{15}N]^2}{[^{14}N_2][^{15}N_2]} = 4.00$$

The corrected composition of the gas atmosphere and the percentage contribution of ^{14}N and ^{15}N at the end of the experiment are given in Table 6.3.

TABLE 6.3. Chemical Composition of the Gas Atmosphere

Component	Contribution of component (mole%)	Component	Percentage contribution to total N
$^{15}N_2$	8.51	^{15}N	77.7
$^{14}N^{15}N$	4.98	^{14}N	22.3
$^{14}N_2$	0.71		
O_2	12.1		
Ar	65.2		
CO_2	8.5		

Table 6.4 presents data on the quantity and the rate at which N_2 is fixed by Alnus glutinosa based on the experiment presented in Figure 6.3. The same calculation can be applied to the nodulated Phaseolus vulgaris (Fig. 6.3) tested concomitantly with the Alnus glutinosa root nodules. In both experiments the atmosphere, temperature (20°C), and other environmental conditions were the same. In 12 hours the bean nodules (4.8 mg N) and the roots (1.9 mg N) contained respectively 1.04 and 1.90 atom % excess ^{15}N. This implies an amount of nitrogen fixed from the atmosphere present in the nodule of (22.3/77.7 x 1.04) + 1.04 = 1.34 % N equivalent to 1.34/100 x 4.8 = 0.064 mg N, and in the root (22.3/77.7 x 1.9) + 1.9 = 2.4 % N equivalent to 2.4/100 x 1.9 = 0.046 mg N.

Total quantity of nitrogen fixed by the nodules of bean is then 0.064 + 0.046 = 0.110 mg N or 1.9 µg N/h·mg nodule N. The Alnus glutinosa root nodules with a fixation of 1.07 µg N/h·mg N (Table 6.4) were about one-half as active as the bean nodules.

6.6.2.2 Acetylene Reduction Tests

As with leguminous plants, the acetylene reduction test is one of the easiest methods for determination of the N_2-fixing capacity of nonlegume root nodules. See IV.12 for a description of this method. Stewart et al. (65) and Sloger and Silver (66) were the first to use the acetylene reduction test for the nonlegumes. The former found rates of 0.08 and 0.01-0.03 nmoles C_2H_4/min·mg fresh wt of nodule for Alnus (species?) and Comtonia peregrina, respectively. Later Silver and Mague (30) assessed N_2-fixing capacity of Myrica cerifera in the field in Florida. In one experiment they reported a reduction of 1080-1820 nmoles C_2H_2/h·mg N and in another experiment with one Myrica tree in the field, values of 700, 337, and 248 nmoles C_2H_4/h·mg N were found, respectively for young, medium-aged, and old tissue. They also tested Casuarina cunninghamiana root nodules with the C_2H_2 method, but gave no figures; they indicated that the reducing capacity in Casuarina proceeded undiminished for about 4 hours (30).

Sloger presented more precise data in his thesis (67): Casuarina cunninghamiana root nodules reduced 0.1 µmoles C_2H_2/h·g fresh wt; C. equisetifolia,

TABLE 6.4. Fixation of N_2 by *Alnus glutinosa* Root Nodules in a 12-Hour Period[a]

^{15}N excess	Fraction of $^{14}N_2$ fixed	Fraction of total N fixed	Total N_2 fixed by root nodules	μg N_2 fixed/h·mg nodule N	μg N_2 fixed/ h·g fresh wt nodules
0.99 atom %	22.3/77.7 × 0.99 = 0.29%	0.99% + 0.29% = 1.28%	1.28/100 × 5.8 mg N = 0.074 mg N	74/(5.8 × 12.0) = 1.07 Mg N	74/(2 × 12) = 3.08

[a]Fresh weight, 2.0 g; N content, 5.8 mg.

0.2-0.7 μmoles C_2H_2/h·g fresh wt (or 0.004-0.017 μmoles C_2H_2/h·μmole N); and Myrica cerifera, 1.6-3.4 μmoles C_2H_2/h·g fresh wt (or 0.033-0.067 μmoles C_2H_2/h·μmole N). Russell and Evans (68) found considerable variation in activity in the nodules of one species of Alnus rubra with field collected nodules less active than those from laboratory-grown plants; nodules from plants grown in the laboratory reduced 0.08-0.13 nmoles C_2H_2/min·mg fresh wt, while field nodules reduced only 0.022-0.026 nmoles C_2H_2/min·mg fresh wt.

Akkermans (69) measured acetylene reduction by excised and attached root nodules in an Alnus glutinosa grove in Holland. Like others, he found a large variation in activity of field-grown nodules. Most had an activity below 1000 nmoles C_2H_4/h·g nodule dry wt, but others had an activity of 1000-70,000 nmoles C_2H_4/h·g nodule dry wt. Nodules from young plants grown in water culture had a higher activity ranging from 9,000 to 90,000 nmoles C_2H_4/h·g dry wt. His values (Fig. 6.13 and Table 8 in ref. 69) indicate, however, a fixation of only 200-320 nmoles C_2H_2/h·g dry wt with rarely a rate of 950 nmoles/h·g dry wt. The very high activities are reported in the summary of his study, but they could not be found in any of the experimental data. Other experiments with root nodules of Hippophaë rhamnoides showed a maximal C_2H_2-reducing rate of 7000 nmoles C_2H_4/h·g dry wt nodule (Fig. 15 in ref. 69).

Our own measurements (Table 6.5) are more or less consistent with those reported in the literature. Our highest activity was found in a Casuarina rumphiana tree growing in the Botanical Gardens, Bogor, Java, Indonesia, which gave C_2H_2 reduction values of 58-67 nmoles/min·mg nodule N or 0.22-0.25 nmoles/min·mg fresh wt. (see also Fig. 6.4).

It is generally accepted that the acetylene concentration applied is not very critical. For N_2-grown Clostridial cells, acetylene saturation occurs between 2.5 and 10% (vol/vol) acetylene and even 50% (vol/vol) acetylene was not inhibitory (70). But we found a striking optimum for acetylene reduction by Alnus glutinosa root nodules of 17-month age at 20% (vol/vol) acetylene with the activity considerably less at 10% (vol/vol) and distinctly inhibited at 30% (vol/vol). (Figure 6.5).

TABLE 6.5. Acetylene Reduction by Nonleguminous Plants

Plant species	nmoles C_2H_2 reduced/min·mg nodule N	nmoles C_2H_2 reduced/min·mg fresh wt nodule	nmoles C_2H_2 reduced/min·mg dry wt nodule	nmoles N_2 fixed/min·mg N nodules[a]
Alnus glutinosa	12.3–20.7	0.074–0.124	0.27–0.45	8.2–13.8
Casuarina rumphiana	58.1–67.3	0.22–0.25	0.79–0.91	38.8–44.8
Myrica cerifera (30)	1. 18.0–30.3 2. 4.1–11.7	1. 0.079–0.13[b]	1. 0.29–0.48[b]	1. 12.0–20.2 2. 2.8–7.8
M. javanica	10.1–10.5	0.041–0.043	0.150–0.156	6.8–7.0

[a] Calculated according to the conversion factor $C_2H_2:N_2 = 3$.
[b] Converted on the basis of N content of fresh wt = 0.44 and fresh wt/dry wt = 3.6.

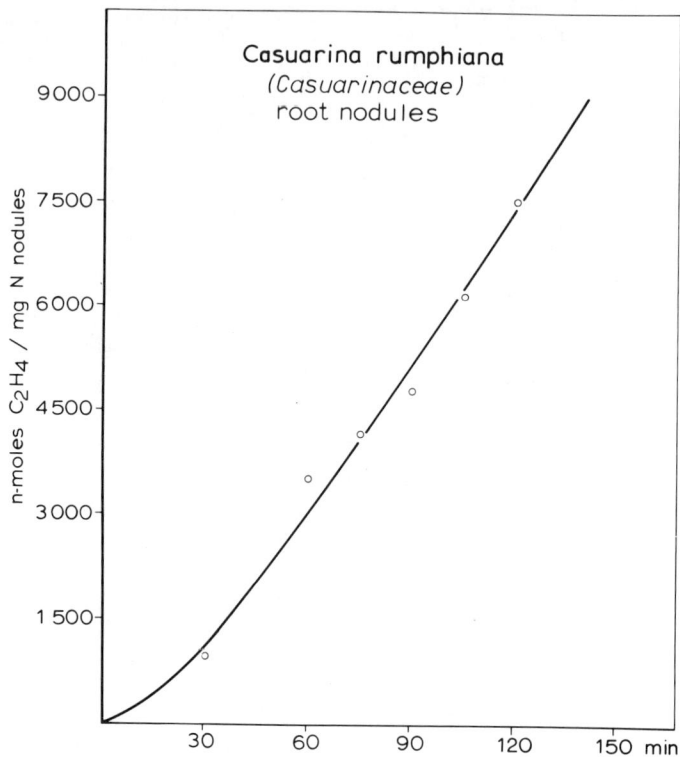

Fig. 6.4 Acetylene-reducing activity of Casuarina rumphiana root nodules tested immediately after excision in air with added acetylene [5.3% (vol/vol)] under the ambient conditions of temperature and light. Field experiment, Botanical Gardens, Bogor, W. Java, Indonesia.

Nodulated plants of Trema cannabina var. scabra produced about 11.5 µmoles C_2H_4/h·g fresh wt. of nodule, while detached root nodules gave about one-half of this activity and nodule-free roots were inactive (218). It is of interest that field surveyed nodules were of much lower activity (C_2H_2-reduction assay) than nodules from green-house grown seedlings (personal communication M. J. Trinick).

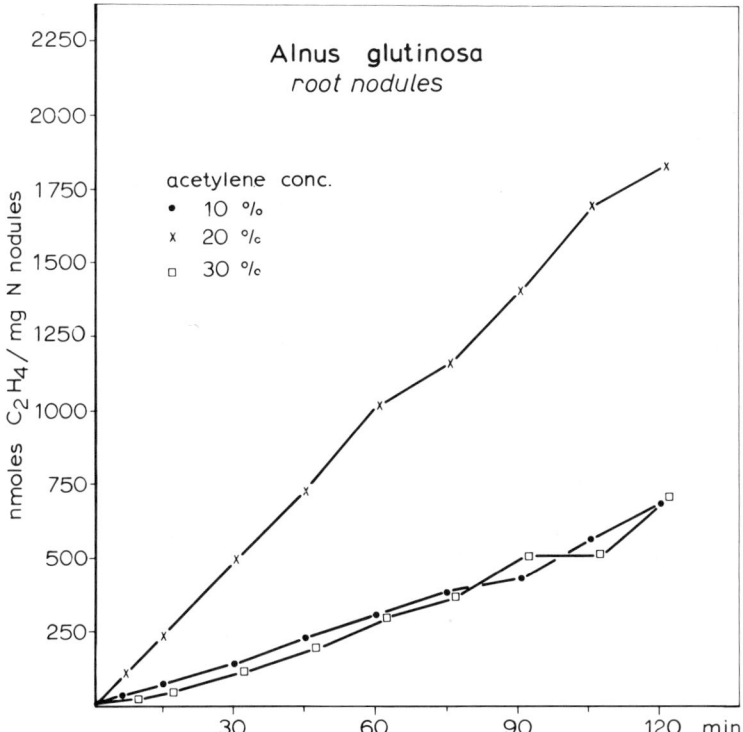

Fig. 6.5. Acetylene-reducing activity of root nodules of 17-month-old <u>Alnus glutinosa</u> plants grown on nitrogen-free nutrient solution. The excised root nodules were incubated at 20° C in air with 10, 20, or 30% (vol/vol) acetylene.

6.6.3 Experiments with Nodule Breis

6.6.3.1 $^{15}N_2$ Tests

Sloger and Silver (59) exposed nodule homogenates of <u>Myrica cerifera</u> to an atmosphere containing 20.5% (vol/vol) $^{15}N_2$ (96.6 atom % excess ^{15}N), 21.8% (vol/vol) O_2 and 57.7% (vol/vol) He. With 5 g of nodules homogenized in cacodylate buffer (pH 7) some $^{15}N_2$ fixation was found ranging from 0.021-0.026 atom % excess ^{15}N when tested in the presence of $Na_2S_2O_4$ for 4 hours, but no enrichment was observed without dithionite. Dinitrogen fixation by the homogenate was, however, only one-thirtieth of that by whole nodules where 0.588 atom % excess ^{15}N was found. However, with Alnus glutinosa root nodules under $^{15}N_2$ we have failed to detect N_2 fixation in the presence of a reducing agent. Our hemogenates darkened rapidly, probably due to the oxidation of phenol compounds by traces of oxygen, even in the presence of dithionite. It is possible that these phenolic compounds inhibit the N_2 fixation process as has been shown in soybean (71).

6.6.3.2 Acetylene Reduction Experiments

No acetylene reduction tests with nodule breis of non-legumes are reported. Trials performed by us with Alnus glutinosa root nodules showed no evidence of C_2H_2 reduction following complete homogenization. Incompletely homogenized breis showed a small activity probably produced by some intact plant cells. As in the $^{15}N_2$ experiments, the homogenates darken presumably from development of phenolics which probably inhibit N_2 fixation. It is quite possible that other non-leguminous plant material with less phenolic compounds will give better results.

6.7 EVIDENCE OF N_2 FIXATION IN THE GUNNERA SYMBIOSIS

6.7.1 Growth Experiments with Whole Plants

Silvester and Smith (15) reported a growth experiment in New Zealand with Gunnera dentata in sand culture supplied with nitrogen-free nutrient. These plants were grown for 10 weeks in a controlled room. They thrived and showed no deficiency symptoms, while radish seedlings grown under the same conditions developed severe nitrogen deficiency symptoms 14 days after germination. Growth rate for the Gunnera plants was 2.9 g/100 g·day, while nitrogen increased from 0.783 mg/plant to 6.84 mg/plant. From these results, the authors estimated a N_2 fixation rate of up to 7.2 g N/m^2·annum (= 72 kg N fixed/ha·yr).

N$_2$ Fixation in Gunnera Symbiosis 225

6.7.2 ^{15}N$_2$ Tests

Silvester and Smith (15) exposed rooted rosettes of Gunnera arenaria in flasks in the light to a gas mixture containing 10% (vol/vol) N$_2$ (95 atom % excess ^{15}N). After an exposure period of 1.5-9 hours, the ^{15}N concentration was measured in the node clusters, leaves, internodes, and roots of the plants. The node clusters, which contain the Nostoc glands, were rapidly labeled, with a slower incorporation of ^{15}N into other parts of the host suggesting that fixed nitrogen is readily transferred to the host. After 9 hours, a small but measurable quantity of ^{15}N was detectable in the ethanol insoluble fraction indicating that fixed nitrogen had been incorporated into host protein.

6.7.3 Acetylene Reduction Tests

Silvester and Smith (15) also exposed tissue slices of the species Gunnera arenaria to an atmosphere of 10% (vol/vol) acetylene. They found ethylene production to be linear with time and an inhibition of ethylene formation by hydrogen gas. In 8 hours 145 nmoles C$_2$H$_4$ were produced per milligram algal protein and in the presence of 10% (vol/vol) H$_2$, there was an inhibition of acetylene reduction of 53-69%. This result requires additional documentation since H$_2$ does not inhibit C$_2$H$_2$ reduction by any other N$_2$-fixing system.

Becking (41, 42, 275, 276) studied N$_2$-fixing capacity of Gunnera macrophylla, a component of the montane vegetation of Java and other islands in Indonesia. This species had already been studied morphologically with regard to the Nostoc glands by a number of authors (72-75), but its N$_2$-fixing capacity has not been assessed. At the habitat site in Cibeureum, Cibodas Nature Reserve, Mt. Gedeh, Java, Indonesia, average plant density was 2-3 plants/m^2. Each petiole base of this plant bears three large glands, one central and two lateral, colonized by a Nostoc species. An average number of 15 leaves per plant could be counted, which means that each plant had 45 active Nostoc glands. Several in situ measurements were performed at the ambient temperature of 18°C using 20 Nostoc glands per experiment. The glands were randomly chosen from several plants and cut out with surrounding tissue from the stem of the plants. The tissue slices were immediately (within 15 min) exposed to an atmosphere containing 8-10% acetylene. A C_2H_2-reduction rate for these 20 Nostoc glands of 6200-10,800 nmoles/h was found. Accepting a conversion factor of 3.0 moles C_2H_2 reduced per mole N_2 reduced (IV.12), a N_2 fixation rate of 2.0-3.6 µmoles N_2/h. A rate of 9.5-16.6 kg N fixed/ha·annum is calculated for a density of 2 plants/m^2 and 10 light h as is usual at this altitude. These values are much lower than those reported by Silvester and Smith (15) for Gunnera arenaria in New Zealand.

6.8 PHYSIOLOGICAL ASPECTS OF N_2 FIXATION BY ROOT-NODULE SYSTEMS

6.8.1 Isolation of Nitrogenous Compounds from Nodules

Leaf et al. (77) determined the free amino acids and related soluble N compounds in Alnus glutinosa by means of chromatographic procedures. Citrulline was the predominant amino acid accompanied by smaller amounts of aspartic, glutamic, and λ-amino butyric acids, arginine, and others. When N_2 labeled with ^{15}N was applied, the highest atom % ^{15}N was found in glutamic acid, followed by citrulline or aspartic acid. Ammonia contained a smaller portion of ^{15}N than these compounds and arginine showed only a very small enrichment. When citrulline was degraded to ammonia and ornithine, the liberated ammonia was richer in ^{15}N than even glutamic acid.

In Myrica gale they (78) found aspartic acid to be the dominant amino acid. Glutamine and various other amino acids were present in smaller amounts. The highest enrichment with ^{15}N was in the amide-N of glutamine and the next highest in the amide-N of asparagine and amino-N of glutamic acid. These data support the view that the fixed nitrogen passes through ammonia before entry into organic compounds.

Wheeler (79) studied the free amino acids in the root nodules of nine species of nonleguminous plants belonging to seven genera. In species of Myrica (M. gale, M. cerifera, M. pilulifera, M. cordifolia) and also in Hippophaë rhamnoides, Elaeagnus angustifolia, Ceanothus velutinus var. laevigatus, and Casuarina cunninghamiana, the amino acid pattern resembled that observed in the nodules of Myrica gale

(78), asparagine being the dominant amino acid in terms of nitrogen content, with substantial amounts of glutamine often also present. In Alnus inokumai, citrulline was prominent, as previously shown for the nodules of A. glutinosa (77), though now amides were also present in substantial quantities. In nodules of Coriaria myrtifolia glutamic acid, glutamine, and arginine were the chief amino acids.

6.8.2 Influence of Photosynthate

In a physiological study Wheeler (80) found that maximal rates of N_2 (C_2H_2) fixation in the root nodules of first-year Alnus glutinosa and Myrica gale plants growing under natural illumination but at constant temperature were attained about midday. The nitrogen content of the sap exuding from the stumps of decapitated plants and the level of soluble nitrogen in the nodules were also highest around midday, the same being true of the rate of respiration in detached nodules. The hypothesis that the midday period must have the maximum ingress of photosynthates into the nodules proved not to be valid, since the analyses of "soluble" and "reserve" carbohydrate in the nodules showed no accumulation during the midday period. In later experiments Wheeler (81) followed the translocation of ^{14}C-labeled photosynthates to the nodules of first-year alder plants growing under the same conditions as mentioned previously. It was shown that there is a maximum influx of new photosynthates at the time of the midday peak in fixation. Analysis of fluctuations in the levels of the main sugars present in the nodules at different times of day, and a study of the effects of interrupting supplies of photosynthates to the root system by stem ringing, suggest that a substantial part of the nodule carbohydrate is unavailable for fixation and that maximal rates of fixation are only attained when quantities of new photosynthates are entering the nodules. No evidence was obtained for an autonomous element in the midday maximum in N_2 fixation when plants were kept in continuous darkness; a decline in acetylene reduction to an insignificant rate 24 hours after darkening was accompanied by a decline in nodule sugars, notably in the amount of sucrose present. The rate of acetylene reduction by the root nodules of Alnus glutinosa reached a maximum at about

25°C, and at this temperature, was some 6 times faster than at 15°C.

6.8.3 Transfer of Nitrogenous Compounds from Nodule to Shoot

Bond (82) showed that in Alnus glutinosa plants a substantial enrichment in ^{15}N was detectable in the shoot within 6 hours after the beginning of the exposure of the root nodules to $^{15}N_2$. By ringing the nodulated plants over a short zone at the shoot base, thereby removing the tissues external to the xylem, the upward movement of the fixed nitrogen was not affected. He concluded that the fixed nitrogen, which is probably in organic form, can be translocated in the xylem by the transpiration stream, and that this is its normal route.

Stewart (83) made a quantitative study of N_2 fixation and its transfer from the nodule to the remainder of the plant in Alnus glutinosa plants during their first season. Fixation per plant reached a maximum in late August (30.5 mg N_2 fixed per plant in 12 days), but fell rapidly with the onset of autumn. Fixation per unit nodule dry weight was greatest in young nodules and of the same order as that of nodulated legumes. Throughout the growth season there was a steady transfer from the nodules of some 90% of the N_2 fixed.

From Wheeler's (80) experiments with decapitated Alnus glutinosa plants, it is evident that the transfer of fixed nitrogen from the nodules occurs without delay as the total amount of nitrogen in the sap is highest for plants decapitated close to midday. However, although the mechanism of transfer of fixed nitrogen from the nodules is rapid, a temporary accumulation of fixed nitrogen in the nodules occurs at the period of most rapid fixation as shown by the analyses.

6.8.4 Mineral Requirements of the Dinitrogen-fixing Symbiosis

Nonleguminous plants growing on dinitrogen have more or less the same requirements for growth as other plants (see III.6.6.1) except for certain trace elements. Two of these, molybdenum and cobalt, are of special interest since one is a component of nitrogenase and the other is required for growth of N_2-fixing organisms.

6.8.4.1 Molybdenum

Becking (84, 85) determined the molybdenum content of the various parts of an Alnus glutinosa tree in an alder grove at Wageningen, The Netherlands. The highest molybdenum levels (Table 6.6) were found in the seeds and the root nodules. An effort was also made to obtain molybdenum-deficient alder plants. Because it is difficult and laborious to produce molybdenum deficiency in plants in water culture, use was made of certain molybdenum-deficient, iron-rich, low-moor, peat soils found in heather regions in the provinces of Groningen and Drenthe in The Netherlands.

TABLE 6.6. Nitrogen and Molybdenum Contents of Parts of a 12-m High Alnus Glutinosa Tree in an Alder Grove at Wageningen, The Netherlands[a]

Part of plant	Nitrogen (% of dry matter)	Mo (ppm of dry matter)[b]
Leaves (young)	3.84	0.25
Stems and branches	0.92 (0.73-1.11)	0.18
Roots	1.76	0.28
Nodules	2.37	1.71
Seeds	3.69	2.10

[a] The soil was a sandy loam (pH = 5.3) containing approx. 0.2 ppm available Mo [Aspergillus niger method, Hewitt and Hallas (274)].
[b] Determined chemically with potassium rhodanide, after reduction of any hexavalent molybdenum by stannous chloride.

Alnus glutinosa plants were grown in this soil in in small glass trays. Parallel tests were carried out with cauliflower (Brassica oleracea) as a molybdenum-indicator plant, since cauliflower is extremely sensitive to molybdenum deficiency producing characteristic symptoms of molybdenum deficiency. In one series of jars 150 μg molybdenum was added to 500 g of the molybdenum-deficient soil per jar. The large differences in the growth of the alder plants supplied with molybdenum and those without molybdenum are shown in Plate 6.7.

Plate 6.7. Six-month old <u>Alnus glutinosa</u> plants growing in pots with molybdenum-deficient soil. All plants are inoculated and nodulated. Left: two plants without added molybdenum; right: two plants with 150 μg Mo as $Na_2MoO_4 \cdot 2H_2O$ per pot. Note the pale-colored leaves of the molybdenum-deficient plants (x 0.11)

The molybdenum-deficient plants had yellowish leaves in contrast to the dark green leaves of the plants provided with molybdenum (84, 85). Analyses of the various parts of these plants showed lower molybdenum contents in the molybdenum-deficient plants than in the plants supplied with molybdenum (Table 6.7). The highest molybdenum concentration was in the root nodules and was distinctly higher than in the roots bearing the nodules.
Another phenomenon was observed in these experiments. In the healthy plants only one or a very few "crown nodules" developed on the root near the stem. In contrast, molybdenum-deficient plants had many small root nodules equally distributed over the entire root system. Similar symptoms are known for leguminous

TABLE 6.7. Molybdenum Contents of Leaves, Stems, Roots, and Nodules of Alnus Glutinosa Plants Grown in Mo-deficient Soil in the Absence or Presence of Added Molybdenum[a]

Plant tissue	Molybdenum content (ppm of dry matter)	
	I[b]	II[c]
Leaves	0.01	0.27
Stems	0.14	1.89
Roots	0.24	2.62
Nodules	2.00[d]	17.30

[a] The molybdenum-deficient soil (pH = 5.2) from Groningen contained approx. 0.006 ppm available Mo [Aspergillus niger method, Hewitt and Hallas (274)].
[b] I: no added Mo.
[c] II: 150 µg Mo (as $Na_2MoO_4 \cdot 2H_2O$) supplied per pot containing 500 g of soil.
[d] Approx. value since the sample was very small.

plants infected with an ineffective Rhizobium strain or for effective Rhizobium strains on molybdenum-deficient legumes. Thus, as in legumes, there is in nonlegumes a control by molybdenum of the production of root nodules.

The essentiality of molybdenum for nonleguminous nitrogen fixation was observed in Casuarina cunninghamiana and Myrica gale as well as Alnus glutinosa by Hewitt and Bond (86) and Bond and Hewitt (87). These authors grew nodulated Casuarina and Alnus plants on dinitrogen and nonnodulated plants with nitrate-nitrogen in water culture with and without molybdenum. In both plant species a molybdenum requirement for normal growth was shown. In addition the nitrate-grown plants without molybdenum showed abnormal visual features in the shoot and an accumulation of unreduced nitrate in the tissues. Evidence was obtained in the Alnus plants that molybdenum is of greater significance to nodulated plants than to those utilizing nitrate-nitrogen. In similar experiments with Myrica gale (87), the essentiality of molybdenum for growth on N_2 was demonstrated and it was shown that most of the molybdenum accumulated in the root nodules.

6.8.4.2 Cobalt

A cobalt requirement for N_2 fixation was demonstrated in Alnus glutinosa, Casuarina cunninghamiana, and Myrica gale (88, 89). Lack of cobalt resulted in severe nitrogen-deficiency symptoms. No cobalt requirement was observed in nonnodulated plants supplied with combined nitrogen. In the Casuarina plants supplied with Co, the highest cobalt concentration was found in the root nodules (6.8 ppm of dry matter) followed by the roots (4.0 ppm Co of dry matter). In the Myrica plants this was reversed with 7.1 ppm Co in the root dry matter and 2.9 ppm Co in the nodule dry matter. Root nodules from plants supplied with cobalt were relatively rich in vitamin B_{12} analogs. Bond et al. (90) found by the Euglena gracilis assay that the vitamin B_{12} content of various nonlegume root nodules from plants supplied with cobalt was similar to that of legume nodules. Root nodules of Alnus glutinosa, Casuarina cunninghamiana, Myrica gale, and Hippophaë rhamnoides contained about 130-300 ng vitamin B_{12} per gram of fresh tissue, although some values were considerably lower. The results varied with the extraction procedure. The vitamin B_{12} content of Coriaria myrtifolia root nodules was much lower, i.e., 28-53 ng/g fresh tissue. Roots of these nonleguminous plants contained negligible amounts of vitamin B_{12}.

Also Kliewer and Evans (91) showed the presence of vitamin B_{12} coenzyme in the root nodules of legumes and of Alnus rubra (= Alnus oregona). However, the published B_{12} coenzyme values were incorrect due to omissions of a dilution factor (personal communication, Kliewer and Evans). The correct value for the root nodules of a 2-yr-old Alnus rubra plant should be 62 nmoles, not 174 nmoles, per gram of fresh tissue.

6.8.5 Effect of Gaseous Environment on Dinitrogen Fixation

Various gases are known to influence, or inhibit, N_2 fixation in legumes and free-living diazotrophs (23, 92, 93). Some of these gases have been tested on symbiotic N_2 fixation by nonlegumes.

6.8.5.1 Dinitrogen

Dinitrogen fixation by root nodules is influenced by the N_2 concentration of the environment. Burris et al. (94) calculated the apparent Michaelis constant for sliced soybean nodules. By statistical methods referring to many experiments a K_m of 0.02 ± 0.004 atm N_2 was obtained. Maximum N_2 fixation was observed above 0.10 atm N_2. In contrast, Bond (95) reported a maximum N_2 fixation at 0.25 atm N_2 for whole excised root nodules of Alnus glutinosa. Sloger (67), who did an identical experiment as Burris et al. (94) with excised whole nodules of Myrica cerifera, calculated a K_m of 0.069 ± 0.004 atm N_2. This value is about 3 times higher than that reported for leguminous root nodules.

6.8.5.2 Oxygen

MacConnell (96) provided different levels of oxygen to nodulated Alnus glutinosa plants in water culture and observed that the number of nodules was progressively reduced as the oxygen tension was lowered from the normal 21% (vol/vol). Nodulated alder plants were considerably more sensitive to oxygen level than nonnodulated plants. Therefore, oxygen supply seems to be of special importance in the development and function of alder root nodules. Later, Bond (97) showed with the ^{15}N method than N_2 fixation was small at low oxygen concentrations, but rose with increasing oxygen supply and attained a maximum at about 12% (vol/vol) oxygen in Hippophaë rhamnoides and 20-25% (vol/vol) in Casuarina cunninghamiana and Myrica gale. Results with Alnus glutinosa closely resembled those obtained with Hippophaë rhamnoides; the optimum fixation occurred at 12% (vol/vol) O_2. Application of 40% (vol/vol) O_2 virtually eliminated N_2 fixation in the root nodules of Alnus glutinosa, but not in Casuarina cunninghamiana, Myrica gale, and Hippophaë rhamnoides (95). Sloger (67) tested the effect of O_2 tension on N_2 fixation ($^{15}N_2$) in Myrica cerifera. In agreement with the results obtained by Bond (97) he found optimal N_2 fixation at 20-30% (vol/vol) O_2, a reduction at 10% (vol/vol) O_2, and a strong inhibition by 40% (vol/vol) O_2 (Figure 6.6).

6.8.5.3 Hydrogen

Bond (98) found an inhibition of N_2 fixation by hydrogen in Alnus glutinosa, Casuarina cunninghamiana, and

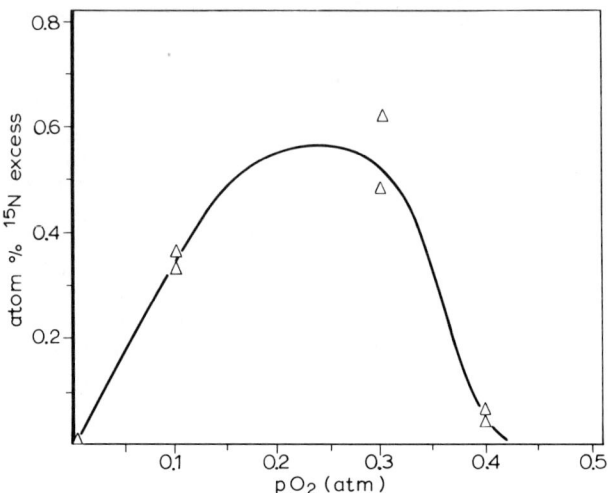

Fig. 6.6 Effect of O_2 tension on N_2 fixation by detached root nodules of <u>Myrica cerifera</u>. After Sloger (67).

<u>Myrica gale</u> similar to that in peas and soybeans. The application of 60% (vol/vol) hydrogen to the gas mixture resulted in an inhibition of fixation of 77-86% in the legumes and 84-88% in the nonlegumes. Exposure of <u>Alnus glutinosa</u> nodules to 20 and 60% (vol/vol) hydrogen inhibited N_2 fixation 52 and 80%, respectively.
 Hoch et al. (99) reported an evolution of hydrogen by leguminous root nodules (soybean), which was inhibited by N_2 and N_2O. Intact root nodules of the nonlegumes, <u>Alnus glutinosa</u>, <u>Alnus rugosa</u>, and <u>Elaeagnus commutata</u> did not evolve hydrogen when tested in an atmosphere of 80% (vol/vol) He and 20% (vol/vol) O_2 (24).

6.8.5.4 Carbon Monoxide

The effect of carbon monoxide on N_2 fixation by nonlegumes was studied in detached root nodules of <u>Alnus glutinosa</u>, <u>Casuarina cunninghamiana</u>, and <u>Myrica gale</u> by Bond (98) with the $^{15}N_2$ method. Parallel tests with peas were included. The inhibition of N_2 fixation by 0.1 and 1.0% (vol/vol) carbon monoxide in the gas mixture was 31-66 and 98%, respectively. A simultaneous experiment with peas showed an inhibition of 79 and 99%, respectively. Thus, carbon-monoxide inhibition of N_2 fixation was similar in nonlegumes and in legumes although comparisons at lower CO concentrations would be more useful. As will be pointed out in III.6.8.5, most workers have been unable to detect hemoglobin-like compounds in nonleguminous root nodules. Therefore, the mechanism of action of carbon monoxide in root nodules of nonlegumes is unknown.

6.8.6 Presence of Hemoglobin, Poly-β-Hydroxybutyrate and Growth Substances in Root Nodules

6.8.6.1 Hemoglobin

Hemoglobin is always present in active N_2-fixing leguminous root nodules and is considered to play an important role in the N_2 fixation process in legumes (100, II.8, III.10).
 Smith (101) could not demonstrate hemoglobin in Alnus glutinosa root nodules and this was confirmed (102) for root nodules of Alnus glutinosa, A. incana, Hippophaë rhamnoides, and Myrica gale. Bond reported that hemoglobin was not detected spectroscopically by Smith (cf. ref. 103, p. 457, footnote) in root nodules of Myrica gale and that the intensive red color of the bog-myrtle root nodule is due to anthocyanin pigments as also observed in Alnus and Casuarina (104).
 In a study of the presence of hemoglobin in nodular tissue from Alnus glutinosa, Myrica gale, Hippophaë rhamnoides, and Casuarina cunninghamiana, Davenport (105) observed by spectroscopic examination, especially in Casuarina, an absorption band at 563 nm due to deoxygenated hemoglobin. This hemoglobin was not extractable by aqueous solvents in contrast to simultaneously tested leguminous hemoglobin. Apparently it was firmly bound to cell debris and could not be removed. Alnus and Myrica root nodules also contained a bound form of hemoglobin, similar to that in Casuarina.
 Hemoglobin was not detected in root nodules of several Alnus species (Becking, unpublished). The red color of these nodules and root tips, especially in nitrogen-deficient plants, was caused by pigments of the anthocyanin type as already observed by Bond (103, 104). Also Moore (24) did not detect hemoglobin with

a microspectroscope in nodule slices of Alnus rugosa, Elaeagnus commutata, Shepherdia canadensis, and Hippophaë rhamnoides.

Although root nodules of Trema are pink in gross appearance in cross-section (218), hemoglobin has not been detected [personal communication, C. A. Appleby, (257)]. It is of interest that the Trema endophyte nodulates three species of legumes [Vigna sinensis, Macroptilium atropurpureus (formerly Phaseolus atropurpureus), and M. lathyroides (formerly P. lathyroides)], and these root nodules contained hemoglobin.

6.8.6.2 Poly-β-hydroxybutyrate

Free-living diazotrophs such as Azotobacter spp. and Beijerinckia spp., as well as many other bacteria, are rich in the storage substance poly-β-hydroxybutyrate. The presence of this substance has been demonstrated in the root nodules of leguminous plants but could not be detected in root nodules of Alnus glutinosa and A. incana (107).

As will be described in III.6.10, Becking et al. (108) observed in electron micrographs of ultrathin sections of the bacteria-like cells of the Alnus glutinosa endophyte lipid inclusions which were rather resistant to thin sectioning. It is quite possible that these lipid globules consist of poly-β-hydroxybutyrate and that this substance is not readily assessed by chemical methods because of its low quantity and the scarcity of bacteria-like endophytic cells in the root nodule tissue.

6.8.6.3 Growth Stimulating and Inhibiting Substances

Root nodules have generally been associated with the production of growth substances by either the tissue or the endophyte. Auxin-like compounds are known to be involved in root hair curling in legumes (230-232). Although a similar phenomenon has been noted in non-legumes (5-7), the substances responsible for the effect have not been identified.

Dullaart (233) examined by spectrophotometric methods the indole-3-acetic acid (IAA) content of the acid ether-soluble fraction of methanol extracts of root nodules of Alnus glutinosa. The root nodule concentration of IAA (0.6-1.0 mg/kg fresh wt) was some

5-10 times higher than that of roots devoid of nodules. This pattern is reminiscent of the earlier report of Silver et al. (110) for field collected Alnus serulata root nodules and roots assayed for IAA by a biological method, although the latter reported much higher levels.

A related substance, indol-3-carboxylic acid (ICA), was found to be present at a much lower concentration (0.05-0.15 mg/kg fresh wt) exclusively in the roots (233). A similar phenomenon has been observed in the root nodules and roots of the legume, Lupinus luteus (234). It may be significant that ether extracts of tomato tumor tissue contained both IAA and ICA, whereas only ICA was detected in normal tissue (235).

The reduced IAA content of root nodules of Myrica cerifera and Casuarina cunninghamiana has been correlated with the negative geotropic growth of the rootlets at the apices of nodular lobes (110). The characteristics of the nodulation type in these species is discussed in III.9.1, as well as the effects of gibberellic acid and cytokinins. Thus far no cytokinin determinations have been performed with the root nodules of nonlegumes.

Root exudates of legumes have been reported to stimulate the growth of Rhizobium as well as other soil bacteria in the rhizophere (236, 237). The occurrence of a red pigment in the root nodules of Alnus glutinosa has been reported (238). It was bactericidal against a number of common pathogenic bacteria when excreted into the environment, and it has been suggested that the substance is a "streptomycin-like" compound excreted by the actinomycetal endophyte.

6.9 ROOT NODULES OF NONLEGUMES

6.9.1 Structure

Root nodules of nonleguminous plants can be divided into two main groups: the Alnus and Myrica/Casuarina types; and the unusual Trema type.

Under field conditions the nodule clusters of nonleguminous plants may attain considerable size. Alnus root nodules may reach a diameter of 5-6 cm, or that of a tennis ball. Equivalent sizes are reported by Mowry (109) and certain botanists for the so-called "rhizothamnia" of the Casuarina species developed under natural conditions. Field grown Ceanothus velutinus collected in Oregon, U. S. A. (Plate 6.2) contained root nodules up to 5-6 cm in diameter. The root nodules of nonleguminous plants in water or sand culture are considerably smaller, but it should be remembered that these plants are in general much younger than field specimens. In the field these nodules are perennial for many years and they increase in size every season.

6.9.1.1 Alnus Type

The Alnus type of root nodule is found in the genus Alnus of the Betulaceae and, in addition, in the nodulated species of the families Elaeagnaceae, Rhamnaceae, Coriariaceae, and in the Rosaceae. All these nodules are modified lateral roots with an arrested or very slow-growing apical meristem. The nodular lobes are usually dichotomously branched, producing a root nodule of coralloid appearance. The Alnus type root nodule is depicted in Plate 6.1, B & C showing Alnus glutinosa root nodules. Plates 6.2 and 6.3 present the root nodules of Ceanothus velutinus (Rhamnaceae) and Dryas drummondii (Rosaceae); the latter root nodules are similar in appearance, although they belong to different and quite unrelated taxonomic groups.

Sometimes transitional nodule structures between the two nodule types occur and a species belonging normally to one of the nodulation types may develop a nodule structure of the other type. This, however, occurs only under aberrant conditions, e.g., when Alnus rubra seedlings originating from the northwest coast of North America are inoculated with the endophyte of the European Alnus glutinosa (5), or when the South American Alnus jorullensis is inoculated with the Alnus glutinosa endophyte (33). Usually in Alnus the apical meristems of the nodule lobes have only a restricted growth, but in this case some of the apical meristems, but not all nodule lobes like in Myrica and Casuarina root nodules, produce negative geotropic rootlets, as found in the Myrica/Casuarina nodulation type (Plate 6.1, A). These abnormal nodules produced by apparently the wrong endophytes are also characterized by inefficient or poor N_2 fixation. On the other hand Comptonia peregrina of the Myricaceae may sometimes show the Alnus nodulation type, since root nodules of this species as depicted by Ziegler (4) showed no outgrowth of the apices of the nodular lobes.

6.9.1.2 Myrica/Casuarina Type

Root nodules of Casuarina spp. and Myrica spp., but probably not Comptonia asplenifolia (see III.6.9.1.1), show the characteristic of this type with the apex of each root-nodule lobe producing a normal, but negative geotropic root. Therefore, these root nodules become covered with upward growing rootlets (Plate 6.1). The negative geotropic growth may be associated with an abnormal auxin pattern of these roots. Indeed, Silver et al. (110) could not detect auxin in the negative geotropic nodule roots of Myrica cerifera and Casuarina cunninghamiana, while nonnodulated roots of Myrica cerifera exhibited a normal positive geotropic curvature and an auxin content within the anticipated range of 10 mg IAA/kg tissue fresh wt. The authors attributed the absence of measurable amounts of endogenous auxin in the negative geotropic nodule roots to the presence of a very active indolacetic acid oxidase system; and the activity of this enzyme was far less in nonnodulated root tissue.

6.9.2 Formation

As in legumes, nodule formation can be studied by the Fåhraeus technique (111), i.e. using a glass slide and following root hair growth under the microscope at low magnification. The initially perfectly straight root hairs become deformed and curved at the site of nodule formation by the action of the endophyte (Plate 6.8, A & B). The curling of the root hairs is most likely due to excretion of auxins or related growth substances by the endophyte as reported for Rhizobium (112, 113).

In those cases where the primary infection could be found and the infection thread could be followed through the root hair, the endophyte entered the plant at the root hair tip. Here a pocket was formed by invagination of the plant cell wall (Plate 6.9). Electron microscopy revealed that the outer cytoplasmic membrane was folded and that this proceeded inward with the infection thread. As in legumes, the growth of the infection thread is associated with the host-cell nucleus. During its extension within the root hair

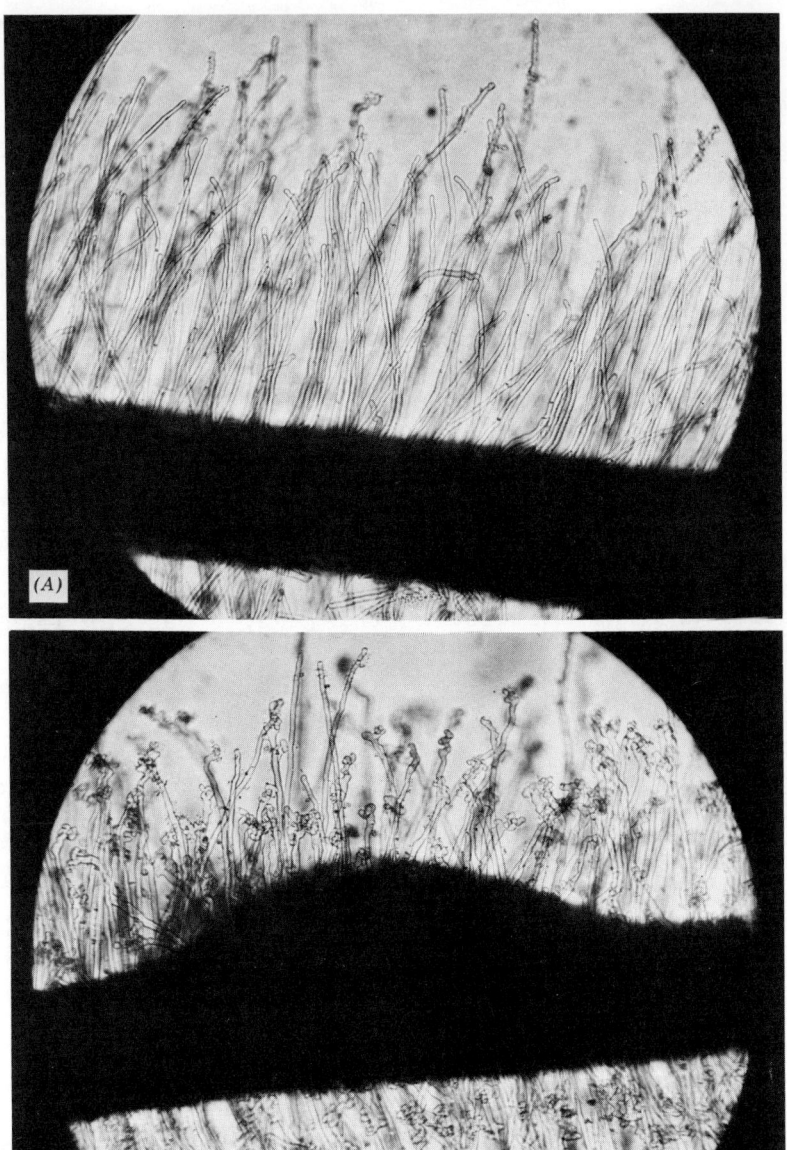

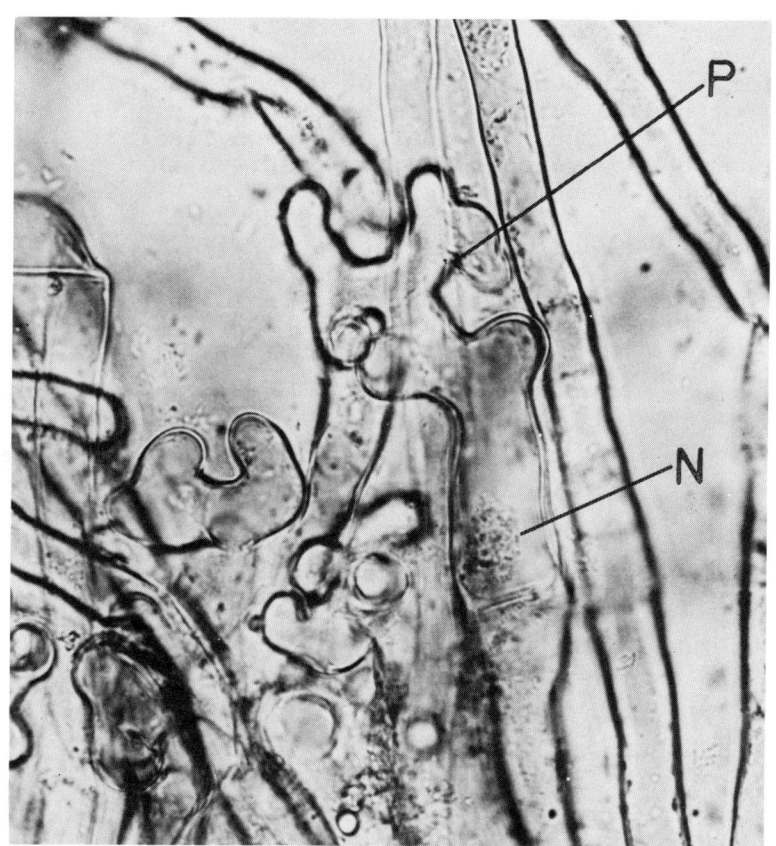

Plate 6.9. Tip of root hair of <u>Alnus glutinosa</u> deformed by the action of the endophyte. At the tip of the root hair a pocket (P) is formed by invagination of the cell wall of the root hair, and from this an infection thread develops inside the host cell accompanied in its growth by the nucleus (N) of the root hair. Living unstained preparation, phase contrase (x 990).

Plate 6.8. Root-nodule development in <u>Alnus glutinosa</u> plants. (<u>A</u>) Alder roots with perfectly straight unicellular and bicellular root hairs. (<u>B</u>) Alder root hairs showing curling due to the production of growth substances by the endophyte in the rhizosphere. A root nodule will develop at the place where root hairs become curved. Living, unstained preparation (x 84.5).

cell the infection thread is accompanied by the host nucleus (see N of Plate 6.9).

In the very early stage of development, the site of root nodule formation is apparent only as a slight thickening of the main root. The longitudinal section of such a very young root nodule showed that even at this stage, when the root nodule was only a few cell layers thick, the host cells already contained vesicles (Plate 6.10). Host cells with bacteria-like endophyte cells were not observed. In addition, the thin section revealed that the root nodule was a modified lateral

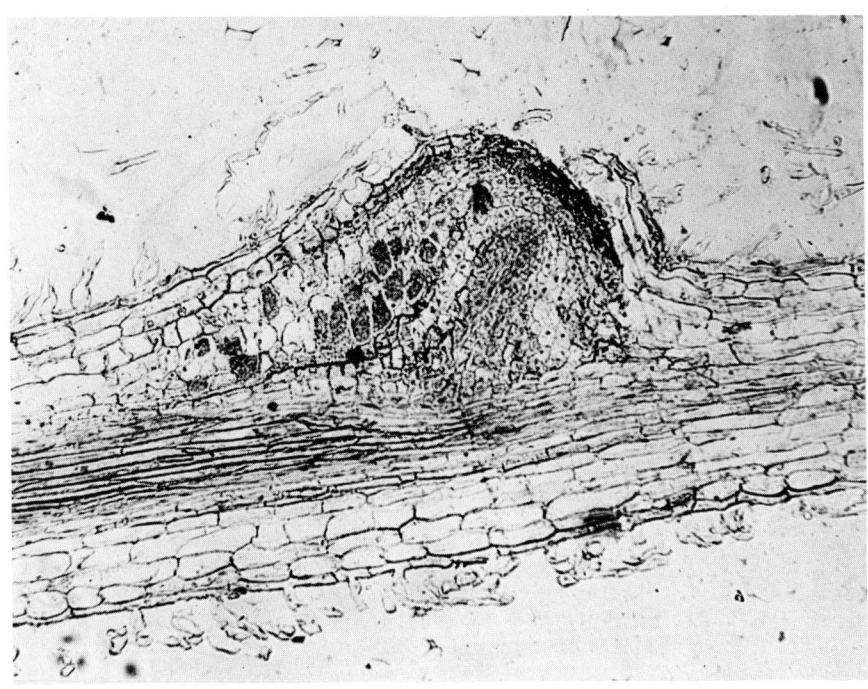

Plate 6.10. Longitudinal section through a very young root nodule of <u>Alnus</u> <u>glutinosa</u>. A "pre-nodule" is formed by the presence of the endophyte in the cortical parenchyma cells of the root. Subsequently the final nodule is formed by the outgrowth of the side-root primordium taking over the endophyte infection. Fixed preparation, stained with erythrosin and methylene green dyes. (x 139.)

root originating from protoxylem or protophloem strands within the pericycle of the parent root. Two endodermic layers were present: one of the parent root and the other of the newly formed lateral root producing the root nodule. This was made particularly evident by the application of erythrosin and methyl-green dyes, since the two endodermic layers stained green whereas the other tissue and the endophyte stained red [(6) and Plate 6.8]. It is noteworthy that the endophyte is originally present in the cortical parenchymatous tissue of the parent root and it is only taken over by the newly developing root nodule at the later stage. During the further growth of the root nodule, the endophyte continues to be transferred from the host cells containing the endophyte to those near the apex produced by the apical meristem of the nodular lobe.

Gibberellic acid has been observed to reduce root nodulation in some legumes (239, 240), but such effects have not yet been studied in nonlegumes.

Cytokinins are reported to produce pseudo-nodulation in excised root cultures of tobacco roots (241). Phillips and Torrey (242, 243) found that cytokinin was released by two Rhizobium species into the culture medium and that it produced polyploid divisions when tested on cultured pea-root segments. The substance was identified as a zeatin-like compound (244) which induced cell proliferation in soybean callus tissue and increased 34 fold over the 12-week culture period. There was clearly a specific dependence of the cells on cytokinin for DNA synthesis for the peak of cytokinin activity was observed at the beginning of the growth phase in pea-root callus culture (245, 246).

Cytokinin production in Alnus may be related to the delayed senescence observed in alder leaves which generally show no yellowing in autumn, but drop green and become black on the forest floor. A chlorophyll retention effect in senescence is a well-known effect of cytokinins (247-253). Although no cytokinin determinations of alder tissues have been made, an appreciable cytokinin content may be expected. A prominent role of cytokinin in root nodule formation is also likely since a profuse production of psuedonodules in sterile non-inoculated Alnus glutinosa was observed when kinetin or 2-iso-pentenyl-adenine was added to the rooting medium (254).

6.9.3 Cytology

Cross sections of the apex of a nodule lobe of alder showed that the endophyte is not distributed arbitrarily in the corticl parenchyma cells of the host. The endophyte is not present in the cortical parenchymatous layers near the epidermis and the parenchymatous layers near the stele. This zone formation is probably the result of a physiological interaction between host tissue and the endophyte. All cortical parenchyma cells displayed rather big nuclei, an indication that this tissue, or a part of it, is tetraploid. Transverse sections through the endophytic layer of very young root nodules revealed very large host nuclei in the process of division. Tetraploidy has also been claimed for the nodular tissue of legumes (114-117). Alnus glutinosa has 28 or 56 chromosomes (118, 119); other Alnus species 28, 42, or 56 chromosomes (120). In Alnus glutinosa root nodule tissue mitotic divisions were observed and separate chromosomes could be counted (Plate 6.11). However, tetraploidy could not be ascertained, because of natural polyploidy observed in some of the Alnus glutinosa plants tested. In this case tetraploidy can only be demonstrated by examining root tips and nodular tissue of the same plant.

However, there is some indirect evidence that Alnus glutinosa nodular tissue is probably polyploid. In plant tissue, a close relation exists between cell size and nuclear size. Sections of root nodule tissue showed that infected cortical parenchyma cells are about twice as large as uninfected cells in the same tissue (108, Fig. 6.2). This observation favors the hypothesis of a doubling of chromosomes in the infected tissue. On the other hand, Kodama (258, 259) counting chromosome numbers in three Japanese species of alder, found no difference in chromosome number in the root nodule cells and in the root tip cells. However, he examined only the apical meristem of the root nodules and not the infected cells.

Trema root nodules (Plate 6.12, A & B) are structurally very different from leguminous root nodules since they contain a central vascular bundle. Therefore, they are modified roots and in this respect closely resemble the actinomycetal root nodules of other non-legumes. Cross sections (Plate 6.12, B) through nodules often show a horseshoe-shaped bacteroid area of infected parenchyma and a vascular bundle more near the

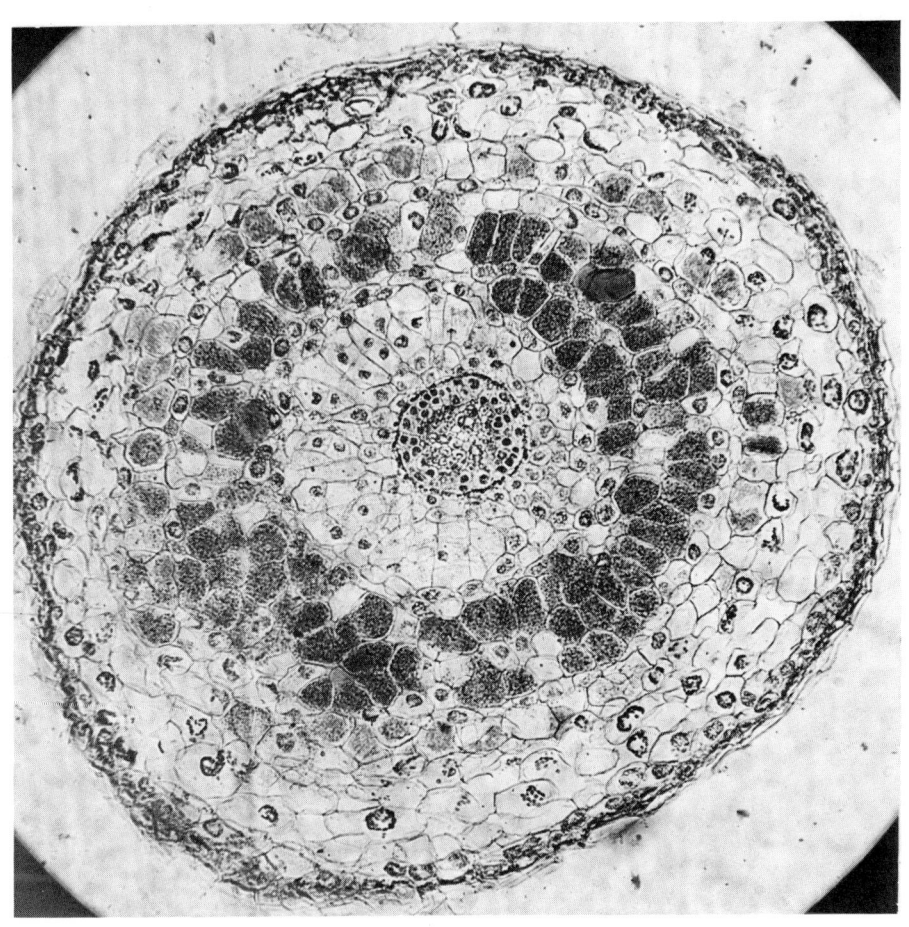

Plate 6.11. Transverse section through a root-nodule lobe of
<u>Alnus</u> <u>glutinosa</u>. In some cortical parenchyma cells big nuclei in
the process of mitosis and individual chromosomes are visible.
Fixed preparation, stained with Heidenhain's iron-hematoxylin
(x 630).

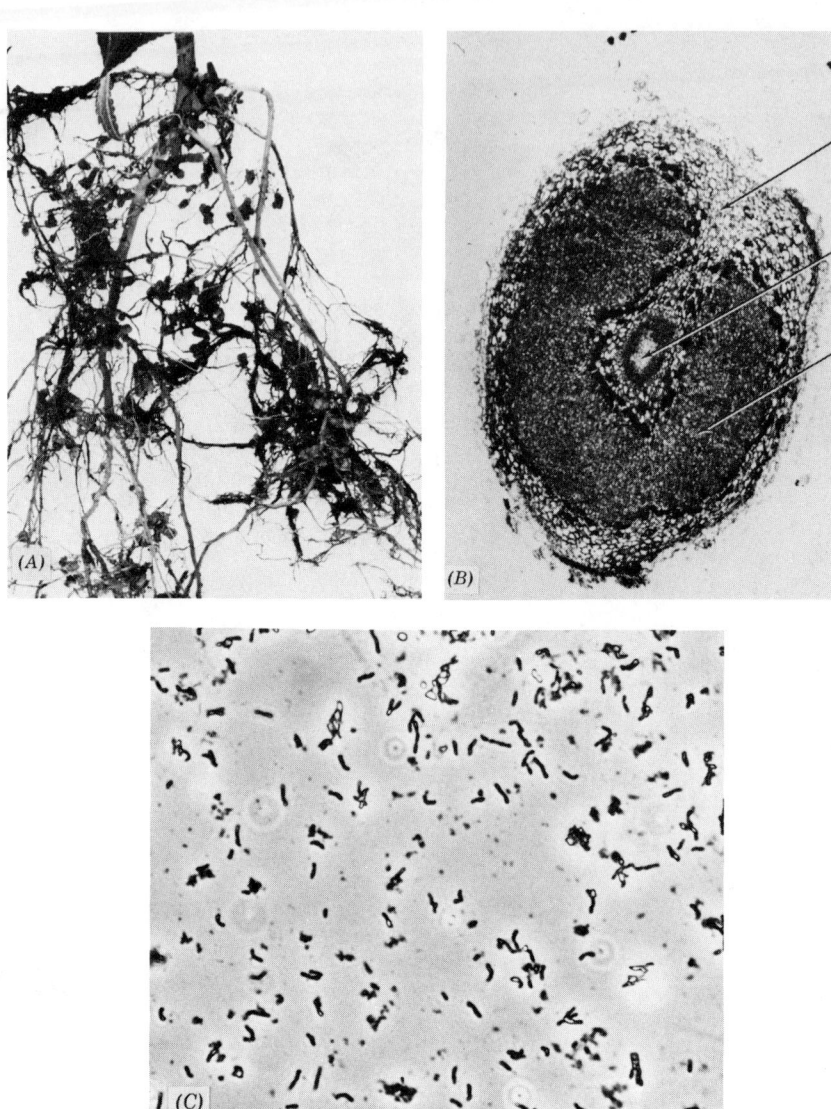

Plate 6.12. <u>Trema cannabina</u> var·<u>scabra</u> de Wit. (A) Upperpart of root system with the <u>Rhizobium</u> inhabited root nodules (x 0.6). (B) Cross-section through a root nodule showing cortical parenchyma cells (P), vascular tissue (V) and innermost the bacteroid area (C) (x 41). (C) <u>Rhizobium</u> strain NGR 231 ("cowpea-type") isolated from <u>Trema</u> root nodule. Seven-day-old culture on <u>Rhizobium</u> nutrient medium (yeast mannitol agar) (x 670). (Photographs by M. J. Trinick.)

edge (218). However, Trema root nodules resemble those of the "cowpea-type" of legume in that they are large, pink, loosely attached, and elongate with branching.

Developments in understanding the physiology and biochemistry of Trema cannibina may be slow because it is rather difficult to grow. Germination of seeds took about 6-8 months for greenhouse conditions, and about 2 additional months for the seedlings to reach a height of 5 cm, after which growth became much faster. So far vegetative propagation from leaf or stem cuttings have been unsuccessful (personal communication, M. J. Trinick).

6.10 ENDOPHYTE OF THE ROOT NODULE SYSTEM

6.10.1 Morphology and Cytology

In root nodule host cells of nonleguminous plants, three morphological forms of the endophyte can be distinguished: hyphae, vesicles, and bacteria-like cells. The hyphae, usually only 0.3-0.5 µm in diameter, are the primary form of the endophyte. These very thin hyphae moreover scarcely show light refraction and, therefore, are hardly visible with optical microscopy of living tissue, but are readily visible in ultra-thin sections viewed by electron microscopy. Some species may show, however, thicker hyphae like Myrica (Table 6.8). Using light microscopy the hyphae can also be made visible to a certain degree by fixation and staining procedures. Hyphae occur in young host cells and in mature host cells only in the central part of each cell. The hyphae grow from host cell to host cell through the plant's cell wall. They bear in a later stage vesicles or club-shaped terminal structures, which are spheres or elongated bodies, lying near the plant cell wall and being absent in the center of the host cell. The spherical vesicles are mostly 3.0-5.0 µm in diameter and the club-shaped structures in some endophyte species may reach a length up to 12 µm. Close inspection of each vesicle or club-shaped body shows that they are the terminal swelling of a hypha.

The bacteria-like endophyte cells are small particles, sometimes also called granulae, being usually only 0.5-1.0 µm in diameter and filling the host cells

TABLE 6.8. Differentiation of Species in the Genus Frankia

Actinomycete species	Host species	Hyphae (μm)
1. F. casuarinae	Casuarina	0.3-0.5
2. F. brunchorstii	Myrica and Comptonia (?)	1.2-2.8
3. F. alni	Alnus	0.3-0.5
4. F. elaeagni	Elaeagnus, Hippophaë, Shepherdia	0.3-0.5
5. F. ceanothi	Ceanothus	0.3-0.4
6. F. discariae	Discaria	0.3-0.4
7. F. coriariae	Coriaria	0.4-0.7
8. F. dryadis	Dryas	0.5-0.8
9. F. purshiae	Purshia	0.3-0.5
10. F. cercocarpi	Cercocarpus	0.3-0.5

[a] EM, electron microscopy; LM, light microscopy.
[b] Becking, unpublished.

Terminal swelling of Hyphae (µm)		Polyhedral-shaped cells (µm)	Studied by[a]
Spherical	Clubshaped		
-	Length: 3-4; diameter at widest point: 0.6-1.5	0.4-1.0	EM[b], LM
-	Length: 7.5-12.5; diameter at widest point: 1.6-2.4	1.5-2.5	EM, LM
3.0-5.0; rarely 6.0-8.0	-	0.5-1.5	EM, LM
2.5-4.0; rarely 1.8-2.2	-	0.3-0.9	LM
1.5-3.0; in average approx. 2.0-2.5	In younger stages with slowly tapering base	unknown	EM[b], LM
Approx. 4.0	-	unknown	LM
-	Length: 9-12 diameter at widest point: 1.2-1.3	unknown	EM[b], LM
-	Length: 1.5-5.0; diameter at widest point: 1.5-2.0	1.0-2.0	EM[b], LM
-	Length: 2.2-5.5; diameter at widest point: 2.2-4.4 µm	unknown	LM
-	Length: 4; diameter at widest point: 3	unknown	LM

entirely. This type of endophyte cell is only found in a few host parenchyma cells, dispersed among more numerous vesicle or club-shaped structure filled host cells. When released from the tissue artificially by squashing, but also observed as such in the ultra-thin sections by electron microscopy, they appeared to be irregular polyhedrons. The angular surfaces suggest that they have been tightly packed in the host cell like seeds in a pomegranate.

Vesicles and bacteria-like cells were never found together in the same host cell as shown in many non-leguminous root nodules (Alnus, Myrica, Elaeagnus, Dryas, etc.). Host cells containing vesicles and hyphae were always alive as shown optically and by electron microscopy; and in the electron micrographs host-cell organelles such as cytoplasm, cytoplasmic membranes surrounding the endophyte, endoplasmic reticulum, plant mitochondria, and the host-cell nucleus could be clearly observed. In contrast, bacteria-like cells are always found in host cells lacking host-cell organelles. Apparently, the latter host cells are dead (108).

Light and electron microscopy showed that bacteria-like cells can be released from the dead host cells. The endophyte particles were liberated into the adjacent host cell or into the soil environment by dissolution or rupture of the plant-cell wall. At present the further fate of the bacteria-like cells in the soil cannot be discussed in other than a speculative approach. They may, like bacterial spores, represent a resting stage, with their thick cell walls enabling survival in the soil. So far, however, nothing is known about the status of the endophyte of nonleguminous plants in soil, except that they must survive because sterile seedlings of these plants may nodulate in soil.

With respect to cytology, it is noted that electron microscopy confirmed the identity of the endophyte as an actinomycete. The hyphae have the same dimensions as an actinomycete and they also possess membranous bodies, the so-called "plasmalemmosomes" or "mesosomes," characteristic of actinomycetes, mycobacteria, and some gram positive bacteria. These structures are sometimes regarded as the bacterial equivalent of mitochondria (12), but they originate from invaginations of the outer cytoplasmic membrane (110). They play a prominent role in cell division and cell-wall formation

(122). The endophyte is certainly not a fungus because it differs in hyphal diameter and in the absence of a nuclear membrane. The endophyte's nuclear material has the form of nucleoids typical of bacteria.

Electron micrographs also revealed that the vesicles or club-shaped structures have a very complex internal structure with many incomplete cell walls and many cytoplasmic membranes and plasmalemmosomes. Vesicles dominate a nodular tissue which actively fixed N_2 by the ^{15}N method. Exposing nodule slices to acetylene, to compare the same fresh weight of nodular tissue (representing also about the same number of host cells) rich in vesicle structures to that poor in vesicle structures, demonstrated that the former produced more $N_2(C_2H_2)$ fixation. This supports the hypothesis that the vesicle structures are mainly associated with N_2 fixation.

In a later stage, the vesicles or club-shaped structures are completely digested by the surrounding active host cytoplasm leaving a shrunken and distorted empty structure. The main source of nitrogen must be from the living actively functioning vesicle, since the amount of nitrogen obtained from resorption of the vesicle structures by the host cytoplasm cannot account for the observed total nitrogen increase of nonlegumes over a period of time.

Vesicles or club-shaped structures cannot produce bacteria-like cells, since they have only incomplete internal cell walls. Therefore, bacteria-like endophyte cells must originate directly from hyphae. The bacteria-like endophyte cells are bounded at a young stage by a thin cell-wall. Their distribution within the host cell in separate groups suggests that each group probably originates from the division of a single or very few hyphae. In a subsequent developmental stage the bacteria-like cells are surrounded by a thick cell-wall and they contain lipoid reserve material suggesting that they are in a resting stage. These endophyte cells lack plasmalemmosomes. In contrast to vesicles and club-shaped structures, older stages of these bacteria-like cells are not dissolved because the host cells are dead. They probably play a role in the reproduction and dispersal of the endophyte.

Some nodule-bearing nonlegumes are very difficult to nodulate in water culture and only produce root nodules when grown in habitat soils. A correlation exists between the presence of bacteria-like cells and nodulation, for nonlegumes, in which bacteria-like cells are unknown (Table 6.8) or very infrequent, do not nodulate in water culture. It strengthens the opinon that the bacteria-like cells are connected with reproduction and dispersal.

As expected the cytology of Trema root nodules infected with their rhizobial endophyte is quite unlike that of the nonlegumes infected with actinomycetes. Light and electron microscopy have revealed that most host cells in the Trema root nodule are filled with intracellular infection threads. About 95% of the infected host cells contained infection threads, and the endophyte was released into the host cytoplasm in only 5% of the host cells. The released endophyte is enclosed by an envelope, but in contrast to legumes where usually only one or a few bacteria are enclosed by a membrane, a very large number of cells are enclosed by one sac (228). Because of the high frequency of infection threads in Trema root nodules, it is suggested that the infection thread continues to grow within the host tissue, spreading from one host cell to the other. Moreover, in view of the fact that the high specific activity of nitrogenase in Trema nodules is at the same level as that of legumes, it is likely that the endophyte in the infection thread stage is able to fix dinitrogen (228).

6.10.2 Classification

Becking (123) has proposed a classification of these symbiotic actinomycetes of nonleguminous root nodules. This classification is mainly based on cross-inoculation groups, as is the Rhizobium classification. As far as possible these physiological differences are linked with morphological characteristics such as the presence of vesicles or club-shaped structures, the dimension of hyphae, and vesicular structures (Table 6.8).

This classification, which proposed a new family, Frankiaceae, with a single genus Frankia, contains imperfections. Ten species of Frankia are associated with special hosts, which show no cross-inoculation abilities. But within some of the proposed species there is sometimes complete incompatibility or degrees of incompatibility. For instance, Bond (124) found no cross inoculation possible between the endophyte of Coriaria myrtifolia and C. japonica, when the endophyte of C. myrtifolia was applied to C. japonica. All degrees of poor, moderate, and satisfactory nodulation and N_2 fixation occur in the various Alnus species with respect to their reciprocal cross inoculations (125). Moreover, the symbiotic performance of the combination of Myrica cerifera with the M. gale endophyte, Myrica rubra, M. cordifolia, and M. californica in reciprocal combinations and with endophytes of other Myrica species showed poor N_2 fixation although in some cases there was good nodulation (125). However, it was found inadvisable to subdivide the proposed species without more data about the endophytes themselves. On the other hand, the barriers of incompatibility between the cross-inoculation groups may be not strict (unsurmountable). Recently, Rodriguez-Barrueco (personal communication and Ref. 260) observed that an Alnus glutinosa inoculum can produce nodulation in Myrica gale, but that the reciprocal combination of Myrica gale inoculum tested on Alnus glutinosa did not give root nodulation. The latter observation is an argument that both endophytes are taxonomically different. Moreover the root nodulation produced in Myrica by alder inoculum is aberrant from the normal type in Myrica (personal communication Rodriguez-Barrueco).

The current classification comprises the following ten Frankia species: F. casuarinae, F. brunchorstii, F. alni, F. elaeagni, F. ceanothi, F. discariae, F. coriariae, F. dryadis, F. purshiae, and F. cercocarpi (123). These are all obligate symbionts, but with a non-obligatory stage (resting stage) in soil. The morphological descriptions in Table 6.8 result from a compilation of the work of many authors on the following species: Casuarina cunninghamiana (126), Myrica gale (127) and M. cerifera (30), Alnus glutinosa (5, 108, 127), Hippophaë rhamnoides (127, 128), Elaeagnus macrophylla and E. angustifolia (128), Ceanothus sanguineus and C. velutinus (129, 130), and Coriaria arborea (131). Recently, the actinomycetal structures of the Cercocarpus montanus, C. paucidentatus (132) and Purshia tridentata (133) root nodules have been described. The table also includes recent electron microscopic observations of various plant species by the author. These are marked with a superior b.

The taxonomic position of these actinomycetes within the Actinomycetales is unclear. The segmentation of the vegetative mycelium into elongate or coccoid elements is a feature of the Actinomycetaceae; on the other hand, the terminal swelling on the mycelial hyphae may be regarded as degenerated conidia (adapted to the N_2-fixation process), a feature of the Streptomycetaceae. The classification of the Actinomycetales is mainly based on the major constituents present in cell-wall preparations (134). According to this scheme the Actinomycetales are divided into six main groups with major differences in lysine, aspartic acid, glycine, L-DAP (DAP = 2,6-diaminopimelic acid), meso-DAP, arabinose, and galactose content. Based on our analyses, the Alnus endophyte contained meso-DAP and, therefore, is related to the cell-wall types of groups II, III, and IV. It also contained arabinose and galactose, which indicates a relation to cell-wall type IV (Mycobacterium/Nocardia), but the presence of glycine points also toward cell-wall type II (Micromonospora). As in the Alnus endophyte and in most other Frankia species, the vesicle at the tip resembles closely the single terminal spore of Micromonospora. Finally, the cell-wall preparations of Alnus endophyte contained lysine and aspartic acid, which are typical for the mainly anaerobic or microaerophilic actinomycetes of cell-wall type VI, representing the Actinomyces bovis type. In view of the different combinations and lack of consistent relationship to a defined type, a special family Frankiaceae was prepared for the Frankia species. The presence of DAP in the endophyte of Casuarina cunninghamiana, C. equisetifolia and some other Casuarina species could not be demonstrated (Becking, unpublished), and Quispel (261) reported the same for the Myrica gale endophyte.

6.10.3 Isolation

Many attempts have been made to isolate the endophyte of nonleguminous plants, and there are many claims of the successful isolation of the endophyte starting with the observation of Von Plotho (135) for the Alnus endophyte and followed by Uemura (136), Danilewicz (137), and Pommer (138, 139). In Hippophaë rhamnoides root-nodule actinomycete isolations are reported by Niewiarowska (140), in Ceanothus velutinus by Wollum et al. (130), in Coriaria arborea by Allen et al. (131), and in Purshia tridentata by Webster et al. (64). However, Koch's postulates have never been fulfilled, and all reinfection tests of the endophyte on host plants were negative.

Quispel (141) first prepared from surface-sterilized Alnus root nodules a crushed nodule inoculum that remained free of contaminant growth after incubation in a favorable medium. Plants inoculated with this inoculum were grown aseptically, and no contaminants could be detected on the resulting nodulated plants in which N_2 fixation was still shown. In agreement with these findings, the author (142) was unable to grow the alder endophyte in pure culture, and he suggested that it was an obligate symbiont. As in Quispel's experiment, no outgrowth from externally sterilized and aseptically sliced root nodules was observed, when the tissue slices were placed on agar plates of the conventional bacteriological media. But these tissue slices still produced good nodulation when they were crushed and the suspension applied to alder seedlings grown aseptically in test tubes. Also no outgrowth of the endophyte was observed when the externally sterilized tissue parts were plated on agar plates of the complex media used for the in vitro culture of plant tissues (5, 142). This observation opened the possibility of cultivating parts of non-leguminous root nodules in plant tissue culture. In contrast, tissue cultures of leguminous root nodules containing the endophyte were not obtained until recently (143), since Rhizobium can grow outside the tissue in the medium and the plant tissue is then quickly overgrown by the bacterium.

In our experiments with alder root nodules small tissue fragments, 1-2 mm diameter, from externally sterilized root nodules were placed on a complex nutrient medium containing mineral nutrients, sucrose, and agar, and supplemented with coconut milk (150 ml/ 1), α-naphthalene-acetic acid (0.1 mg/l), and calcium panthotenate (2.5 mg/l) (142). These tissue fragments were incubated first for several weeks on this medium or on nutrient agar containing 0.7% Difco Bacto yeast extract to test for visible "sterility." Contamination was usually high, but some of the surface-sterilized tissue parts produced callus tissue which eventually grew into callus tissues of 4-5 cm in diameter (Plate 6.13, A & B). The callus tissue still contained the

Plate 6.13. Root-nodule callus tissue of Alnus glutinosa containing the endophyte growing in vitro on a complex organic medium used for the cultivation of plant tissue. (A) Young root-nodule explant. (B) The same root-nodule tissue after several months of growth. (x 0.9)

endophyte in the center, and when this tissue was macerated and applied to sterile alder seedlings, nodules were produced. Explants from the new callus tissue were made onto medium of the same composition.

A cytological study of the callus-producing fragments of nodules revealed that transmission of the endophyte from the original tissue to the newly formed callus tissue is difficult. In thin sections, a distinct transitional zone was observed between the original root-nodule tissue and the callus tissue (5). Only very rarely was the endophyte transmitted to the newly formed callus tissue turning most explants free of endophyte. In rare cases in which the endophyte was transmitted to the callus tissue, hyphal structures were observed in the plant cells with large vacuoles typical of callus tissue, but vesicle structures of the endophyte were never observed. The hyphal structures did not solely occur within the host tissue but also in the spaces between the host cells.

When callus tissue of explants containing the endophyte was ground and added to sterile alder seedlings in test tubes, nodulation usually did not occur, and within a few months these plants showed severe symptoms of nitrogen deficiency. In a few plants nodules developed but these nodules were ineffective (142, Fig. 4). A microscopic study of these nodules showed that the endophyte was present within the host cells, but neither vesicles nor bacteria-like endophyte cells were observed. It is interesting that ineffective root nodules occasionally produced by nodular material of normal plants show the same internal features. This observation supports the hypothesis that N_2 fixation is only associated with the vesicles.

Thus, although it was possible to cultivate the alder endophyte in root-nodule callus tissue and to propagate the endophyte in "pure culture," the endophyte lost the property of producing root-nodules effective in N_2 fixation. Also when alder root-nodule tissue cultures containing the endophyte growing in test tubes (Plate 6.13) were exposed to acetylene, no $N_2(C_2H_2)$ fixation could be detected.

Although the _Alnus_ endophyte like other actinomycetal endophytes of nonlegumes could not be obtained in pure culture on artificial media and only slight growth could be obtained within host tissue _in vitro_, this does not necessarily infer that the endophyte is an obligate symbiont. The alder endophyte can multiply in soil (262), and electron microscopy of the purified endophyte growing in association with sterile alder seedlings revealed that the microbe was able to grow in the mucilaginous layer of the root (263). It is evident that the microsymbiont can multiply outside the host plant when diffusable root substances or unidentified growth factors in the soil are present.

It is likely that when the nutritional requirements of the endophyte are fully understood it will be cultured like common bacteria on artificial media.

In contrast to the difficulty encountered with the actinomycetal endophytes, nodule squashes of Trema root nodules (see Plate 6.12, C, page 246) plated on yeast-mannitol agar commonly used for the isolation of Rhizobium yielded abundant growth. The isolated culture (NGR 231) was a gram negative, non-spore forming, motile rod measuring 1.7-3.5 x 0.7-0.9 µm. The flagellation was subpolar. All of its cultural features were typical of the "cowpea-type" Rhizobium, and the serological evidence supports this (218, 228).

6.11 ENDOPHYTE OF THE GUNNERA SYMBIOSIS

The endophyte of Gunnera species are blue-green algae of the genus Nostoc. Some authors (10, 11, 144) have isolated the endophytic alga, which is a Nostoc punctiforme capable of fixing dinitrogen in pure culture. Miehe (74) and Schaede (12) studied the penetration and the development of these algae within the host tissue. Miehe (76) noted that the algae were already present in the slime substance on the leaves and above special glands producing slime at the periphery of the stem. These glands consist of thin-walled host cells and are called "phycorrhizas." Via the younger stage of these so-called "phycorrhizas," the algae penetrate the stem passing through the intercellular spaces to a special tissue called "phycome." At this site the algae dissolve the middle lamellae of the host cells and penetrate into the host cytoplasm. According to Miehe (74) the "phycome" structure is a deformed root primordium. Once in the host cytoplasms the algae multiply rapidly and soon the whole host cell is completely filled with algal cells. Schaede (12) confirmed to a large extent Miehe's observations, but was of the opinion that the host-cell division in the "phycorrhiza," which resulted in the formation of many thin-walled cells, was induced or at least stimulated by the endophyte.

Baas-Becking (75) pointed out that the "phycome" structures or "wrats" in Gummera macrophylla are not abortive root primordia as Miehe supposed since roots always develop at the base of these "wrats." Moreover, these "phycomes" form outside the endodermis of the

Discussion and Conclusions

root, while roots always develop from the pericyclic part of the stele. "Phycomes" or "wrats" are, therefore, independent structures more or less functioning as specialized glands. Baas-Becking (75) suggested, on the basis of anatomical evidence, i.e., that the "algal cells" terminate close to the endodermis and the pericycle is only one or two cells thick allowing these parenchymatous cells to communicate with the sieve tubes, that the symbiont probably exerts a chemical influence. This root-producing substance might be a growth substance of the auxin type. Indeed, alcoholic as well as ether extracts of the "phycome" contained an auxin-like substance, while algae-free tissue did not. For this reason he proposed the tentative hypothesis that Gunnera is an auxoheterotroph and that the symbiosis with Nostoc is obligatory (75).

Plate 6.14, A shows Gunnera macrophylla in its natural habitat at Cibeureum, Mt. Gedeh, Java (1650 m), while Plate 6.14, B & C shows the stem of this Gunnera species with the "phycome" structures and the fanwise spreading of the algal colonies from the periphery. Most authors (11, 144, 145) regard the symbiotic Nostoc as similar to the free-living Nostoc punctiforme, but Reinke (9) pointed to a more specific relation and named it Nostoc gunnerae Reinke.

The general physiology of the Nostoc within the Gunnera tissue has been elucidated by Silvester (146). The endophytic algae of the Nostoc glands of G. albocarpa do not contain the photosynthetic pigment phycocyanin. Also tests for oxygen production by the endophytic algae were negative, and there was no inhibition of the light stimulated N_2 fixation by the application of DCMU (3,4-dichlorophenyl-1, 1-dimethylurea). These results suggest that the endophytic algae living within the Gunnera tissue are really symbiotically dependent on their host, since they have lost Photosystem II, responsible for the transfer of electrons from water and evolution of oxygen.

6.12 DISCUSSION AND CONCLUSIONS

The previous sections have discussed the physiology of the nonleguminous root-nodule symbioses as well as the algal symbiosis in Gunnera. These diazotrophs may contribute substantially to the nitrogen status of

Plate 6.14. Gunnera macrophylla Blume. (A) Natural vegetation of Gunnera macrophylla growing at the Cibeureum waterfalls (1650 m), Nature reserve Cibodas, Mt. Gedeh, W. Java, Indonesia. (x 0.07), (B) Stem of the plant showing the position of the Nostoc glands in relation to the petiole base (x 0.56), (C) Cross section through the stem showing the fan-like outgrowth of Nostoc from the periphery of the gland in the colorless parenchymatous host tissue. (x0.56)

soils (see also IV.4). Under field conditions (Table 6.9) an N_2 fixation rate of 60-100 kg N/ha·yr is reached, and some plant species like Casuarina equisetifolia and Alnus rubra may fix up to 200 kg N/ha·yr.

Compared to leguminous plants, the nonlegumes are a more diverse group with specializations and adaptations for extreme environmental conditions. Nonlegumes with N_2-fixing ability are found in all types of environmental conditions from near the Artic Circle (e.g., Dryas) to the tropics (Casuarina spp.). They

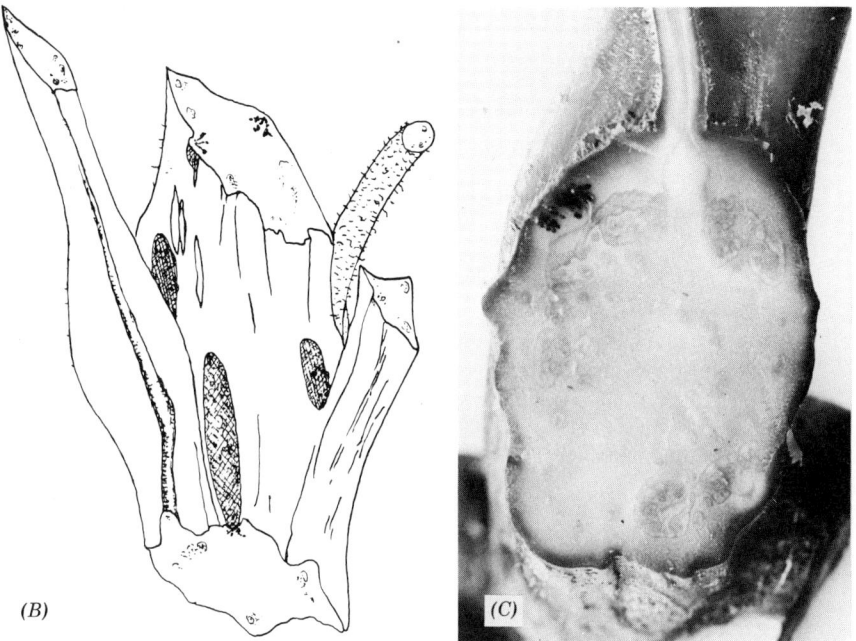

(B) (C)

occur in the temperate regions (Alnus), subtropical regions (Coriaria, Casuarina) as well as in the extremely wet tropical monsoon climate (Casuarina, and some Myrica species). On the other hand, they occur also in regions with temporary very dry conditions, being a prominent component of the chaparral vegetation (e.g., Ceanothus spp., Purshia spp., Cercocarpus spp.) of the northwest coast of North America extending from Canada to Mexico. Nonleguminous plants are more diverse in growth habit than the common legumes used in agriculture. Within this group are tree-like species, Alnus and Casuarina, shrub-like species, Ceanothus and some Myrica species, and some herbaceous species like Coriaria as well as real herbs Dryas. Likewise, growth may vary from the tree-like form as Alnus and Casuarina to the small prostate form as Ceanothus prostratus and Dryas spp., with an ultimate height of about 10 cm and dense mats on the

TABLE 6.9. Field Measurements of N_2 Fixation in Some Nonlegumes

Nonlegume	N_2 fixation (kg N/ha·yr)	Reference
Casuarina equisetifolia	229	63
Myrica cerifera	3.4	30
M. rubra (3-15 yr)	15-25[a]	22
Alnus crispa (50 yr old)	61.5	16
A. crispa	40.0	211
A. crispa (5 yr old)	157	37, 163
A. glutinosa (0-7 yr)		
(1 plant/m^2)	28	148
(5 plants/m^2)	100	149
A. glutinosa (12 yr old)	26-48	212
A. glutinosa	58	69
A. incana	40	213
A. rubra	139	156, 157
A. rubra	up to 300	214, 273
A. rugosa	193	215
A. rugosa (natural stand)	85	216
Hippophaë rhamnoides		
(0-3 yr old)	27	65
(13-16 yr old)	179	
H. rhamnoides		
(1-2 yr old)	2	69
(10-15 yr old)	15	
Ceanothus spp. (natural shrub community)	60	61
Dryas drummondii	18-36	19
D. drummondii and some Shepherdia canadensis	61.5	217
Gunnera dentata	72	15
G. macrophylla	14-25	41, 42

[a] Data from comparison of Myrica-pine stands and pure pine stands; N_2 fixation determined by subtracting total amount N pine stand from total amount N Myrica-pine stand.

Discussion and Conclusions 263

ground. Some species are adapted to specialized ecological sites like Myrica cordifolia, which stabilizes loose sand in sand dunes. Others like Gunnera species are very moisture-loving, occurring preferentially in marshes or places of seepage.

This large variation in tolerance to ecological niches make nonlegumes very suitable for soil improvement in areas where legumes cannot be used. These possibilities of soil enrichment and revegetation of extreme ecological habitats should be explored and adaptation of nonlegumes to these ecological patterns employed. Tree-like species like Alnus can be exploited in forestry as legumes are in agriculture; this position of Alnus is already recognized for Alnus glutinosa and A. incana in Europe (147-152), where Alnus has been grown in association or interplanted with Pinus sylvestris, Picea abies, or Populus species. The same relation of Alnus rubra with conifers is practiced at the northwest Pacific coast of North America (25, 26, 153-157). Also many indigenous Alnus species have this role in Japan as shown by studies of Uemura (22). In tropical countries Casuarina species are used for the same purpose (126, 134, 158-162).

In contrast to most legumes, many nonlegumes can be used for soil improvement, restoration of vegetation, and colonization of habitats of nonagricultural and nonforestry soils. In fact, in nature these nonlegumes already have this function in plant colonization and vegetation development. At extreme ecological sites such as raw mineral deposits exposed after glacier recession Alnus crispa subsp. sinuata and Dryas spp. are the first plants to colonize the site (16, 31, 163). They form the natural chain in the chronosequence of vegetation adding 60-150 kg N/ha·yr, which makes possible the development of other vegetation. Alnus sieboldiana fulfills the same function in primary succession on volcanic deposits in Japan (164), as well as Myrica javanica, which in Indonesia initiates the establishment of vegetation on lava flows (17).

Therefore, many nonlegumes can be successfully used to revegetate eroded areas, landslides, or bare soil including even the most compact and hard soil unsuitable for growth of other plants. The European montane species Alnus viridis has been introduced by Benecke (165) to reforest eroded soils at high elevations in New Zealand. Although Alnus species do not occur in New Zealand, it was not necessary to introduce the Alnus endophyte (166). This is rather curious because of the high degree of species specificity or adaptation of endophytes to certain host species.

In Europe black wastes from anthracite mining are now regularly colonized by Alnus species, and in the United States the same has been done with Comptonia asplenifolia (28). Also city refuse wastes are now revegetated by Alnus (167). Therefore nonleguminous plants are useful not only for the enrichment of soil with nitrogen, but also in the colonization of bare soil of devasted areas. Such a role can never be occupied by leguminous plants, because their ecological niche is much narrower and their nutrient requirement for elements other than nitrogen is much higher. A detailed discussion of the ecological aspects of these symbioses appears elsewhere in this volume (IV.4).

6.13 REFERENCES

1. Allen, E. K., O. N. Allen, and L. J. Klebesadel, Proc. Alaskan Sci. Conf., 14th, Anchorage, Alaska, 54 (1964).
2. Rayner, M. C., Mycorrhiza, New Phytologist No. 15, Wheldon and Wesley Ltd., London, 1927.
3. Engler, A., in Syllabus der Pflanzenfamilien, 12th ed., H. Melchior, Ed., Berlin, Vol. 2, 1964.
4. Ziegler, H., Mitt. Deutsch. Dendrol. Ges., 61, Jahrb. 1959/60, 28 (1962).
5. Becking, J. H., Ann. Inst. Pasteur, (Suppl.), 111, 211 (1966).
6. Becking, J. H., Proc. Symp. Nitrogen in Soil, Groningen, 1967; Nitrogen, Dutch Nitrogenous Fert. Rev., 12, 47 (1968).
7. Becking, J. H., Plant Soil, 32, 611 (1970).
8. Hutchinson, J., The Families of Flowering Plants, 2nd ed., Clarendon, Oxford, Vol. 1, Dicotyledons, 1959.
9. Reinke, J., Morphologische Abhandlungen, Verlag Wilhelm Engelmann, Leipzig, 1873.
10. Harder, R., Z. Bot., 9, 145 (1917).
11. Winter, A. G., Beitr. Biol. Pflanz., 23, 295 (1935).
12. Schaede, R., Planta, 39, 154 (1951).

References

13. Willis, J. C., A Dictionary of the Flowering Plants and Ferns, Revised by H. A. K. Airy-Shaw, 8th ed., Cambridge University Press, London, 1973.
14. Batham, E. J., Trans. Proc. Roy. Soc. New Zealand, 73, 209 (1943).
15. Silvester, W. B., and D. R. Smith, Nature (London), 224, 1231 (1969).
16. Crocker, R. L., and J. Major, J. Ecol., 43, 427 (1955).
17. Van der Pijl, L., Ann. Jard. Bot. Buitenzorg, 48, 129 (1938).
18. Bond, G., and P. Montserrat, Nature, 182, 474 (1958).
19. Lawrence, D. F., R. E. Schoenike, A. Quispel, and G. Bond, J. Ecol., 55, 793 (1967).
20. Rodriguez-Barrueco, C., Bot J. Linn. Soc., 62, 77 (1969).
21. Bond, G., in Nitrogen Nutrition of the Plant, Symposium 1967, E. A. Kirkby, Ed., Univ. of Leeds, pp. 1-8, 1970.
22. Uemura, S., Plant Soil, Special Vol., 349 (1971).
23. Wilson, P. W., in Handbuch der Pflanzenphysiologie, W. Ruhland, Ed., Springer Verlag, Berlin, Vol. 8, pp. 9-47, 1958.
24. Moore, A. W., Can. J. Bot., 42, 952 (1964).
25. Tarrant, R. F., L. A. Isaac, and R. F. Chandler, J. Forestry, 49, 914 (1951).
26. Tarrant, R. F., Forest Sci., 7, 238 (1961).
27. Goldman, C. R., Ecology, 42, 282 (1961).
28. Schramm, J. R., Trans. Am. Phil. Soc. N. S., 56, Part I (1966).
29. Silver, W. S., J. Bacteriol., 87, 416 (1964).
30. Silver, W. S., and T. Mague, Nature (London), 227, 378 (1970).
31. Lawrence, D. B., Development of Vegetation and Soil in Southeastern Alaska with Special Reference to the Accumulation of Nitrogen, Final Rep., Off. Naval Res. Project No. 160-183, Washington (1953).
32. Schoenike, R. E., Proc. Minn. Acad. Sci., 25 and 26, 55 (1957-1958).
33. Rodriguez-Barrueco, C., Phyton (Buenos Aires), 23, 103 (1966).
34. Dommergues, Y., Agrochimica, 7, 335 (1963).
35. Jansen van Ryssen, F. W., and N. Grobbelaar, S. Afr. J. Sci., 66, 22 (1970).
36. Bond, G., Ann. Rev. Pl. Physiol., 18, 107 (1967).
37. Bond, G., Plant Soil, Special Vol., 317 (1971).

38. Grobbelaar, N., J. M. Strauss, and E. G. Groenewald, Plant Soil, Special Vol., 325 (1971).
39. Killick, D. J. B., Bothalia, 10, 5 (1969).
40. Van Steenis, C. G. G. J., The Mountain Flora of Java, E. J. Brill, Leiden, 1972.
41. Becking, J. H., Nitrogen Fixation in Rice Soils and Some Natural Ecosystems in Indonesia, Rep., Netherlands Foundation for the Advancement of Tropical Research (WOTRO), 1972.
42. Becking, J. H., Fortschr. Bot., 1, 377 (1973).
43. Good, R. D. O., New Phytol., 29, 170 (1930).
44. Burke, W. D., N. Z. J. Bot., 1, 377 (1963).
45. Morrison, T. M., Nature (London), 189, 945 (1961).
46. Oliver, W. R. B., Rec., Dominion Mus. Wellington, 1, 21 (1942).
47. Gross, H., Quartär, 9, 3 (1957) (transl., H. E. Wright, 1958).
48. Baker, R. G., Bull. Geol. Soc. Am., 76, 601 (1965).
49. Watts, W. A., Proc. 7th Congr. Int. Ass. Sedimentary Res., 7, 89 (1967).
50. Bond, G., in Symbiotic Associations, P. S. Nutman, and B. Mosse, Eds., Cambridge University Press, London, pp. 72-91, 1963.
51. Hewitt, E. J., Sand and Water Culture Methods, 2nd Commonwealth Agr. Bur. (1966).
52. Hoagland, D. R., and W. C. Snyder, Proc. Am. Soc. Hort. Sci., 30, (1933).
53. Russell, S. A., and H. J. Evans, Forest. Sci., 12, 164 (1966).
54. Aprison, M. H., and R. H. Burris, Science, 115, 264 (1952).
55. Aprison, M. H., W. E. Magee, and R. H. Burris, J. Biol. Chem., 208, 29 (1954).
56. Magee, W. E., and R. H. Burris, Plant Physiol., 29, 199 (1954).
57. Bond, G., J. Exp. Bot., 6, 303 (1955).
58. Bond, G., Ann. Bot., N. S., 21, 513 (1957).
59. Sloger, C., and W. S. Silver, in Non-heme Iron Proteins: Role in Energy Conversion, A. San Pietro, Ed., Antioch Press, Yellow Springs, Ohio, pp. 299-302, 1965.
60. Ziegler, H., and R. Hüser, Nature (London), 199, 508 (1963).
61. Delwiche, C. C., P. J. Zinke, and C. M. Johnson, Plant Physiol., Lancaster, 40, 1045 (1965).
62. Harris, G. P., and T. M. Morrison, Nature (London), 182, 1812 (1958).

63. Stevenson, G., Nature (London), 182, 1523 (1958).
64. Webster, S. R., C. T. Youngberg, and A. G. Wollum II, Nature (London), 216, 392 (1967).
65. Stewart, W. D. P., and M. Pearson, Plant Soil, 26, 348 (1967).
66. Sloger, C., and W. S. Silver, Bact. Proc., 64 (1967).
67. Sloger, C., Nitrogen Fixation by Tissues of Leguminous and Non-leguminous Plants, Ph. D. Thesis, University of Florida (1968).
68. Russell, S. A., and H. J. Evans, Assessment of Nitrogen Fixation by Alnus rubra by use of the Acetylene Reduction Technique, Rep., Pac. N. W. Forest Range Exp. Sta. (1970).
69. Akkermans, A. D. L., Nitrogen Fixation and Nodulation of Alnus and Hippophaë under Natural Conditions, Ph. D. Thesis, Leiden (1971).
70. Hardy, R. W. F., R. D. Holsten, E. K. Jackson, and R. C. Burns, Plant Physiol., 43, 1185 (1968).
71. Koch, B., H. J. Evans, and S. Russell, Plant Physiol., Lancaster, 42, 466, (1967).
72. Treub, M., Med. Kruidk, Arch., 2^e Ser., 3, 404 (1882).
73. Merker, P., Flora (Marburg) N. R., 47, 211 (1889).
74. Miehe, H., Flora (Jena), N. F., 17, 1 (1924).
75. Baas-Becking, L. G. M., Biol. Jaarb. Dodonaea, 14, 93 (1947).
76. Stewart, W. D. P., G. P. Fitzgerald, and R. H. Burris, Arch. Mikrobiol., 62, 336 (1968).
77. Leaf, G., I. C. Gardner, and G. Bond, J. Exp. Bot., 9, 320 (1958).
78. Leaf, G., I. C. Gardner, and G. Bond, Biochem. J., 72, 662 (1959).
79. Wheeler, C. T., and G. Bond, Phytochem., 9, 705 (1970).
80. Wheeler, C. T., New Phytol., 68, 675 (1969).
81. Wheeler, C. T., New Phytol., 70, 487 (1971).
82. Bond, G., J. Exp. Bot., 7, 387 (1956).
83. Stewart, W. D. P., J. Exp. Bot., 13, 250 (1962).
84. Becking, J. H., Nature (London), 192, 1204 (1961).
85. Becking, J. H., Plant Soil, 15, 217 (1961).
86. Hewitt, E. J., and G. Bond, Plant Soil, 14, 159 (1961).
87. Bond, G., and E. J. Hewitt, Nature (London), 190, 1033 (1961).
88. Bond, G., and E. J. Hewitt, Nature (London), 195, 94 (1962).

89. Hewitt, E. J., and G. Bond, *J. Exp. Bot.*, **17**, 480 (1966).
90. Bond, G., J. F. Adams, and E. H. Kennedy, *Nature (London)*, **207**, 319 (1965).
91. Kliewer, M., and H. J. Evans, *Nature (London)*, **194**, 108 (1962).
92. Bergersen, F. J., R. H. Burris, and P. W. Wilson, in *Recent Advances in Botany, Lectures and Symposia, IX Int. Botan. Congr.*, D. L. Bailey, Ed., Montreal 1959, University of Toronto Press, Canada, pp. 589-593, 1961.
93. Mortenson, L. E., in *The Bacteria*, I. C. Gunsalus and R. Y. Stanier, Eds., Academic Press, New York, Vol. 3, pp. 119-166, 1962.
94. Burris, R. H., W. E. Magee, and M. K. Bach, *Ann. Acad. Sci. Fenn.*, **60**, 190 (1955).
95. Bond, G., in *Utilization of Nitrogen and its Compounds by Plants, 13th Symposium Soc. Exp. Biol.*, H. K. Porter, Ed., Cambridge University Press, London, pp. 59-72, 1959.
96. MacConnell, J. T., *Ann. Bot. (London)*, N. S., **23**, 261 (1959).
97. Bond, G., *Z. Allg. Mikrobiol.*, **1**, 93 (1961).
98. Bond, G., *J. Exp. Bot.*, **11**, 91 (1960).
99. Hoch, G. E., H. N. Little, and R. H. Burris, *Nature (London)*, **179**, 430 (1957).
100. Bergersen, F. J., *Bact. Rev.*, **24**, 246 (1960).
101. Smith, J. D., *Biochem. J.*, **44**, 585 (1949).
102. Egle, K., and H. Munding, *Naturwiss.*, **38**, 548 (1951).
103. Bond, G., *Ann. Bot. (London)*, N. S., **15**, 447 (1951).
104. Bond, G., in *Nutrition of the Legumes, Proc. Univ. Nottingham*, E. G. Hallsworth, Ed., 5th Easter School Agr. Sci., Butterworths, London, pp. 216-231, 1958.
105. Davenport, H. E., *Nature*, **186**, 653 (1960).
106. Thornton, H. G., *Rep. Rothamsted Exp. Sta.*, 1939-45, 74 (1946).
107. Schlegel, H. G., *Flora*, **152**, 236 (1962).
108. Becking, J. H., W. E. de Boer, and A. L. Houwink, *Antonie van Leeuwenhoek, J. Microbiol. Seriol.*, **30**, 343 (1964).
109. Mowry, H., *Soil Sci.*, **36**, 409 (1933).
110. Silver, W. S., F. E. Bendana, and R. D. Powell, *Physiol. Plant.*, **19**, 207 (1966).
111. Fåhraeus, G., *J. Gen. Microbiol.*, **16**, 374 (1957).

112. Kefford, N. P., J. Brockwell, and J. A. Zwar, Aust. J. Biol. Sci., 13, 456 (1960).
113. Sahlman, K., and G. Fåhraeus, Kgl. Lantbruks-Högskol. Ann., 28, 261 (1962).
114. Wipf, L., Bot. Gaz., 101, 51 (1939).
115. Wipf, L., and D. C. Cooper, Proc. Nat. Acad. Sci. U. S., 24, 87 (1938).
116. Wipf, L., and D. C. Cooper, Am. J. Bot., 27, 821 (1940).
117. Trolldenier, G., Arch. Mikrobiol., 32, 328 (1959).
118. Wetzel, G., Bot. Arch., 25, 258 (1929).
119. Woodworth, R. H., J. Arnold Arboretum (Harvard Univ.), 12, 206 (1931).
120. Darlington, C. D., and A. P. Wylie, Chromosome Atlas of Flowering Plants, 2nd ed., George Allen and Unwin Ltd., London, 519 pp., 1961.
121. Giesbrecht, P., Zentr. Bakteriol. Parasitenk. 1. Abt. Orig., 179, 538 (1960).
122. Imaeda, T., and M. Ogura, J. Bacteriol., 85, 150 (1963).
123. Becking, J. H., Int. J. Syst. Bacteriol., 20, 201 (1970).
124. Bond, G., Nature (London), 193, 1103 (1962).
125. Mackintosh, A. H., and G. Bond, Phyton (Buenos Aires), 27, 79 (1970).
126. McLuckie, J., Proc. Linn. Soc. N. S. W., 48, 194 (1923).
127. Schaede, R., Planta, 19, 389 (1933).
128. Hawker, L. E., and J. Fraymouth, J. Gen. Microbiol., 5, 369 (1951).
129. Furman, T. E., Am. J. Bot., 46, 698 (1959).
130. Wollum, A. G. II, C. T. Youngberg, and C. M. Gilmour, Soil Sci. Soc. Am., Proc., 30, 463 (1966).
131. Allen, J. D., W. B. Silvester, and M. Kalin, N. Z. J. Bot., 4, 57 (1966).
132. Hoeppel, R. E., and A. G. Wollum II, Can. J. Bot., 49, 1315 (1971).
133. Krebill, R. G., and J. M. Muir, Northwest Science, 48, 266 (1974).
134. Lechevalier, M. P., N. N. Gerber, H. Lechevalier, M. Higgins, and J. Hegyi, The Separation of Actinomycetes into Genera by Morphological and Chemical Methods, Actinomycetes Workshop, Am. Soc. Microbiol., New Brunswick, N. J. (1967).
135. Von Plotho, O., Arch. Mikrobiol., 12, 1 (1941).
136. Uemura, S., Rep. Govt. Forest Exp. Sta., Tokyo, 62, 41 (1952).

137. Danilewicz, K., Acta Microbiol. Polon., 14, 321 (1965).
138. Pommer, E. H., Flora (Jena), 143, 603 (1956).
139. Pommer, E. H., Ber. Deut. Bot. Ges., 72, 138 (1959).
140. Niewiarowska, J., Acta Microbiol. Polon., 10, 271 (1961).
141. Quispel, A., Acta Bot. Neerland., 3, 495 (1954).
142. Becking, J. H., Nature (London), 207, 885 (1965).
143. Holsten, R. D., R. C. Burns, R. W. F. Hardy, and R. R. Hebert, Nature (London), 232, 173 (1971).
144. Jönsson, B., Bot. Notis., 1 (1894).
145. Pringsheim, E. G., Arch. Protistenk., 38, 126 (1918).
146. Silvester, W. B., in Nitrogen Fixation and the Biosphere, Int. Synth. Meet. I. B. P., Sect. PP-N, Edinburgh 1973, P. S. Nutman, Ed., Cambridge University Press, England, pp. 521-538, 1976.
147. Virtanen, A. I., Acta Chem. Fenn. B., 6, 57 (1936).
148. Virtanen, A. I., Physiol. Plant., 10, 164 (1957).
149. Virtanen, A. I., Commun. Inst. Forestry Fenn. 55.22, 1 (1962).
150. Kohnke, H., J. Forestry, 39, 333 (1941).
151. Stone, E. L., Proc. Nat. Joint Comm. Fert. Appl., 31, 81 (1955).
152. Van der Meiden, H. A., Handboek voor de Populierenteelt (Handbook for Poplar Culture), 3rd ed., Konink. Ned. Heidemaatschappij, Arnhem, Holland, 1960.
153. Johnson, H. M., J. Forestry, 15, 981 (1917).
154. Worthington, N. P., R. H. Ruth, and E. E. Matson, U. S. For. Serv., Misc. Publ., 881, (1962).
155. Bollen, W. B., C. S. Chen, K. C. Lu, and R. F. Tarrant, Res. Bull., No. 12, Forest Res. Lab. School of Forestry, Oregon State University, Corvallis (1967).
156. Tarrant, R. F., K. C. Lu, W. B. Bollen, and J. F. Franklin, Pacific North-West Forest Range Exp. Sta. Res. Paper PNW-76, 1969).
157. Tarrant, R. F., and J. M. Trappe, Plant Soil, Special Vol., 335 (1971).
158. Rao, K. A., Yearbook 1923, Madras Agr. Dep., 60 (1924).
159. Aldrich-Blake, R. N., Oxf. Forestry Mem. No. 14 (1932).
160. Parker, R. N., Indian Forestry, 58, 362 (1932).
161. Jagoe, R. B., Malayan Agr. J., 32, 77 (1949).
162. Puri, G. S., Indian Forestry, 84, 74 (1958).

163. Lawrence, D. B., Am. Sci., 46, 89 (1958).
164. Tezuka, Y., Jap. J. Bot., 17, 371 (1961).
165. Benecke, U., Plant Soil, 33, 30 (1970).
166. Benecke, U., Plant Soil, 30, 145 (1969).
167. Guldemond, J. L., Tijdschr. Nat. Raad Landb. Onderzoek T. N. O., 2, (1972).
168. Hooker, J. D. and B. D. Jackson, Index Kewensis, 1, Oxford (1893).
169. Kamerling, E., Ned. Tijdschr. Natuurk., 71, 73 (1915).
170. Shibata, K., and M. Tahara, Bot. Mag., Tokyo, 31, 157 (1917).
171. Uemura, S., Bull. Govt. Forest Exp. Stat. Meguro, 167, 59 (1964).
172. Janse, J. M., Ann. Jard. Bot. Buitenzorg, 14, 53 (1897).
173. Allen, E. K., and O. N. Allen in Microbiology and Soil Fertility, G. M. Gilmour and O. N. Allen, Eds., 25th Ann. Biol. Coll., Corvallis, Oregon, pp. 77-106, 1965.
174. Narasimhan, M. J., Indian Forestry, 44, 265 (1918).
175. Youngken, H. W., Am. J. Pharm., 87, 391 (1915).
176. Chevalier, A., Mém. Soc. Nat. Sci. Nat. Math., Cherbourg, 32, 85 (1900-1902).
177. Brunchorst, J., Bot. Inst., Tübingen, Untersuch., 2, 151 (1886-1888).
178. Shibata, K., Jahrb. Wiss. Bot., 37, 643 (1902).
179. Arzberger, E. G., Missouri Bot. Gard. 21st Ann. Rep., 60 (1910).
180. Roberg, M., Jahrb. Wiss. Bot., 86, 344 (1938).
181. Kellerman, K. F., Yearbook U. S. Dep. Agr., 213 (1911).
182. Wilson, E. H., J. Arnold Arboretum, 2, 25 (1920).
183. Rabotnov, T. A., and Y. A. Mednis, Priroda, Mosk., 6, 94 (1936).
184. Meyen, J., Flora (Jena), 12, 49 (1829).
185. Woronin, M., Mem. Acad. Imp. Sci. St. Petersbourg, Ser. VII, 10 (1866).
186. Karavayev, M. N., Bot. Zh. SSSR, 44, 100 (1959).
187. Castellanos, A., Lilloa, 10, 413 (1944).
188. Bushnell, O. A., and W. B. Sarles, Soil Sci., 44, 409 (1937).
189. Nobbe, F., E. Schmid, L. Hiltner, and E. Hotter, Landw. Versuchs-Sta., 41, 138 (1892).
190. Spratt, E. R., Ann. Bot., 26, 119 (1912).
191. Milovidov, P. F., Zentbl. Bakt. Parasitkde., II Abt., 73, 58 (1928).

192. Fred, E. B., I. L. Baldwin, and E. McCoy, Root Nodule Bacteria and Leguminous Plants., University of Wisconsin Stud. Sci., 5 (1932).
193. Roberg, M., Jahrb. Wiss. Bot., 79, 472 (1933-1934).
194. Oersted, A. S., (reported by E. Warming, 1876), Bot. Tidsskr., Ser. 3, 1, 108 (1865).
195. Warren, J. A., U. S. Dep. Agr. Bur. Pl. Indus. Circ., 70 (1910).
196. Nobbe, F., and L. Hiltner, Naturwissen. Z. Land-u. Forstw., 2, 366 (1904).
197. Beal, W. J., Bot. Gaz., 15, 232 (1890).
198. Quick, C. R., J. Forestry, 42, 827 (1944).
199. Hellmers, H., J. S. Horton, G. Juhren, and J. O'Keefe, Ecology, 36, 667 (1955).
200. Jepson, W. L., A Flora of California, No. 2, Berkeley, California (1936).
201. Sarauw, G. F. L., Bot. Tidsskr., 18, 127 (1893).
202. Gasparrini, G., Atti Accad. Sci. Fis. Mat., Napoli, 6, 221 (1851).
203. Hellmers, H., and J. M. Kelleher, Forest Sci., 5, 275 (1959).
204. Morrison, T. M., and G. P. Harris, Nature (London), 182, 1746 (1958).
205. Morrison, T. M., and G. P. Harris, Proc. N. Z. Ecol. Soc., 6, 23 (1959).
206. Wagle, R. F., and J. Vlamis, Ecology, 42, 745 (1961).
207. Vlamis, J., A. M. Schultz, and H. H. Biswell, J. Range Mgmt., 17, 73 (1964).
208. Schnegg, H., Flora, 95, 161 (1902).
209. Reinke, J., Nachr. K. Gesell. Wiss., No. 4, Georg-Augusts-Universität, Göttingen (1872).
210. Mattfeld, J., Ostenia, 102 (1933).
211. Crocker, R. L., and B. A. Dickson, J. Ecol., 45, 169 (1957).
212. Delver, P., and A. Post, Plant Soil, 28, 325 (1968).
213. Ovington, J. D., J. Ecol., 44, 171 (1956).
214. Zavitkovski, J., and M. Newton, Plant Soil, 35, 257 (1971).
215. Daly, G. T., Can. J. Bot., 44, 1607 (1966).
216. Voigt, G. K., and G. L. Steucek, Soil Sci. Soc. Amer. Proc., 33, 946 (1969).
217. Lawrence, D. B., and L. Hulbert, Bull. Ecol. Soc. Amer., 31, 58 (1950).
218. Trinick, M. J., Nature (London), 244, 459 (1973).

References

219. Sabet, Y. S., Nature (London), 157, 656 (1946).
220. Mostafa, M. A. and M. Z. Mahmoud, Nature (London), 167, 446 (1951).
221. Allen, E. K., and O. N. Allen, Proc. Soil Sci. Soc. Amer., 14, 179 (1950).
222. Farnworth, R. B. and M. A. Clawson, Agron. Abstr., 96 (1972).
223. Clawson, M. A., R. B. Farnworth, and M. Hammond, Agron. Abstr., 138 (1972).
224. Hutchinson, J., The Genera of Flowering Plants, Vol. 1 and 2 Dicotyledones, Clarendon, Oxford (1964).
225. Becking, J. H., in The Biology of Nitrogen Fixation, A. Quispel, Ed., North-Holland Publishing Co., Amsterdam, pp. 583-613, 1974.
226. Miguel, C., and C. Rodriguez-Barrueco, Plant Soil, 41, 521 (1974).
227. Bond, G., in Nitrogen Fixation and the Biosphere, Intern. Synthesis Meet., I. B. P., Sect. PP-N, Edinburgh 1973, P. S. Nutman, Ed., Cambridge University Press, England, pp. 443-474, 1976.
228. Trinick, M. J., in Internal Symposium on N_2 Fixation, W. E. Newton and C. J. Nyman, Eds., Pullman, Washington, June 1974, pp. 507-517, 1976.
229. Becking, J. H. in Proceedings IVth Intern. Conference Global Impacts of Applied Microbiology (GIAM IV), São Paulo, Brazil, July 1973 (1976).
230. Thimann, K. V., Proc. Natl. Acad. Sci. U. S., 22, 511 (1936).
231. Thimann, K. V., Trans. 3rd Comm. Intern. Soc. Soil Sci. A, 24 (1939).
232. Chen, H. K., Nature (London), 142, 753 (1938).
233. Dullaart, J., J. Exp. Botan., 21, 975 (1970).
234. Dullaart, J., Acta Botan. Neerl., 16, 222 (1967).
235. Clarke, G., M. H. Dye, and R. L. Wain, Nature (London), 184, 825 (1959).
236. Rovira, A. D., Soils and Fertilizers, 15, 167 (1962).
237. Peters, R. J. and M. Alexander, Soil Sci., 102 380 (1966).
238. Seidel, K., Naturwiss., 59, 366 (1972).
239. Thurber, G. A., J. R. Douglas, and A. W. Galston, Nature (London), 181, 1082 (1958).
240. Fletcher, W. W., J. W. S. Alcorn, and J. C. Raymond, Nature (London), 184, 1576 (1959).
241. Arora, N., F. Skoog, and O. N. Allen, Proc. Soil Sci. Soc. Amer., 46, 610 (1959).

242. Phillips, D. A., and J. G. Torrey, Physiol. Plant., 23, 1057 (1970).
243. Phillips, D. A. and J. G. Torrey, Pl. Physiol. (Lancaster), 49, 11 (1972).
244. Short, K. C., and J. G. Torrey, Pl. Physiol. (Lancaster), 49, 155 (1972).
245. Short, K. C., and J. G. Torrey, J. Exp. Bot., 23, 1099 (1972).
246. Libbenga, K. R. and J. G. Torrey, Amer. J. Bot., 60, 293 (1973).
247. Mothes, K., L. Engelbrecht, and O. Kulajewa, Flora, 147, 445 (1959).
248. Mothes, K., Naturwiss, 47, 337 (1960).
249. Engelbrecht, L., and K. Conrad, Ber. Deut. Botan. Ges., Sitzungsber., 27 Januar 1961, 64, 42 (1961).
250. Engelbrecht, L., and K. Conrad, Ber. Deut. Botan. Ges., Sitzungsber, 24 Mai 1961, 74, 42 (1961).
251. Conrad, K., Flora, 151, 345 (1961).
252. Mothes, K., in Régulateurs Naturels de la Croissance Végétale, Centre National de la Recherche Scientifique, Paris, pp. 131-140, 1964.
253. Srivastava, B. I. S., Int. Rev. Cytol., 22, 349 (1967).
254. Rodriguez-Barrueco, C., and F. Bermudez de Castro, Physiol. Plant., 29, 277 (1973).
255. Nutman, P. S., Biol. Rev., 31, 109 (1956).
256. Fåhraeus, G., and H. Ljunggren, in The Ecology of Soil Bacteria, T. R. G. Gray and D. Parkinson, Eds., Liverpool University Press, pp. 396-421, 1968.
257. Appleby, C. A., Comment at International Symposium on Nitrogen Fixation, Pullman, Washington, June 1974, and in Proc., W. E. Newton and C. J. Nyman, Eds., Washinton State Univ. Press, pp. 274-292, 1976.
258. Kodama, A., Bot. Mag. (Tokyo), 80, 230 (1967).
259. Kodama. A., J. Sci. Hiroshima Univ. Ser. B., 13, 261 (1970).
260. Rodriguez-Barrueco, C., and G. Bond, in Nitrogen Fixation and the Biosphere, Int. Synth. Meet. I. I. B. P., Sect. PP-N, Edinburgh, September 1973, P. S. Nutman, Ed., Cambridge Univ. Press, pp. 561-565, 1975.
261. Quispel, A. in The Biology of Nitrogen Fixation, A. Quispel, Ed., North-Holland Publ., Co., Amsterdam, pp. 499-520, 1974.
262. Rossi, S., Ann. Inst. Pasteur, 106, 505 (1964).

References

263. Becking, J. H., in *International Symposium on Nitrogen Fixation*, Pullman, Washington, June 1974, W. E. Newton and C. J. Nyman, Eds., Washington State Univ. Press, pp. 581-591, 1976.
264. Holsten, R. D., R. C. Burns, R. W. F. Hardy, and R. R. Hebert, *Nature (London)*, 232, 173 (1971).
265. Child, J. J. and T. A. LaRue, *Pl. Physiol. (Lancaster)*, 53, 88 (1974).
266. Child, J. J., *Nature (London)*, 253, 350 (1975).
267. Hall, W. C., G. B. Truchelut, C. L. Leinweber, and F. A. Herrero, *Physiol. Plant.*, 10, 306 (1957).
268. Radin, J. W., and R. S. Loomis, *Pl. Physiol. (Lancaster)*, 44, 1584 (1969).
269. Reid, M. S., and H. K. Pratt, *Nature (London)*, 226, 976 (1970).
270. LaRue, T. A. G., and O. L. Gamborg, *Pl. Physiol. (Lancaster)*, 48, 394 (1971).
271. Gamborg, O. L., and T. A. G. LaRue, *Pl. Physiol. (Lancaster)*, 48, 399 (1971).
272. Klevenskaya, I. L., in *The Recent Achievements in the Study of Biological Nitrogen Fixation*, E. N. Mishustin, Ed., Publ. House "Nauka," Moscow, 1971.
273. Zavitkovski, J., and M. Newton, *Plant Soil*, 35, 257 (1971).
274. Hewitt, E. J., and D. G. Hallas, *Plant Soil*, 3, 366 (1951).
275. Becking, J. H., in *Nitrogen Fixation and the Biosphere*, International Synthesis Meeting, I. B. P., Section PP-N, Edinburgh, September 1973, P. S. Nutman, Ed., Cambridge Univ. Press, pp. 539-550, 1976.
276. Becking, J. H., in *International Symposium on Nitrogen Fixation*, Pullman, Washington, June 1974, W. E. Newton and C. J. Nyman, Eds., Washington State Univ. Press, pp. 556-580, 1976.
277. Schaede, R., *Die Pflanzlichen Symbiosen.*, 3rd ed., Gustav Fischer Verlag, Stuttgart, 1962.

CHAPTER 7

Rhizobium:
General Microbiology

J. M. VINCENT

Department of Microbiology
University of Sydney
Sydney, Australia

7.1. Introduction, 280
    7.1.1. General Description, 280
    7.1.2. The Meaning of "Strain," 280
    7.1.3. Types, 281
    7.1.4. Specific Designations, 281
7.2. Morphology and Cytology, 281
    7.2.1. The Cultivated Cell, 281
    7.2.2. Comparison between Cultured Rhizobia and Bacteroids, 284
    7.2.3. Life Cycles and Endospores, 285
    7.2.4. The Cell Surface, 286
7.3. Cultural Characteristics, 286
    7.3.1. Growth on Solid Media, 286
    7.3.2. Indicators, 287
    7.3.3. Colonial Variants, 287
    7.3.4. Growth on Other Media, 288
    7.3.5. The "Autoplaque" Phenomenon, 288
    7.3.6. Sensitivity of Rhizobia, 289
7.4. Metabolic Requirements, 289
    7.4.1. Nutrition, 289
        7.4.1.1. Carbon Source, 289
        7.4.1.2. Nitrogen Source, 291
        7.4.1.3. Vitamin Requirements, 291
        7.4.1.4. Mineral Elements, 292
    7.4.2. Other Metabolic Requirements, 294
        7.4.2.1. Oxygen, 294
        7.4.2.2. Hydrogen Ions, 295
        7.4.2.3. Temperature, 295

7.5. Metabolic Activities, 295
    7.5.1. Enzymes and Processes, 295
        7.5.1.1. Carbon Compounds, 295
        7.5.1.2. Nitrogen, 299
        7.5.1.3. Other Enzymes, 300
        7.5.1.4. Comparison of Vegetative Cells and Bacteroids, 301
    7.5.2. Products, 301
        7.5.2.1. Protons and Hydroxyl Ions, 301
        7.5.2.2. Biologically Active Organic Compounds, 302
        7.5.2.3. Extracellular Gums, 303
        7.5.2.4. Poly-β-hydroxybutyric Acid, 304

7.6. Factors Affecting Survival, 305
    7.6.1. General Considerations, 305
    7.6.2. Temperature, 306
    7.6.3. Moisture, 307
    7.6.4. Inorganic Ions, 308
    7.6.5. Organic Toxic Agents, 310
        7.6.5.1. Dyes, 310
        7.6.5.2. Pesticides and Hormones, 310
        7.6.5.3. Antibiotics, 312
    7.6.6. Inhibition by Amino Acids, 315
    7.6.7. Miscellaneous Toxic Effects, 315

7.7. Interbiotic Relationships between Rhizobia and Other Microorganisms, 316
    7.7.1. General, 316
    7.7.2. Antagonism by Fungi, 317
    7.7.3. Antagonism by Actinomycetes, 318
    7.7.4. Antagonism by Eubacteria, 318
    7.7.5. Antagonism among Rhizobia, 318

7.8. Bacteriophage and Lysogeny, 320
    7.8.1. General, 320
    7.8.2. Properties of Rhizobial Phages, 320
    7.8.3. Phage-Host Interaction, 321
    7.8.4. Host Range, 322
    7.8.5. Relationships between Phage Sensitivity and Other Properties in the Rhizobial Host, 322
    7.8.6. Morphology, 324
    7.8.7. Serology, 324
    7.8.8. Lysogeny, 324
    7.8.9. Genetics of Rhizobiophage, 325
    7.8.10. Rhizobiocins, 325

7.9. The Antigenic Structure of Rhizobia, 328
    7.9.1. General Considerations, 328
    7.9.2. Review of Techniques, 328

Contents

- 7.9.3. Serological Relationships among the Root Nodule Bacteria, 330
- 7.9.4. Flagellar Antigens, 330
- 7.9.5. Somatic Antigens, 331
  - 7.9.5.1. Minimal Antigenic Constitution, 331
  - 7.9.5.2. Somatic Antigens in Bacteroids, 332
  - 7.9.5.3. Chemical Nature of Somatic Antigens, 332
  - 7.9.5.4. Differential Immunogenic Response to Somatic Antigens, 333
- 7.9.6. Other Antigens, 334
  - 7.9.6.1. Vi-like Antigens, 334
  - 7.9.6.2. Internal (Group) Antigens, 334
  - 7.9.6.3. The Question of Extracellular Gum, 335
- 7.9.7. The Reproducibility of Serological Reactions, 336
- 7.9.8. Relationship of Antigenic Constitution to Other Properties, 337
- 7.10. Taxonomy, 339
  - 7.10.1. Present Taxonomic Position, 339
    - 7.10.1.1. Family Rhizobiaceae, 339
    - 7.10.1.2. Rhizobium Frank, 1889, 339
    - 7.10.1.3. Species, 340
    - 7.10.1.4. Host Specificity as Basis for Speciation, 340
  - 7.10.2. New Taxonomic Approaches to Rhizobial Classification, 341
    - 7.10.2.1. Information from Numerical Taxonomy, 342
    - 7.10.2.2. DNA-Base Composition and Homology, 343
  - 7.10.3. Other Relevant Information, 346
    - 7.10.3.1. Rhizobial Phage, 346
    - 7.10.3.2. Antigenic Constitution, 347
    - 7.10.3.3. Patterns of Isoenzymes, 348
  - 7.10.4. Apparent Taxonomic Relatedness within the Genus Rhizobium, 349
    - 7.10.4.1. Fast Growers, 349
    - 7.10.4.2. Slow Growers, 350
  - 7.10.5. Wider Relationships, 350
  - 7.10.6. Recognition and Characterization of Rhizobium, 351
- 7.11. References, 354

7.1 INTRODUCTION

7.1.1 General Description

The genus Rhizobium, as at present defined, contains those bacteria able to form morphologically defined nodules on the roots of a member of the family Leguminosae. It also includes cutlures which are no longer invasive, but which have an authentic history of origin from an invasive strain. The bacteria which meet these requirements show the following characteristics:

Gram negative rods (0.5-0.9 µm x 1.2-3.0 µm), occurring singly or in pairs; generally motile when young having peritrichous, polar or subpolar flagella; often with prominent granules of poly-β-hydroxybutyrate; without endospores.

Aerobic chemorganotrophs; best growth at 25°-30° on complex media, notably with yeast extract; poor growth on peptone agar; growth on litmus milk slow, evantually resulting in acid or alkaline reaction, sometimes with a cleared upper "serum zone," indicative of slight proteolytic activity; no coagulation; many able to use nitrate, ammonia, or an amino acid as sole source of nitrogen; dinitrogen utilized primarily in symbiosis with a legume host.

"Fast growers," likely to have a mean generation time of the order of 2-4 hours, give little detectable growth on agar media in 24 hours, but generally form relatively large (2-4 mm diameter), gummy, colorless or white colonies, in 3-5 days.

"Slow growers," likely to have a mean generation time of 6-8 hours, yielding small colonies ($\leqslant$1 mm) after 7-10 days; generally colorless, white, or cream-colored, rarely pink; gum less abundant than in the fast growers, dense and sticky.

7.1.2 The Meaning of "Strain"

Every rhizobial culture not known to have had a common clonal history with another culture constitutes a separate strain. The identity and history of a strain needs to be maintained between laboratories and countries, and should not be confused, or lost, by any system of local labeling that may be adopted for immediate convenience. The experience of different

Morphology and Cytology

groups can then be used for direct comparisons; information will be cumulative, waste of effort reduced.

Cultures of the one original strain but with different clonal histories may well have diverged significantly; those derived from single colony reisolates are likely to reveal differences not apparent in a massed, to some extent genetically heterogeneous, culture. Careful records need, therefore, to be kept of clonal history and of any colony reisolates.

7.1.3 Types

A given assembly of strains can be grouped according to any convenient characteristic, e.g., "serotypes" according to shared serological reactions. Each typing system is likely to result in the fresh sorting of a given collection of strains. Members so typed are not necessarily otherwise related, and it is a mistake to extrapolate from such groupings to predict other aspects of strain behavior. The situation is different when the typing characteristic is being employed for the recognition of strains used under defined experimental conditions (as in serotyping of strains supplied in a mixed inoculum).

7.1.4 Specific Designations

The commonly listed species (Rhizobium leguminosarum, R. lupini, etc.) will be used in this account without prejudice to the question of their taxonomic validity. The term "cowpea rhizobia" will be used for the large miscellaneous collection, generally of slow growers, which are obtained from the many hosts (even non-legumes!) additional to those to which the currently named species are adapted. In other cases of present taxonomic difficulty the same noncommittal device will be resorted to as, for example, the "Lotus rhizobia."

7.2 MORPHOLOGY AND CYTOLOGY

7.2.1 The Cultivated Cell

The younger cells stain evenly with simple basic stains [except that strains of Rhizobium leguminosarum and R. trifolii commonly contain metachromatic granules (1)]. Older, generally longer, cells have unstained granules of poly-β-hydroxybutyrate (PHB), which can occupy such a large part of the cell as to cause the classical "banded" appearance, due to compression of the cytoplasmic material (see Plate 7.1). They have also militated against the definition of nucleoid areas by staining and light microscopy. The polymer, though not lipid in the true sense, stains with lipophilic dyes and can be extracted with chloroform. Its high refractive index makes it conspicuous with phase microscopy.

Young cells are motile and sufficiently well equipped with flagella for these to have been demonstrated with light microscopy of stained preparations. There is some confusion, however, when it comes to distinguishing the detailed arrangement of the flagella. This difficulty is illustrated in a study of fast-growing rhizobia from Lupinus densiflorus (2) in which most of the cells had a single subpolar flagellum, but in others the position varied from what was taken to be truly polar to lateral. Apparently some cells carry both peritrichate and subpolar flagella, and of these, the former are more easily detached (3). The question of the arrangement of flagella in the rhizobia requires a comprehensive reinvestigation with the electron microscope.

Abundant water-soluble gum (exopolysaccharide), which is characteristic of the fast-growing rhizobia, appears to have no morphological role. However, some strains have a capsule-like structure that becomes more prominent with age (4) but seems to bear no relationship to the gum (5). Small colony (nongummy) variants may produce more encapsulated cells than the parent gummy strain, and fluorescein-labeled antibody,

---

Plate 7.1. Thin sections of Rhizobium trifolii TA1 from a 24 hr, shaken broth culture using a yeast extract-mannitol medium. Poly-β-hydroxybutyrate (b), polyphosphate (P), and glycogen-like (g) inclusions are present. The region between plasma membrane and cell wall membrane contains electron-dense material, prominent in the inset. The inset is an enlargement of the area outlined showing the prominent rigid layer (e.g., arrow head) and two circular membrane profiles (*) associated with the plasma membrane. (x 8800.) Electron micrograph supplied by Dr. P. J. Dart, Rothamsted Experiment Station.

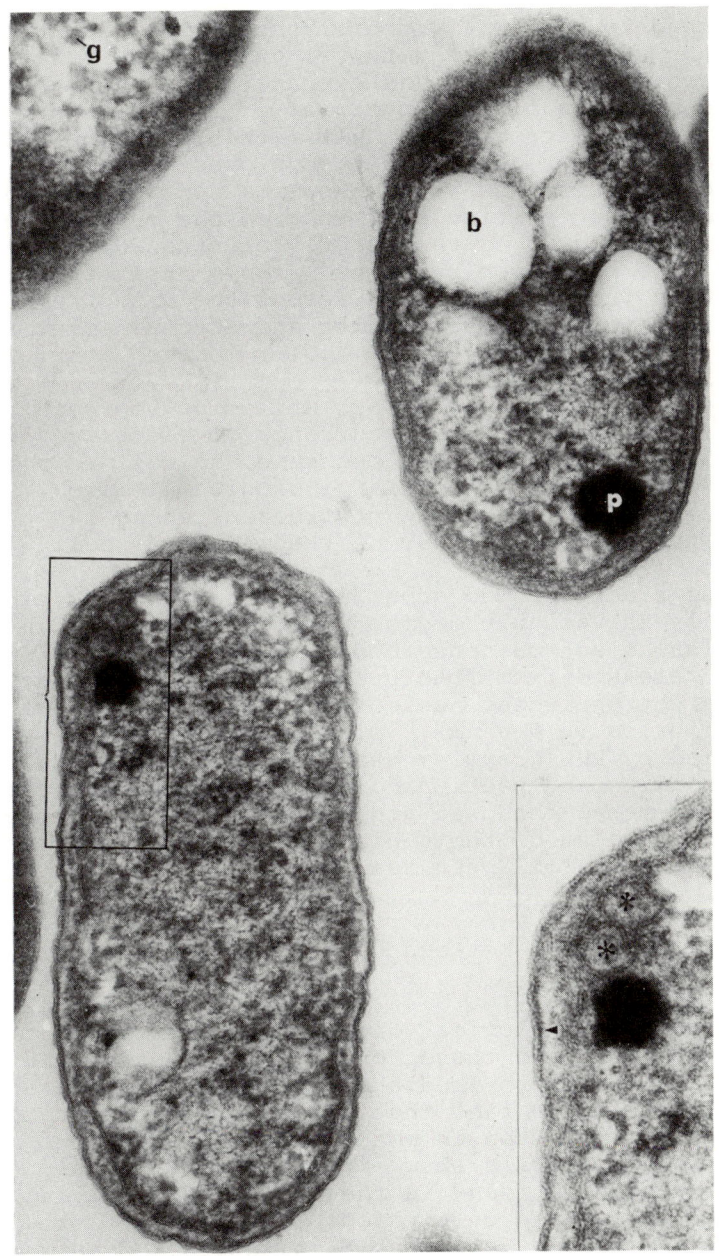

developed to cell wall lipopolysaccharide, has reacted with the capsules; not with the gum. These structures may then be interpreted as extracellular accumulation of the normal antigenic material of the cell wall.

Electron microscopy of ultra-thin sections shows typically gram negative cell envelopes (6). The earlier preparations (6-8) showed a conspicuous inter-membrane zone (particularly towards one or both poles) which appeared to contain fairly electron-dense material. The condition might be largely an artifact, resulting from difficulties associated with preparing these organisms for sectioning and electron microscopy, but might be related to the "polar bodies" attributed to Palacios-DeBarrao (7).

More recent preparations, utilizing improved techniques, confirm the occurrence of electron dense material between wall and membrane, as well as revealing additional intracellular inclusions (Plate 7.1 and Ref. 275). The technique of freeze-etching has also been applied to the study of the rhizobial cell surface (276).

The finest ground substance containing the ribosomes is in variable degree aggregated into roughly spherical electron-dense bodies (15-80 nm) similar to those released from fractured cells. A roughly central electron-lucent nucleoid zone has been resolved into fibrils (2.5-6 nm). Older cells are largely occupied by PHB granules that add to the difficulties of embedding and sectioning and are visible particularly in metal-shadowed and negatively stained preparations because of their high elevation relative to the generally collapsed cell (6).

7.2.2 Comparison between Cultured Rhizobia and Bacteroids

The marked morphological differences between cultivated rhizobia and those found in the root nodule have for a long time provoked a good deal of comment and speculation. The fact that, occasionally in ordinary media and more often in media containing various plant extracts, some cells show irregularities roughly similar to those found in the nodule has encouraged some to regard the cultivated pleomorphic forms as a suitable model for studying N_2 fixation. Consequently the term "bacteroid" has been used interchangeably for abnormal cultivated and nodule-inhabiting forms. Golebiowska (10) concluded that true bacteroids were quite different in their appearance to involution forms produced in variously dosed media. Jordan (11), although retaining the term "bacteroid" for cultivated, as well as nodule, forms, generally qualifies it with "artificially produced" when in culture. The caution expressed in the same article as to the relationship between the two types of cells is well taken but not always observed. This superficial analogy has led to considerable confusion. In the present account "bacteroid" will be restricted to those cells of distinctive form and physiology, which make up the bulk of the rhizobial population in the root nodule. This is a restriction on earlier practice (12) but seems justified until detailed comparisons between the two forms permit the question of analogy to be resolved.

7.2.3 Life Cycles and Endospores

The early descriptive morphology of Rhizobium (12) emphasized the diverse forms of bacteroids and the atypical morphology of cells cultivated in modified media or under otherwise suboptimal conditions of growth. Observations between bacteroids and cultivated rhizobia were freely transposed, and small structures observed and confidently assessed close to, or beyond, the resolving limits of the light microscope. Uneven staining, almost certainly due to PHB granules, was interpreted as evidence for the internal production of "swarmers," as were small "motile" bodies released from lysed cells, most probably due to the same granules showing vigorous Brownian movement.

The position became more confusing when gram positive endospore-forming rods were brought into the picture (13, 14). This claim has received no support from any worker thoroughly conversant with the rhizobia and the safeguards needed in demonstrating the nodulating capacity of a culture. Moreover, in the light of what is now known about the cell wall structure of the gram positive and gram negative cells, it is very difficult to envisage such a transition as would be entailed in the relationship put forward by Bisset and Hale. Equally difficult to accept is the claim that Medicago sativa was nodulated by a pure gram positive spore-forming organism as well as by the normal separately cultivated gram negative form (15).

An attempt to find heat-resistant endospores in 164 strains, mostly recent isolates, gave completely negative results (16). Media that could be expected to promote endospore production were used, but no strain survived 60° for 5 min; numerical taxonomy data clearly separate the rhizobia from Bacillus.

7.2.4 The Cell Surface

The surface of rhizobia has been studied by whole cell agglutination, by the precipitation reaction of solubilized surface antigens (5), and by electrophoresis (17). Four slow-growing species and one strain of R. trifolii revealed only acid groups at the surface, whereas the other four fast-growing strains contained some basic groups as well. These observations agree with the common occurrence of uronic acid in the wall antigens of rhizobia (5), and, in the case of the two strains of R. trifolii that have been compared, the relationship between the charge, characteristic of the whole cell (18), and the electrophoretic property of the isolated antigen (18). The walls of R. trifolii contained glucose, rhamnose, glucosamine, muramic acid, glutamic acid, and diaminopimelic acid, as well as the usual range of amino acids found in gram negative walls; in this case, lysine, aspartic acid, glycine, serine, valine, methionine, leucine, and tryptophane (19). Other constituents have been demonstrated in wall (lipopolysaccharide) antigens (5).

7.3 CULTURAL CHARACTERISTICS

7.3.1 Growth on Solid Media

A large part of the growth of most fast-growing rhizobia after 3-5 days consists of extracellular gum which may, in some cases, flow over the larger part of the agar surface. The growth of R. leguminosarum, R. trifolii, and R. phaseoli is often water clear, or has differentiated, more opaque, areas set in a clear surroundings. Some are evenly mucoid throughout. Colonies of R. meliloti do not generally develop the excessive gumminess of the R. leguminosarum type, and the growth is often more evenly opaque, even chalky white rather than clear or milky. The colonies of the fast growers are often convexly elevated and may have a "pearl-like" appearance when viewed from the underside. Others are relatively flat; some tend to flow along the inoculation line.

Colonies of slow-growing strains do not become apparent for several days; they require about 10 days to reach maxim size and do not produce the free-flowing gum that is so common with the fast-growing forms. The pink pigmented colonies of isolates from Lotononis bainesii are normal and fully invasive (20). Isolates from other species of Lotononis are not necessarily pink.

7.3.2 Indicators

Congo red incorporated in the agar medium (final concentration, 0.0025%) has been used as a means of distinguishing rhizobia (which seldom absorb any appreciable amount of the dye) from many other bacteria, including agrobacteria, which take the dye up strongly. The distinction is not absolute as this property is one of several sometimes shared between Agrobacterium and Rhizobium, most commonly R. meliloti. Dye adsorption may also be affected by nature of the medium and conditions of cultivation (21).

The incorporation of bromothymol blue, final concentration 0.0025%, provides a useful indicator of relatively small changes in acidity or alkalinity. Many non-rhizobial contaminants will also be revealed in this medium by their more drastic effect on pH.

7.3.3 Colonial Variants

Mutation to different colony forms is not uncommon, often in the direction of a small (ca. 1 mm) nongummy colony. In the author's experience these small colony variants, though fully invasive, have invariably lost their N_2-fixing capacity (22). On the other hand, most pigmented colonies that have been reported as variants of Rhizobium have been noninvasive, a fact which leaves some doubt as to their authenticity as rhizobia except when good corroborative evidence is available (23).

Some investigators have applied the "smooth-rough" terminology to describe one form of colonial variation in Rhizobium. Jensen, for example (24), described such variants in R. meliloti as opaque, firm almost cartilaginous, with a surface that became deeply wrinkled. A coherent pellicle formed on liquid media. In Jensen's experience such "rough" variants produced nodules and from these the same form could be recovered. The "rough" form reacted with antiserum developed to the smooth, parent culture. Others have noted loss of invasiveness with their rough variants. The small colony variant of R. trifolii, noted above (22), is not associated with any of the appearance, or instability in liquid culture, to be expected in a rough variant.

7.3.4 Growth on Other Media

The usual yeast extract media are representative of those based on plant extract, which are suitable for the growth of Rhizobium. Most strains grow poorly on glucose peptone agar but some R. meliloti, again like Agrobacterium, make a reasonable degree of growth on this medium.

Growth in liquid medium is generally evenly turbid and may become extremely viscous due to the production of gum. The different rapidity with which the colonies develop on agar is reflected in their generation time in liquid media, which for slow growers is twice to four times that of the fast growers (25).

Reactions in litmus milk have often been taken to be distinctive between R. meliloti (acid reaction) and the slightly alkaline reaction found with other species. The production of a "serum zone" appears as a strain, rather than a species, characteristic.

7.3.5 The "Autoplaque" Phenomenon

Another phenomenon that calls for some comment is the "autoplaque" condition (26). The signs of the condition are progressive erosion of older growth on agar media and the formation of discrete clear areas strongly reminiscent of phage plaques. However, no transmissible agent was demonstrated in filtrates from affected areas, and there was no sign of phage, or phage pieces, when the material was viewed with the electron microscope. The condition appears to be one of self-lysis, triggered off under certain undefined circumstances, in certain regions of growth on the surface of solid media. No similar effect could be demonstrated in liquid media. This could well have been the condition noted in 1908 by Loew and Aso (12), and one which can easily be ascribed to contamination or phage when encountered in freshly isolated cultures. The frequent appearance of lysis in streakings from nodules, attributed to phage, may have been due to this condition. Its occurrence seems to be partly dependent on strain and partly on the nature of the medium (28).

7.3.6 Sensitivity of Rhizobia

The fact that rhizobia are sensitive to some conditions less likely to affect most other bacteria should be mentioned. Extremely small amounts of heat-charred residues on reused glass petri dishes, effects of residual ethylene oxide in gas-sterilized petri dishes, and the unfavorable effect of a second heat sterilization of media are examples. A preferred carbohydrate source may need to be sterilized separately by membrane filtration (as in the case of galactose for the Lotononis rhizobium).

7.4 METABOLIC REQUIREMENTS

7.4.1 Nutrition

7.4.1.1 Carbon Source

The fast-growing species can use many sugars, polyols, and organic acids. The slow growers are more specialized in their requirement and commonly prefer pentoses (12, 29, 30).

Table 7.1 combines data for carbon compound utilization with other data based on changed pH. They agree very well with early respirometric studies (29). The regular fast growers (R. leguminosarum, R. trifolii, R. phaseoli, and R. meliloti) made good use of almost all the C sources. Dextrin (not shown in the table) was used by only 12.5%. The fast growers from

TABLE 7.1. Utilization of Carbon Compounds: Proportion (%) of Strains Responding to Carbon Source[a]

No. of strains / Carbon source	Growth response					Changed pH		
	Fast growers	Slow growers				Slow growers	Fast growers	
	40 (i)	36 (ii)	14 (iii)	5 (iv)	36 (v)	12 (vi)	11 (vii)	4 (viii)
Glucose	100	67	70	60	70-89	67	100	100
Galactose	100	72	79	60	67-92	75	100	100
Fructose	100	86	100	100	47-53	67	100	100
Arabinose	100	98	100	100	56-75	100	100	100
Xylose	100	92	100	80	23-36	83	100	100
Rhamnose	100	0	7	0		8	91	100
Maltose	98	14	17	0	17-81	50	73	100
Sucrose	100	0	0	0	0	0	92	100
Lactose	90	6	17	0	0-36	0	18	75
Trehalose	100	0	0	0				
Raffinose	88	8	0	0				
Mannitol	93	80	79	60	58-69	0	54	100
Dulcitol	90	0	0	0		0	0	25
Fumarate	93	11	100	0		100	100	100
Malate	93	30	28	0	0	100	100	100
Succinate	85	14	7	0	6-44	100	100	75
Citrate	95	56	64	60	0	100	100	75
Pyruvate	80	30	64	20	17	100	100	75

Notes on rhizobia: (i) Pooled data for R. trifolii (16); R. leguminosarum (14); R. phaseoli (5); R. meliloti (5). (ii) Cowpea rhizobia. (iii), (vi) Slow R. lupini. (iv), (v) R. japonicum. (vii) Fast-growing cultures from Lotus and Anthyllis. (viii) Fast-growing cultures from Lupinus densiflorus.
[a]Source of data: i-iv (31); v (32); vi-viii (2).

Lotus and Lupinus densiflorus showed results that generally paralleled the "orthodox" fast growers, with the dulcitol result exceptional. The slow growers were less regular in all their responses. Failure to use rhamnose, sucrose, trehalose, raffinose, and dulcitol appeared to be a consistent feature; maltose, lactose, and the organic acids varied a good deal between strains.

Metabolic Requirements   291

7.4.1.2 Nitrogen Source

The conclusion that the free-living Rhizobium is incapable of utilizing dinitrogen (12, 33) has been recently invalidated for several rhizobia growing under newly defined conditions (299-301).
   Many strains can utilize nitrate or ammonium as the sole or supplementary source of nitrogen. In these circumstances care has to be taken to avoid an inhibitory change of pH in a lightly buffered medium, due to selective uptake of anionic or cationic source of N. Nitrite may also be used (34).
   Strains unable to use inorganic combined N are generally satisfied by the addition of a single amino acid. Glutamic acid is generally acceptable (35), but some strains and substrains prefer other amino acids. A suitable carbon compound, such as α-ketoglutarate (32, 34, 36) may replace the need for amino acid, provided a source of combined inorganic N is also available. Vitamin-free casein hydrolyzate was superior to any tested combination of amino acids for the growth of R. japonicum, perhaps due to a peptide growth factor. Peptone is a poor source of N for rhizobia although it may substitute for yeast extract for the growth of the Lotononis rhizobium (which in several ways is a rather peculiar organism). Urea was utilized by a large collection of representative strains and species and most showed a moderate degree of growth on biuret (37). The exceptions were strains of R. meliloti and Agrobacterium radiobacter.
   The inhibitory and unfavorable selective action of higher levels of amino acids will be dealt with separately in III.7.6.6.

7.4.1.3 Vitamin Requirements

Responses to thiamin and biotin were obtained with R. trifolii early in the study of bacterial growth factors (25, 38). In a larger survey it was found that 35 of 63 representative rhizobial strains needed the addition of one or more vitamins to the basal medium (39). Species appeared to differ in their detailed requirements. The R. leguminosarum, R. trifolii, R. phaseoli group was the most demanding in that 26 of 31 strains required the addition of one or more of biotin, thiamin, and calcium pantothenate. Calcium pantothenate, generally with biotin, thiamine, or both was the most common requirement in this group (25 strains); one strain required only biotin. <u>Rhizobium meliloti</u>, <u>R</u>. <u>lupini</u>, and the cowpea rhizobia, when they needed any vitamin (9 of 23), were satisfied with biotin. None of six strains of <u>R. japonicum</u> needed any of the vitamins used in these tests. The less-demanding vitamin requirement of slow growers was also shown in the case of a larger collection of R. japonicum, for which only biotin, out of the 10 vitamins tested, had any effect, with 3 of 39 strains stimulated and 4 inhibited (32). There appears, however, to be an additional, undefined factor, present in yeast extract needed for the continued cultivation of slow growers.

Nicotinic acid, pyridoxin, folic acid, p-amino benzoic acid, inositol, B_{12}, and riboflavin have generally been without effect. In fact, the synthesis of vitamin B_{12} has been recorded by rhizobia (40) and is the reason for their cobalt requirement. Agar itself contains other factors able to stimulate growth of some strains, even in the presence of added yeast extract (41).

The seemingly strange observation that low concentrations of chloracetic and chloropropionic acids accelerated the growth of some strains of <u>R</u>. <u>leguminosarum</u> could possibly be explained in terms of inhibition of the formation of metabolites that could otherwise compete with β-alanine in the synthesis of pantothenate (42).

A very interesting development has been the finding that some ineffective mutants of <u>R</u>. <u>leguminosarum</u> and <u>R</u>. <u>trifolii</u>, selected as resistant to the metabolic inhibitors D-alanine and D-histidine, were auxotrophic for vitamins or other growth factors, and that some back mutations to prototrophy restored effectiveness (43).

7.4.1.4 Mineral Elements

Reported approximate optimal concentration of iron has ranged from 0.005-0.2 m<u>M</u> (44).

A more general requirement for divalent cation (Mg or Ca) can be met by a total concentration of 0.5 m<u>M</u>. The distinctive effects due to a shortage of each ion specifically became apparent only when their separate contributions were well below that level (Mg, 0.1 m<u>M</u>; Ca, 0.025 m<u>M</u>). The viability of the culture was reduced when either element was deficient; in the case of Mg, the effect was particularly striking. Morphologically the outcome of each deficiency was distinctive. Magnesium-limited cells were elongated and sometimes branched; Ca-limited were irregular, swollen, and roughly spherical. Unlike spheroplasts, however, they were osmotically stable (35, 45-49).

The abnormal shape of the Ca-deficient rhizobia was suggestive of a specific role for this element in normal wall structure. This was supported by: its relative concentration in walls of cells grown under conditions of Ca limitation; the failure of Mg to overcome the wall deficiency, although it increased in the cell (19, 50); the immediate susceptibility of Ca-deficient cells to the action of lysozyme, without need to pretreat with EDTA; the outward diffusion of internal group antigens from Ca-deficient cells (51); and additional capacity to absorb antibody (taken to indicate a less closely knit bacterial surface) (52). The structural role envisioned for Ca could involve the binding of otherwise free -COOH groups, either in the peptidoglycan layer, between it and other wall constituents, or within acidic groups of the lipopolysaccharide antigen. It would appear, however, that any such site of Ca binding would be located in an electrophoretically inaccessible region (53).

Strontium, but neither Ba nor Mg, was a less efficient substitute for Ca for growth, normal morphology, and as a protectant against lysozyme (50, 54).

A peculiar effect of apparent sensitivity to Ca in a substrain of R. trifolii appears to have resulted from a drop in pH brought about by the Ca-stimulated organism in a weakly buffered defined medium. Because the parent strain did not reduce the pH to the same growth-inhibiting level, its Ca stimulation remained uncomplicated (55).

The need for trace amounts of cobalt for the synthesis of vitamin B_{12} (cobalamin) has been clearly defined. Maximum growth of R. japonicum was obtained with approximately 0.01 μM (56). Fast-growing species also showed a marked Co response, R. meliloti being specially dependent on an essential, though very small, amount of this element (57). The increment in growth response was greater at 0.03 μM added Co than at 3 μM, although B_{12}-coenzyme synthesis was greatest at the higher concentration (58). The method lends itself to a microbiological assay of Co (59). More recently, it

has been found that inhibition of rhizobial growth by
some other heavy metals (Ni and Cu) can be prevented
by sufficient Co (0.01 µM) (60).
  More detailed studies with cell-free extracts led
to the conclusion that Co deficiency in R. meliloti
prevented sufficient synthesis of vitamin B_{12} for the
normal functioning of methyl-malonyl-coenzyme A
mutase (61). More recently a more subtle effect has
been indicated due to Co deficiency leading to excessive production of ribonucleotide reductase apoenzyme
(62). This enzyme is dependent on B_{12} coenzyme (hence
on Co) for its activity, and it is postulated that
failure to provide the normal product, which represses
the production of apoenzyme, leads to apoenzyme
accumulation.
  The need for low concentrations of zinc and manganese has also been demonstrated with several rhizobial
species. Zinc response was maximal at 0.1-1.0 µM; Mn,
in the range 0.1-10 µM (63).
  R. trifolii and R. meliloti show restricted growth
when potassium is omitted from a defined medium and a
clear linear response in cell yield up to 0.006 mM.
A small response was also obtained with low concentrations of added Na. Salts of both elements are toxic
at higher concentrations, although strains and species
differ in their tolerance (64). The omission of K in
the defined medium used by Bergersen (35) causes a
heavy demand for biotin, presumably as a substitute
source of this element (65).

7.4.2 Other Metabolic Requirements

7.4.2.1 Oxygen

It is common experience for the rhizobia to respond to
increased aeration; none of a large collection of
strains grew anaerobically (1). However, excellent
growth was obtained at less than 0.01 atm O_2, although
no strain grew in its complete absence (25). In other
work it was found that O_2 affected the growth rate of
R. meliloti only at a low level (29 ml O_2/l·hr) and
comparable growth curves were obtained over a 40-fold
range of available O_2 (66). Some strains are able to
grow to a considerable depth in shake culture.

7.4.2.2 Hydrogen Ions

A comprehensive set of data (1) has been condensed and arranged to show the proportion of strains in each group able to grow at each specified pH level (Table 7.2). All strains grew at pH 5.5-7.5. Acid tolerance was, in decreasing order, slow growers (R. lupini, R. japonicum, cowpea) > (R. leguminosarum, R. trifolii, R. phaseoli) > R. meliloti. Tolerance of high pH was in reverse order. These results are not unlike the early quoted ranges (12), but the latter indicated rather greater acid tolerance on the part of R. lupini (pH 3.2) and a uniform tolerance of high pH to a degree not borne out in the more recent data.

7.4.2.3 Temperature

Earlier data (12) include reports of growth over the range 0-50°C, with the optimum in the range, 20-28°C. The highest figure is well into the lethal range for the rhizobia and seems quite improbable. Maximal growth temperature for a large collection of rhizobia of tropical and temperate legumes has since been determined (67). Among temperate strains, those of R. meliloti were the most tolerant (36.5-42.5°C) being 8° higher on the average than those of R. leguminosarum and R. trifolii (31-38°C). The collection from tropical legumes (cowpea miscellany) ranged from the lowest to the highest (30-42°C). In another survey 9 of 79 strains (8 of them fast growers) were able to grow at 4°C; 8 (all R. meliloti) grew at 39°C (1).

7.5 METABOLIC ACTIVITIES

7.5.1 Enzymes and Processes

7.5.1.1 Carbon Compounds

Acid production (2, 12) and growth (31) have been used to obtain some preliminary information as to an organism's carbon metabolism, but generally to establish taxonomic relatedness. Many carbon sources are acted on by the faster-growing rhizobia, which seem to have more pathways at their disposal than do the slow growers.

TABLE 7.2. Proportion of Strains of Rhizobia Showing Growth over a Range of pH[a]

Species	No. of Strains	pH								
		3.5	4.0	4.5	5.0	8.0	8.5	9.0	9.5
R. leguminosarum } R. trifolii } R. phaseoli	37	0	0.05	0.65	1.00	1.00	0.81	0.19	0
R. meliloti	11	0	0	0.09	0.82	1.00	1.00	1.00	0.91
R. lupini, R. japonicum, "cowpea"	31	0.06	0.35	0.84	1.00	0.90	0.65	0.03	0

[a] Collated from Graham and Parker (1).

Key enzymes of the Emden-Meyerhof, pentose phosphate, and the Entner-Doudoroff pathways have all been demonstrated with the fast-growing rhizobia (68), although a large survey has indicated that fructose-1-6-diphosphate aldolase (critical for the Emden-Meyerhof pathway) was relatively insignificant and sporadic (69). The activity, or inactivity, of 6-phosphogluconate dehydrogenase is an important point of departure between the fast and slow growers because it constitutes the essential link between the Entner-Doudoroff and phosphate pentose pathways. With the virtual absence of this enzyme, the slow-growing rhizobia (Table 7.3), are left essentially dependent on the Entner-Doudoroff glycolytic sequence.

R. japonicum has been investigated in some detail as a representative slow grower. The results substantiate its dependence on the Entner-Doudoroff pathway when using glucose and, largely, though not entirely, on this system, when supplied with gluconate (70). In the second case there appears to be a supplementary "Acetobacter-like" direct oxidation of gluconate via 2-keto- and 2,5-diketo-gluconate, then as α-ketoglutarate into the tricarboxylic cycle. This supplementary pathway has been suggested as a means of securing 4-C compound replenishment when, apparently, an absence of isocitrate lyase prevents the usual production from acetate (71). Otherwise the evidence favors the functioning of the tricarboxylic acid cycle.

Polyol entry into the glycolytic cycle has been demonstrated as due to inducible dehydrogenases by which mannitol and sorbitol produce fructose, and arabitol, the analogous xylulose (72). The two C_6 and the C_5 polyols induce the same two enzymes, one of which is substrate specific for mannitol and arabitol, the other for sorbitol.

Other sugars probably convert to fructose-6-phosphate (mannose via mannose-6-phosphate; pentoses by the C_2-C_4 split) or to glucose-6-phosphate (directly from glucose by means of UDP-controlled epimerization of galactose). Disaccharides require hydrolysis or phosphorylysis. An alternative pathway to pyruvate and glyceraldehyde-3-phosphate (analogous to the Entner-Doudoroff sequence for glucose) might also be available to rhizobium. The block in the pentose-phosphate pathway characteristic of the slow growers could well explain the common superiority of pentose sugars for these rhizobia in that these sugars would be brought directly into the system.

TABLE 7.3. Enzymatic Differentiation Between Fast- and Slow-growing Rhizobia[a]

Group	Growth rate[c]	Specific enzyme activity[b]				
		G6PD[d]	6PGD[e]	Isocitrate D[f]	Malate D[g]	ED[h]
A. Fast						
R. meliloti	0.22 ± 0.01	241 ± 33	110 ± 16	336 ± 33	788 ± 171	27.2 ± 5.2
R. trifolii	0.23 ± 0.02	341 ± 110	153 ± 24	497 ± 67	1890 ± 263	15.7 ± 4.0
R. legumi-nosarum	0.14 ± 0.02	225 ± 65	56 ± 31	300 ± 84	442 ± 112	19.0 ± 9.0
R. phaseoli						
Fast-growing Lotus and cowpea	0.15 ± 0.01	178 ± 77	76 ± 23	330 ± 155	808 ± 143	14.5 ± 4.5
B. Slow						
R. japonicum	0.034 ± 0.004	24.9 ± 3.6	<0.5-1.1	241 ± 32	1085 ± 167	2.83 ± 0.2
R. lupini	0.05 ± 0.007	50.4 ± 10.1	<0.5	317 ± 45	1454 ± 228	1.74 ± 0.44
Cowpea	0.04 ± 0.004	36.2 ± 7.0	<0.5-3.0	248 ± 49	1258 ± 35	2.06 ± 0.65

[a]Calculated from data given by Martinez-De Drets & Arias (69).
[b]nmoles substrate reduced or formed/min·mg protein.
[c]Generations/hr in exponential phase.
[d]Glucose-6-phosphate dehydrogenase.
[e]6-Phospho-gluconate dehydrogenase.
[f]Isocitrate dehydrogenase.
[g]Malate dehydrogenase.
[h]Production of pyruvate from 6-phosphogluconate.

7.5.1.2 Nitrogen

Nitrogenase has recently been demonstrated in cultured rhizobia by the use of media containing carefully selected carbon sources and the establishment of sufficiently reduced oxygen tension (299-301). Nitrate can be reduced either as an electron acceptor or as a source of assimilable ammonia. Some strains utilize urea or biuret (37). More has been done with the bacteroid than with the cultured rhizobia to determine metabolic processes involving N.

Proteolysis, as judged by gelatin liquefaction, is weak, although the "serum zone" formed in milk by some species seems to involve some degree of proteolysis, as indicated by an increase in soluble N and amino N (12).

R. meliloti has been shown to accumulate an internal pool of amino acid (using ^{14}C-labeled histidine or glutamate). This accumulation against a concentration gradient is energy dependent; movement from the pool depends on protein synthesis (73). Transamination in rhizobia was also demonstrated with the synthesis by cell-free extracts of glutamic acid from α-ketoglutaric acid, when glycine, L-histidine, D-aspartic acid, or D-alanine acted as $-NH_2$ donors (74, 75). There were no indications of specialized pathways with the cultured rhizobia. D-alanine aminotransferase appears to be concerned with the provision of D-amino acids for cell wall synthesis (277).

High levels of aspartate aminotransferase and alanine aminotransferase were detected in rhizobia grown in the presence of ammonium salts, as well as in bacteroids (76). Concentration of these, as well as of glutamate dehydrogenase, was influenced by the N source in the growth medium. While each amino acid failed to stimulate the production of its own transferase, each enzyme was stimulated by ammonia, as well, though to a lesser degree, by nitrate. The combination of NH_4^+ and α-ketoglutarate was best. Enzyme induction was inhibited by the inclusion in the medium of actinomycin D (1 μg/ml), puromycin (2.5 μg/ml), or cycloheximide (5 μg/ml).

Multiple forms of aspartate aminotransferase and alanine aminotransferase have been detected in

different species of rhizobia, and differences have
been noticed in the binding of pyridoxal-5-phosphate
to partially purified aminotransferases from several
rhizobium strains; for example, the coenzyme was
easily dissociated from the aspartase aminotransferase
of strains of R. japonicum (77).

7.5.1.3 Other Enzymes

β-Hydroxybutyrate Dehydrogenase. This enzyme, which
catalyzes the reversible oxidation of β-hydroxybutyrate to acetoacetate, is likely to play an important
role in both the cultured bacterium and in the bacteroid, in which the polymerized form of this substrate
is an important energy reserve. Multiple forms of the
enzyme have been detected (78). Comparisons with two
strains of R. japonicum showed identical electrophoretic patterns for each pair between cultured bacteria
and the bacteroids from the nodule.

Ribonucleotide Reductase. Rhizobia are the second of
two genera recognized as possessing a coenzyme B_{12}-
dependent ribonucleotide reductase (62). The apoenzyme is produced in excess in cells grown with Co
deficiency, evidently due to failure to have normal
repression which is dependent on the deoxy product.
Note that Co also has a role in providing B_{12}
coenzyme-dependent methyl-malonyl-coenzyme A mutase
(61).

Esterases. About 90% of the rhizobial strains contained an alkaline phosphatase (79). Activity was
higher in the fast-growing strains but varied in
mutant forms (80). Acid phosphatase was less active
and was produced by fewer strains.

Esterase and phosphatase isoenzymes have been
studied in Lotus rhizobia and bacteroids of variable
effectiveness (81). No difference could be found in
the cultured bacteria that related to effectiveness,
but such a difference was found between the bacteroids of an effective strain (characterized by slow-
moving esterases and acid phosphatases) and of a
partially effective strain (fast-moving bands). A
similar approach has been used chiefly for taxonomic
purposes (82).

7.5.1.4 Comparison of Vegetative Cells and Bacteroids

Apart from the lack of nitrogenase in ordinarily cultured cells and its presence in bacteriods, the most striking metabolic difference between the two forms of rhizobia is in the nature of their hemoproteins (see II.8). Both forms of R. japonicum contain nonautoxidizable cytochrome c and b; other cytochromes ($a-a_3$ and autoxidizable b) are unique to the cultured cell; the bacteroids uniquely contain a soluble autoxidizable CO-reactive cytochrome c, another hemoprotein and other pigments (33). Earlier it had been shown that a decrease in pO_2 reduced the level of cytochromes a and b and increased cytochrome c, i.e., moved the hemoproteins of the cultured bacteria in the direction of those of the bacteroid (83). The bacteria of ineffective nodules had some of the properties of those that had been cultivated. More recently R. japonicum cultivated under strict anaerobic conditions with NO_3^- as terminal electron acceptor developed a cytochrome pattern similar to that of effective bacteroids (33, 278). Cytochrome patterns of R. leguminosarum also differ between the cultured and bacteroid forms (279).

Since effective, but not ineffective, nodules had been reported to contain coproporphyrin, its excretion from cultured effective and ineffective rhizobia was studied under conditions that maximized the effect (84). No relationship was found between effectiveness and the production of coproporphyrin, even by including viomycin-resistant, ineffective mutants in the comparison.

7.5.2 Products

7.5.2.1 Protons and Hydroxyl Ions

Early records showed that the fast growers were most likely to produce excess H^+, while the slow growers were responsible for an alkaline reaction (12), a generalization which later work supported (30). The reaction was, of course, subject to modification according to the nature of the C source or other media constituents. Norris (85) has completed a most comprehensive survey of a very large collection of rhizobia (717 strains) whose original hosts were widespread over the family Leguminosae, and which fell into the same generally distinct groups (85). This is summarized in Table 7.4.

7.5.2.2 Biologically Active Organic Compounds

Rhizobium leguminosarum produced indole acetic acid (IAA) and gibberellin-like substances (86) which were also found in small amounts in other rhizobia and in Agrobacterium (87). Compounds related to IAA, such as indole-3-lactic acid, indole-3-glycolic acid and tryptophol, also have been found (89-91, 280, 281).

TABLE 7.4. Production of Acid and Alkaline End Points by Strains of Rhizobia Growing on Yeast Mannitol Agar (85)

Taxonomic position of host	No. of strains	With final reaction (%)	
		Acid	Alkaline
Fam. Caesalpinaceae	11	0	100
Fam. Mimosaceae[a]	57	35	65
Leucaena	16	94	6
Fam. Papilionaceae			
Tribe Dalbergieae	19	0	100
Tribe Galegeae[b]	102	42	58
Tribe Genistae	39	13	87
Tribe Hedysareae	112	14	86
Tribe Loteae[c]	24	58	42
Lotus	18	55	45
Tribe Phaseoleae[d]	215	29	71
Phaseolus vulgaris	55	96	4
Tribe Podalyrieae[e]	25	40	60
Tribe Sophoreae	2	0	100
Tribe Trifolieae	97	96	4
Tribe Viceae	14	93	7

[a] 15 of the 20 acid producers isolated from Leucaena.
[b] Most acid producers from Astragalus, Psorelea, and Sesbania (39 of 43); all of 47 Indigofera isolates produced alkaline reaction.
[c] Most acid producers (10 of 14) from Lotus — mainly from L. corniculatus.
[d] 53 of 63 acid producers from Phaseolus vulgaris.
[e] 6 of 10 acid producers from Pultanea.

7.5.2.3 Extracellular Gums

Glucose and, to a lesser extent galactose, are major constituents of the exopolysaccharides of the fast-growing species. Uronic acids, including 4-methyl-0-glucuronic acid, are relatively abundant in R. trifolii, R. leguminosarum, and R. phaseoli, but deficient or barely detectable in R. meliloti. Pyruvy- and acetyl-substituted groups are also common, particularly in those species which contain large amounts of uronic acid. The agrobacterial exopolysaccharides conform closely to those of R. meliloti (92-100, 282).

Most recently there has been a comprehensive survey of 18 strains (4 species) of fast growers as well as 3 strains of Agrobacterium tumefaciens (Table 7.5) (101). This provides strong support for the R. leguminosarum, R. trifolii, R. phaseoli grouping on the one hand, and an apparent relationship between R. meliloti and Agrobacterium on the other. The first group shares the larger content of uronic acid, and a larger amount of pyruvate and acetyl than the uronic acid-deficient R. meliloti-Agrobacterium group.

TABLE 7.5. Quantitative Analyses of Exopolysaccharides of Rhizobium and Agrobacterium[a]

Species[b]	Component (%)				
	Glucose	Galactose	Uronic acid	Pyruvate	Acetyl
R. leguminosarum [7]	50	10	17	13	9
R. trifolii [4]	50	11	17	12	10
R. phaseoli [3]	53	9	17	12	9
R. meliloti [4]	77	9	1	5	7
Agrobacterium tumefaciens [3]	78	10	0.5	7	4

[a] Condensed from Zevenhuisen (101).
[b] Bracketed: number of strains.

Several workers have failed to detect 4-methyl-0-glucuronic acid in the uronic acid-containing gums where, according to earlier findings, they would be expected to occur. This failure could be due to a technical difficulty associated with its demonstration. Two apparently discrepant results need additional consideration. The first is a failure to show the generally established relationship between rhizobial species and content of uronic acid (102); the second is the finding of mannose widely distributed among the polysaccharides of both fast and slow growers (103). This sugar was, in fact, reported for the same strain of R. meliloti (SU47) in which it had not been found previously under conditions where it would have been readily detected (95). It seems significant that in both the discrepant cases, the polysaccharide fraction had been precipitated out of yeast-containing liquid media and that, in the first case, surfactant or alkali, had been allowed to act on the cells prior to their separation from the supernatant (283). Some progress has been made in determining the detailed structure of the exopolysaccharides of R. meliloti (284, 285).

The extracellular polysaccharides of the slow-growing rhizobia are more difficult to prepare without complications due to extraction procedures; glucose is common to all (103-106), but the strains differ a great deal otherwise in their components. Galactose and mannose have been commonly reported, less often, rhamnose. Uronic acid is strain variable, and some of it might have been in the form of 4-methyl-0-glucuronic acid. Additional and unidentified components have been noted with some strains. The detection in the gum fractions of typical cell wall components might well relate to the greater difficulty encountered in separating the gum of the slow growers from the bacteria themselves.

A particulate enzyme preparation from R. japonicum, which was able to synthesize glucan (mainly β-1,2; some β-1,3 and β-1,6), appeared to be identical with one previously isolated from Agrobacterium tumefaciens (107).

7.5.2.4 Poly-β-hydroxybutyric Acid

The cells of cultivated rhizobia, and the bacteroids of some species, have long been noted for their

relatively large nonstaining areas. This has led to
confusion with "vacuoles," "endospores," and projected
life cycles. The areas concerned have a high refractive index, stain with Sudan Black (108), and have
been identified as chloroform-soluble polymerized
β-hydroxybutyrate. In fact, earlier workers (12)
recognized that the chloroform-soluble, lipid-like
material they obtained in quantity from R. meliloti
was neither a true glyceride nor a sterol, but a
polymer of a monohydroxy acid: either hydroxybutyric
or hydroxyvaleric. The correctly identified polymer
was reported for Rhizobium, among other bacteria (109,
110), and, in more detail, with various rhizobia (6,
111), where it can account for half of the cell dry
weight. It probably constitutes an osmotically convenient, neutral food reserve. The polymer is prominent in the bacteroids of some nodules (110, 112) but
is relatively rare in the bacteroids of fully effective nodules involving a faster-growing species, such
as R. trifolii. The polymer molecular weight of poly-
β-hydroxybutyrate of 11 different bacterial genera
ranged from 1000-250,000 (probably reflecting preparative differences), but the same basic molecule
appeared to be involved in all cases (113).

7.6   FACTORS AFFECTING SURVIVAL

7.6.1   General Considerations

Survival of rhizobia in various situations (in culture,
on the seed, and in the soil) is of great practical
importance, but other chapters will be concerned with
these aspects. The present account will endeavor to
consider the basic phenomena, as far as possible in a
quantitative fashion.
Reports of survival over very long periods have
often been based on the demonstration of one or a few
survivors. However some early quantitative data (12)
showed only a tenfold drop in a culture of R. meliloti
after 14 years: a phenomenal result which few, if
any, other workers have been able to match. Surprisingly too, some rhizobia survived several years in
well water (though not in distilled water or 0.0008%
$CaCl_2$) (114). A variable number of R. meliloti was
found in sterilized soil after 30-45 years' storage;
R. trifolii and R. leguminosarum after 14 years' (115).

Although there appears to be a contradiction between such reports and the frequent finding, based on quantitative determinations, that rhizobia do in fact die rapidly under most conditions, the explanation seems generally to rest in the capacity of the cell once dried to persist in a metabolically dormant state. A substrate like soil or peat is likely to enhance such survival, and lyophilized cultures seem also to last well. There is no unequivocal evidence in support of an endospore or any other spore form as the explanation for long survival of rhizobia.

7.6.2 Temperature

Survival of rhizobia is considerably better at low temperatures, and despite a few reports to the contrary (13), the main body of evidence is that the rhizobia are susceptible to moderately elevated temperatures. Early data were to the effect that 2-3 min at 60° were sufficient to kill all of an unspecified number; other data quoted 10 min at about this temperature, and others a decimal reduction time of 80 sec at 50°C during the first 5 min exposure. Twenty-two of twenty-four rhizobia survived 50°C for 10 min (1), but none of a large collection survived 5 min at 60°C; no endospores were produced and heat resistance could not be increased by training or by modification of the medium on which the cultures were grown prior to testing (16). On the other hand, almost 1% survivors at 70°C has been reported, though none at 80°C (116). The associated report of "spore-like" structures, which neither conformed to the classical definition of a spore nor correlated with the number of heat-resistant cells, is difficult to interpret and needs confirmation.

Rhizobial species differ markedly in their tolerance of moderately elevated temperatures. Rhizobium meliloti was more tolerant than R. leguminosarum, R. trifolii, and R. phaseoli; the tropical miscellany was more variable (67, 117). The maximum temperature at which growth of different strains of R. meliloti occurred ranged from 36.5-42.5°C and of isolates from tropical legumes from 30-42.5°C. Results for cultures stored at 40°C in sandy soil (67) yielded the following decimal reduction times:

Factors Affecting Survival

R. trifolii (2 strains), R. leguminosarum (1): 2.1-2.7 hours
R. meliloti (2):   5.3-7.7 hours
Cowpea rhizobia (4):   2.9-8.3 hours

There are some indications of local adaptation to elevated temperatures, e.g., some Egyptian isolates from Trifolium alexandrinum grow at 35°C, whereas no isolates from other collections were able to do so (118), and native rhizobia of the cowpea miscellany, isolated from the hot inland region of New South Wales, survive higher temperatures than strains from the cooler tablelands (119). It is significant, however, that the tolerances of R. meliloti (all introductions into Australia) did not reflect the region from which they were isolated.

Storage at low temperatures resulted in decimal reduction times for agar cultures of 18 weeks at 5°C and 8 weeks at 25°C and for liquid cultures, 8-21 weeks at 2°C and 23-40 weeks at -15°C, according to strain and substrain (120).

7.6.3 Moisture

There were early reports of poor survival when rhizobia were allowed to dry out on various substrates such as cotton, filter paper, glass, and seeds (12). In fact, like other gram negative bacteria, the rhizobia are likely to die rapidly under drying conditions. Most rapid death will occur during the stage of water removal; once dried, the survivors are likely to persist for a much longer period.

The addition of 10% sucrose protected clover rhizobia when dried at normal atmospheric pressure on glass (120). A comparison of other sugars and sorbitol showed that maltose was an even better protectant extending the decimal reduction time from <1 to 70 hours, according to conditions and experiment (121). Glucose, lactose, and cellobiose were inferior; raffinose and sorbitol, as well as sucrose, were intermediate in the protection they provided. Death was most rapid during the first 24-27 hours, when the main loss of water occurred; cells surviving that phase died at about one-tenth the rate during the next 16 days. However even at this stage a similar proportionate benefit accrued to the maltose-protected cells. The reason for the superiority of maltose over the other protectants is not apparent. It could not be attributed to any control on the rate at which water was lost from the suspended cells or to any difference in the level of residual water. Nor could it be attributed to preferred assimilibility, osmotic effect, the presence or absence of a carbonyl group, or solubility. Other less-protective substances shared these properties. The superiority of the α-glucoside (maltose) over the β-glucoside (cellobiose) indicated the importance of detailed molecular configuration.

The nature of the suspending medium was also important in lyophilization (120). Sucrose was superior to sorbitol, mannitol, lysine, an amino acid mixture, milk, and yeast mannitol broth; its use has secured between 12 and 52% initial survival of R. trifolii. Survival with lyphilization from broth itself was generally <1%. It has since been found advantageous to add 5% peptone to the suspending medium. Storage survival data showed an interesting difference among substrains of R. trifolii where both small-colony (nongummy) variants survived after lyophilization relatively poorly [decimal reduction time (d.r.t.), 7 weeks] compared with the several larger colony forms (d.r.t., 31-91 weeks).

Differences have been shown in the initial death during lyophilization of R. meliloti, due to the age of the cells being dried (122). Cells of a 24-hour culture averaged 30% survivors compared with 10% for 96-hour cultures. The cells were washed and suspended in a protective medium (5% dextran, 7.5% sucrose, 1% sodium glutamate) before drying. Longer term survival at 30°C was quite poor, however, with d.r.t. of 2-3 weeks compared with another result of approximately 20 weeks for lyophilized culture held at the same temperature (123). The inclusion of Na^+ in the suspending medium (as the glutamate) might have been responsible for the difference (see below).

Not unexpectedly the survival of R. meliloti in aerosols at 20°C was maximal at high relative humidity and minimal at low (124).

7.6.4 Inorganic Ions

Heavier metals (Hg, U, Ni, Cu) had the usual toxic effect (12), but the Co status of the organism modified the reaction of R. meliloti to Ni and Cu (60). On the other hand, the inhibitory effect of Zn at 100 µ$\underline{M}$ was not reduced by a high level of Co. Neither Cu nor Mn was toxic up to the maxima tested (10 and 100 µ$\underline{M}$, respectively). Continued cultivation in the presence of 16 m$\underline{M}$ Mn resulted in marked loss of symbiotic effectiveness without any adaptation to Mn (125), even following 10 successive transfers in a medium high in Mn (Fig. 7.1).

Salts of calcium and alkaline ions have long been known to be inhibitory at a high enough concentration: $CaCl_2$ (0.14 $\underline{M}$); KCl (0.32 $\underline{M}$); NaCl (0.6 $\underline{M}$); and LiCl (0.024 $\underline{M}$) (12). Experience in the author's laboratory resulted in similar figures for Na and K (0.3 $\underline{M}$ for R. trifolii; 0.5-0.6 $\underline{M}$ for R. meliloti). Indications of greater resistance of R. meliloti compared with R. trifolii and other rhizobial species are borne out by other reports (1, 126). In the second case, however, it was found that the two strains of R. meliloti, which were less affected by Na^+ (as Na_2HPO_4), were more susceptible to relatively low concentrations of $CaCl_2$ (approximately 0.015 $\underline{M}$). Other reports

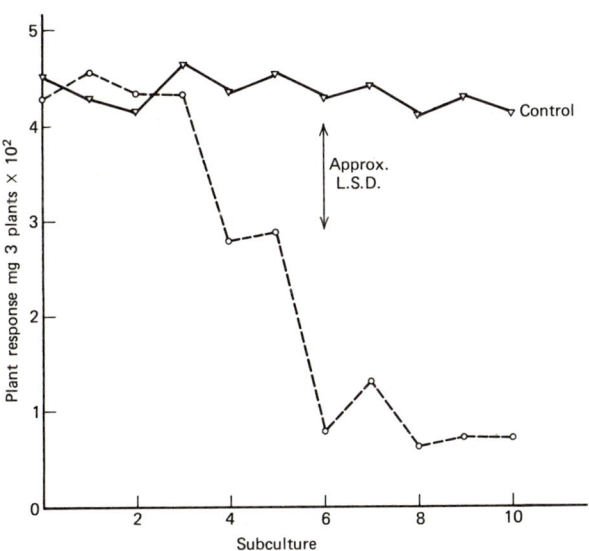

Fig. 7.1. Decline in effectiveness of R. trifolii with continued cultivation in presence of 16 m$\underline{M}$ manganese.

illustrate differences between species and strains of rhizobia in their salt tolerance (118, 286, 287).

7.6.5 Organic Toxic Agents

7.6.5.1 Dyes

Rhizobia, like other gram negative bacteria, are relatively resistant to cationic dyes such as crystal violet, although this varies between strains over a wide range from 200 to 10 ppm (12). The alkali-producing slow growers appeared to be more sensitive to crystal violet (sensitive to 6.6 ppm) than the acid-producing fast growers (resistant to 1000 ppm (7).

7.6.5.2 Pesticides and Hormones

The toxicity of pesticides has received a good deal of attention in both strictly laboratory investigations and under more practical use conditions. The latter will be dealt with in IV.11. From a basic point of view experiments with these biologically potent materials are often difficult or impossible to assess and compare. These difficulties arise when the actual growth-limiting concentrations cannot be determined (as where the result is simply in terms of an inhibition zone likely to be complicated by relative solubility and diffusibility), and where the potency of the chemical is complicated by the presence of other substances (such as carrier or solvent). The task of assigning chemically meaningful names to the numerous proprietary labels, so often quoted without identification, can also be a forbidding one.

All fungicides are likely to be toxic to rhizobia to some degree, so that where the seed has to be both treated against fungal infection and inoculated with a rhizobia it becomes a matter of achieving a balance between the two effects by the careful selection of fungicide and possible manipulation of the inoculation technique to lessen the risk to the rhizobia. Such basic points as have emerged will be treated in this section. Compounds containing toxic metals (Hg, Cu, and Zn) always constitute a serious risk, as do halogen-substituted aromatic compounds, although reports differ as to degree and according to detailed formulations (127-134). Various fungicides based on thiuram (tetramethylthiuram disulphide) are generally least toxic to rhizobia, and this advantage has in fact translated to field use.

Insecticides have not been studied in any detail in the laboratory, but there are clear indications of toxic effects with inoculated seed (120, 135).

Hormone herbicides are generally without effect on rhizobia up to concentrations far in excess of those that have a marked effect on the plant itself (136). Details as to inhibiting levels are confused by the fact that species, and strains within species, differ considerably in their sensitivity to even one herbicide, and because the result is likely to be affected by conditions and time of incubation. The quoted data show how pointless it is to arrive at any generalization as to the relative sensitivity of the different rhizobial species to one herbicide, when the inhibitory dose within a species ranged from 50 to 4000 ppm for R. trifolii, 400-8000 ppm for R. leguminosarum, and 100-3000 ppm for R. lupini. A comparison (137) of potency for the one strain by the one group of investigators is more meaningful. Toxicity to the

Herbicide	Inhibiting concentration (14 days' incubation, ppm)
2,4,5-Trichlorophenoxyacetate (2,4,5-T)	50
2,4-Dichlorophenoxyacetate (2,4-D)	200
Methylchlorophenoxyacetate (MCPA)	200
Methylchlorophenoxybutyrate (MCPB)	500
2,4-Dichlorophenoxybutyrate (2,4-DB)	500

rhizobia was increased with additional halogen substitution and decreased when butyric substituted for acetic as the aliphatic acid portion of the molecule.

Gibberellic acid at up to 1000 ppm had no direct effect on the growth of any of three strains of R. trifolii or one of R. meliloti (138).

IAA produced by rhizobia on a tryptophane-containing medium, was able to delay the initiation of new growth but was without effect on exponentially dividing cells. The degree of prolongation of the lag phase depended on the ratio of IAA to the number of bacteria (139).

7.6.5.3 Antibiotics

The rhizobia, like other gram negative bacteria, are more likely to be susceptible to wide spectrum antibiotics. The effect of such antibiotics cuts across specific boundaries so that strains and substrains within the one species show a considerable range of sensitivity and resistance. Tests with a large collection of R. japonicum and with the faster-growing species of R. trifolii, R. leguminosarum, and R. meliloti showed that the tetracycline antibiotics were generally the most active (140, 141). In a collection of 46 strains of the faster-growing species and 27 slow growers, the latter were generally more resistant to streptomycin, aureomycin, and penicillin G (142), but other reports indicate that generalizations cannot be made. Strain-to-strain variation seems to be the most characteristic feature. Most attention has recently been given to a group of antibiotics for which resistant laboratory-selected strains almost always show parallel complete loss of N_2-fixing effectiveness. Twenty-nine of thirty-three mutant clones, from 11 normally effective strains of R. leguminosarum, R. trifolii, and R. meliloti, which were selected for resistance to viomycin were uniformly ineffective (143). Neomycin resistance gave a similar result and full cross-resistance was recorded between the two antibiotics. Acquisition of resistance to kanamycin conferred resistance to neomycin but not to viomycin, and like resistance to polymyxin, was only rarely associated with loss of effectiveness. No loss of effectiveness was found with streptomycin-resistant mutants. On the other hand, loss of nodulating capacity has been reported in clones of R. trifolii resistant to streptomycin at levels varying from 10-100 ppm (144). Transformation of resistance also meant simultaneous transfer of noninfectiveness. Since the genes associated with nitrogen fixation (nif), as well as antibiotic resistance, are plasmid associated, it is perhaps not surprising that there is some connection between symbiotic effectiveness and antibiotic resistance (see III.12.3.4).

The ineffectiveness of viomycin-resistant mutants of R. meliloti has been confirmed (145), there being one-step mutants of intermediate resistance and highly resistant two-step mutants. Representative figures are:

Status of substrain of Rhizobium meliloti, R21	Minimal Inhibiting Concentration (M.I.C.) of Viomycin (ppm)
Susceptible to viomycin	5
Resistant — 1 step	13
Resistant — 2 step	45

Effectiveness was completely lost at the first step. Sensitivity to glycine (M.I.C. of 2.6×10^{-2} $\underline{M}$) and to D-alanine (M.I.C. of 1.6×10^{-2} $\underline{M}$) was the same with all substrains, and no difference could be demonstrated in the nature of the antigenic surface, either by agglutination or gel diffusion.

Resistance to viomycin in this strain did not confer neomycin resistance and the further addition of divalent and monovalent inorganic ions (Mg^{2+}, Na^+, K^+, Li^+) prevented the inhibitory action of antibiotic (146). However this effect was not limited to viomycin or to the rhizobia; it was interpreted in terms of competition for sites of action. The synthesis of cellular protein by intact cells was inhibited in only the sensitive strain, but this difference was lost with cell-free preparations (147). Resistance was also reduced when the cellular surface was modified with Tris or TES buffer or with EDTA. The difference between the intact resistant cell and the cell-free systems, or modified cell, led to the conclusion that resistance was likely to be due to selective reduction in the permeability of cells towards viomycin. No large-scale changes could be detected in other permeability conditions as gauged by the uptake of glucose and glutamate or by leakage from loaded cells. The proposed selective change in permeability was suggested to result from a greater (30% or more) accumulation of phospholipid at the expense of neutral lipid in the cell envelope. Viomycin appeared to complex with phospholipid in vitro (148).

Kanamycin-resistant mutants of $\underline{R. trifolii}$ and $\underline{R. meliloti}$ were found that were ineffective in N_2 fixation, no significant physiological difference appeared to account for the loss of effectiveness (149).

It is implied in these papers that the apparent single gene mutation responsible for resistance to these antibiotics was also directly responsible for lost effectiveness. However account has to be taken of the fact that resistant fully effective strains to all of these antibiotics can also occur without specific selection. The difference might be due to a different mechanism for resistance in the two cases, or it might result from the action of antibiotic on an extra-chromosomal effectiveness factor. Naturally resistant strains, not having been exposed to the antibiotic, would then retain their effectiveness.

Some peculiar effects of toxic products of the leguminous host adversely affecting rhizobia have been reported. Juice from soybean allowed the soybean and cowpea rhizobia to grow but sap from clover and lucerne was bactericidal to these and other nonhomologous rhizobia. The toxic factor concerned was heat labile and not toxic to a range of other soil organisms (150).

A toxic water-soluble seed coat factor has also been demonstrated (151, 152, 288). Expressed in terms of decimal reduction times (d.r.t.) on beads, seeds, and extract-treated beads, the following was obtained (121) during the early stages of exposure with cells of R. trifolii suspended in 9% maltose:

Treatments		
Relative humidity	Test surface	D.r.t. (hours)
0%	Beads	16
	Seeds	6
	Beads + seed extract[a]	6
100%	Beads	No death
	Seeds	10
	Beads + seed extract[a]	4

[a] From Trifolium subterraneum, var. Mt. Barker.

One such substance isolated from the seed of white clover (Trifolium repens) has been identified as myricetin (3,5,7,3',4',5'-hexhydroxyflavanol) (153).

7.6.6 Inhibition by Amino Acids

The first case of an amino acid having a toxic effect on rhizobium was noted with 100 ppm glycine (25). Tolerance to this amino acid could be progressively increased by cultivation in 1000-3000 ppm of glycine. Loss of infectivity accompanied increased glycine resistance, and was often preceded by loss of effectiveness.

More recent studies have shown that rhizobia, initially sensitive to relatively low levels of glycine and some other amino acids (notably the DL- or D-form) can be selected in the course of continuous transfer to grow at much higher concentrations. R. meliloti could, for example, be "adapted" in this way to tolerate 750-1500 ppm of such amino acids as glycine, L-cysteine and D-alanine (154) and D-methionine (155). Cultures adapted to such high concentrations variously lost effectiveness and sometimes invasiveness. There was an amino acid-species-strain interaction in that some amino acids affected more, some less, strains and there were differences between R. meliloti and R. trifolii (156). Results with progressive loss of effectiveness with adaptation to increasing levels of methionine (157) and selected data given in Table 7.6 illustrate the rather complex situation that has been encountered. Loss of symbiotic capacity associated with "adaptation" to an initially inhibitory amino acid has an apparent similarity to the situation with selected resistance to antibiotics such as neomycin and viomycin, but available data do not give much indication as to what common mechanism might be involved.

7.6.7 Miscellaneous Toxic Effects

Several biotin-independent strains of rhizobia were inhibited by 0.001-1 ppm of nicotinic acid (158). The inhibitory effect, interpreted as due to interference by nicotinic acid in biotin synthesis was countered by the addition of 0.001-0.1 ppm biotin but not by any other growth factor. Some strains of R. japonicum were inhibited by biotin itself (159, 160).

TABLE 7.6. Influence of Exposure to Higher Levels of Some Amino Acids on Symbiotic Capacity of R. trifolii[a]

Amino acid	Max. concn[b] (ppm)	Percent gain or loss in host benefit with strain		
		200	201	210
Glycine	300	-92	-97	-94
L-Alanine	300	-96	-93	-22
D-Alanine	300	-95	-93	-95
L-Histidine	200	-93	-28	+5
D-Histidine	200	-95	-94	-94
L-Phenylalanine	200	-96	-10	+6
D-Phenalanine	200	NN	-93	NN
L-Arginine	300	-21	-16	+13
L-Cysteine	250	-18	0	+9

[a] Representative data from Strijdom and Allen (154); gain (+) or loss (-) as a result of 11 successive weekly subcultures on increasing dosage of amino acid, compared with host benefit. NN, no nodule produced.
[b] Maximum ppm amino acid in the last culture prior to plant test.

7.7 INTERBIOTIC RELATIONSHIPS BETWEEN RHIZOBIA AND OTHER MICROORGANISMS

7.7.1 General

Not unexpectedly microorganisms found in soil and on the root surface often affect rhizobia when they are tested together in laboratory culture or recovered after associated growth on the root. This effect may be favorable or unfavorable (161): 24% of 246 bacteria and actinomycetes isolated from near clover were antagonistic to one or more of the five test strains of R. trifolii (162); 16% were stimulatory (163).

Antagonism towards rhizobia by other microorganisms may be direct, predation (feeding by protozoa, myxobacteria or, Bdellovibrio) (164), or indirect by rendering conditions less favorable for rhizobial growth, e.g., by change of pH, competition for a limiting metabolite, or production of a toxic or inhibitory antibiotic. Antibiotic effects between rhizobia and other microorganisms have been most investigated. The topic has been comprehensively reviewed up to 1964 (165) in an account that usefully emphasizes the importance of the composition of the medium and the nature of the test organism.

7.7.2 Antagonism by Fungi

Only 6 of 107 fungi were inhibitory against any of five strains of R. trifolii in one investigation (162), 8 out of 95 in another (166). However the situation was quite different in fungal isolates obtained from newly cleared soils where the larger part (34 of 47) of the penicillium isolates were strongly bactericidal or bacteriostatic (166). Difficulties with rhizobia in these soils were attributed to this relatively large representation of antibiotic-producing fungi. Results, including actinomycetes and eubacteria for comparison, are summarized in Table 7.7.

TABLE 7.7. Antibiotic Activity of Soil Isolates towards R. trifolii (166)

Nature of Antagonist	Newly cleared soils		Established clover soils	
	Tested	No. of species Producing antibiotic	Tested	No. of species Producing antibiotic
a. Fungi				
Penicillium	47	34	19	5
Aspergillus	17	2	8	2
Nonsporing phycomycetes	2	2		
Phoma	5	1	5	0
Ascomycetes	4	0	5	1
Others	43	0	58	0
b. Actinomycetes	44	5	12	3
c. Eubacteria	15	1	12	1
	167	45	119	12

7.7.3 Antagonism by Actinomycetes

There is obviously a close connection between results obtained when actinomycetes are tested directly in culture vs. rhizobia, and those obtained when the antibiotics they produce are substituted for the growing organism. A detailed account has been provided of the sensitivity of five strains each of R. meliloti, R. lupini, and R. japonicum and four strains of R. trifolii to 19 collection cultures of actinomycetes and three fresh soil isolates (165). Rhizobium meliloti and R. japonicum were most affected, R. trifolii least. Other tests showed an antibiotic effect with a minority of soil actinomycetes (166-168).

7.7.4 Antagonism by Eubacteria

The most inhibitory bacteria of those tested by Krasil'nikov and Korenyako (161) were a strain of Pseudomonas and one of Achromobacter. In a later survey, involving 1091 isolates from the root surface of clover, there were 76 identified bacterial antagonists of which about a quarter were Pseudomonas and practically all of these were strongly inhibitory (169). Table 7.8 is a condensation of these results.

This distribution differs from earlier findings of others with spore formers (12, 38, 170-172) and actinomycetes (165, 167, 168, 170, 173-175), and probably reflects the ability of the organisms studied in the later report to colonize the root surface.

7.7.5 Antagonism among Rhizobia

Medium that has already supported the growth of Rhizobium can be thereby rendered unfavorable for fresh growth of the same organism (176). This effect is not simply a matter of nutrient exhaustion but is evidently due to the accumulation of a toxic factor. More recently inter-rhizobial inhibition due to the production of antibiotic has been reported, and this was the most common cause of antagonism found in an extensive survey of 270 cultures (177, 178). Antagonism was recorded as: "antibiotic" when the factor was dialyzable and was not UV induced; "bacteriocin-like" when it was not dialyzable and did not produce

TABLE 7.8. Generic Representation of Bacteria Antagonistic to R. trifolii (169)

Genus	Number of antagonists	Number strongly antagonistic[a]
Pseudomonas	20	15-16
Flavobacterium	11	2-7
Xanthomonas	9	1-2
Achromobacter		
Alcaligenes	4	2
Erwinia	3	1-2
Aerobacter		
Corynebacterium		
Arthrobacter	26	5-10
Brevibacterium		
Bacillus	3	1
Streptomyces	3	0
Nocardia		
Unidentified	4	0-1
	83	28-40

[a] Versus two strains of R. trifolii.

phage plaques when diluted; and "phage" when it was not dialyzable, UV induced from the test strain and produced phage plaques.

The causes of antagonism were distributed amongst the 116 antagonistic cultures as follows:

Nature of antagonist	Number of strains	%
Antibiotic	95	35
Bacteriocin	12	4.5
Phage	9	3.5
None	154	57
Total	270	

All three factors, tested against an augmented battery of 11 strains, varied from restricted to a wide range of activity. None of the antibiotic effects encountered in this later survey was as potent as that of the first report.

7.8 BACTERIOPHAGE AND LYSOGENY

7.8.1 General

The earlier work with the bacteriophage of rhizobium dates from 1923 and was followed by many reports of its frequent occurrence in nodules, on the root, and in soil supporting the plant and the rhizobial symbionts. By 1950 bacteriophage had been demonstrated for all the main groups of rhizobia, particularly the faster growers (38). Less attention had been given to the slow growers, but R. japonicum phage was subsequently detected in all samples of rhizosphere soil and nodules (179). The frequent occurrence of rhizobial phage is indeed a common phenomenon where the appropriate legume occurs regularly. Trifoliphage was detected in all 10 extracts from clover roots and nodules, and in all the 10 soils in which clover had been grown (180). Trifoliphage was particularly abundant in 100-yr-old clover plots and had a particularly wide host range (14 of 16 cultures of R. trifolii being susceptible). No phage was found in six samples of soil that were without legumes.

7.8.2 Properties of Rhizobial Phages

Rhizobiophage lends itself to most of the techniques applicable to other bacterial viruses, although special media may be required to lessen interference due to excessive gum. Quantitative determination of rhizobiophage requires care with technique and the condition of the culture used to establish the indicator lawn. It can also be affected by erratically reversible attachment to host debris, an effect which may be lessened by the use of low-salt suspending medium (181). Plaque morphology is a strain characteristic, except that a confusing range of plaque size can occur if the phenomenon of reversible inhibition is operating. The physiological age of the culture used to establish the indicator lawn may also affect the size and, consequently, the general appearance of the plaque.

Host-controlled modification (182, 183) may pose a technical problem in that the quantitative recovery of phage on a particular indicator strain may be affected by the strain on which the phage was grown.

The occurrence and possible ecological significance of rhizobial phage in soil raise questions concerned with multiplication and survival.

Stock melilotiphage preparations maintained a constant titre for two years at 4°C, but this declined more rapidly as storage temperature was increased and phage strains differed greatly in their sensitivity (184). These melilotiphages ranged from some that were rapidly destroyed at 50°C to one which had a d.r.t. of 680 min at 60°C. Some survived well from pH 5.7 to 8.8 but were sensitive at pH 4.7 and 9.9. Maximum yield of phage was in the pH range 6.8-7.8, probably due to these conditions most favoring the host.

A collection of 25 rhizobiophages varied considerably in their resistance to osmotic shock and, in some degree, to ultraviolet irradiation (185).

7.8.3 Phage-Host Interaction

The course of events between rhizobiophage and its host is similar to that of other phage systems. Adsorption by the specific susceptible rhizobium is constant over a wide range of phage concentration (186). It is affected by divalent cations and sometimes complicated by debris inhibition. On the one hand, approximately 80% adsorption has been reported after 1-min exposure, on the other only 20% after 10 min. Adsorption constants have ranged from 1×10^{-8}/min to 2×10^{-9}/min. Latent period has ranged from 70 to 295 min; burst size, 11-490 (184, 186-188).

The effects of several outside agents on phage multiplication have been studied. These include a nonspecific action of rhizobial exopolysaccharides, the effect of ribonuclease (probably via its influence on bacterial metabolism), and inhibition by chymotrypsin during a short critical early period in the infection process (188). Bacteriophage modified the rhizobial surface, as indicated by increased electrophoretic mobility of the host, not immediately, as would be expected if this were the direct result of adsorption, but about the middle of the latent period of phage multiplication (189).

Regrowth of a resistant form regularly occurs after lysis of the susceptible majority of an exposed rhizobial population. In the case of virulent phage this will represent selection of a minority mutant form; where the phage is temperate it may be due to lysogenization of some of the host cells. Phage-resistant mutants could be demonstrated in single colony isolates from cultures without any prior exposure to a particular bacteriophage. Some of the regrowth, however, yielded fully susceptible progeny, which may have represented back-mutations, reversion from an imperfectly lysogenized culture, or "escapes" due to some protective product of the host cell's metabolism (186). Some information is now available on the influence of infection by phage on host RNA and DNA metabolism (289).

7.8.4 Host Range

The range of hosts susceptible to a particular rhizobial phage is variable. In some cases it is limited to relatively few strains within a single host species; in others it ranges widely and may, to a variable extent, cross taxonomic boundaries within the Rhizobiaceae (180, 186, 190-193).

7.8.5 Relationships between Phage Sensitivity and Other Properties in the Rhizobial Host

Reports as to relatedness between phage and serological specificities differ (179, 180, 194). In the case of the groups of trifoliphage defined on morphological grounds, some were restricted to one or few serotypes; others were able to attack a wide range of hosts with distinct serological specificities (195). Loss of adsorbing capacity associated with a changed antigenic surface as a result of lysogenic modification (196) is another example of connection between the two properties.

Susceptibility or resistance of stock cultures to phage is a poor predictor os a strain's N_2-fixing capacity, and the coexistence of sensitive and resistant forms in the one nodule has been readily demonstrated (186, 190, 197, 198). Associated change in the effectiveness of cultures that had become resistant after exposure to phage was not apparent in earlier investigations but was later clearly demonstrated in detailed studies of trifoliphages (199). In the case of rhizobial cultures the phage-resistant survivors yielded a large proportion of changed colonial forms. Some were stable over a 3-yr period (6-7 subcultures), but most reverted to the parent strain colony after one or sometimes two plant passages. The most striking change was the relatively large proportion of ineffective substrains among the resistant survivors. There was no relationship between changed colonial form and changed effectiveness, and ineffectiveness persisted in subculture and after plant passage. Change from an originally ineffective to the effective condition occurred much less often. The emphasis as to the likely ecological significance of the frequent association between rhizobia and their phages in the soil and the host plant has moved from the direct "lucerne fatigue" proposition of Demolon and Dunez (38) to consideration of phage as a modifier and selector of the rhizobial genotype.

Coexistence of phage and susceptible rhizobia has been investigated under the varied conditions provided by liquid culture, vermiculite, and soil (180). In the first, the most artificial condition, phage persisted to a high titre and all the bacteria that survived were resistant. The same result was obtained when the nutrient solution was added to soil and vermiculite. Without such addition, however, the rhizobia in the soil all remained susceptible and no phage could be recovered a month after inoculation. The result with vermiculite was intermediate. Clones of the resistant cells recovered from those treatments having liquid medium were all ineffective in N_2 fixation. In a further ecological study three combinations of virulent phage and susceptible bacterium were allowed to interact for a period of 24 weeks in liquid culture (200). The rhizobia declined from an initial population of about 100,000/ml during the first three days whereas the phage increased. In one case the bacterial decline continued to extinction; in the other two a population fully resistant to the phage developed. The resistant cells still adsorbed the specific rhizobiophage. None of these mutations was detected without bacteriophage.

7.8.6 Morphology

It has been possible to arrange 28 trifoliphages in 11 distinct morphological groups on the basis of the nature of the tail appendages and dimensions (195). The morphology of these rhizobiophages conformed generally to the patterns and detailed substructure of other bacteriophages, although some of them (baseplate of CT6, tail appendages of CT1 and CT3) have only been reported once previously (Plate 7.2). Trifoliphage WT1 which showed some contractile activity resembled P11 BNV6 of Agrobacterium, which was considered to be noncontractile (201).

Similar morphological features have been found in bacteriophage of other rhizobial species (289-291).

7.8.7 Serology

Wide serological relationships have been found among phages of R. trifolii, R. leguminosarum, R. meliloti, and R. lupini, as revealed by neutralization with trifoliphage and lupiniphage antisera (202). These results cross the taxonomic boundaries of the respective hosts and contrast with specificities revealed within the group of trifoliphages studied by Barnet (195).

7.8.8 Lysogeny

The occurrence of this condition in Rhizobium was first reported in the reaction between a particular strain of R. trifolii (SU298), as lysogen, and a related strain (SU297) as indicator (203). This case is complicated since SU297 serves as a source of "inducing factor," essential for the release of phage from the lysogen (SU298), which in this case is not inducible by UV or other agents. The "inducing factor" is itself a temperate phage (i) carried in the lysogenic state by SU297, inducible by UV and propagatable on indicator strain, R. trifolii, NU18 (204). The situation for SU298 is complicated further by the fact that it behaves as a double lysogen and, when "induced" by phage i, produces not only phage 7 but a smaller amount of phage 8; which lyses SU298 itself. Phage 7 is a typical temperate phage which can be used to lysogenize R. trifolii SU297 causing lysogenic conversion, which is reflected in antigenic changes in the bacterial surface and loss of capacity to adsorb the phage (196).

Many other cases of lysogenicity have been reported with R. meliloti, R. trifolii, and R. leguminosarum, the last being multiple lysogens (193, 205-208). Studies with R. meliloti that involved initial, substitute, or double lysogenization showed complicated patterns of changed sensitivity (209).

7.8.9 Genetics of Rhizobiophage

Temperate rhizobial phages are beginning to be used for genetic studies; streptomycin resistance and lysine dependence in the host rhizobium have been transduced, but five such transducing melilotiphages failed to transfer effectiveness (210).

A useful start has been made with the more detailed study of the nucleic acid and chromosome mapping of the bacteriophage of several rhizobial species (211, 212, 289, 292-294). The phage nucleic acid appears to be double stranded DNA with exposed terminals (at least in the case of the lupiniphage). Molecular weight ranged from 27-100 M daltons (with a melilotiphage yielding a small fraction as well). Base ratios ranged from (G + C) % of 52-62, i.e., close to those found in the host rhizobia.

7.8.10 Rhizobiocins

Bacteriocins were produced by 34 of 136 strains tested in a collection that ranged over a wide range of fast- and slow-growing rhizobia (213); bacteriocin-like substances have also been demonstrated in a large collection of R. trifolii (178). Tail-like particles produced by a presumed strain of R. lupini (but fast-growing with mean generation time of 70 min) appeared to be responsible for irreversible inactivation of a

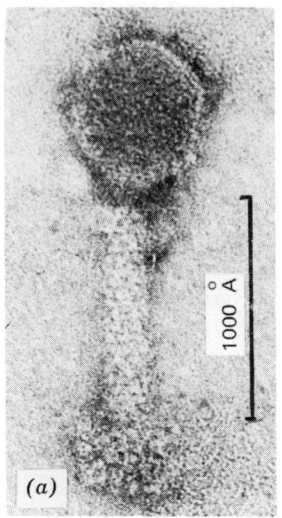
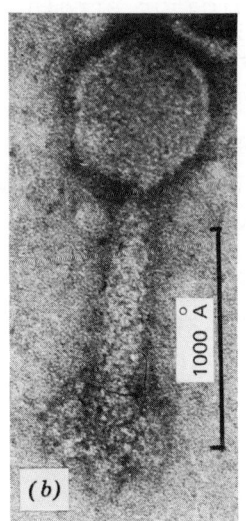
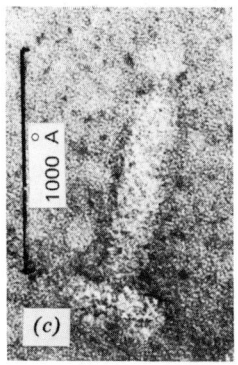
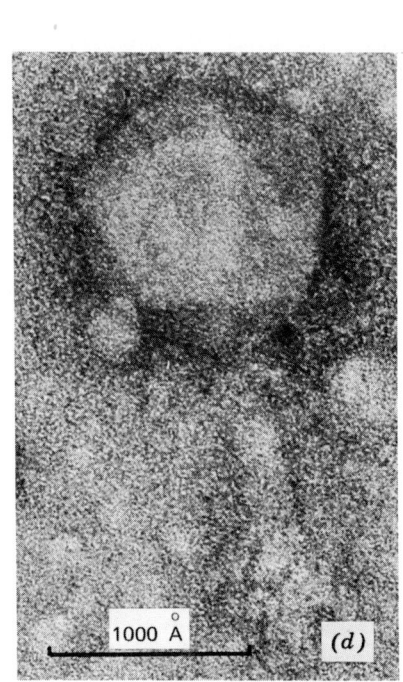
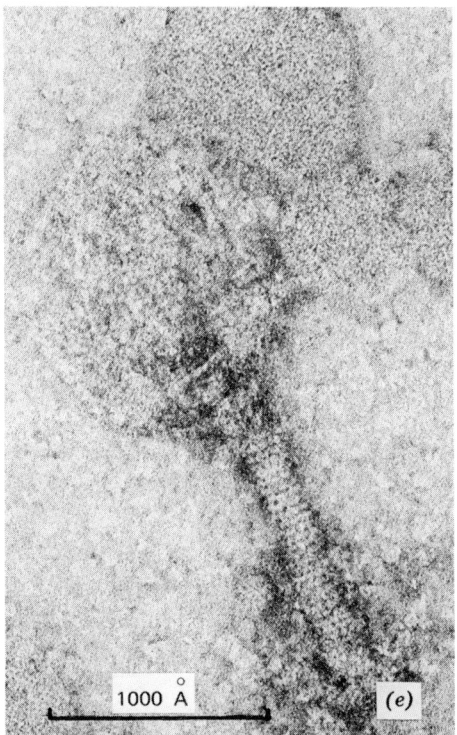

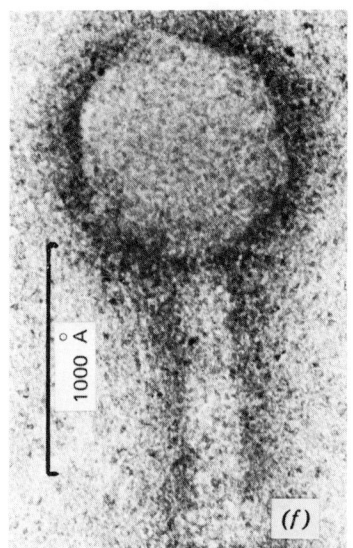
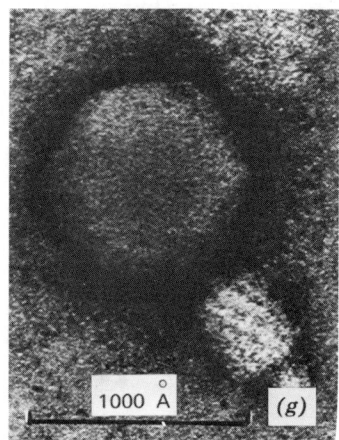
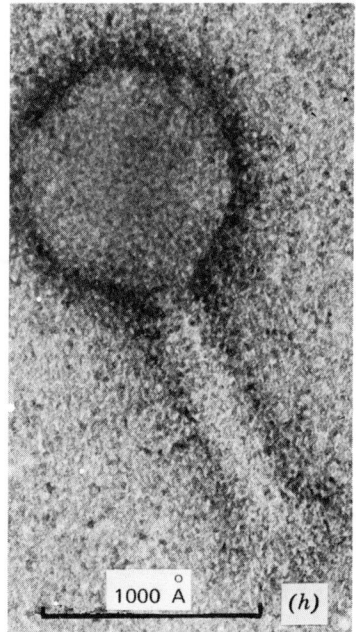

Plate 7.2 The morphology of some trifoliphages (195): CT, contractile tail; WT, wide slightly contractile tail; CT1, with different orientations of base plate (a, b); detached tail (c); CT3 with intact (d) and disrupted head (e); CT6, showing disc-like baseplate (f) and contracted tail (g); WT1 (h). (Photographs, Dr. Y. M. Barnet, School of Microbiology, University of New South Wales.)

related isolate. They also inhibited the development of a virulent phage in the latter. The apparently contractile tails were 123 nm long, with a base plate carrying 6 fibers, each 32 nm. These, together with phage-like particles having a head and short tail) obtained from R. trifolii (295), have been interpreted as incomplete or defective phage material able to exert a bacteriocin-like action.

7.9 THE ANTIGENIC STRUCTURE OF RHIZOBIA

7.9.1 General Considerations

The rhizobia lend themselves to various serological techniques: agglutination, precipitin reaction, complement fixation, antibody absorption, and the use of fluorescein-labeled antibody (215, 216). Most work so far has been with agglutination and the precipitin-reaction techniques, the latter using the Ouchterlony method of immune gel diffusion. Such serological methods have been applied to rhizobia to provide information about their chemical structure and taxonomy and to identify strains (see IV.13).

7.9.2 Review of Techniques

The earlier work was almost entirely restricted to agglutination, and was largely concerned with establishing the validity or otherwise of the current groupings according to the host of origin (12). Generally, cultures allocated to different nodulating groups were serologically distinctive, in that cross reactions outside these groups were not very common, but there were many cases of non-cross-reaction among members of the same group. Work to this stage suffered from two weaknesses: too few strains were being studied, and no distinction was drawn between heat-labile flagellar antigens and heat-stable somatic antigens.

The value of distinction between flagellar and somatic agglutination was demonstrated, particularly with fast growers (217-219). More recently immune gel-diffusion was successfully applied to the rhizobia (220). See Plate 7.3 for typical gel diffusion patterns. Because agglutination depends on exposed

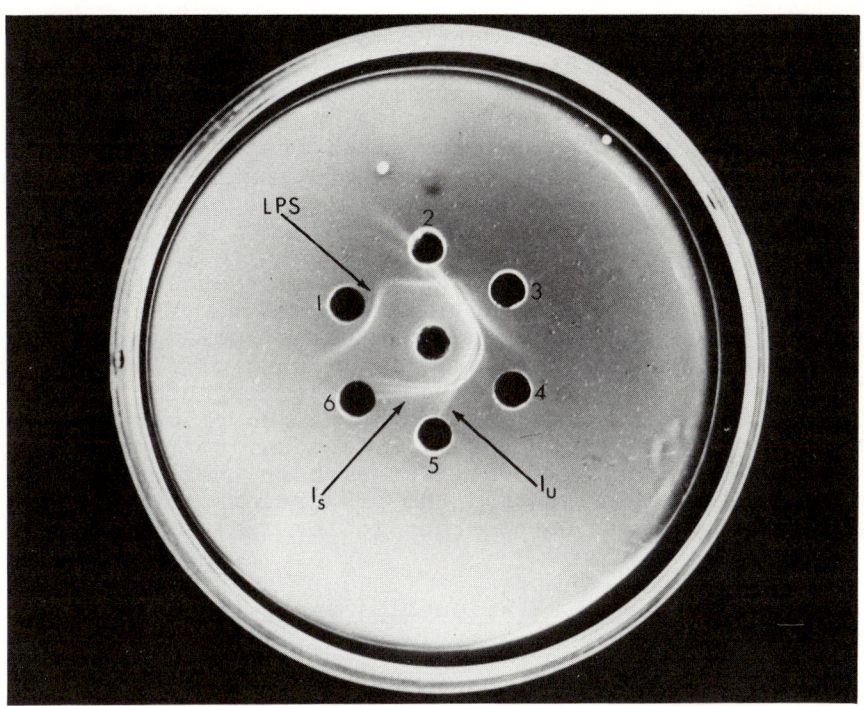

Plate 7.3. Gel diffusion patterns with rhizobial antigens. Center well: Antiserum to R. trifolii TA1. Outer wells: (1) lipopolysaccharide (LPS) somatic antigen from R. trifolii TA1; (2) whole cells of R. trifolii TA1; (3) broken cells of R. trifolii TA1; (4) broken cells of R. trifolii SU297/31 (non-cross-agglutinating); (5) broken cells of R. meliloti U45; (6) broken cells of R. japonicum CB756 (slow-growing). LPS, strain-specific lipopolysaccharide line; I_s, shared internal antigen lines; I_u, unshared internal antigen lines.

antigens located on the cell surface or on flagella, it yields information as to location. Gel diffusion depends on soluble or solubilized antigens which, according to the treatment given in their extraction or use, may coincide with those responsible for agglutination or represent others which, being deeper-seated or obscured, cannot participate in this reaction. Antibody absorption can be used with both methods as a means of obtaining more analytical information about the identity or nonidentity of cross-reacting strains (51, 216, 217, 221, 222); quantitative absorption has permitted more detailed analysis of surface antigens (52, 196).

Fluorescein-conjugated antibody has some specialized use for the identification of somatic-antigen specific rhizobia either in mixed populations or directly in nodules (223-224).

Reactions between a given rhizobial strain and a particular antiserum are generally reproducible. Those involving the flagella require some care to ensure they are sufficient and active; somatic reactions can be repeated routinely without difficulty. However variability in immunological response from rabbit to rabbit using a single antigen preparation must not be ignored. This can be seen, not only in variable homologous titer, but also in the response to minor shared antigens and the proportions of the form of antibody molecule (whether IgG or IgM).

7.9.3 Serological Relationships among the Root Nodule Bacteria

The most comprehensive survey of overall cross-reactivity involved 113 strains of a range of rhizobial species tested for flagellar and somatic agglutination with 58 rhizobial antisera and 16 against Agrobacterium. The rhizobia and agrobacteria fell into three groups: (i) R. leguminosarum, R. trifolii, and R. phaseoli; (ii) R. meliloti, Agrobacterium radiobacter, and A. tumefaciens; (iii) R. japonicum, R. lupini, and cowpea rhizobia. Cross-reactions were absent between these groups but common within them (225).

7.9.4 Flagellar Antigens

Less work has been done with flagellar antigens than with the somatic, partly because they are less reliable in their demonstration (though giving abundant flocculent agglutination when the test suspension is suitable) and partly because they are less strain specific. Twelve strains of R. trifolii provided two flagellar groups, 16 strains of R. meliloti, three (226). Larger collections required the postulation of more flagellar antigens (218, 227-229).

7.9.5 Somatic Antigens

Somatic antigens are capable of bringing about close cell-to-cell agglutination in the presence of specific antibody, a reaction that leads to the complete aggregation of the cell suspension in a fine compact deposit. They are stable to 100°C for 30 min. Some withstand 120°C but the agglutinability of others is very much affected by this extreme temperature (230, 231). The capacity of a cell suspension of R. japonicum to stimulate antibody production in the rabbit was generally reduced by heat treatment (231). Somatic antigen is liberated sufficiently from cells suspended in saline to produce a band nearest and concave to the antigen well in gel diffusion. In the case of R. japonicum and other slow growers and of R. meliloti, the development of this line is enhanced by heating at 100°C for 20 min (222, 232, 233). Multiple lines within the somatic band seem to represent different molecular or aggregate forms of the same specificity. Cross-reacting antigens that share some, but not all, determinate groups, show the expected spur formations. In the case of R. meliloti, a further line which appears to correlate with a strain-specific agglutinogen is found close to the antiserum well (233).

7.9.5.1 Minimal Antigenic Constitution

Nine antigenic determinants had to be invoked to describe the observed cross-reactions with 12 strains of R. trifolii, 15 in a similar study of 16 strains of R. meliloti (226, 228, 229). Larger collections showed that even these arrays did not suffice. The heterogeneity revealed in a Czechoslovakian collection (234) and in one from South Africa (235) would also need many determinants to be postulated. Multiplicity of somatic antigens is also apparent in the several detailed investigations of R. japonicum (221, 236-238).

Various workers have grouped their strains according to patterns of reactivity against a range of antisera. This is useful as an interim measure but is likely to obscure more fundamental relationships that require detailed antigenic analysis.

7.9.5.2 Somatic Antigens in Bacteroids

Suspensions prepared directly from the nodules due to R. japonicum almost invariably duplicated the antigenic characteristics of the cultured form (231, 239). This offers considerable experimental opportunities; it is also of considerable interest that the bacteroid form should be so much like the cultured bacterium in the fine detail of its antigenic surface. It must also mean that the bacteroids are readily freed of the plant-originated membrane envelope that surrounds them in situ. Fluorescein-conjugated antibody has been used directly with nodule smears (223), and the contents of larger nodules have been directly tested in gel diffusion (240, 241).

7.9.5.3 Chemical Nature of Somatic Antigens

Lipopolysaccharide (LPS) antigens of two serologically distinct strains of Rhizobium trifolii have been investigated in some detail (5). Several features were similar to other gram negative bacteria, including firmly bound lipid, 2-keto-3-deoxy-octonate (KDO), glucose, mannose, and, in one strain, a heptose while the very low phosphorus content and presence of glucuronic acid was unusual (cf. K-antigens of encapsulated Escherichia coli). The two lipopolysaccharides differed significantly in detailed composition (presence or absence of heptose), and electrophoretic behavior. The markedly anionic lipopolysaccharide of strain TA1 reflected the surface characteristics of the cells (17, 18), was possibly responsible for the gelatinous nature of some preparations, and could also relate to capsules, observed in old cultures of the same strain (4).

    The strain-specific lipopolysaccharide of these and other strains was fully antigenic with rabbits, and highly active in the gel-diffusion precipitin test. Lines were clearly developed with 0.1-1.6 µg LPS per well; they aligned with the somatic line of whole cells, and the LPS specifically absorbed antibody responsible for this line. Antisera to the isolated LPS had a high agglutination titer, showed normal somatic lines with the homologous cell suspensions, as well as with the LPS itself, but failed to show any additional lines due to internal antigens from broken cells (see below). They also lacked the minor, difficult to absorb, antibody for which calcium-deficient cells were more efficient absorbants than calcium adequate (52).

The sugars of lipopolysaccharides from a wider collection of rhizobia have been reported on more briefly (242). Constituent identifications, based on paper chromatography and not confirmed with eluted samples, differed significantly in detail from other work. There were no consistent specific or generic differences. Other strains have been reported to contain galactose, a range of pentoses, and, in the smooth forms, rhamnose but not fucose. Neither of the two methyl pentoses was recorded for rough variants (243). Analyses have also been reported with other members of family Rhizobiaceae [Agrobacterium (242, 244), Chromobacterium (245)] and for Xanthomonas. The The last shares with the R. trifolii an abundance of uronic acid; in this case galacturonic not glucuronic acid (5).

7.9.5.4 Differential Immunogenic Response to Somatic Antigens

A survey of 151 strains of R. trifolii showed the expected correspondence between high agglutination titer (>800) and the development of the somatic ("a") line in gel diffusion. Multiplicity of antigenic determinants, predicted on the basis of cross-agglutination and antibody absorption, was shown by spurs between homologous and adjoining heterologous lines. There were, however, antisera with which up to 60% of high-titer cross-agglutinating strains failed to produce a somatic line. In these cases most of the antibody responsible for agglutination was in the IgM form, which is reputedly more effective in agglutination than precipitation. Moreover, experience with different rabbits treated with the same antigen preparation showed that the proportion between IgM and IgG could vary significantly between animals. Even the homologous agglutination titer was in the order of 80% due to IgM when the animal had been immunized with isolated antigen. Such antisera had sufficient IgM antibody to give a distinct inner (IgM) gel-diffusion line due to precipitation with LPS, as well as the more usual somatic line, nearer the antigen well, due to IgG. The occurrence and nature of precipitin reactions between these high-titer IgM antisera and cross-agglutinating strains showed interesting degrees of relatedness (246).

7.9.6 Other Antigens

7.9.6.1 Vi-like Antigens

A "vi-like" condition occasionally interfered with somatic agglutination of unheated cells of R. trifolii tested at 52°C. Antiserum to heated cells showed the normal somatic reaction. The heat-labile interfering antigen itself caused agglutination when the temperature of the test was reduced to 20°C (247).

7.9.6.2 Internal (Group) Antigens

Gel diffusion of mechanically broken cells reveals additional precipitation lines, closer to the antiserum well than the slowly diffusing somatic antigen. Similar lines are obtained with frozen, lyophilized and Ca-deficient (hence leaking) cells (51). These have been termed "internal" antigens (in the sense that they are not available to act as agglutinogens).

The internal antigens are of considerable interest in that they reveal similarities within and differences between fast and slow growers. Plate 7.3 shows gel diffusing patterns due to surface and internal antigens of representative rhizobia and includes the more specific somatic antigen of Rhizobium trifolii (identifying with isolated lipopolysaccharide) and internal lines, fully shared with a second strain of the same species, partly with R. meliloti, but not with the slow grower (CB756). More detailed results (248, 249) are summarized in Table 7.9, which also shows evidence of relationship between agrobacteria and the fast rhizobia.

There also appear to be additional antigens near the surface of R. trifolii that become available to the animal for the production of antibody but, apparently blocked by cross-linking in Ca-adequate rhizobia are not as efficient for absorbing these antibodies as are the same antigens in Ca-deficient cells (52).

TABLE 7.9. Internal Group Antigens of Rhizobia

Group	Number of gel-diffusion lines with antiserum to:	
	R. trifolii (248)	R. japonicum (249)
A. Fast growers	TA1	D344
1. R. leguminosarum R. trifolii R. phaseoli	3-4	0-(1s)[b]
2. R. meliloti Fast growers from Leucaena and Lotus	1-2[a]	0-(1s)[b]
Agrobacteria		
B. Slow growers		
R. japonicum R. lupini "Cowpea" "Lotus"	0	1-2

[a] A UV-induced ineffective mutant of R. meliloti gave no line; parent strain gave 2.
[b] (1s): a thin spurred line, mostly not shared with the corresponding homologous line.

7.9.6.3 The Question of Extracellular Gum

This term is used to describe the material that is produced as a water-soluble exudate (exopolysaccharide) by most of the fast-growing rhizobia, but is apparently in a less freely diffusible condition in slow growers. The usual product is likely to be considerably contaminated by other highly active compounds, which may be hard to separate from the precipitated gum.

The rhizobial exopolysaccharide shares with many other polysaccharides the ability to precipitate with certain antisera to pneumococci (99), but there is no unequivocal evidence of its own immunogenicity or identity with any immunogen of the rhizobial cell. Gel-diffusion lines obtained with gum preparations of R. trifolii have been clearly identified with those produced by specific agglutinating lipopolysaccharides (5) or with internal antigens released from Ca-deficient or broken cells (51). These lines are obtained as well with thoroughly washed cells as with a gummy suspension, as well with a nongummy variant as with the gummy parent culture. Any step towards the purification of the gum reduces its line-producing capacity, and a positive reaction requires about 10,000 times as much gum as LPS (2 mg gum per well compared with 2×10^{-4} mg isolated LPS). The "exopolysaccharide reaction" was completely blocked by absorption of the antiserum with specific lipopolysaccharide (in the case of the somatic line), or with thoroughly washed and broken cells (in the case of internal lines). On the other hand, gum was unable to block the reaction between the cell antigens and the homologous antiserum. The sharing of hexosamines between an exopolysaccharide preparation (0.1%) and somatic antigen (7.6%) (250) provides further evidence of an amount of contamination by LPS sufficient to account for the development of lines by gel diffusion against whole cell antisera.

The capacity of rhizobia to react with widely cross-reactive antisera to pneumococci is another matter. This has been shown against the antiserum to type 27 pneumococcus (99, 251) as a property common to unmodified polysaccharide of a strain of R. meliloti and several of R. trifolii. The same polysaccharides take on wide cross-reactivities, some shared with Klebsiella, when the sugar residues are freed of pyruvate. This haptenic behavior of Rhizobium gum assists in characterization of the exopolysaccharide.

7.9.7 The Reproducibility of Serological Reactions

The antigenic stability of cultures after plant passage has been amply established; in fact, with few exceptions, the bacteroids themselves show their antigenic relatedness to the cultured bacteria from which they were derived (see III.7.9.5.2).

A comparison after 10 years of subculturing (226) generally showed a high degree of antigenic stability. This contrasted with changes in cultural and symbiotic properties, unaccompanied by marked loss or change of antigenic specificity. In the course of 30 years of experience any variation in antigenic constitution of

R. trifolii has been minor, but the situation with R. meliloti is less satisfactory, in that a good many of the distinctive reference cultures found in the earlier work have subsequently become progressively more widely cross-reactive (233). The reason for the changing antigenic reactivity of R. meliloti is not yet apparent, but it is possibly significant that the few early cases of wide cross-reactivity reported in 1941 included cultures that had already been in older established collections, and that recently collected field isolates were practically all in the specific agglutinating condition. Variation of a different kind can be encountered if one works with a large number of single-colony reisolates. In the author's experience nine strains of R. trifolii and R. meliloti yielded 4-22% clones that were nonreactive with the homologous antisera. Plant tests showed the tested cultures to be authentic rhizobia and the cultures derived from negative-reacting colonies in turn provided a mixture of reactive with a minority of nonreactive colonies. A similar short-term phasic variation has been reported with some other gram negative bacteria. It has not been studied in detail, but some account needs to be taken of the phenomenon when the technique is being used for quantitative typing.

There can be considerable variation between rabbits in the production of antibodies to certain ("minor") cross-reacting antigens. This may involve failure to produce a minor antibody or its restriction to the IgM form. Animals also vary in the level of antibody they produce to more specific, major, antigens but this is seldom to a degree that seriously interferes with the usefulness of a particular batch of antiserum. This animal-based variability is unfortunate and demands certain precautions if interpreting results based on a single animal. Its recognition and interpretation is not helped if the operator simply pools antisera to obtain a crude average product.

7.9.8 Relationship of Antigenic Constitution to Other Properties

Strains that are related, or apparently identical serologically, can be entirely unrelated in other characteristics; those related on other grounds can be serologically distinct. The validity of this often repeated, but frequently disregarded observation, is shown once again in a detailed investigation of "serogroup 123" of R. japonicum (252).

Some strains which have similar patterns of serological cross-reactivity occur frequently within and between geographical regions. In fact many serotypes can be obtained from one locality; those from different nodules of the same plant commonly differ from each other (227-229, 253). However certain broad relationships have been encountered. Arctic isolates differ from those common in temperate zones (38), and a collection of Australian and overseas cultures of R. trifolii divides on the basis of frequency of certain antigenic groups, even though those obtained in Australia would have originated from accidental imports generally in the last century. There was also some relationship between serotype and host of origin in the case of R. meliloti (217, 229). Group "internal" antigens are shared by strains from widely separated geographical regions.

There is no consistent relationship between ability to invade a host and the nature of rhizobial agglutinogens. The lack of fundamental relationship between the antigenic constitution and invasiveness is shown clearly in cases where loss of the latter property has occurred, leaving specific agglutinability unchanged (23, 199, 254). In one such case no antigenic difference could be found even when the invasive and noninvasive forms were fully studied by means of quantitative antibody absorption. Nor is there any evident relationship between effectiveness in N_2 fixation and antigenic constitution; changes in effectiveness can occur with little, if any, change in serological properties (219, 255).

Experience in regard to bacteriophage susceptibility has been conflicting. On the one hand, there have been cases where one bacteriophage was able to attack rhizobia of quite different antigenic constitution (219, 195), and on the other the presence of certain surface antigens appeared necessary to provide receptor sites (194). Demonstrable modification of surface antigens as a result of lysogenic conversion was associated with failure of the bacteriophage to adsorb on cells that had been lysogenized by the same strain; in this case receptor sites were evidently associated with a somatic antigen (196).

The cultural appearance of rhizobia can change markedly without corresponding changes in antigenic constitution. Even saline-unstable ("rough") mutants retain the ability to react with specific antisera (24). Variant cultural forms, including small-colony, nongummy variants of R. trifolii are unchanged or retain at least the major antigens of the parent strain (22).

7.10 TAXONOMY

7.10.1 Present Taxonomic Position

7.10.1.1 Family Rhizobiaceae

In the 7th edition of Bergey's Manual (256), Rhizobium was one of three genera which made up the Family Rhizobiaceae within the Order Eubacteriales. The current revision reduces the genera to Rhizobium and Agrobacterium, both generally capable of inciting cortical hypertrophies on plants, but largely distinguished on the basis of their nature. Whereas morphologically organized nodules (generally restricted to plants in the legume family) are characteristic of Rhizobium, disorganized galls are produced on many kinds of plants by Agrobacterium. Some differences in metabolic capacities and requirements have also been invoked for inter-generic distinction.

7.10.1.2 Rhizobium Frank, 1889

Strictly speaking, Phytomyxa Schroeter, 1886, has priority and is the valid generic name. However the Judicial Commission of the International Committee on Nomenclature of Bacteria has recently validated Rhizobium, in view of long-standing usage and the confusion that would result from a change of generic name (258).

The general characteristics of the genus have been outlined in III.7.1.1, but its most distinctive feature is the capacity of its members to invade a leguminous plant, so as to cause the production of a morphological and metabolic relationship that may result in the fixation of N_2. Most of the rhizobia in the nodule become modified into "bacteroids" and occur singly or in groups within an enveloping plant-cell membrane.

7.10.1.3 Species

The type species is R. leguminosarum Frank, which is currently applied to the rhizobia that show an invasive preference for host genera like Pisum and Vicia. This and other species (in the currently used sense) are best recognized, on the basis of host preferences:

Species	Preferred hosts
Rhizobium leguminosarum	Pisum, Vicia, Lathyrus, Lens.
R. trifolii	Trifolium
R. phaseoli	Phaseolus vulgaris, R. multiflorus (P. angustifolia)
R. meliloti	Medicago, Melilotus, Trigonella
R. lupini	Lupinus, Ornithopus
R. japonicum	Glycine max.

The first four are fast, the last two generally slow growers. The position of the large number of strains having other host preferences (e.g., the "cowpea miscellany" and the Lotus rhizobia) has yet to be defined, particularly since the cowpea type may infect a non-legume (see III.6.2).

7.10.1.4 Host Specificity as Basis for Speciation

As useful as host preference is from a practical point of view and as a taxonomic character to be given proper weight, accumulated experience has made its shortcomings painfully obvious. The greatest difficulty is with the vast and steadily increasing collection of rhizobia that, for want of a better place, get dumped in the "cowpea miscellany," and which show confusing degrees of cross invasiveness on the one hand, and equally confusing instances of specificity on the other. The separate taxonomic validity of R. lupini and R. japonicum is suspect; some "cowpea rhizobia" nodulate Phaseolus vulgaris, the classical host of R. phaseoli, and some strains of R. phaseoli nodulate other species ordinarily host to "cowpea rhizobia." The Lotus rhizobia pose their own problems. Those from Lotus corniculatus are fast-growing rhizobia;

Taxonomy   341

those from other species of Lotus would not be too
much out of place with R. lupini.
   The slow-growing rhizobia have generally been
recorded as having polar or subpolar flagella, whereas
lateral (peritrichate) flagellation is regarded as the
typical characteristic of fast growers. This distinction is blurred by the observation that fast-growing
strains, isolated from Lupinus densiflorus and more
compatible with Lotus corniculatus and Anthyllis
vulneraria than with slow-grower hosts (Ornithopus
sativus and Lotus uliginosus) are dominantly subpolar
(2). The situation is further confused by the possibility of lateral flagella being more easily fractured
than subpolar (3).
   The "defined" species are not without their own
problems. Certain isolates from African clover are
ill adapted to the Mediterranean clover species and
vice versa (259). Within R. meliloti there are invasiveness subgroups based on different Medicago species,
fortunately with a species like M. sativa acting as a
more widely invasible "bridging" host (229).
   Another aspect of host specificity and rhizobial
species is shown in a critical evaluation of the root
hair deforming capacity of the bacterial partner. A
likely close relationship between R. leguminosarum
and R. trifolii was shown by the capacity of strains
of R. trifolii to cause the marked curled condition of
root hairs of Pisum and of R. leguminosarum to produce
this result with Trifolium (260, 261). This is a
response generally specific to a nodule-producing
association, and its occurrence in these cases, short
of the full invasive process, can be taken as further
evidence of relatedness between these two species.
Other, evidently more distantly related rhizobia (but
not agrobacteria) caused less specific root-hair
responses (branching and/or "moderate curling"). The
reaction of Ornithopus (a normal host for R. lupini)
to a rhizobium isolated from Anthyllis (262) is of
interest in view of the relationship proposed for this
group.

7.10.2 New Taxonomic Approaches to Rhizobial
Classification

Numerical taxonomy, based on the Adansonian principle
of using a large number of unweighted characters with

a sufficient sampling of strains, lends itself to comprehensive surveys and is likely, at least to provide guidelines for subsequent more detailed investigation.

Information on deoxyribonucleic acid base-pair ratios (guanine + cytosine) ÷ (adenine + thymine), and DNA homology have provided more detailed data for restricted areas of the problem.

7.10.2.1 Information from Numerical Taxonomy

Investigation of 83 widely representative rhizobial strains and others that included Agrobacterium (263) led to these conclusions:

a. Strains of R. leguminosarum, R. trifolii and R. phaseoli (with the exception of one strain of the last) formed a homogeneous, interrelated group. This merited consolidation into the species R. leguminosarum.
b. The fast growers, expecially R. meliloti, grouped with the agrobacteria. The incorporation of the latter as a species of Rhizobium was suggested.
c. Slow-growing rhizobia formed a group distinct from all the fast growers, and merited separate generic status.

Reexamination of these data, and the application of different clustering techniques, supported the division between fast- and slow-growing rhizobia, but indicated that Agrobacterium should remain a separate taxon, though more closely related to the fast-growing rhizobia than to the slow (264). Decisions as to specific ranking within the fast and slow groups seemed to need more information than was then available. The groups from the technique of "flexible sorting," and their interrelatedness, are indicated in Table 7.10.

A further study of the family Rhizobiaceae, though with relatively few strains of each genus, indicated separate species status for R. meliloti and R. leguminosarum (the latter, a composite of other fast growers) (257). Some of the agrobacteria would be better regarded as species of Rhizobium, and the slow growers elevated to generic rank.

A more complicated analysis and clustering technique has recently been used to examine an earlier

TABLE 7.10. Groups of Rhizobia and Agrobacteria by "Flexible" Sorting (264)

1. R. trifolii (16), R. leguminosarum (11),
    R. phaseoli (2), R. spp. (7).

2. R. meliloti (11), R. phaseoli (2),
    R. trifolii (1), R. leguminosarum (1),
    R. spp. (2), Agrobacterium (2).

3. Agrobacterium (16).

4. R. trifolii (2), R. leguminosarum (1),
    R. japonicum (1), R. lupini (1),
    R. spp. (2).

5. R. phaseoli (1), R. lupini (9),
    R. japonicum (5), R. spp. (9).

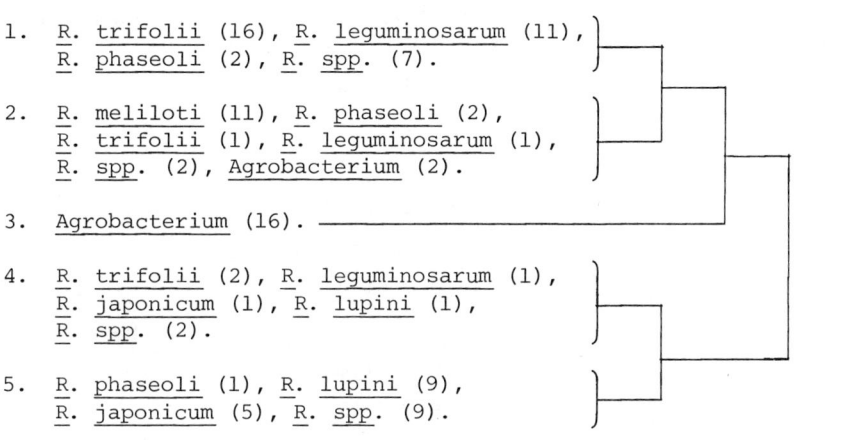

suggestion that the rhizobia and the agrobacteria are related to Arthrobacter (265). Although the tests that were used, by omitting morphological and gram stain characters as well as reaction with plant host, could be expected to lessen distinction between these gram negative bacteria, and the gram positive, wholly saprophytic arthrobacters, it was found that the last were, in fact, clearly separate from rhizobia and the agrobacteria, which were somewhat related to each other. Otherwise the work is of limited value for the rhizobia.

7.10.2.2 DNA-Base Composition and Homology

A compilation of published data on DNA-base composition for bacteria generally (266) emphasizes the relative homogeneity of rhizobia, and points to affinities between them and the pseudomonads. The G + C content for most rhizobia (Table 7.11) ranged from 59 to 65%, with a distinction between those that have been classed as peritrichously flagellate (59.1-63.1; coinciding with almost all of the fast growers) and those where the flagella appeared to be subpolar (61.6-65.5, all slow growers) (3). A sample of 41 strains fell within the range of 58-64% without relationship to the present specific designations (267). A large

TABLE 7.11. DNA-base Composition of Rhizobia (3)[a]

Host group	(G + C)% according to flagellation	
	Peritrichate	Subpolar
Trifolium	60.4-61.1	
Pisum	59.7-62.5	
Phaseolus vulgaris	60.6-62.3	
Medicago, etc.	62.3-62.5	
Lotus	60.6	
Lupinus	59.1-61.8	63.3
Glycine max		63.1-64.8
Vigna	59.9-61.8	61.6-65.5
Others	60.5-63.1	64.3

[a] Note high G + C content of isolate from Lotononis bainesii (68-69%) (296).

collection of R. japonicum yielded (G + C)% between 61.5 and 64.1, and significant differences could be established within this restricted range (238). Frequency distribution (Fig. 7.2) indicates, that while most strains occur about the mode, in what could be taken as a normal distribution, there is an apparent skewness towards the bottom of the range (61.5-61.8). The failure of the (G + C)% to correlate with a phenotypic characteristic such as nutrient requirement is a reminder of the lack of sensitivity of the method. Comparison of the base composition and theoretical DNA homology showed a linear relationship between the two, with the possibility of subgroups existing within the species (296). The pink colony form isolated from Lotononis bainesii has a much higher G + C content (68-69%) than other rhizobia (296).

Deoxyribonucleic acid hybridization with DNA from R. leguminosarum showed marked homology with R. trifolii and distinct nonhomology with R. japonicum. Other hybridization experiments supported the three rhizobial groupings (R. leguminosarum—R. trifolii—R. phaseoli; R. meliloti; slow growers) and the likelihood of relatedness between rhizobia and the

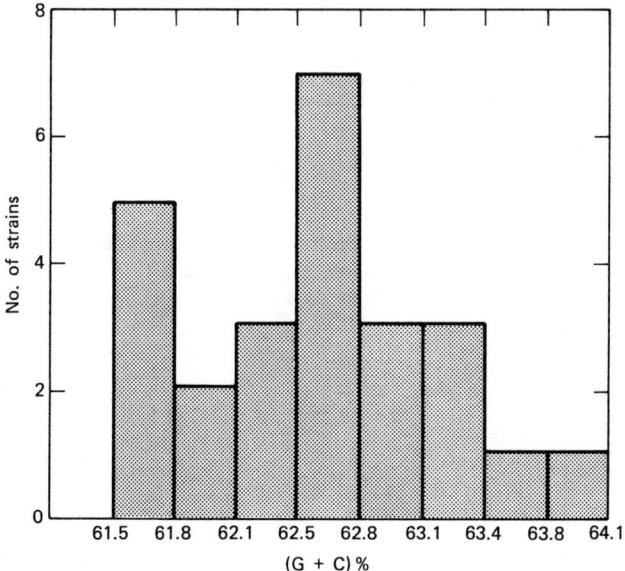

Fig. 7.2. Distribution of strains of R. japonicum on basis of (G + C)%.

agrobacteria (268). One can look at the location of each nomen species in relation to its homology with DNA of A. tumefaciens and with the DNA of R. leguminosarum (Fig. 7.3). The three species of Rhizobium are clearly separated, particularly in their relationship to the reference DNA of R. leguminosarum. In this regard R. meliloti is no better than the agrobacteria (omitting atypical A. pseudotsugae, which appears to be quite out of place); R. japonicum would appear to be even less related to R. leguminosarum. The comparison using A. tumefaciens DNA as reference distinguishes three of the species of that genus as a group distinct from the rhizobia. One Agrobacterium (A. rhizogenes) locates closer to the rhizobia than to the typical members of its own genus. R. japonicum, along with Chromobacterium is close to the two representatives of the Pseudomonodales. These attempts to obtain information about the rhizobial genotype have been more useful in demonstrating the broader groups

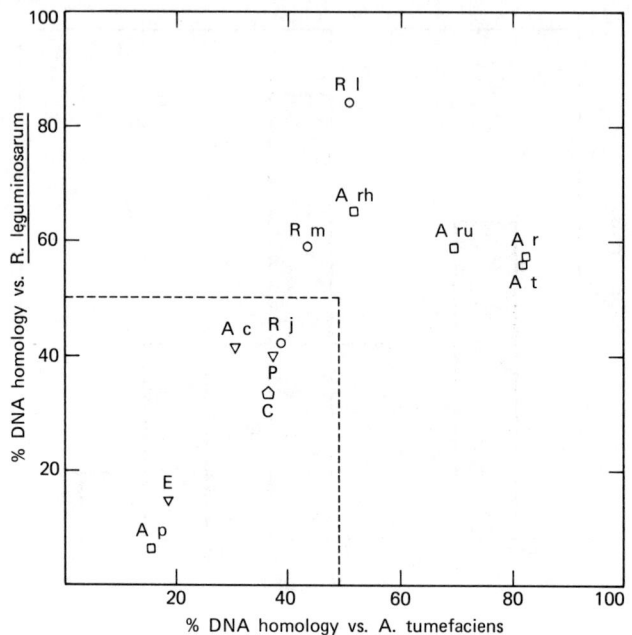

Fig. 7.3. Location of nomen species in relation to homology with DNA of Rhizobium leguminosarum and Agrobacterium tumefaciens. Rl, R. leguminosarum; Rm, R. meliloti; Rj, R. japonicum; At, A. tumefaciens; Ar, A. radiobacter; Arh, A. rhizogenes; Aru, A. rubi; Ap, A. pseudotsugae; Ac, Acetobacter; E, Escherichia; C, Chromobacterium; P, Pseudomonas. Homologous strain values omitted.

than in predicting finer phenotypic behavior, some of which, such as symbiotic specificity, are of immediate practical concern.

7.10.3 Other Relevant Information

7.10.3.1 Rhizobial Phage

Although some rhizobial phages are strain or species specific their usefulness for taxonomic analysis is limited. Phages for Lotus rhizobia obtained from L.

corniculatus were in variable degree (22-77%) able to lyse rhizobia obtained from the same group of hosts (269). In fact their uniform failure to affect rhizobia from Lotus uliginosus fits in well with other evidence pointing to a distinct dichotomy within the lotus rhizobia that coincides with respective L. corniculatus and L. uliginosus affinities. With the exception of a strain of R. meliloti, all other susceptible rhizobia were isolated from a few specialized legumes that included Anthyllis, Astragalus (again relevant to affinities of some lotus rhizobia). Forty-five fast-growing rhizobia, 10 of R. lupini, and 123 of cowpea and other hosts were all negative.

Much wider cross-reactivity between phage for fast- and slow-growing rhizobia has also been reported (270). Neutralization with antiphage serum also showed close serological relationships between phages of R. lupini and R. trifolii (202). An earlier report from the same workers dealt with a collection of more specific phage (192).

7.10.3.2 Antigenic Constitution

A large collection of rhizobia surveyed for flagellar and somatic antigen activity (225) were distributed (225) into three self-contained groups: (i) R. trifolii, R. leguminosarum, and R. phaseoli; (ii) R. meliloti; (iii) R. lupini, R. japonicum, and cowpea rhizobia. However, generally most of the cross-reactions within each group were negative, even when flagellar and somatic reactions were combined to maximize evidence for cross-reactivity. It is, therefore, evident that the surface antigens of rhizobia, though very valuable for the recognition of strains, are limited in their usefulness for species recognition.

Internal antigens, on the other hand, show broader group specificities. Lines produced with fast-diffusing thermolabile antigens in the gel diffusion of broken cells against an appropriate antiserum (e.g., of R. trifolii) reveal a degree of relatedness amongst the fast growers and non-cross-reactivity between the fast and slow growers. Closer relationship among members of the R. leguminosarum—R. trifolii—R. phaseoli than between them and other fast growers was also indicated as is a relationship between R. meliloti and Agrobacterium. Conversely, antisera to R. japonicum revealed common internal lines with a wide range of slow-growing rhizobia, which indicated a degree of relatedness, readily distinguished from a lower order interaction with some fast-growing strains (Table 7.9). Fast-growing strains from Lotus and Leucaena grouped with R. meliloti and the agrobacteria, occupying an intermediate place between the R. leguminosarum—R. trifolii group and the slow growers. An extension, in the author's laboratory, of this approach with antisera to a slow-growing cowpea strain (CB756) and to a fast-growing Cicer isolate has substantiated the validity of such groups. Interestingly, Lotononis bainesii isolations again stand out from other rhizobia by their failure to react with antisera to either fast or slow growers. The same approach has been used with antisera to broken cells of Agrobacterium; this also gave cross-reactions with broken cell antigen of fast-growing rhizobia (including one from Lupinus densiflorus) (271).

The technique of internal antigen analysis holds considerable promise as a relatively simple taxonomic tool and for positive identification of rhizobia in the absence of reaction with more strain-specific surface antigens.

7.10.3.3 Patterns of Isoenzymes

The distribution of esterases of 52 strains (six species) of Rhizobium has been examined after electrophoresis on cellulose acetate (82). The five fast-growing species (R. trifolii, R. leguminosarum, R. phaseoli, R. meliloti, and rhizobia from Lotus corniculatus) yielded esterase patterns related in some degree to the fastest-moving line found in all of the 18 strains of R. trifolii. Agreement with the R. trifolii pattern was closest with R. leguminosarum and the fast lotus strains, rather less with R. phaseoli and R. meliloti. The esterase of R. japonicum was observed in only one of five strains and was unaligned with those of the fast-growing strains. A similar comparison with agrobacteria and R. meliloti (298) again pointed to a degree of relatedness between the genera.

A similar approach has been used with 3-hydroxybutyrate dehydrogenase (78) and aspartase aminotransferase (272). Some tentative conclusions were drawn as to the sharing of some bands within and between the different taxonomic groupings. However, dissimilarity

Taxonomy

between strains in the same species was often as great, or greater, as between strains in different species, or even between the fast- and slow-growing rhizobia (e.g., those involving R. leguminosarum, R. phaseoli, R. lupini, and R. japonicum). The strains of R. trifolii and R. meliloti were more homogeneous in the tested sample. There was no indication of greater dissimilarity in comparisons between fast and slow strains than comparisons within these two groups. These attempts at a taxonomy based on enzyme patterns have made only a limited contribution towards the clarification of rhizobial relationships.

7.10.4 Apparent Taxonomic Relatedness within the Genus Rhizobium

A major acceptable distinction is between those members of the genus that make relatively fast growth on yeast mannitol agar and lower the pH to a moderate degree and those that grow very slowly and produce an alkaline rather than an acid end-point on the same medium. This has been accepted in the current revision of Bergey's Manual.

7.10.4.1 Fast Growers

These comprise Rhizobium leguminosarum, R. trifolii, R. phaseoli, R. meliloti, as presently defined, as well as fast growers able to nodulate various members of the cowpea group of hosts, notably rhizobia effectively nodulating Lotus corniculatus, Lupinus densiflorus, Cicer and Leucaena Leucocephala (273).

At least three subgroups can be further postulated within the fast growers:
1. (a) R. Leguminosarum—R. trifolii
   (b) R. phaseoli
2. R. meliloti
3. Other fast growers from hosts such as Lotus, Lupinus, Cicer, and Leucaena.

Rhizobium phaseoli seems less related to the other two members of that subgroup than the latter are to each other. It has in common with subgroup 3 the capacity to nodulate hosts compatible with the slow growers. Further investigation will probably justify subdivision within 3. above.

7.10.4.2 Slow Growers

The slow growers include two reasonably well-defined subgroups, R. lupini and R. japonicum, but have besides a vast array of as yet poorly defined rhizobia with variable host specificities. There is a good deal of evidence to lead one to the conclusion that the slow growers do in fact merit elevation to separate generic status. This would, however, be opposed by some on the ground that it would lead to confusion, and pose some difficulty in finding an acceptable generic name. None the less, a decision along these lines seems now to be required.

There are rhizobia that cross the usually sharp boundaries between the hosts for fast and those for slow growers. One example is the fast-growing R. phaseoli, which preferentially nodulates Phaseolus vulgaris, but occasionally invades other species belonging to the "cowpea miscellany" (e.g., Macroptilium lathyroides or M. atropurpureum) also regarded formerly as a species of Phaseolus. Conversely some slow "cowpea rhizobia" are able occasionally to nodulate P. vulgaris. The Lotus rhizobia are even more striking and divide into the fast-growing strains (having a marked affinity for Lotus corniculatus, Lupinus densiflorus, Anthyllis vulneraria, but are also able to nodulate Ornithopus sativus) and slow growers, which are more likely to form effective nodules with Lotus uliginosus and L. pedunculatus.

If, as seems likely, the future sees some combinations of present species (as for example, R. leguminosarum to serve for the present R. trifolii, and perhaps R. phaseoli as well) there will be a case for retaining a degree of taxonomic distinction to reflect the significant and marked symbiotic specificities that would then exist within the consolidated species. Perhaps R. leguminosarum symbiotype trifolii would be the nomenclature for clover rhizobia, and R. japonicum symbiotype vignae for a typical cowpea strain.

7.10.5 Wider Relationships

The long-recognized close relationship between Agrobacterium and many rhizobia (12) is well established, particularly in the case of R. meliloti. Examples of relatedness, besides those already noted, are:

nonutilization of urea and biuret by R. meliloti and Agrobacterium (which differentiates them from most other rhizobia)(37); shared susceptibility to Bdellovibrio (104); and, between R. japonicum and A. tumefaciens, the possession of a large amount of phosphatidyl choline (238). There are, however, probably sufficient differences to justify the distinction at the generic level. A range of tests permits reasonably safe distinction between them in addition to nodule formation in the case of Rhizobium.

There is evidence of similarities between a considerable collection of strains of R. japonicum (presumably representative of slow-growing rhizobia generally) and the pseudomonads. These include dependence on the Entner-Doudoroff and other shared catabolic pathways (69, 238), DNA-base ratio and homology data and the occurrence of large amounts of phosphatidyl choline, rather than neutral lipids.

This proposition would take at least the slow-growing rhizobia out of the order Eubacteriales. Then a decision would have to be made for the fast growers, which have been generally regarded as having peritrichous flagella (though not abundantly endowed), and which share only some of the points of similarity raised by Elkan (238). The fast growers appear to have all three usual methods of carbohydrate catabolism (68, 274). DNA-base ratios of this group agrees reasonably well with the pseudomonads; DNA homology is not so convincing. The large amount of phosphatidyl choline reported for Agrobacterium offers indirect support for this as does a report of transformation between a fast-growing rhizobium and a soil Pseudomonas (275).

7.10.6 Recognition and Characterization of Rhizobium

The decision as to whether a culture is or is not a Rhizobium commonly depends on a plant test, as do most decisions as to its kind. Certain conformable and contradictory characteristics are worth noting however (216).

Some distinction between groups of rhizobia and the recognition of Agrobacterium is likely to be possible on the basis of cultural and biochemical characteristics (1) (Table 7.12). Typical R. japonicum—R. lupini—cowpea rhizobia produce very small colonies

TABLE 7.12. Cultural and Biochemical Characteristics of the Major Groups of Rhizobia and Distinction from Agrobacterium (Collated from Graham and Parker, ref. 1)[a]

	R. trifolii, R. leguminosarum, and R. phaseoli	R. meliloti	R. japonicum, R. lupini, and cowpea	Agro-bacterium
i. Colony size (mm)	2+	2+	≤1	2+
ii. Raffinose used	0.84	0.73	0.19	0.89
iii. Citrate used	0.03	0.27	0	0.72
iv. Growth, pH 4.5	0.65	0.09	0.83	0.88
v. Growth, pH 9.5	0	0.91	0	0.56
vi. Growth, 2% NaCl	0	0.45	0	0.61
vii. Growth, 39°C	0	0.72	0	0.06
viii. Response to thiamine	0.65	0	0	0.11
ix. Response to pantothenate	0.89	0	0.03	0
x. Response to biotin	0.59	0.36	0.23	0
xi. H_2S produced	0	0.81	0	1.00
xii. Penicillinase produced	0.08	0.09	0.77	0.11
xiii. Precipitate from Ca glycerophosphate	0	0.55	0	1.00

[a]Entries, except in the case of (i), represent the proportion of rhizobial strains in each group, or species, showing the characteristic.

TABLE 7.13. Nodulation Specificity of Rhizobial Species[a] (Collated from Graham and Parker, ref. 1)

Test host	R. legumi-nosarum	R. trifolii	R. phaseoli	R. lupini	R. japonicum	"Cowpea" rhizobia	R. meliloti
Vicia sativa	0.85	0.1	0	0	0	0	0
Trifolium repens	0.4	0.7[b]	0	0	0	0	0
Phaseolus vulgaris	0.15	0.05	1.0	0.2	0.7	0.5	0
Macroptilium lathyroides[c]	0	0	0.8	0	0.3	0.2	0
Ornithopus sativus	0	0	0	1.0	0.3	0.05	0
Glycine max	0	0	0	0.1	1.0	0.3	0
Vigna sinensis	0	0	0	0.2	0.3	0.5	0
Medicago sativa[d]	0	0	0	0	0	0	1.0

[a] Entries represent the proportion of rhizobial strains in each species nodulating the specified host.
[b] Partly reflects poor compatibility of isolates from African clovers; those from T. ambiguum are also poorly compatible with Mediterranean clovers.
[c] Formerly Phaseolus; M. atropurpureum is more widely susceptible to slow-growing rhizobia including those, such as isolates of Lotononis and Leucaena, which are very host specific.
[d] Whereas M. sativa appears to be always invasible by R. meliloti — other species of Medicago are likely to be more strain specific. Some strains isolated from Leucaena nodulate M. sativa.

and utilize citrate poorly; they commonly produce penicillinase, and do not cause a precipitate in calcium glycerophosphate medium. Rhizobium meliloti often shares with Agrobacterium tolerance to high pH and NaCl, the production of H_2S, and, sometimes, precipitates from Ca glycerophosphate. By way of distinction, Agrobacterium more often utilizes citrate, is more acid tolerant, but less tolerant of 39°C, and regularly produces H_2S and precipitates from Ca glycerophosphate.

Plant tests that permit reasonable distinction between present species (1) are summarized in Table 7.13 (appearing on p. 353). On some occasions strains of R. phaseoli could be expected to show rather more cross reactivity with other hosts of the clover-pea-vetch-bean group. However, the three species, or symbiotypes, in this group can be generally distinguished by their preferred host where the reaction often results in a much more effective association than in other seemingly less compatible cases. Similarly among the slow growers, separation into the present subgroups is indicated by the nature of the preferred host. R. meliloti can almost always be readily distinguished from the other rhizobia, as well as from the agrobacteria, by its nodulation of Medicago sativa.

7.11 REFERENCES

1. Graham, P. H., and C. A. Parker, Plant Soil, 20, 383 (1964).
2. Abdel-Ghaffer, A. S., and H. L. Jensen, Arch. Mikrobiol., 54, 393 (1966).
3. De Ley, J., and A. Rassel, J. Gen. Microbiol., 41, 85 (1965).
4. Dudman, W. F., J. Bacteriol., 95, 1200 (1968).
5. Humphrey, B., and J. M. Vincent, J. Gen. Microbiol., 59, 411 (1969).
6. Vincent, J. M., B. Humphrey, and R. J. North, J. Gen. Microbiol., 29, 551 (1962).
7. Jordan, D. C., and I. Grinyer, Can. J. Microbiol., 11, 721 (1965).
8. Jordan, D. C., and W. H. Coulter, Can. J. Microbiol., 11, 709 (1965).
9. Allen, E. K., and O. N. Allen, in Encyclopedia of Plant Physiology, W. Ruhland, Ed., Springer Verlag, Berlin, pp. 48-118, 1958.

References

10. Golebiowska, J., Poste pow Nauk Rolniczych, 20, 151 (1960).
11. Jordan, D. C., Bact. Rev., 26, 119 (1962).
12. Fred, E. B., I. L. Baldwin, and E. McCoy, Root Nodule Bacteria and Leguminous Plants, University of Wisconsin, Madison, 1932.
13. Bisset, K. A., and C. M. F. Hale, J. Gen. Microbiol., 5, 592 (1951).
14. Bisset, K. A., J. Gen. Microbiol., 7, 233 (1952).
15. Heurtos, M. G., and G. T. Dominguez, Microbiol. Espan., 10, 305 (1957).
16. Graham, P. H., C. A. Parker, A. E. Oakley, R. T. Lange, and I. J. V. Sanderson, J. Bacteriol., 86, 1353 (1963).
17. Marshall, K. C., Aust. J. Biol. Sci., 20, 429 (1967).
18. Humphrey, B. A., and J. M. Vincent, J. Bacteriol., 98, 845 (1969).
19. Humphrey, B., and J. M. Vincent, J. Gen. Microbiol., 29, 557 (1962).
20. Norris, D. O., Aust. J. Agr. Res., 9, 629 (1958).
21. Hahn, N. J., Can. J. Microbiol., 12, 725 (1966).
22. Vincent, J. M., Proc. Linn. Soc. (N. S. W.), 87, 8 (1962).
23. Almon, L., and I. L. Baldwin, J. Bacteriol., 26, 229 (1933).
24. Jensen, H. L., Proc. Linn. Soc. (N. S. W.), 67, 98 (1942).
25. Wilson, P. W., The Biochemistry of Symbiotic Nitrogen Fixation, University of Wisconsin Press, Madison, 1940.
26. Barnet, Y. M., and J. M. Vincent, Aust. J. Sci., 32, 208 (1969).
27. Vandecaveye, S. C., and H. Katznelson, J. Bacteriol., 31, 465 (1936).
28. Barnet, Y. M., Ph. D. Thesis, University of New South Wales (1968).
29. Neal, O. R., and R. H. Walker, J. Bacteriol., 30, 173 (1935).
30. Ishizawa, S., J. Sci. Soil (Tokyo), 24, 163 (1953).
31. Graham, P. H., Antonie von Leeuwenhoek, 30, 68 (1964).
32. Elkan, G. H., and I. Kwik, J. Appl. Bacteriol., 31, 399 (1968).
33. Bergersen, F. J., Ann. Rev. Plant Physiol., 22, 121 (1971).
34. Jordan, D. C., and C. L. San Clemente, Can. J. Microbiol., 1, 659 (1955).

35. Bergersen, F. J., *Aust. J. Biol. Sci.*, **14**, 349 (1961).
36. Jordan, D. C., *Can. J. Bot.*, **30**, 693 (1952).
37. Jensen, H. L., and M. J. Schrøder, *J. Appl Bacteriol.*, **28**, 473 (1965).
38. Allen, E. K., and O. N. Allen, *Bact. Rev.*, **14**, 273 (1950).
39. Graham, P. H., *J. Gen. Microbiol.*, **30**, 245 (1963).
40. Burton, M. O., and A. G. Lochhead, *Can. J. Bot.*, **30**, 521 (1952).
41. Hattingh, M. J., and H. A. Louw, *S. Afr. Med. J.*, **39**, 359 (1965).
42. Jensen, H. L., and M. Schrøder, *Arch. Mikrobiol.*, **58**, 127 (1967).
43. Schwinghamer, E. A., *Can. J. Microbiol.*, **15**, 611 (1969).
44. Thorne, D. W., and R. H. Walker, *Soil Sci.*, **42**, 231 (1936).
45. Ferry, P., H. Blachère, and M. Obaton, *Ann. Agron.*, **II**, 219 (1959).
46. Norris, D. O., *Nature (London)*, **182**, 734 (1958).
47. Norris, D. O., *Aust. J. Agr. Res.*, **60**, 651 (1959).
48. Vincent, J. M., and J. R. Colburn, *Aust. J. Sci.*, **23**, 269 (1961).
49. Vincent, J. M., *J. Gen Microbiol.*, **28**, 653 (1962).
50. Vincent, J. M., and B. Humphrey, *Nature (London)*, **199**, 149 (1963).
51. Humphrey, B. A., and J. M. Vincent, *J. Gen. Microbiol.*, **41**, 109 (1965).
52. Vincent, J. M., and B. A. Humphrey, *J. Gen. Microbiol.*, **54**, 397 (1968).
53. Humphrey, B. A., K. C. Marshall, and J. M. Vincent, *J. Bacteriol.*, **95**, 721 (1968).
54. Humphrey, B. J., and J. M. Vincent, *Nature (London)*, **212**, 212 (1966).
55. Vincent, J. M., and C. H. Jancey, *Nature (London)*, **195**, 99 (1962).
56. Lowe, R. H., H. J. Evans, and S. Ahmed, *Biochem. Biophys. Res. Commun.*, **3**, 675 (1960).
57. Lowe, R. H., and H. J. Evans, *J. Bacteriol.*, **83**, 210 (1962).
58. Kliewar, M., and H. J. Evans, *Arch. Biochem. Biophys.*, **97**, 428 (1962).
59. Kliewar, M., R. Lowe, P. A. Mayeux, and H. J. Evans, *Plant Soil*, **21**, 153 (1964).
60. Wilson, D. O., and H. M. Reisenauer, *Plant Soil*, **32**, 81 (1970).

References

61. De Hertogh, A. A., P. A. Mayeux, and H. J. Evans, J. Biol. Chem., 239, 2446 (1964).
62. Cowles, J. R., H. J. Evans, and S. A. Russell, J. Bacteriol., 97, 1460 (1969).
63. Wilson, D. O., and H. M. Reisenauer, J. Bacteriol., 102, 729 (1970).
64. Steinborn, J., and R. J. Roughley, J. Appl. Bacteriol., 37, 93 (1974).
65. Sherwood, M. T., J. Appl. Bacteriol., 33, 708 (1970).
66. Ertola, R. J., L. A. Mazza, A. P. Balatti, C. M. Cuevas, and R. Daguerre, Soil Sci., 108, 373 (1969).
67. Bowen, G. D., and M. M. Kennedy, Queensland J. Agr. Sci., 16, 177 (1959).
68. Katznelson, H., and A. C. Zagallo, Can. J. Microbiol., 3, 879 (1957).
69. Martinez-De Drets, G., and A. Arias, J. Bacteriol., 109, 467 (1972).
70. Keele, B. R., P. B. Hamilton, and G. D. Elkan, J. Bacteriol., 101, 698 (1970).
71. Johnson, G. V., H. J. Evans, and T. M. Ching, Plant Phys., 41, 1330 (1966).
72. Martinez De Drets, G., and A. Arias, J. Bacteriol., 103, 97 (1970).
73. Jordan, D. C., Can. J. Microbiol., 5, 131 (1959).
74. Jordan, D. C., J. Bacteriol., 65, 220 (1953).
75. Jordan, D. C., Can. J. Microbiol., 1, 743 (1955).
76. Fottrell, P. F., and P. Mooney, J. Gen. Microbiol., 59, 211 (1969).
77. Ryan, E., F. Bodley, and P. F. Fottrell, Plant Soil, Special Vol., 545 (1971).
78. Fottrell, P. F., and A. O'Hora, J. Gen. Microbiol., 57, 287 (1969).
79. Zelazna, I., Acta Microbiol. Polon., 11, 329 (1962).
80. Lorkiewicz, Z., I. Zelazna, and B. Przybojewska, Acta Microbiol. Polon., 14, 225 (1965).
81. Moustafa, E., and R. M. Greenwood, N. Z. J. Sci., 10, 548 (1967).
82. Murphy, P. M., and C. L. Masterson, J. Gen. Microbiol., 61, 121 (1970).
83. Appleby, C. A., and F. J. Bergersen, Nature (London), 182, 1174 (1958).
84. Hendry, G. S., and D. C. Jordan, Can. J. Microbiol., 15, 242 (1969).
85. Norris, D. O., Plant Soil, 22, 143 (1965).
86. Oplistilova, K., and V. Vancura, Rost. Vyroba, 36, 734 (1963).

87. Katznelson, H., and S. E. Cole, Can. J. Microbiol., 11, 733 (1965).
88. Bulard, C., B. Guichardon, and J. Rigaud, Ann. Inst. Pasteur, 105, 150 (1963).
89. Rigaud, J., C. R. Acad. Sci., Paris, 262, 100 (1966).
90. Rigaud, J., Arch. Mikrobiol., 72, 297 (1970).
91. Hartmann, T., and K. W. Glombitza, Arch. Mikrobiol., 56, 1 (1967).
92. Hopkins, E. W., W. H. Peterson, and E. B. Fred, J. Am. Chem. Soc., 52, 3659 (1930).
93. Schluchterer, E., and M. Stacey, J. Chem. Soc., 776 (1945).
94. Haworth, N., and M. Stacey, Ann. Rev. Biochem., 17, 97 (1948).
95. Humphrey, B. A., and J. M. Vincent, J. Gen. Microbiol., 21, 477 (1959).
96. Humphrey, B. A., Nature (London), 184, 1802 (1959).
97. Graham, P. H., Antonie van Leeuwenhoek, 31, 349 (1965).
98. Dudman, W. F., J. Bacteriol., 88, 640 (1964).
99. Dudman, W. F., and M. Heidelberger, Science, 164, 954 (1969).
100. Amarger, N., M. Obaton, and H. Blachère, Can. J. Microbiol., 13, 99 (1967).
101. Zevenhuizen, L. P. T. M., J. Gen. Microbiol., 68, 239 (1971).
102. Clapp, C. E., and R. J. Davis, Soil Biol. Biochem., 2, 109 (1970).
103. Bailey, R. W., R. M. Greenwood, and A. Craig, J. Gen. Microbiol., 65, 315 (1971).
104. de Leizaola, M., and R. Dedonder, C. R. Acad. Sci., 240, 1825 (1955).
105. Skrdleta, V., Rostl. Vyroba, 39, 23 (1966).
106. Keele, B. B., G. H. Elkan, and R. W. Wheat, Bact. Proc., G52, 30 (1967).
107. Dedonder, R. A., and W. Z. Hassid, Biochem. Biophys. Acta, 90, 239 (1964).
108. Burdon, K. L., J. Bacteriol., 52, 665 (1946).
109. Smithies, W. R., N. E. Gibbons, and S. Bayley, Can. J. Microbiol., 1, 605 (1955).
110. Forsyth, W. C. G., A. C. Hayward, and J. B. Roberts, Nature (London), 182, 800 (1958).
111. Hayward, A. C., W. G. C. Forsyth, and J. B. Roberts, J. Gen. Microbiol., 20, 510 (1959).
112. Schlegel, H. G., Flora, 152, 236 (1962).

113. Lundgren, D. G., R. Alper, C. Schnaitman, and R. H. Marchessault, J. Bacteriol., 89, 245 (1965).
114. Albrecht, W. A., and T. M. McCalla, J. Am. Soc. Agron., 29, 76 (1937).
115. Jensen, H. L., Nature (London), 192, 682 (1961).
116. Gillberg, B. O., Arch. Mikrobiol., 62, 328 (1968).
117. Hofer, A. W., and H. B. Little, Appendix to N. Y. State Exp. Stat. Bull., 772 (1956).
118. Essawi, T. M. E., and A. S. Abdel Ghaffar, J. Appl. Bacteriol., 30, 354 (1967).
119. Wilkins, J., Aust. J. Agr. Res., 18, 299 (1967).
120. Vincent, J. M., in Nutrition of the Legumes, E. G. Hallsworth, Ed., Butterworths, London, pp. 108-123, 1958.
121. Vincent, J. M., J. A. Thompson, and K. O. Donovan, Aust. J. Agr. Res., 13, 258 (1962).
122. Amarger, N., M. Jacquemetton, and G. Blond, Arch. Mikrobiol., 81, 361 (1972).
123. McLeod, R. W., and R. J. Roughley, Aust. J. Exp. Agr. An. Husb., 1, 29 (1961).
124. Won, W. D., and H. Ross, Appl. Microbiol., 18, 555 (1969).
125. Holding, A. J., and J. F. Lowe, Plant Soil, Special Vol., 153 (1971).
126. Steinborn, J., and R. J. Roughley, J. Appl. Bacteriol., 37, 93 (1974).
127. Milthorpe, F. L., J. Aust. Inst. Agr. Sci., 11, 89 (1945).
128. Ruhloff, M., and J. C. Burton, Soil Sci., 72, 283 (1951).
129. Hofer, A. W., Soil Sci., 86, 282 (1958).
130. Jakubisiak, B., and J. Golebiowska, Acta Microbiol. Polon., 12, 196 (1963).
131. Kecskes, M., and J. M. Vincent, Agrokem. Talajtan, 18, 57 (1969).
132. Kecskes, M., and J. M. Vincent, Agrokem. Talajtan, 18, 461 (1969).
133. Afifi, N. W., et al., Arch. Mikrobiol., 66, 121 (1969).
134. Diatloff, A., Aust. J. Exp. Agr. An. Husb., 46, 562 (1970).
135. Goss, O. M., and W. A. Shipton, J. Agr. (W. A.), 6, 659 (1965).
136. Fletcher, W. W., and J. W. S. Alcorn, in Nutrition of the Legumes, E. G. Hallsworth, Ed., Butterworths, London, pp. 284-288, 1958.

137. Fletcher, W. W., P. B. Dickenson, and J. C. Raymond, Phyton, 7, 121 (1956).
138. Fletcher, W. W., J. W. S. Alcorn, and J. C. Raymond, Nature (London), 184, 1576 (1959).
139. Dullaart, J., C. A. Wijffelman, and J. Haveman, Antonie van Leeuwenhoek, 37, 219 (1971).
140. Vintika, J., and H. Vintikova, For. Soc. Agr. Sci., 7, 349 (1958).
141. Skrdleta, V., Ved. Pr. Vysk Ust. Rostl. Vyroby, Prahe-Ruzyne, 177 (1965).
142. Graham, P. H., Aust. J. Biol. Sci., 16, 557 (1963).
143. Schwinghamer, E. A., Can. J. Microbiol., 10, 221 (1964).
144. Zelazna-Kowalska, I., Plant Soil, Special Vol., 67 (1971).
145. Hendry, G. S., and D. C. Jordan, Can. J. Microbiol., 15, 671 (1969).
146. Jordan, D. C., Y. Yamamura, and M. E. McKague, Can. J. Microbiol., 15, 1005 (1969).
147. Alexander, D., D. C. Jordan, and M. McKague, Can. J. Biochem., 47, 1092 (1969).
148. MacKenzie, C. R., and D. C. Jordan, Biochem. Biophys. Res. Commun., 40, 1008 (1970).
149. Damery, J. T., and M. Alexander, Soil Sci., 108, 209 (1969).
150. Thorne, D. W., and P. E. Brown, J. Bacteriol., 34, 567 (1937).
151. Thompson, J. A., Nature (London), 187, 619 (1960).
152. Bowen, G. D., Plant Soil, 15, 155 (1961).
153. Fottrell, P. F., S. O'Connor, and C. L. Masterson, Irish J. Agr. Res., 3, 247 (1964).
154. Strijdom, B. W., and O. N. Allen, Can. J. Microbiol., 12, 275 (1966).
155. Hamdi, Y. A., Arch. Mikrobiol., 63, 227 (1968).
156. Strijdom, B. W., and O. N. Allen, Phytophylactica, 1, 147 (1969).
157. Hamdi, Y. A., Plant Soil, 31, 111 (1969).
158. Abdel-Ghaffar, A. S., and H. L. Jensen, IX Int. Congr. Microbiol. Moscow, pp. 87-96 (1966).
159. Elkan, G. H., and I. K. Gwat, Bact. Proc., 4 (1967).
160. Bunn, C., J. J. McNeill, and G. H. Elkan, J. Bacteriol., 102, 24 (1970).
161. Krasil'nikov, N. A., and A. I. Korenyako, Mikrobiologiya, 13, 39 (1944).
162. Hattingh, M. J., and H. A. Louw, S. Afr. J. Agr. Sci., 9, 239 (1966).

163. Hattingh, M. J., and H. A. Louw, S. Afr. J. Agr. Sci., 9, 453 (1966).
164. Parker, C. A., and P. L. Grove, J. Appl. Bacteriol., 33, 253 (1970).
165. Van Schreven, D. A., Plant Soil, 21, 283 (1964).
166. Holland, A. A., and C. A. Parker, Plant Soil, 25, 329 (1966).
167. Thornton, G. D., J. B. Alencar, and F. B. Smith, Proc. Soil Sci. Soc. Am., 14, 188 (1949).
168. Damirgi, S. M., and H. W. Johnson, Agron. J., 58, 223 (1966).
169. Hattingh, M. J., and H. A. Louw, Can. J. Microbiol., 15, 361 (1969).
170. Abdel-Ghaffar, A. S., and O. N. Allen, Trans. Int. Congr. Soil Sci., 4th Amsterdam, pp. 93-96 (1950).
171. Virtanen, A. I., and H. Linkola, Acta Chem. Fenn. Sci., B21, 12 (1948).
172. Afrikyan, E. K., and V. G. Tumanyan, Soils Fert., 22, 285 (1959) (Abstr.).
173. Landerkin, G. B., and A. G. Lockhead, Can. J. Res., Sect. C, 26, 501 (1948).
174. Afrikian, E. G., Ann. Inst. Pasteur, 105, 123 (1963).
175. Wieringa, K. T., Ann. Inst. Pasteur, 105, 417 (1963).
176. Demolon, A., Rev. Gen. Bot., 58, 489 (1952).
177. Schwinghamer, E. A., and R. P. Belkengren, Arch. Mikrobiol., 64, 130 (1968).
178. Schwinghamer, E. A., Soil Biol. Biochem., 3, 355 (1971).
179. Kowalski, M., G. E. Ham, L. R. Frederick, and I. C. Anderson, Bact. Proc., A20, 4 (1967).
180. Kleczkowska, J., Can. J. Microbiol., 3, 171 (1957).
181. Barnet, Y. M., and J. M. Vincent, J. Gen. Virol., 12, 313 (1971).
182. Schwinghamer, E. A., Aust. J. Biol. Sci., 18, 333 (1965).
183. Schwinghamer, E. A., Can. J. Microbiol., 12, 395 (1966).
184. Parker, D. T., and O. N. Allen, Can. J. Microbiol., 3, 651 (1957).
185. Kowalski, M., R. Staniewski, and M. Paraniak, Acta Microbiol. Polon., 12, 180 (1963).
186. Kleczkowska, J., J. Bacteriol., 50, 81 (1945).
187. Staniewski, R., M. Kowalski, and K. Gorkowska, Acta Microbiol. Polon., 12, 184 (1963).

188. Kleczkowski, J., and A. Kleczkowski, J. Gen. Microbiol., 10, 285 (1954).
189. Kleczkowski, J., and A. Kleczkowski, J. Gen. Microbiol., 21, 308 (1959).
190. Laird, D. G., Arch. Mikrobiol., 3, 159 (1932).
191. Conn, H. J., E. J. Bottcher, and C. Randall, J. Bacteriol., 49, 359 (1945).
192. Kowalski, M., and R. Staniewski, Acta Microbiol. Polon., 8, 253 (1959).
193. Schwinghamer, E. A., and D. J. Reinhardt, Aust. J. Biol. Sci., 16, 597 (1963).
194. Marshall, K. C., and J. M. Vincent, Aust. J. Sci., 17, 68 (1954).
195. Barnet, Y. M., J. Gen. Virol., 15, 1 (1972).
196. Barnet, Y. M., and J. M. Vincent, J. Gen. Microbiol., 61, 319 (1970).
197. Almon, L., and P. W. Wilson, Arch. Mikrobiol., 4, 209 (1933).
198. Wilson, P. W., E. W. Hopkins, and E. B. Fred, Soil Sci., 32, 251 (1931).
199. Kleczkowska, J., J. Gen. Microbiol., 4, 298 (1950).
200. Kleczkowska, J., Plant Soil, Special Vol., 47 (1971).
201. Boyd, R. J., A. C. Hildebrandt, and O. N. Allen, Arch. Mikrobiol., 73, 47 (1970).
202. Staniewski, R., M. Kowalski, and I. Lomanska, Acta Microbiol. Polon., 12, 187 (1963).
203. Marshall, K. C., Nature (London), 177, 92 (1956).
204. Barnet, Y. M., Ph. D. Thesis, University of New South Wales, 1968.
205. Davies, R. J., Bact. Proc., 10 (1958).
206. Szende, K., and F. Ordogh, Naturwissenschaften, 17, 404 (1960).
207. Ordogh, F., and K. Szende, Acta Microbiol. Acad. Sci. Hung., 8, 65 (1961).
208. Kowalski, M., Acta Microbiol. Polon., 15, 119 (1966).
209. Staniewski, R., and M. Kowalski, Acta Microbiol. Polon., 14, 231 (1965).
210. Kowalski, M., Plant Soil, Special Vol., 63 (1971).
211. Szende, K., Plant Soil, Special Vol., 81 (1971).
212. Sik, T., and L. Orosz, Plant Soil, Special Vol., 57 (1971).
213. Roslycky, E. B., Can. J. Microbiol., 13, 431 (1967).
214. Lotz, W., and F. Mayer, J. Virol., 9, 160 (1972).

215. Graham, P. H., in Analytical Serology of Microorganisms, Kwapinski, Ed., John Wiley and Sons, New York, Vol. II, pp. 353-378, 1969.
216. Vincent, J. M., A Manual for the Practical Study of the Root-nodule Bacteria, I. B. P. Handbook, Blackwell Scientific Publications, Oxford and Edinburgh, 1970.
217. Vincent, J. M., Proc. Linn. Soc. (N. S. W.), 66, 145 (1941).
218. Vincent, J. M., Proc. Linn. Soc. (N. S. W.), 67, 82 (1942).
219. Kleczkowski, A., and H. G. Thornton, J. Bacteriol., 48, 661 (1944).
220. Dudman, W. F., J. Bacteriol., 88, 782 (1964).
221. Date, R. A., and A. M. Decker, Can. J. Microbiol., 11, 1 (1965).
222. Dudman, W. F., Appl. Microbiol., 21, 973 (1971).
223. Trinick, M. J., J. Appl. Bacteriol., 32, 181 (1969).
224. Schmidt, E. L., R. O. Bankole, and B. B. Bohlool, J. Bacteriol., 95, 1987 (1968).
225. Graham, P. H., Antonie van Leeuwenhoek, 29, 281 (1963).
226. Purchase, H. F., J. M. Vincent, and L. M. Ward, Proc. Linn. Soc. (N. S. W.), 76, 1 (1951).
227. Hughes, D. Q., and J. M. Vincent, Proc. Linn. Soc. (N. S. W.), 67, 142 (1942).
228. Purchase, H. F., and J. M. Vincent, Proc. Linn. Soc. (N. S. W.), 74, 227 (1949).
229. Purchase, H. F., J. M. Vincent, and L. M. Ward, Aust. J. Agr. Res., 2, 261 (1951).
230. Drozanska, D., Acta Microbiol. Polon., 13, 69 (1964).
231. Means, U. M., and H. W. Johnson, Appl. Microbiol., 16, 203 (1968).
232. Skrdleta, V., Antonie von Leeuwenhoek, 35, 77 (1969).
233. Humphrey, B. A., and J. M. Vincent, Microbios, 13, 71 (1975).
234. Drozanska, D., Acta Microbiol. Polon., 15, 323 (1966).
235. Scheffler, J. G., and H. A. Louw, S. Afr. J. Agr. Sci., 10, 161 (1967).
236. Koontz, F. P., and J. E. Faber, Soil Sci., 91, 228 (1961).
237. Skrdleta, V., Plant Soil, 23, 43 (1965).
238. Elkan, G. H., Plant Soil, Special Vol., 85 (1971).

239. Means, U. M., H. W. Johnson, and R. A. Date, J. Bacteriol., 87, 547 (1964).
240. Skrdleta, V., Folia Microbiol., 14, 32 (1969).
241. Skrdleta, V., Zbl. Bakt., 123, 1II (1969).
242. Graham, P. H., and M. O'Brien, Antonie van Leeuwenhoek, 34, 326 (1968).
243. Lorkiewicz, Z., and R. Russa, Plant Soil, Special Vol., 105 (1971).
244. Manasse, R. J., and W. A. Corpe, Can. J. Microbiol., 13, 1591 (1967).
245. Corpe, W. A., and M. R. J. Salton, Biochim. Biophys. Acta, 124, 125 (1966).
246. Humphrey, B., and J. M. Vincent, Microbios, 7, 87 (1973).
247. Vincent, J. M., Aust. J. Sci., 15, 133 (1953).
248. Vincent, J. M., and B. A. Humphrey, J. Gen. Microbiol., 63, 379 (1970).
249. Humphrey, B., J. M. Vincent, and V. Skrdleta, Arch. Mikrobiol., 89, 79 (1973).
250. Stevenson, F. J., C. E. Clapp, and J. A. E. Molina, Proc. Soil Sci. Soc. Am., 34, 759 (1970).
251. Heidelberger, M., W. F. Dudman, and W. Nimmich, J. Immunol., 104, 1321 (1970).
252. Gibson, A. H., W. F. Dudman, R. W. Weaver, J. C. Horton, and I. C. Anderson, Plant Soil, Special Vol., 33 (1971).
253. Loos, M. A., and H. A. Louw, S. Afr. J. Agr. Sci., 7, 135 (1964).
254. Vincent, J. M., Proc. Linn. Soc. (N. S. W.)., 79, iv (1954).
255. Vincent, J. M., Nature (London), 153, 496 (1944).
256. Breed, R. S., E. G. D. Murray, and N. R. Smith, Eds., Bergey's Manual of Determinative Bacteriology, 7th ed., Williams and Wilkins, Baltimore, 1957.
257. Moffett, M. L., and R. R. Colwell, J. Gen. Microbiol., 51, 245 (1968).
258. Lessel, E. F., Int. J. Syst. Bacteriol., 20, 11 (1970).
259. Norris, D. O., and L. t' Mannetje, E. Afr. Agr. Forestry J., 29, 214 (1964).
260. Haack, A., Zentbl. Bakt. Parasitkde, Abt. II, 117, 343 (1964).
261. Yao, P. Y., and J. M. Vincent, Aust. J. Biol. Sci., 22, 413 (1969).
262. Jensen, H. L., Arch. Mikrobiol., 59, 174 (1967).

263. Graham, P. H., J. Gen. Microbiol., 35, 511 (1964).
264. t'Mannetje, L., Antonie van Leeuwenhoek, 33, 477 (1967).
265. Skyring, G. W., and C. Quadling, Can. J. Microbiol., 15, 141 (1969).
266. Hill, L. R., J. Gen. Microbiol., 44, 419 (1966).
267. Wu, L., K. F. Gregory, and M. M. Hauser, Am. Soc. Microbiol. Bacteriol. Proc., G8, 19 (1968).
268. Heberlein, G. T., J. De Ley, and R. Tijtgat, J. Bacteriol., 94, 116 (1967).
269. Bruch, C. W., and O. N. Allen, Can. J. Microbiol., 3, 181 (1957).
270. Ziemiecka, J., Ann. Inst. Nat. Agron. (Paris), 1, 65 (1963).
271. Graham, P. H., Arch. Mikrobiol., 78, 70 (1971).
272. Ryan, E., and P. F. Fottrell, J. Gen. Microbiol., 70, 395 (1972).
273. Trinick, M. J., Exp. Agr., 4, 243 (1968).
274. Katznelson, H., Nature (London), 175, 551 (1955).
275. Dart, P. J., and M. Chandler, Fifth European Congress on Electron Microscopy, Symp. No. 8, Manchester, Sept. 1972.
276. MacKenzie, C. R., W. J. Vail, and D. C. Jordan, J. Bacteriol., 113, 387 (1973).
277. Gosling, J. P., and P. F. Fottrell, Biochem. Soc. Trans., 1, 252 (1973).
278. Daniel, R. M., and C. A. Appleby, Biochim. Biophys. Acta, 275, 347 (1972).
279. Kretovich, W. L., V. I. Romanov, and A. V. Korolyov, Plant Soil, 39, 619 (1973).
280. Isono, K., and Y. Mino, J. Jap. Soc. Grassl. Sci., 17, 57 (1971).
281. Rigaud, J., and J. C. Trinchant, Physiol. Plant, 28, 160 (1973).
282. Hepper, C. M., Antonie van Leeuwenhoek, 38, 437 (1972).
283. Humphrey, B., M. Edgley, and J. M. Vincent, J. Gen. Microbiol., 81, 267 (1974).
284. Björndal, H., C. Erbing, B. Lindberg, G. Fåhraeus, and H. Ljunggren, Acta Chem. Scand., 25, 1281 (1971).
285. Zevenhuisen, L. P. T. M., Carbohydrate Res., 26, 409 (1973).
286. Yadav, N. K., and S. R. Vyas, Ind. J. Agric. Sci., 41, 875 (1971).
287. Yadav, N. K., and S. R. Vyas, Folia Microbiol., 18, 242 (1973).

288. Dadarwal, K. R., and A. N. Sen, Ind. J. Agric. Sci., 43, 82 (1973).
289. Ley, A. N., H. R. Warner, and P. L. Kahn, Can. J. Microbiol., 18, 375 (1972).
290. Lotz, W., and F. Mayer, Can. J. Microbiol., 18, 1271 (1972).
291. Krsmanovic-Simic, D., and M. Werquin, C. R. Acad. Sci. Paris, Sér. D, 276, 2745 (1973).
292. Orosz, L., Z. Svab, A. Kondorosi, and T. Sik, Molec. Gen. Genet., 125, 341 (1973).
293. Mayer, F., W. Lotz, and D. Lang, J. Virol., 11, 946 (1973).
294. Atkins, G. J., J. Virol., 12, 157 (1973).
295. Schwinghamer, E. A., C. E. Pankhurst, and P. R. Whitfield, Can. J. Microbiol., 19, 359 (1973).
296. Elkan, G. H., and R. A. Usanis, Int. J. Syst. Bact., 21, 295 (1971).
297. Godfrey, C. A., J. Gen. Microbiol., 72, 399 (1972).
298. Clark, A. G., J. Appl. Bact., 35, 553 (1972).
299. Kuyz, W. G. W., T. A. LaRue, Nature (London), 256, 407 (1975).
300. Pagan, J. D., J. J. Child, W. R. Scowcroft and A. H. Gibson, Nature (London), 256, 406 (1975).
301. McComb, J. A., J. Elliot and M. J. Dilworth, Nature (London), 256, 409 (1975).

CHAPTER 8

Infection and Development of Leguminous Nodules

PETER DART
ICRISAT
Hyderabad, India

8.1. Historical, 368
8.2. Infection, 372
    8.2.1. Rhizobium in the Rhizosphere, 372
    8.2.2. Root Hair Curling, 373
    8.2.3. Host Bacterium Specificity, 376
        8.2.3.1. Wall-degrading Enzymes, 378
        8.2.3.2. Hormones, 380
    8.2.4. Root Hair Invasion and Infection Thread Formation, 382
    8.2.5. Infection Thread Numbers, 391
        8.2.5.1. Effect of Temperature, 392
        8.2.5.2. Combined Nitrogen, 394
        8.2.5.3. Delayed Inoculation, 395
    8.2.6. Infection without Infection Threads?, 395
8.3. Nodule Development, 399
    8.3.1. Nodule Form, 399
    8.3.2. Stimulation of Division in the Root Cortex, 410
        8.3.2.1. Location, 410
        8.3.2.2. Ploidy in Nodule Cells and the Role of Hormones, 412
    8.3.3. Nodule Initiation, 416
        8.3.3.1. Meristematic Activity, 416
        8.3.3.2. Cell Differentiation and Invasion by Rhizobium, 417
        8.3.3.3. Vascular Connections and the Nodule Cortex, 420
    8.3.4. Bacteroid Development, 424
        8.3.4.1. Cross-Inoculation Relationships, 425
        8.3.4.2. Clover, Pea, Medic Groups, 425
        8.3.4.3. Soybean, Phaseolus, Cowpea-type, 428

                8.3.4.4. Bacteroids of Other Cross-
                          Inoculation Groups, 429
                8.3.4.5. Ineffective Nodules, 431
                8.3.4.6. Role of the Plant in Bacteroid
                          Development by the Same
                          Rhizobium Strain, 436
                8.3.4.7. Bacteroid-like Forms
                          in vitro, 437
                8.3.4.8. Bacteroid Zone Size, 438
        8.3.5. Physiological Factors Influencing
                Nodule Initiation and Development, 442
                8.3.5.1. Temperature, 442
                8.3.5.2. Combined Nitrogen, 443
                8.3.5.3. Other Environmental Factors, 444
        8.3.6. Nodulation of Excised Roots, 445
        8.3.7. Rhizobium-Host Interactions in Tissue
                Culture, 446
        8.3.8. Perennial Nodules, 448
8.4. Selected Methods, 449
        8.4.1. The Fåhraeus Slide Technique, 449
                8.4.1.1. To Make Covership Spacing
                          Pieces, 449
                8.4.1.2. To Make Cell, 450
                8.4.1.3. Seed Pretreatment and Sowing, 450
        8.4.2. Culture of Legumes for Nodulation
                Studies, 451
                8.4.2.1. Axenic Culture in Test Tubes, 451
                8.4.2.2. Root-enclosed Assembly, 452
                8.4.2.3. Sand Culture, 453
8.5. Concluding Comments, 454
8.6. Glossary of Common Names Used in Chapter III.8, 455
8.7. References, 457

   8.1 HISTORICAL

Drawings of legume roots in the sixteenth century included nodules, and Dalechamps (1) in 1586 remarked on the abundant nodules on Ornithopodium tuberosum. Malpighi (2) in 1679 believe the tumors on Pisum sativum plants to be insect galls as he observed "worms" inside them. Perhaps these were nematode galls, or nodules secondarily invaded by nematodes or by insect larvae. The concept of nodules as pathological outgrowths, perhaps induced by insect stings or containing small fungi (3), was widely accepted in the early nineteenth century, although Trevinarus (4) believed they were not due to invasion of the root but were a kind of adventitious bud. Others thought they might be distorted roots, absorbing or storage organs.

Following Pasteur's elegant demonstration that some diseases were caused by germs or microorganisms, Woronin (5) in 1867 illustrated the anatomy of Lupinus mutabilis nodules and carefully described the small corpuscles and rod-shaped bodies they contained. When the disintegrating cells from an old nodule were teased into water, some of these bodies moved rapidly and also divided. Although Woronin could not culture them, he was sure that these were vibrio-like bodies, or bacteria, which formed the "diseased," nodule growths.

Ericksson (6) found nodules on a dozen legume species and examined sections of Faba vulgaris (Vicia faba) nodules in detail. His drawings show how they arouse in the inner root cortex with the dense nodule cells containing the vibrio-like bodies joined to the root epidermis by several fine hyphal strands with granular contents, crossing the empty outer cortex cells. The strands were found much branched in the nodule, but although the cells were drawn with granular contents, hyphae and vibrios were not noted in the same cells.

Woronin's (7) description in 1878 of the formation by Plasmodiophora brassicae of the hypertrophy or club foot on Cruciferae roots led to the suggestion that the nodule was similarly induced by a myxomycete (8); the corpuscles (i.e., the bacteroids) being the spores of the plasmodium. The nature of the causative organism and the relation between the two separate forms were hotly debated with much depending on whether the thread form had a wall and whether true bacteria could be pleomorphic (i.e., bacteroids). Frank (9) found "hyphae" and corpuscles in the same cell and believed that the latter were hyphal buds.

Both Eriksson and Frank suggested that the nodule arose by invasion from outside the root (i.e., the organism was not seed borne) but it was left to the masterly studies of Ward (10) with Vicia faba and the Prazmowski (11) with Pisum sativum to describe very accurately the infection of curled root hairs and the growth of the infection thread into the root cortex to initiate the nodule (Plate 8.1). Vicia faba did not nodulate in sterilized soil but did nodulate readily in water culture when inoculated with a dried nodule. The infection thread was described as a "germinal hypha" that ramified through the young nodule before budding off the "yeast cells — gemmules, germs" (i.e., the bacteroids). Ward was unable to culture the organism, but from his and Frank's field observations on many legumes, believed it to be almost universally present in soils.

Two years earlier in 1886, Hellriegel and Wilfarth (12, 13) had demonstrated that legumes nodulated by microorganisms in soil extracts could fix N_2. Beijerinck (14) confirmed that the nodules of several legumes were formed by bacteria named Bacillus radicicola, which he isolated, grew in culture, and used to reinoculate and nodulate Vicia faba. He noted the infection threads in V. faba nodules crossing cells, but believed they were remnants of the spindle left after cell division.

Prazmowski (11) also cultured the organism from P. sativum nodules and formed nodules by reinoculation. He regarded the infection thread as a gelatinous tube excreted by the bacteria for protection. By 1890 Frank (15) had also cultured Rhizobium leguminosarum Frank had produced nodules by inoculations, and believed that the infection tube was formed by the plant cytoplasm and the bacteria to conduct the minute coccoid form of the symbiont into the root cortex where it would grow. This was disputed by Dawson (16, 17), who could not detect cellulose in the thread wall, which she and Ward had demonstrated so clearly. Others (18), however, claimed it was present.

That the infection thread wall is formed by the plant and contains normal wall constitutents was confirmed by McCoy's histochemical study of the infection thread in Medicago sativa nodules (19).

Plate 8.1. a. Mucigel layer at the tip of a Trifolium subterraneum root with embedded root cap cells. b. Young root hairs on T. subterraneum root enclosed by the mucigel layer. c. Double infection thread in T. subterraneum, starting in a lateral branch and then joining in the main part of the root hair. The infection thread expands as it crosses the cell walls (arrows). d. T. parviflorum infection threads, initiated in lateral branches, growing away from the epidermis before aborting. e. This infection thread, initiated in a lateral branch of a T. fragiferum root hair, first grew away from the root before turning to follow the nucleus down into the epidermal cell. f. T. parviflorum infection thread which formed a vesicle (arrow), filled with rhizobia before continuing into the epidermis and aborting. g. Start of an infection thread in a kink in a root hair of T. parviflorum. The newly initiated thread twists as it grows through the curled part of the root hair, straightening in the main part of the hair. Nomarski optics. h. T. parviflorum infection thread with contorted growth occasionally found. Nomarski optics. i. Section showing branching of Pisum sativum root hairs as they grow through the cortex, passing close to the nuclei (n). R. Kumarsinghe kindly provided the material in Plate 8.1d, f, g, i; P. S. Nutman (43) 8.1e, and K. Libbenga (188) 8.1i.

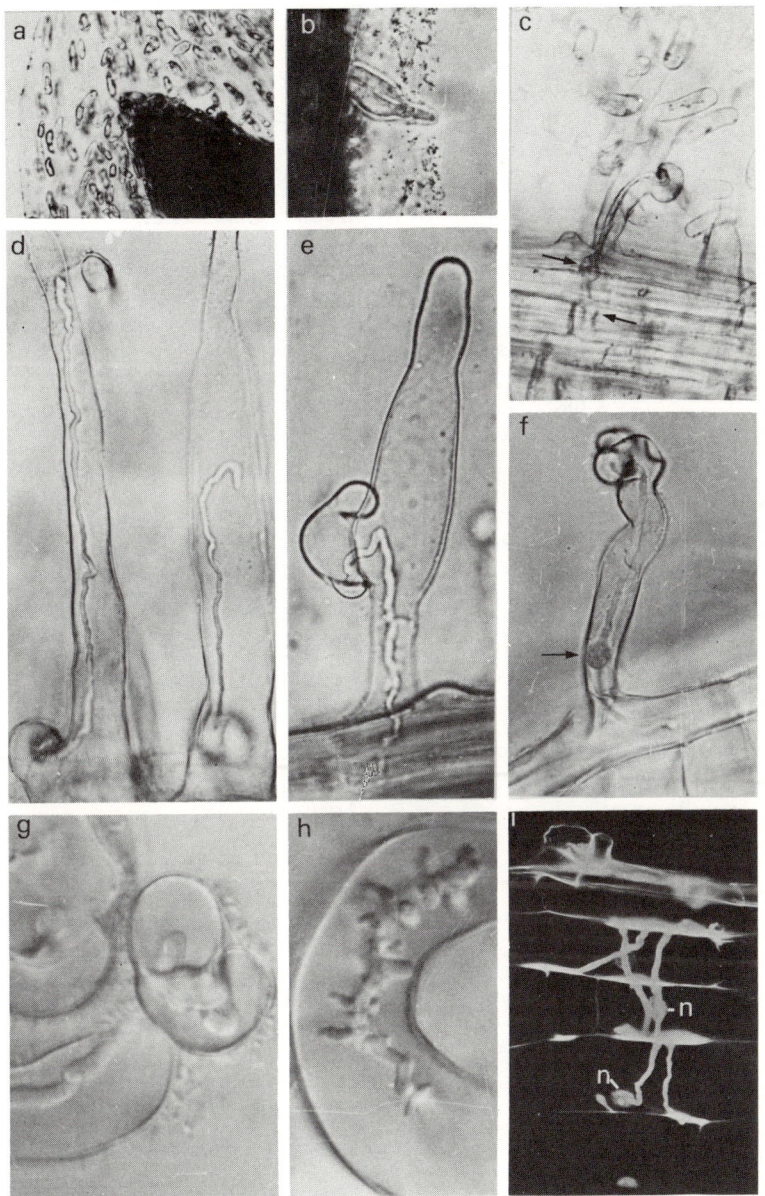

Beijerinck divided the nodule organisms into a fast-growing group from Vicia, Ervum (Lens), Trifolium, Lathyrus, and Pisum, and a slow-growing group from Lotus, Lupinus, Ornithopus, Phaseolus, and Robinia. Nobbe et al. (20) demonstrated clearly by inoculations with cultures that Lupinus luteus did not nodulate with the Pisum sativum, Robinia, Cytisus, and Gleditschia organisms, but would do so with the lupin organism, and that Phaseolus vulgaris and P. sativum did not nodulate with isolates from Lupinus or Robinia. Atkinson (21) found that the organisms from Vicia sativa and Dolichos sinensis would not cross-infect. Kirchner (22) observed that Glycine max growing in the botanical garden at Hodenheim, Germany, was the only one of 100 legume species which were not nodulated, and remained so until inoculated with soil from Japan containing Rhizobium japonicum Kirchner. Hiltner (23) showed that nodule bacteria from Trifolium pratense would not nodulate Medicago sativa. Nobbe and Hiltner in 1896 patented a process of inoculation using a pure culture of Rhizobium growing on gelatine-"Nitragin."

Thus the cross-inoculation group concept — that separate Rhizobium species only nodulate specific, related legumes — came into being. The need to provide inoculants for agriculture prompted wide-scale isolations of rhizobia from different nodules and their testing on the legumes to be grown. By 1932 Fred, Baldwin, and McCoy (24) recognized eight major cross-inoculation groups (alfalfa, R. meliloti; clover, R. trifolii; pea, R. leguminosarum; bean, R. phaseoli; lupin, R. lupini; soybean, R. japonicum; cowpea, Rhizobium sp; Lotus, Rhizobium sp.; and eight minor groups.

8.2 INFECTION

8.2.1 Rhizobium in the Rhizosphere

The distribution of Rhizobium in soils and the numbers developing in the legume rhizosphere affect infection. Factors influencing Rhizobium growth in the soil are described in Chapters III.7 and IV.9.

Because Rhizobium multiplies rapidly in the legume rhizosphere and the distribution along the root is uneven, it is difficult to determine how many rhizobia are needed for infection and nodulation. As few as

10-50 rhizobia on the seed, which later multiply, may be sufficient for maximum nodulation of Glycine max in sand culture (25, 26). In water culture the number of nodules on Vigna radiata was increased only slightly as the initial inoculum was raised from 89 to 8.9×10^4 rhizobia/ml of culture solution (27).

Purchase and Nutman (28) used the ingenious method of restricting multiplication of a nodulating R. trifolii strain in the rhizosphere by adding a large inoculum of a non-nodulating mutant at the same time, and then examined the relation between infection of Trifolium pratense and numbers of rhizobia in the rhizosphere. Early infection and nodulation were promoted by a large population of nodulating rhizobia. The numbers of infections almost doubled with each doubling of the log number of nodulating rhizobia up to the time the first nodule was formed. Thereafter the nodulating population had much less affect on the slower rate of infection that followed. With few nodulating bacteria in the rhizosphere the few infections were still restricted to well separated zones of the root, and not randomly distributed (29). Nodulation of T. pratense increased asymptotically as the number of virulent bacteria in the rhizosphere increased to about 10^4. This was not a proportional increase as each additional nodule required an increasing number of extra rhizobia. Only about 10 bacteria were required to initiate nodulation, although this was delayed if fewer than 2.4×10^4 nodulating rhizobia were present (28).

Soil pH affects the survival of Rhizobium. Acidity also affects root hair curling, infection, and nodulation with the response dependent on the legume species. Nodulation of Trifolium and Medicago spp. at low pH requires a higher level of calcium around the root hairs during their infection. These effects of pH and calcium are discussed in more detail in III.7 and IV.10.

8.2.2 Root Hair Curling

In Trifolium sp., root hairs start to develop some distance from the root tip. Every epidermal cell seems capable of forming a root hair although not all do, particularly in water culture. The developing root hair grows rapidly from the tip at 8-15 µm/hr for

Trifolium glomeratum, with rapid cytoplasmic movement within the hair (29).

The root hair tip appears smooth (30), contains a relatively small amount of cellulose and consists mostly of pectic and hemicellulosic substances. At the tip the cellulose fibrils are arranged in a random network but some 25 μm back from the tip another wall layer, with cellulose fibrils running predominantly parallel to the long axis of the hair, is laid down inside the first formed (31, 32). This appears to have a larger cellulose component and to be much stronger. The surface of uninoculated root hairs usually appears smooth, perhaps because they are covered by a thin cuticle (33) or by a mucilaginous layer (34). Inoculation with *Rhizobium* exposes a fibrillar structure on some root hairs (30, 35).

The root cap cells and probably also the epidermal cells secrete large amounts of polysaccharide (36) and globules of material may also form at the root hair tips (37). This material forms the "mucigel," a coating of variable thickness along the young root which often extends beyond the root hairs and is present on sterile and nonsterile root surfaces (30, 38). Sloughed off root cap cells also lie among the root hairs (Plate 8.1a and b). Rhizobia and other bacteria are found enmeshed within the mucigel in larger numbers than outside (35, 39).

When seeds or small seedlings are inoculated, rhizobia usually colonize the whole root as it grows, but for *Pisum sativum* they seem initially localized to the points of lateral root emergence (40) possibly because the rhizobia are unable to penetrate the prominent mucigel except where it is disturbed by lateral roots, or becuase growth promoting substances are released from the roots at these points.

Uninoculated root hairs are normally straight. Curling or contorted growth and branching of root hairs is the first visible plant response to *Rhizobium* (10, 11, 19, 41-43) (Plate 8.1c-g). For seedlings of *Trifolium* spp. and *Vicia hirsuta* in the presence of different *Rhizobium* strains, from 7 to 37% of the hairs were deformed (43). Deformation was most common on mature hairs usually more than 5 mm from the root tip of *T. glomeratum* (44).

The degree of root hair distortion depends on the plant, the *Rhizobium* strain and the position on the root (19, 43, 45-49). Marked hair curling was also

Infection

found for Trifolium spp. inoculated with Medicago sativa nodule bacteria and vice versa, and with Pisum sativum plants inoculated with Rhizobium trifolii or R. meliloti. No infection threads were found for any non-nodulating association (19). Almost 60% of P. sativum root hairs were deformed and about 23% markedly curled when inoculated with homologous rhizobia, while for noninvasive strains of R. meliloti, R. phaseoli, and cowpea rhizobia the deformed hairs ranged from 55 to 12% with much less than 5% markedly curled. Similarly for Ornithopus sativus homologous rhizobia induced about 35% of deformed hairs with 15% curled, and for heterologous rhizobia less than 3% of hairs were deformed and virtually none curled (47). Yao and Vincent (48) examined these effects in more detail. For T. glomeratum, M. sativa, and Macroptilium atropurpureum markedly curled hairs generally occurred only with homologous strains among which the amount of curling varied. Heterologous strains deformed hairs much less and to varying degrees.

The curling response of Trifolium spp. to R. leguminosarum is greater than for other heterologous strains (19, 47, 48, 50) resembling the response for the homologous strain. Some R. leguminosarum strains nodulate T. pratense and some R. trifolii strains will form a few nodules on P. sativum (50, 51). The cross-infecting strains need not be related antigenically (52). Non-nodulating R. trifolii strains produce only moderate curling on T. glomeratum and T. fragiferum (48, 53). Root hairs are also curled on some plants (e.g., Lotus spp., Vigna unguiculata, Stylosanthes humilis), although infection threads have not been found in the root hairs — only in nodule sections (54).

Root hairs of nonlegumes may increase in length (47) but are not curled or distorted by Rhizobium. However, deformation and curling is apparent in the presence of their suspected actinomycete endophyte (see III.6.9.2).

If rhizobia are grown in liquid culture and then removed by filtration, the resulting sterile filtrate will induce root hair deformation of seedling legumes although both the number of hairs affected and degree of deformation are much less than in the presence of rhizobia and depend on the host and method of growing the plants (19, 23, 41, 48, 55-59). Trifolium spp. root hairs are also deformed by sterile culture filtrates of Medicago sativa nodule bacteria and vice versa (19, 48) but less than with the filtrate from homologous strains.

The deforming substance in culture filtrates is present in a crude extracellular polysaccharide preparation from R. trifolii (53), is heat labile (destroyed by heating for 20 min at 80°), and retained by a dialysis sac. When R. trifolii or its culture filtrate is added to T. repens roots the deforming substance in the medium becomes heat stable (not destroyed by 20 min at 110°) (57). This substance can be adsorbed onto T. pratense roots and eluted with 0.2 N acetic acid or 6 M urea (58). Root hair branching can be induced by a highly diffusible heat stable fraction of the rhizobia-free filtrate, as well as by a dialysable fraction. Treatment of the filtrate with nuclease, periodate, and especially trypsin reduces the braching response (59).

The amount of root hair curling does not seem to be a limiting factor in infection — curling is increased by treating roots with Polymixin B with no effect on infection (44) and nitrate has a differential effect on curling and infection (see III.8.3.5.2).

8.2.3 Host Bacterium Specificity

Serological analysis of rhizobia isolated from nodules of Trifolium glomeratum, T. repens, T. subterraneum, T. incarnatum, Medicago minima, and Glycine max (see IV.13) suggests that a nodule is generally formed by only one strain even though adjacent nodules on a plant given a mixed inoculum may be formed by different strains. However identification of single nodule contents by serology (60, 61), by strain-specific, fluorescent antibody staining (62), and by using inoculum strains with genetic markers such as drug resistance (63, 64) has shown that up to 30% of the nodules on G. max roots (60-62), M. truncatula, T. glomeratum (63), up to 40% on Pisum sativum roots (64), and about 3% of T. subterraneum (65) and M. sativa (63) can contain both inoculum strains in Leonard jar, flask, or tube culture. Generally one stain consistently occupies the majority of the nodules, and in the mixed nodules is the predominant strain numerically (62, 64). Within pea nodules with mixed infections, transfer of the P-group R factor plasmid RP4 may occur so that up to 10% of the rhizobia isolated from such nodules may be nonparental progeny (64). It is not known whether the rhizobia enter the mixed nodules via a single infection thread.

The mechanism for excluding other virulent strains from a nodule is not obvious except for hosts infected through the root-hair infection-thread process where only the bacteria near the tip of the thread multiply. Possibly only a small, single-strain colony of rhizobia is present at the site of infection.

It is difficult to decide whether bacteria other than rhizobia also occur in nodules because there is no way to check the efficiency of surface sterilization. However Beijerinck (14) and others (24) claimed to have isolated saprophytes from nodules. Klebsiella pneumoniae can be isolated from surface sterilized nodules of G. max and Trifolium sp. (66). A fast-growing aerobe was also isolated from G. max nodules along with the slow-growing R. japonicum (67). Bacteriological, serological, and plant infection tests showed that it was not Agrobacterium radiobacter, though probably a member of the Rhizobiaceae. It could also be readily recovered from Vigna unguiculata and Arachis hypogaea nodules (all induced by slow-growing rhizobia) but not from Medicago sativa, Pisum sativum, and Trifolium pratense nodules (induced by fast-growing rhizobia), when a mixed inoculum contained the contaminant. The contaminant could "enter" V. unguiculata nodules 3 weeks after they had formed (68, 69).

Although legume nodules generally seem to harbor only one Rhizobium strain, a given root can certainly form nodules with more than one strain. There are both host and bacterial factors in the competition between strains to form nodules (see IV.9).

Rhizobium strains vary in their ability to form infection threads, to initiate early nodulation, and in the rate at which they form nodules (e.g., ref. 63). Strains ineffective in fixing dinitrogen often form more nodules than effective strains, possibly because differences in nodule physiology influence the susceptibility of the root to successful infections. The effects of Rhizobium genetic factors on nodulation are dealt with in III.12 and the role of host genes in nodulation in III.11.

As more nodule bacteria are isolated from a wider range of plants and their nodulating abilities tested, the concept that most strains and plants fall into closely defined groups within which there is reciprocal nodulation of all hosts by strains isolated from each plant in the group becomes less tenable. Some legumes are more promiscuous than others, nodulating with strains from hosts of more than one cross-inoculation group, e.g., Vigna unguiculata with R. japonicum, R. lupini, and R. phaseoli. Often strains isolated from such promiscuously nodulating plants do not reciprocally nodulate other hosts in that plant's cross-inoculation group (380). A detailed description of cross-inoculation specificities can be found in III.7.

8.2.3.1 Wall-degrading Enzymes

Small differences in the host wall are involved in the resistance or susceptibility of plants to fungal enzymes (71). Hemagglutinins and related glycoproteins are widely distributed in legumes (72). The phytohemagglutinins of Phaseolus vulgaris bind to R. phaseoli. Red blood cells are agglutinated onto P. vulgaris roots and root hairs at discrete sites where infection by Rhizobium might occur (73). Glycine max lectin labeled with fluorescein isothiocyanate, specifically bound to 22 out of 25 R. japonicum strains but not to 23 other strains that do not nodulate soybean (74). This suggests that such proteins may be involved in the initial binding and recognition of specific Rhizobium strains by the host. Glycoproteins also inhibit the activity of polygalacturonases secreted by various fungi in a response that is dependent on both host and fungal species (75).

Plant cell walls are degraded by many fungal and bacterial pathogens which produce cellulases (β-1,4-glucanases), hemicellulases (depolymerizing polysaccharides other than cellulose), pectic glycosidases (often grouped as polygalacturonases and which hydrolytically cleave the α-1,4-glycosidic bonds of pectic substances), and pectic lyases (trans eliminases). Several categories of these enzymes exist, differing in substrate preference and perhaps the point of attack (71, 76). Rhizobium in culture does not use cellulose (14, 19), pectin, or calcium pectate (19, 57, 77-79) as substrates, and apparently does not produce cellulase, pectolytic enzymes, or pectin esterase.

Wall-degrading enzymes are produced during their growth and development by plant cells, e.g., pollen tube "tip" growth and epicotyl extension. Infection but not nodulation of Trifolium glomeratum has been increased by exogenous pectolytic enzymes (44). Some plasticizing of the root hair wall presumably does occur during infection as the lateral branching and curling associated with hair infection often occurs in "mature" root hairs that have stopped elongating (43). Polygalacturonase was found in the medium around the roots of Trifolium repens and Medicago sativa seedlings and in extracts of the roots when the plants were inoculated with virulent, homologous Rhizobium strains. Less enzyme was found associated with uninoculated plants or plants inoculated with an avirulent mutant or a noninfective strain from another cross-inoculation group (57, 80-83). Activity varied with seedling age. Polygalacturonase was also detected when water-soluble, extracellular slime of the homologous virulent strains was added to seedlings.

Fåhraeus and Ljunggren (81) postulated that Rhizobium strains capable of infecting a legume release a specific polysaccharide that induces more pectolytic activity by the root and that this accounts for the cross-inoculation specificities. This activity, by loosening the root hair wall structure, would facilitate infection thread initiation.

The pectolytic enzymes of M. sativa and T. repens are most active at pH 5 (57) or 5.5 (84) with little activity at more acid pHs; this has been suggested as the acid-sensitive stage in infection (84).

Others have been unable to find increased pectolytic activity for Glycine max, T. repens, T. pratense, or M. sativa when inoculated with homologous virulent rhizobia (78, 85-88). Any significantly different activity was usually associated with fungal contaminants.

The composition of the root medium affects cellulase and pectolytic activity; little was produced by T. pratense seedlings grown in distilled water as in Ljunggren's experiments, and much more in a mineral salts solution, and generally more with increased calcium. The pectolytic activities were also generally higher than those detected by Ljunggren. While conditions such as the number of seedlings in a given volume of medium, the medium composition, and the day length all influenced the patterns of early nodulation, no corresponding differences were found in pectolytic activity in either root exudates or root extracts. Seedling age influenced activity whether Rhizobium was present or not. Activities were not apparently related to inoculation or the infecting ability of the Rhizobium strain. Enzyme samples from different assays differed in their relative activity for sodium polypectate and pectic acid, indicating that at least two enzymes were present (87, 88).

The discrepancies between pectolytic enzyme levels in different experiments could be due to subtle differences in technique — whether pectic acid or pectin is used and their type and methoxyl content, particularly since two pectolytic enzymes are involved. Thus cell-wall degrading enzymes are present during seedling growth but the activities as measured to date reflect more the changing growth conditions than effects of Rhizobium.

8.2.3.2 Hormones

Hormones may be involved in infection. Exogenous auxin reduces the minimum turgor needed to increase cell wall area (89). Localized auxin production by Rhizobium at the root hair surface might, thus, soften the wall in such a manner that the characteristic deformation preluding infection could then occur. Molliard in 1912 (90) showed that R. leguminosarum in liquid culture produced a substance that induced cell division in the pericycle of Pisum sativum seedlings and a marked elongation of the cortical cells, particularly opposite the phloem. The nuclei of these cells were also greatly enlarged. Autoclaving at 120° destroyed these activities.

Rhizobium produces auxins, mainly indole acetic acid (IAA) and indole carboxylic acid, in culture from tryptophan (56, 91-94). Auxins are produced by Rhizobium in the rhizosphere of Trifolium pratense and T. subterraneum; this IAA production is not restricted to R. trifolii (92). T. subterraneum var. "Woogenellup" nodulates much better with some Rhizobium strains if the roots are exposed to light, perhaps because of the increased IAA oxidation (95). The poor nodulation by certain strains seems related to their ability to produce IAA in culture from added tryptophan; Woogenellup roots have a higher tryptophan content than other cultivars (217). Addition of IAA can increase nodulation (96), but added auxin usually delays initial nodulation and even prevents some plants nodulating. Anti-auxins can increase the rate of nodulation (92, 97, 98).

Root hairs of T. repens and T. glomeratum were not curled or infections stimulated by adding IAA and tryptophan ($10^{-7} - 10^{-12}$M) although they grew longer (44, 46). As root hairs can be curled by unidentified substances in a dialyzable fraction of the Rhizobium culture filtrate, IAA, if involved in the curling, is not the only substance (46, 58).

Increasing concentrations of exogenous IAA progressively decreased curling of Medicago sativa root hairs in the presence of R. meliloti (99). Caffeic acid, often synergistic with IAA, at $10^{-2} - 10^{-4}$M decreased infection and nodulation but did not affect curling of hairs. Phenoxyacetic hormone herbicides severely reduced infections at 10^{-4}M and less so at 5×10^{-6}M (44). Occasionally concentrations of hormone herbicides in soil are reached that could inhibit infection but not growth of Rhizobium (100, 101).

Ethylene at low concentration inhibits nodulation of excised roots of Phaseolus vulgaris (102) and T. subterraneum grown in flasks on agar (103). Application of 2 ppm ethrel (2-chloroethyl-phosphonic acid), a compound which releases ethylene, to the roots of P. sativum reduced nodulation. The effect was not translocated to the other half of a split root system. Foliar sprays with ethrel also reduced nodulation (104). If ethylene does inhibit subsequent infection and nodule initiation, the response must initially be mediated by internal root levels, as the infection pattern of T. patens plants grown together on slides seems little affected by nodulation of the other plant (105).

Some Rhizobium strains grown in a defined liquid medium produce small amounts of gibberellin-like substances active in bioassays (106). More gibberellin is found in nodules of Pisum sativum, Phaseolus vulgaris, and Lupinus luteus than in the rest of the root on a fresh weight basis (107, 108). Gibberellic acid (GA) at 10^{-8}M reduces the number and rate of infections of T. glomeratum, although root hair growth is not affected (44). Radley (107) suggested that GA could be the substance formed by existing nodules that restricts the development of further nodules (109).

Addition to the agar root medium or foliar sprays of GA reduces nodulation of several plants (110-112).
Rhizobium produces cytokinin in culture (113-115). Infection of T. glomeratum seedlings is severely restricted by exogenous kinetin at 10^{-3} and 10^{-4}M, but at 10^{-7}M the formation of the first nodule was delayed and numbers of infections stimulated with the extension of the initial rapid phase of infection (44).
Exogenous abscisic acid (ABA) at 1.9×10^{-6}M in the root medium inhibits nodule formation on P. sativum; Rhizobium growth and the numbers of root hairs seem unaffected but root hair deformation and infection thread numbers are perhaps stimulated. Abscisic acid probably affects nodulation at a stage subsequent to infection (116).

Colchicine added to liquid medium or soil can increase root hair deformation, infection, and nodulation (117-121). Both diploid and tetraploid varieties of T. pratense responded similarly, but not consistently. Ethylene diamine tetra acetic acid (EDTA) at 10^{-4}M decreased infection of T. glomeratum and delayed the appearance of the first nodule (44). EDTA also locally inhibited nodulation of P. sativum in liquid culture (122). Although hormones affect infection, the mechanisms involved are obscure.

8.2.4 Root Hair Invasion and Infection Thread Formation

Infections seem generally restricted to the root zones with deformed hairs (43), and usually occur in markedly deformed root hairs in the region of most curling, either in a "kink" or a notch near the hair tip (Plate 8.1f and g) or associated with a hair branch (Plate 8.1c-e) (10, 11, 16, 17, 19, 21, 42, 43, 47). Occasionally infections occur in straight root hairs (35, 42, 43). The proportion of infections in lateral branches of hairs depends on the species and is 29-88% for Trifolium spp. and Vicia hirsuta (43). Infection often occurs near the contact point of two adjacent root hairs, sometimes with both hairs becoming infected. Infections in such "touching" hairs amounted to about 4% of infections for P. sativum (47).

Nodules form on adventitious roots of the water plant Neptunia oleracea, which lack root hairs, apparently from infection threads initiated in the epidermal cells (123). <u>Glycine max</u> and <u>Trifolium</u> spp. root epidermal cells, which have not developed root hairs, occasionally form infection threads (43, 124).

The first sign of infection is usually a hyaline spot or swelling in the root hair wall (10, 42). Cytoplasmic streaming increases and the cytoplasm near the infection point increases in opacity, with the host cell nucleus remaining nearby. Rapid cytoplasmic streaming continues and in about 3 hours the refractile sheath of the infection thread becomes visible, though not yet clearly demarked from the cytoplasm. Small areas of callose are found on the root hair wall of <u>Trifolium</u> spp., especially at the tip, becoming less pronounced as the individual hairs age. Inoculation with <u>Rhizobium trifolii</u> increases the callose abundance mainly as prominent deposits in the region where the hyaline spot is found. These deposits are permanent, but do not extend much along the developing thread (125).

The thread grows at about 7 μm/hour for <u>T. glomeratum</u>, similar to the previous growth rate of the hair itself (44). The nucleus, now almost double the size of the uninfected hair nucleus and with a prominent nucleolus, remains close to the tip of the thread "leading" its growth along the major cytoplasmic strand usually centrally placed in the hair. The growing point of the thread usually remains obscured by cytoplasm, and it is not known whether it is enclosed by the thread wall; older parts become more visible as much of the cytoplasm retracts. If the nucleus moves from the infection thread for any length of time, its growth stops, but may sometimes renew when the nucleus returns (126). Should the infection thread start in a lateral branch of the hair, it may either follow the nucleus towards the base of the hair, or (occasionally) grow towards the hair tip away from the root, depending on the direction of movement of the nucleus (42, 43, 47, 126), and may even turn again and grow towards the hair base (Plate 8.1d and e). For <u>Trifolium</u> spp. up to 43% of threads grew away from the epidermis, depending on the species (70).

It is not known how <u>Rhizobium</u> initiates the infection thread. McCoy (19) suggested mechanical rupture with rhizobia entering a break in the root hair wall. Alternatively rhizobia may become trapped within the fold of a growing, deformed hair. This would be an impediment to normal wall growth, and induce wall synthesis to enclose the rhizobia. As the rhizobia are increasing in number and volume, the infection thread growth accommodates for this with the form determined by the wall deposition mechanism of the root hair (127). The root hair and the infection thread grow at their tips, with little cellulose deposited until some distance back from the tip. Rhizobia are usually aligned longitudinally in the thread in a single row (126); thread growth presumably results from bacterial division and growth stretching the thread tip, it's most plastic region, with accompanying deposition of plant cell wall material maintaining the thread wall thickness. Within the nodule the tip of the infection thread is usually the thinnest region, and contains the least electron-dense material.

Infection threads vary in thickness with the host species, but generally the finer the root hair, the finer the thread. The thread wall is usually smooth and the thread usually straight, bending only to follow the root hair wall (Plate 8.1g). Some threads grow irregularly (Plate 8.1h); these usually abort (43). The thread in the curled part of the root hair is often itself curled and may branch and anastomose (Plate 8.1c).

Histochemically the thread wall contains the same components of pectic substances, hemicelluloses, and some cellulose, as the root hair tip (19), and electron micrographs of infection threads in root hairs and nodules show similar structure to the primary wall of plant cells (128-130). The thread has a certain rigidity, as occasional rupturing of the root hair from its epidermal base leaves the infection thread intact and unchanged in its form. Rhizobium is presumed to produce the "zoogloeal matrix" surrounding the bacteria inside the more rigid wall. This matrix stains with the basophilic dye, toluidene blue, in similar fashion to the RNA of the host nucleus, and also with Schiff's reagent and Alcian blue, suggesting the presence of mucopolysaccharides (54).

The start of the infection thread, as Ward (10) noted (Plate 8.2a) may have the appearance of a small tubular invagination of the hair wall that broadens into the infection thread proper. Sometimes a prominent funnel shape marks the thread junction with the root hair wall. Electron micrographs of the inoculated root hair surface show an exposed fibrillar structure on which sit Rhizobium rods and small (ca. 0.4 μm diameter) cocci (30, 35). The latter may be the swarmers originally suggested by Beijerinck (14) as the infecting form of Rhizobium. Small, coccoid, very motile forms are sometimes found in culture (131-133) and in the rhizosphere of legumes (35, 133), which are perhaps small enough to penetrate the gaps in the exposed cellulose meshwork of the root hair. Rhizobia can occasionally be seen swimming freely within the root hair and may multiply to fill the whole hair (126, 134). Rhizobium in culture can produce cellulose fibrils (135), and these are perhaps involved in the affinity of Rhizobium and plant wall, as the structures found attached to "swarmers" on the root surface (35). Rhizobium rods often attach "end on" to the roots of legumes (30, 126) and nonlegumes (136), but other bacteria also attach to surfaces similarly (137).

Rhizobia may have the ability to induce breaks in plant walls, because the infection thread crosses already formed walls on its progress through the cortex, developing as a tube joined at each cell junction to the wall of the cell it is crossing. The already-formed cortical cell wall "dissolves" and joins the infection thread. A characteristic swelling of the intercellular space and often a large accumulation of rhizobia occurs before the thread enters the adjacent cell (Plate 8.1c). As in root hairs, the infection thread and nucleus are closely associated. The infection thread may branch many times, often at intercellular spaces (Plate 8.1i). Within the cortex infection threads usually grow towards the stele. Ultimately within the nodule a gap develops in the thread wall to release the rhizobia, often in a vesicle formed as a swelling of the thread tip (Plate 8.6).

Occasionally two infection threads may start in a hair and very rarely three (124). For Trifolium spp., Vicia hirsuta, and Pisum sativum, up to 12% of infections may be double (43, 47). These usually intertwine and anastomose or only one develops (Plate 8.1c). Infection threads of some hosts (e.g., T. scabrum) may branch repeatedly and this leads to thread growth ceasing (43). Multiple infection threads were common in Phaseolus vulgaris roots (138). Within the root cortex two infection threads from separate root hairs may sometimes join together.

Not all infections give rise to nodules (10). Depending on the host plant and partly on the Rhizobium strain, many infections cease to grow (abort) before

Plate 8.2. a. Drawings of infection threads in Vicia faba roots published by Ward in 1887 (10). Note the similarity of his Plate 13 with Plate 8.1g, and the vesicle in the right-hand thread with Plate 8.1f. b, c, d. Drawings of nodule development in Pisum sativum taken from sections, by Libbenga and Harkes (188). In 8.1b the infection thread has initiated meristematic activity in the inner cortex opposite a protoxylem point, while still growing through the outer cortex.
c. The thread has grown into the nodule initial with many cortex cells now dividing and some penetrated by the threads. d. The zone of dividing cells is now larger, with older cells in the initial now expanding as they cease to divide, and the meristematic region has become localized at the edge of the initial away from the root stele.

Plate 8.3. Diagram of a Pisum sativum nodule showing two vascular connections with the root. Note that the root endodermis is continuous between the nodule traces indicating that the nodule arose in the cortex. a - root stele, e - root endodermis, n - nodule endodermis, v - nodule vascular trace and ve - its associated bundle endodermis, m - nodule meristem, d - differentiating zone, b - bacteroid zone, db - degenerate zone. After Bond (192).

Plate 8.4. a. "Cylindrical" nodules on Trifolium subterraneum. b. T. subterraneum nodules induced to form roots by high-temperature treatment (see Section 8.3.5.1). The nodule expands into a callus-like growth at its apex (e.g., arrow) before several roots develop. Inset shows a single root developing from a T. pratense nodule also after high-temperature treatment. (Photo J. M. Day and P. J. Dart.) c. Cicer arietinum root with two, large, much-branched, nodule masses. d. Large cylindrical nodules, with occasional branching (e.g., arrows) on Cajanus cajan. The detached nodule on the right has been cut in two. e. Round nodules on Centrosema pubescens. The inset shows the surface patterning. f. Collar nodule on Lupinus arboreus. The main nodule on the right has several rounded components.

Plate 8.5. a. *Trifolium subterraneum* nodule showing an obliquely sectioned vascular trace in the cortex, the nodule endodermis (ne), and the bacteroid zone (b). The trace is surrounded by the bundle endodermis and is bordered by densely cytoplasmic, transfer cells. b. Bundle endodermis in *Glycine max* with Casparian thickening (e.g., arrow). Phase contrast. c. Electron micrograph of two transfer cells in *T. subterraneum* nodule showing the elaborate wall ingrowths and associated mitochondria in a dense cytoplasmic matrix containing many polyribosomes. (x19,760.) (M. R. Chandler and P. J. Dart, unpublished.)

Plate 8.6. *Trifolium subterraneum* nodule. a. Meristem showing files of cells with large nuclei and nucleoli. Divisions then occur in other planes, vacuolation increases, and cells become invaded by infection threads (e.g., arrows). b. Oblique section through the differentiating zone showing infection threads (it) crossing cells releasing vesicles of zooglaeal matrix and rhizobia. c. Rhizobia dispersed through the cytoplasm, dividing rapidly, and filling the cells. The cells filled with young bacteroids have peripheral, nonstaining starch deposits (s). d. Infection thread crossing a cell, with a vesicle of zoogaeal matrix and rhizobia (v) released midway. In another cell a vesicle of rhizobia and matrix lies free in the cytoplasm (arrow). e. Infection threads expanding in the intercellular spaces, with one entering three adjacent cells (arrows). This is enlarged in f showing the release of rhizobia from a small gap in the side of the thread. g. Infection thread, and terminal vesicle which has no obvious enclosing wall.

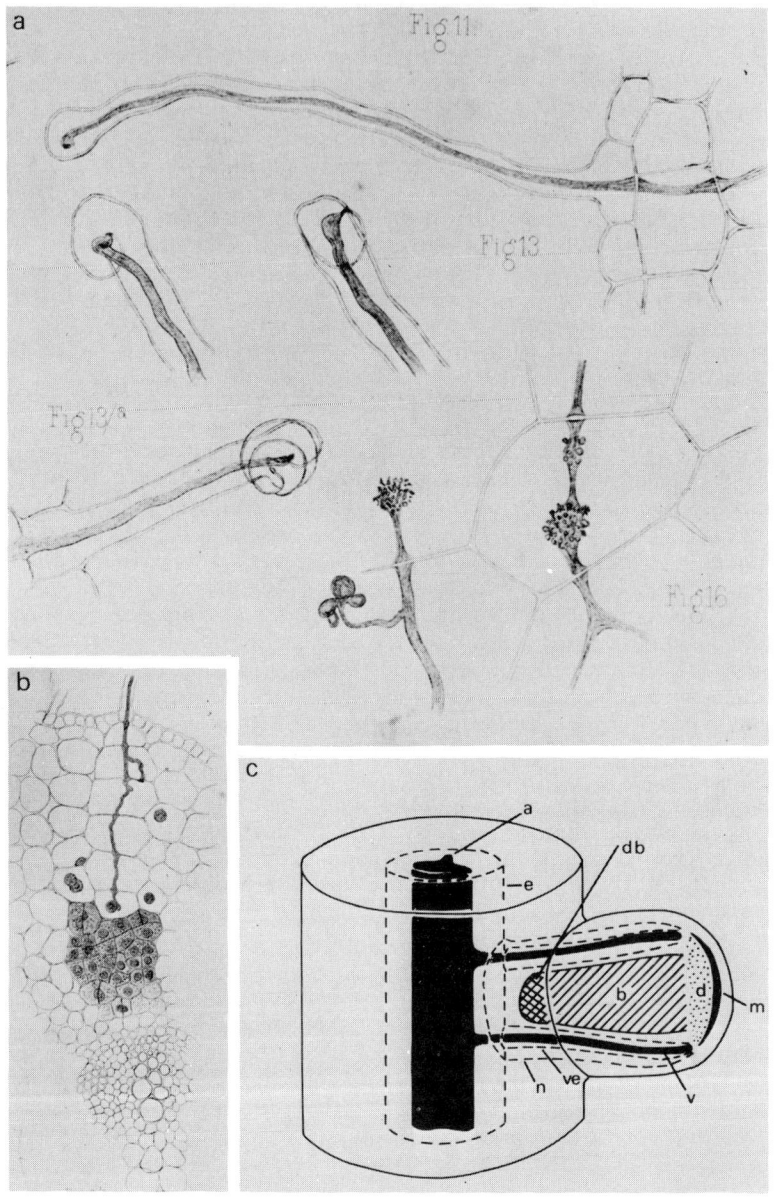

Plate 8.2

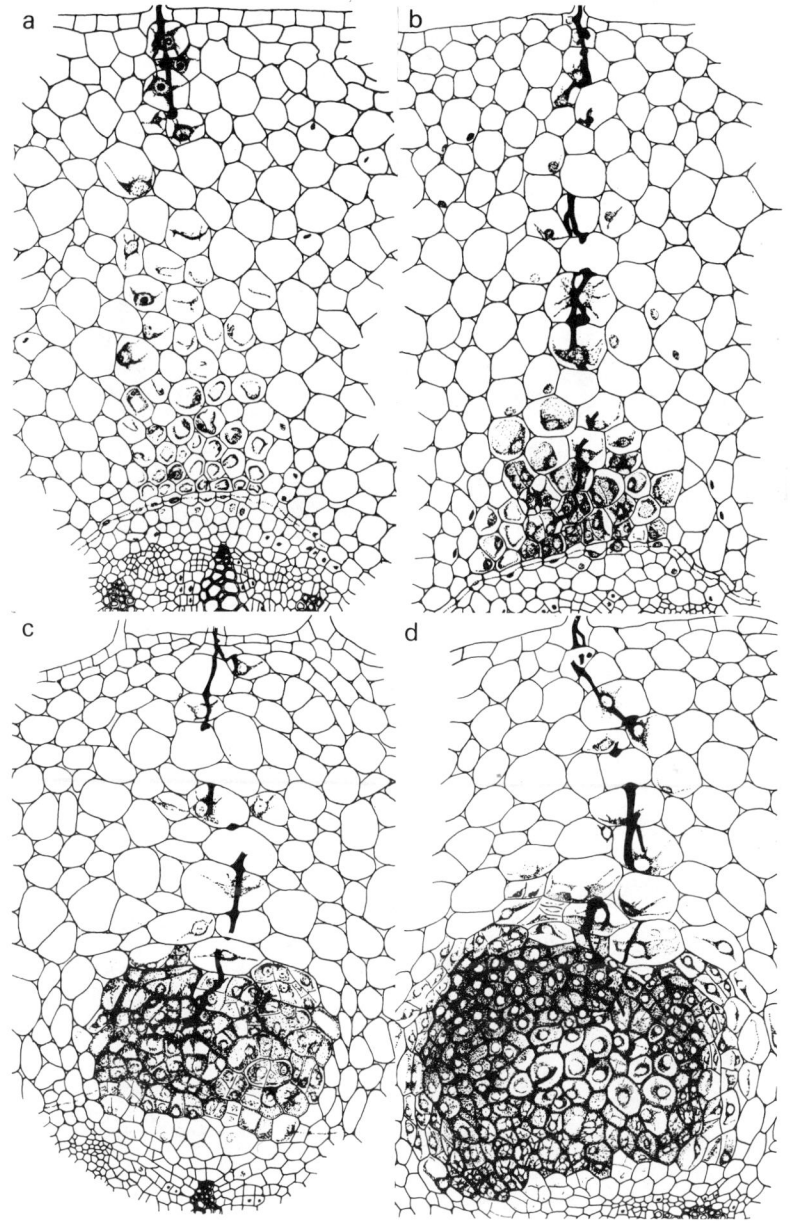

Plate 8.3

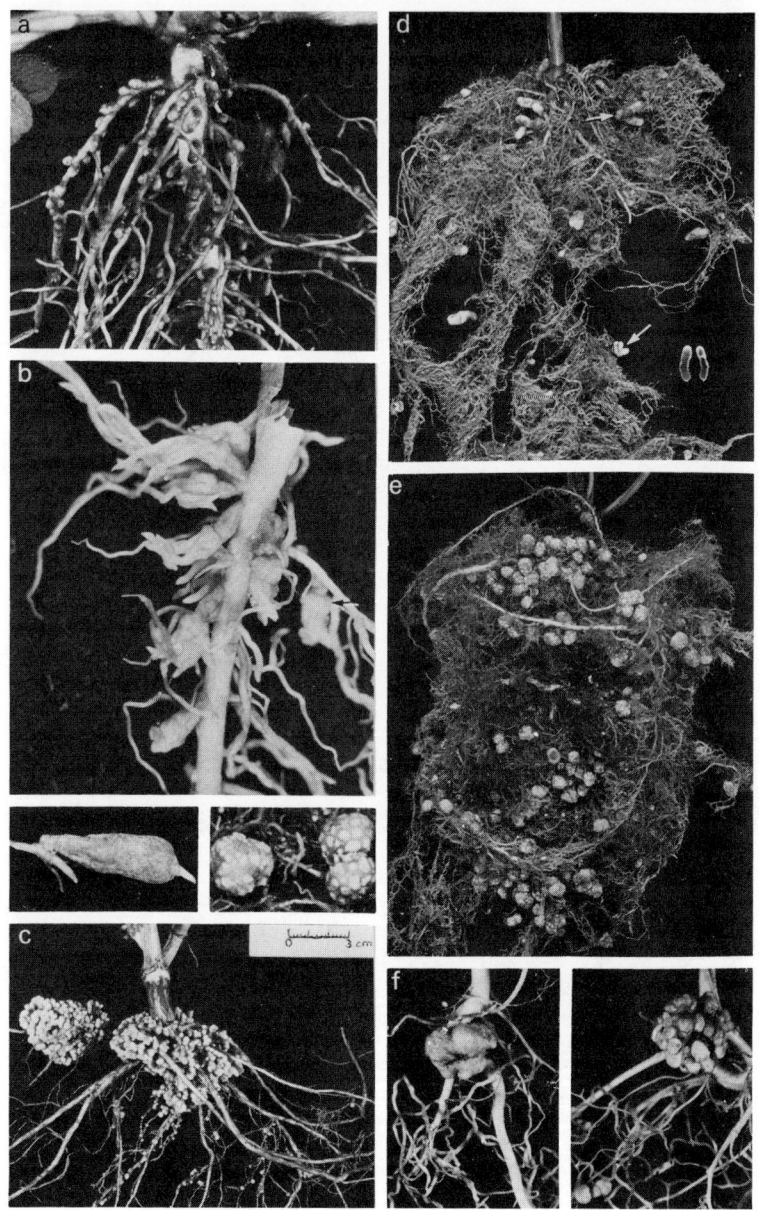

Plate 8.4

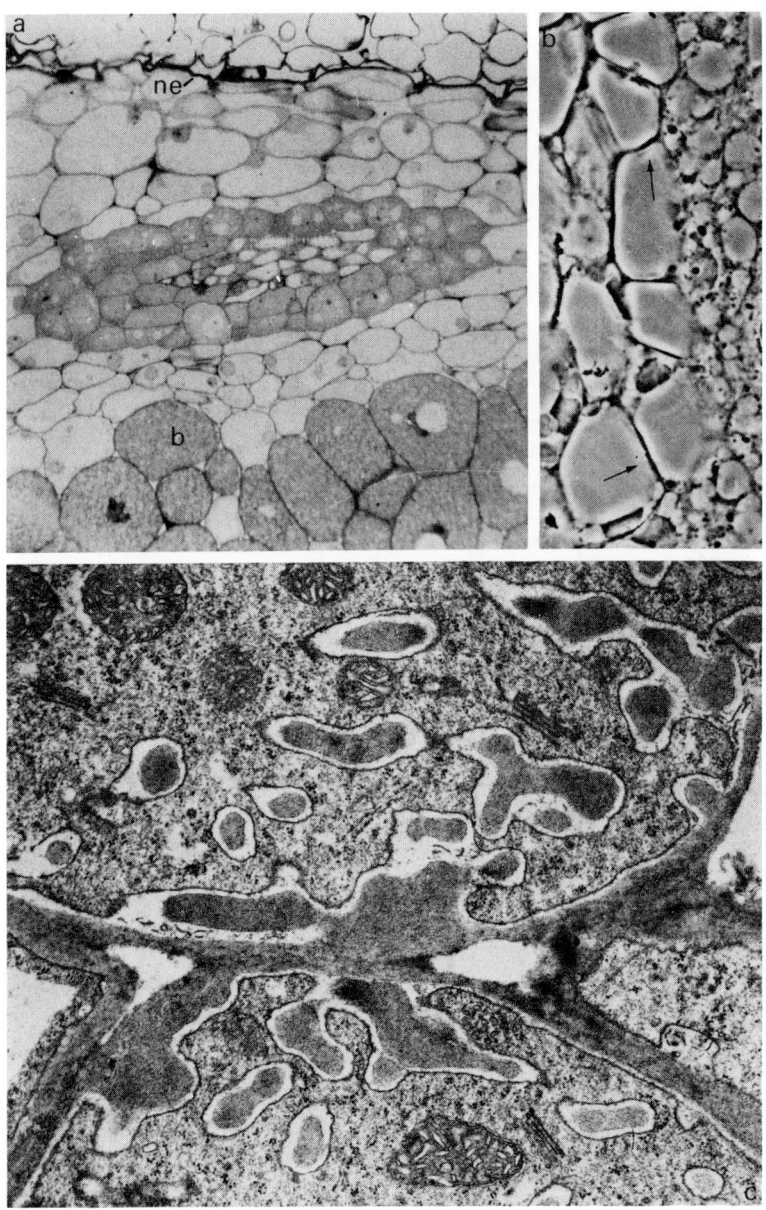

Plate 8.5

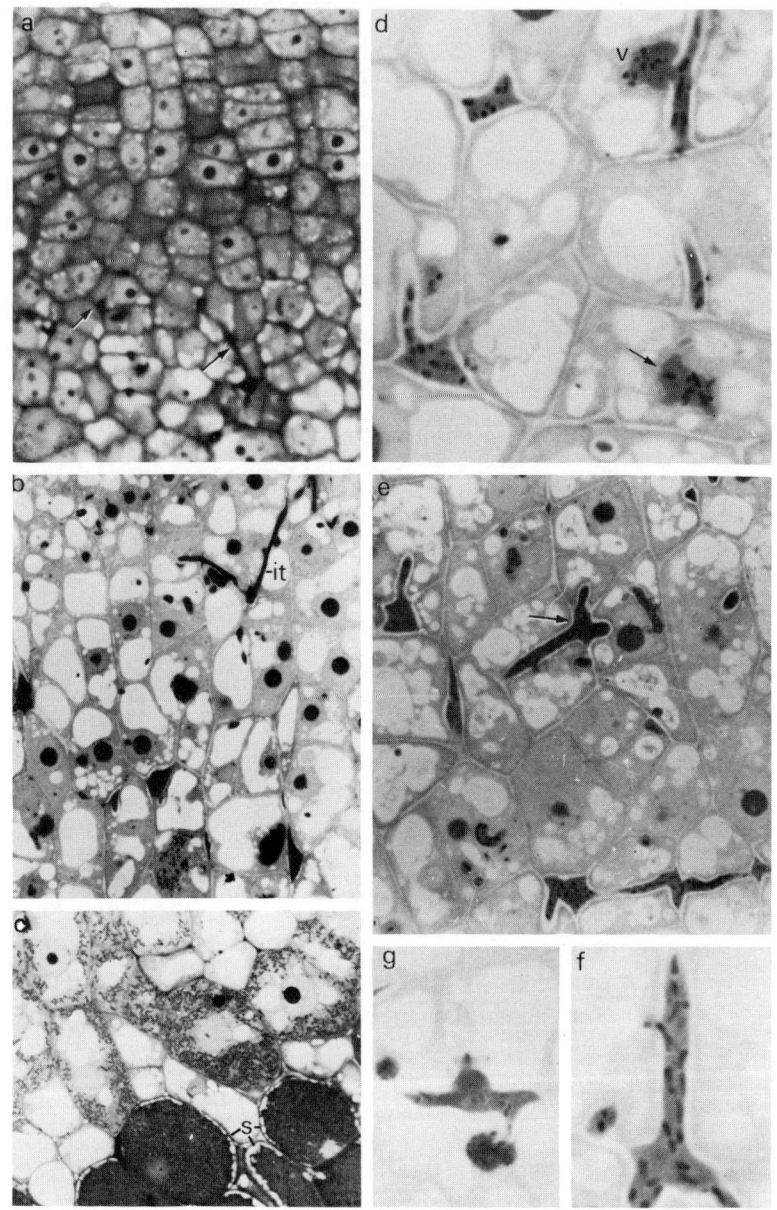

Plate 8.6

Infection 391

reaching the base of the root hair. Others abort at the base of the hair and in the root cortex. The nucleus associated with aborted threads often rounds up, becomes smaller and appears to degenerate (42, 126). A 15-min film of the infection of roots of Trifolium spp. has been made (126). The infection of seedlings with small roots can be followed microscopically if the root is grown between cover slip and slide on a specially made assembly (42, 43).

8.2.5 Infection Thread Numbers

Only a small proportion of root hairs is infected. In about 1000 hairs on a young lateral root of Medicago denticulata Peirce in 1902 (139) found only one infection thread. McCoy (19) found 2.5% of hairs infected in her sample of 600 hairs of M. sativa, and Haack (47) found 2.8% infected on Pisum sativum from a sample of 20,000 hairs. Such counts are very difficult to make. Large samples are necessary because infections are not randomly distributed among the root hairs but usually occur in localized groups (70, 105, 140, 141).

The mean numbers of infected hairs on 12 Trifolium spp. inoculated with the same Rhizobium strain varied from 0.8-82 on 11-day-old plants and was not correlated with root length or seed weight. Rhizobium strains from six different localities induced different numbers of infections on 12-day-old T. fragiferum with means for the strains varying from 25-114 per plant, but with up to a 3-fold variation between replicate plants. Numbers of infections were not related to strain effectiveness. The relative invasiveness of three R. trifolii strains was similar to T. parviflorum, T. fragiferum, and T. glomeratum. Infections begin some 3-20 days after germination, depending on the species. The number of infections increases almost exponentially until the first nodule is formed; then the rate of infection slows markedly. If nodule formation is delayed, most of the root becomes susceptible to infection. The formation of a lateral root also slows the rate of infection in some species (29, 70, 140, 141).

T. subterraneum cultivars differ in the number of nodules formed. Lines bred for sparseness or abundance of nodulation (142) do not differ in the number of infection threads formed by 9 days, and are similar to Tallarook, Mt. Barker, and Cranmore varieties all of which have more infections than Yarloop. Nodule number at 9 days is not related to the number of infections (141).

Infections are not distributed at random along the root but at first are restricted to broad zones down the root initiated by single infections, with later infections within these zones tending to be grouped together. Infections spread both up and down the root tending to reduce the demarcation between zones. Within the broader zones infections are also grouped, with several often within 250 µm. This grouping is not related to any obvious cytological feature. Nodules do not always develop in acropetal succession (63, 70, 105, 140, 141), and their location is unrelated to the initial sites of infection. The number of infections of Trifolium parviflorum, T. patens, and T. glomeratum in soil were fewer than on agar, even though R. trifolii numbers seemed adequate, but distributions were similar and the first nodule reduced the rate of infection in all treatments (143).

McCoy (19) estimated that only 1 in 68 infections gave rise to a nodule on 6-week-old Medicago sativa seedlings, whereas Purchase (45) found few "surplus" infections for 24-to-60-day-old T. pratense. Only 1.4-20% of infected hairs formed nodules on seedlings of T. parviflorum, T. glomeratum, and T. procumbens, and from 4 to 32% for Vicia hirsuta, the ratio depending on the host and Rhizobium strain (43).

Nodule number is to some extent dependent on the root habit, but whether it limits infection or nodule initiation is not known. Excising effective nodules from Trifolium pratense stimulated further nodule production, suggesting that a hormone inhibitor was produced by existing nodules (109). As nodules are powerful sinks for carbohydrate, nodulation may alternatively be limited by the availability of photosynthate. Ineffective strains may form nodules, but since these are not fixing N_2, utilize little carbohydrate. Increasing the CO_2 concentration around the leaves (144) or increasing the light intensity stimulates nodulation of Trifolium spp. (145).

8.2.5.1 Effect of Temperature

Root temperature markedly affects the infection of Trifolium spp. At 7° infection of T. subterraneum was delayed until about 13 days with the first of few nodules forming at 15 days. The much slower rate of infection at 7° was not obviously reduced by nodule formation and did not seem to result from too few rhizobia or root hairs. The time taken to form the first nodule depended on the cultivar (140). Cranmore was not infected at all by 40 days and in the root zone where nodules usually formed, lateral roots were produced instead (146). As the temperature increases, infection and nodulation start earlier, until at 19° infection begins about 3 days after germination and nodules form 3-4 days later. Plants transferred from 7° to 19° at 9 days, before infections had begun, rapidly become infected, and nodule formation began at 10 days. The number of infections equalled those produced by plants which had been at 19° continuously, but again the rate of infection appeared to be slowed by the formation of the first nodule. Although some infections occurred among the oldest root hairs on the transferred plants, there may have been newly formed root hairs within these zones. However, infections generally formed over a much less extensive zone of the root (140, 141).

Holding 24-hour seedlings of Trifolium spp. at 3°C increased the number of infections when the plants were subsequently inoculated and grown at 19°C. The effect was transitory, disappearing on plants more than 7 days old (147).

Infection of T. glomeratum and T. parviflorum was also much reduced at root temperatures below 18° or above 30°, and nodule number even more affected. This was not simply related to root length or leaf area. Temperature had little affect on the marked zonation of infections along the root or thread morphology. Transfer of 4-to-14-day-old seedlings from 6 or 36° to 24° markedly stimulated infection within 2 days, so that a larger number were ultimately formed than on plants kept at 24° continuously. Even as short a period as 6 hours at 24° stimulated the infection of 6-day seedlings otherwise kept at 6°. Infection over the next 4 days proportionally increased with the time spent at 24° (125).

Only at extremes of hot or cold is initial infection restricted or completely inhibited. Later stages in nodule development, perhaps involving growth of the infection thread into the cortex, are more sensitive to temperature.

8.2.5.2 Combined Nitrogen

Nitrate and ammonium (0.1 m$\underline{M}$) in the water culture considerably reduced <u>Medicago sativa</u> root hair production, and their curling by <u>R. meliloti</u> (41, 148). Pretreatment of seedlings with nitrate for 3 days before inoculation reduced curling by about 90%, but only by 30% if supplied at inoculation. Later stages of nodule initiation on the contrary were much more influenced by the nitrate status after inoculation. All stages of infection including thread growth seemed sensitive to nitrate, fewer formed and more aborted and appeared disorganized. Exposure to nitrate for only 24 hours on any one of the first 3 days after inoculation reduced nodulation (149).

Adding IAA alleviates some of the effects of nitrate on curling, thread formation and abortion, and nodulation of <u>M. sativa</u> (99, 150). Nitrate reduced the IAA levels in <u>Rhizobium</u> cultures possibly by oxidation of the IAA by the nitrite produced by <u>Rhizobium</u> from nitrate (151, 152). <u>Rhizobium</u> sp. vary in their rate of nitrite production and IAA breakdown (153). IAA production from tryptophas was not inhibited by nitrate, but was by ammonium, urea, and glycine which do not produce nitrite.

Initial infection of <u>Trifolium glomeratum</u> and <u>T. repens</u> is stimulated by very small nitrate or nitrite additions (10-20 µg N per plant) and delayed by 40 µg N and above. Ammonium and urea 10 µg N have no effect on early infections but 100 µg decreases infection. Urea at 1 mg N per plant prevents infecttions over the first 17 days. Increasing nitrate levels reduce infections in the oldest part of the root and nodules also form further from the hypocotyl. Root hair initiation and growth is inhibited by N additions; nitrite was particularly inhibitory. Combined nitrogen additions to the root tended to delay the formation of the first nodule, but when the nitrate (20 µg N) was taken up through the cotyledons nodulation was not delayed, but infections were still stimulated. There was no evidence of more threads aborting in the hair or growing abnormally as in <u>M. sativa</u>, perhaps because the nitrate levels were lower. The increased infections with small N additions, thus, resulted from a prolongation of the initial rapid phase of infection before the first nodule formed. Levels that inhibited infections also delayed the time at which infections began. The much slower rate of infection after the first nodule had formed was not apparently influenced (57, 70, 154, 155).

The stimulation of infection by nitrate was negated by the addition of glucose but required 10 mg per seedling, even though 10 µg was sufficient to prevent the delay in noudlation by nitrate. Glucose alone decreased infections only with the largest addition (10 mg per plant) but this did not affect nodule formation. Added glucose stimulated Rhizobium growth and may have been sufficient to reduce the nitrate concentrations below that which influcenced infection (155).

8.2.5.3 Delayed Inoculation

For Trifolium pratense, delaying inoculation with the single strain until 12-30 days after sowing increased nodule formation and shortened the time between inoculation and nodulation; even after an 80-day delay plants nodulated (156). Delaying inoculation from 5 to 25 days had a similar stimulation of nodulation for Medicago truncatula, T. subterraneum, Pisum sativum, and Glycine max (54, 142, 157, 158). As inoculation is delayed the older parts of the root apparently become progressively resistant to infection (44) with nodules forming lower down the primary root, then on the upper, first-formed lateral roots, and with long delays on the later-formed laterals. Root hairs of Trifolium spp. become resistant to infection 8-10 days after they are formed. After delayed inoculation, nodules formed much closer together on some lateral roots, with up to 20 times more per unit root length than found on roots inoculated at sowing. It is not known whether more infections are successful, but initial infection is stimulated (155).

When an already-nodulated plant is inoculated after 4-6 weeks with another strain, few nodules are formed by the second strain. A 13-day delay in application of a second inoculum to Glycine max reduced nodulation by this strain to about 20% even though 5×10^{10} rhizobia were added per plant (61).

8.2.6 Infection without Infection Threads?

Infection threads have not been found in root hairs or nodules for several species (Table 8.2b), with the implication that a different mode of infection occurs. Although much is known of the root-hair infection-thread pattern, reliable observations of infection threads in root hairs of different legumes are few (Table 8.1), although they have been seen in sections of nodules of 46 additional species (Table 8.2a).

TABLE 8.1. Observations of Infection Threads in Root Hairs

Plant Species	Reference	Plant Species	Reference
Vicia faba	10, 200	Trifolium repens	
V. hirsuta	15, 43, 70	T. fragiferum	
V. villosa	190	T. scabrum	
V. sativa	47	T. nigrescens	29, 35,
Pisum sativum	11, 16, 47	T. dubium	42-5, 49,
		T. procumbens	70, 105,
Medicago denticulata	139	T. subterraneum	126, 128-
M. sativa	41, 45, 99, 149, 187, 327	T. pratense	30, 134,
		T. incarnatum	140, 141,
		T. parviflorum	146, 147,
Phaseolus oleraceus	17	T. glomeratum	154, 265,
P. vulgaris	138	T. patens	344, 366,
Caragana aborescens	248	T. arvense	367
		T. ornithopodioides	
Cicer arietinum	189, 367	T. angustifolium	
Cajanus cajan	250	T. alexandrinum	
Cyamopsis tetragonoloba	202	Ornithopus sativus	47, 165
		Glycine max	124
Vigna radiata	27		
V. mungo	368		
Sesbiania grandiflora	246		

Also claimed for:	
Lathyrus sativus, Lens culinaris, Melilotus indica, Trigonella foenum-graecum	367
Canavalia gladiata	368
Centrosema pubescens, Crotolaria verrucosa, C. juncea Cyamopsis psoralioides, Desmodium gangeticum, Indigophora endecaphylla, Pongamia pinnata, Sesbania bispinosa, Vigna unguicalata	367

TABLE 8.2. Observations of Nodule Sections

Plant Species	Reference	Plant Species	Reference

a. Nodules containing infection threads, and uninvaded cells in bacteroid zone

Plant Species	Reference	Plant Species	Reference
Vicia faba	6, 171, 200, 369, 370	Medicago denticulata	139
V. sativa	370	M. lupulina	171
V. hirsuta	16, 171	M. sativa	124
V. villosa	124, 190	M. arborea	256
V. tetrasperma	171	M. truncatula	130
Pisum sativum	8, 14, 16, 192, 369-71	Melilotus officinalis	171
Lathyrus aphaca	14	M. alba	124
L. sativus	370	M. indica	367
L. sylvestris	14	Trigonella foenum-graecum	367, 368
L. pratensis	171	Glycine max	124, 194
Coronilla glauca		Phaseolus lathyroides	198
C. iberica		P. vulgaris	54, 370
C. scorpioides		P. coccineus	17
Onobrychis viciifolia		P. oleraceus	
Neptunia oleracea	123	Acacia cornigera	
Ornithopus sativus	165, 372	A. bynoeana	
Astragalus glycyphyllus	54, 256	A. melanoxylon	256
Doryncium hirsutum	256	Albizzia spp.	
Lotus corniculatus	54, 371	Cajanus cajan	202
Carmichaelia australis	303	Canavalia gladiata	368
Cytisus scoparius	256	Caragana tragacathoides	256
Galega officinalis	370, 373	C. arborescens	201, 248
Ulex europeaus	171	Colutea arborescens	256
Trifolium repens	196, 232	C. persica	256
T. squamosum	16	Enterolobium cyclocarpum	201
T. incarnatum	196	Indigofera atropurpurea	256
T. pratense	196, 199, 373	I. macrostachya	
T. alpinum	256	Mimosa dysocarpa	
T. subterraneum	130	Robinia pseudoacacia	169, 256
T. parviflorum	232	R. viscosa	256
Ononis natrix	256		

(Continued)

TABLE 8.2. (Continued)

Plant Species	Reference	Plant Species	Reference

a. Nodules containing infection threads, and uninvaded cells in bacteroid zone

Plant Species	Reference	Plant Species	Reference
Vigna unguiculata	124, 161, 367	Sesbania grandiflora	246
V. radiata	161, 374	Wisteria sinensis	178
V. mungo	161		

b. Nodules where infection threads not found, containing no uninvaded cells in bacteroid zone

Plant Species	Reference	Plant Species	Reference
Lupinus albus	163, 164, 284	Desmodium dillenfi	256
L. mutabilis	163	D. uncinatum	54
L. polyphyllus	171	Genista florida	
L. perennis	163	G. hispanica	
L. angustifolius	54	G. pilosa	
Aeschynomene indica	166	G. siberica	
A. americana	367	G. tinctoria	
Arachis hypogaea	54, 160, 256	Laburnum anagyroidcs	256
Cassia mimosoides	256	Leucaena leucocephala	
Crotalaria vespertilio	375	Piptedenia rigida	
Cytisus capitatus		Sophora moorcroftiana	
C. purgans	256		
C. sessilifolius		S. tomentosa	367
Dalbergia sissoo	367	Spartium junceum	256
D. lanceolaria	367	Stylosanthes sundaica	367

Nodules of Arachis hypogaea arise in the junctions of lateral roots (160), often in well-defined lines running along both primary and secondary roots. Possibly such points provide favorable intercellular sites for Rhizobium multiplication and penetration into the cortex; no infected root-hairs were observed. Intercellular zoogleal strands of rhizobia but not infection threads are found in young A. hypogaea nodules (161). Small indentations of the host cell wall are occasionally found, and these may have been the initial

site of entry. The nodule is recognizable as such
before intracellular rhizobia can be found. Rhizobium
is distributed in the nodules by cell division and
only invaded cells in the bacteroid zone seem to
divide (160, 161); consequently, among the invaded
tissue of the nodule there are no uninvaded cells as
in nodules where rhizobia are distributed through infection threads. R. leguminosarum can be taken up by
Pisum sativum leaf protoplasts during enzymic digestion
of the plant cell wall (162) presumably without
infection thread development.

A similar pattern of "intercellular" infection and
dissemination of rhizobia by cell division probably
exists for other nodules where infection threads have
not been found, e.g., Lupinus sp. [(14, 163, 165) and
Table 8.2b], and where uninvaded cells are absent or
uncommon. Aeschynomena indica forms nodules on the
stem base as well as on the roots. Both stem and root
nodules arise near emerging lateral roots; root hairs
are not present. No infection threads are found in
the nodules, and the rhizobia are spread by cell
division (166). One consequence of such a mode of
infection would perhaps be an ability to form nodules
with a wider range of Rhizobium strains. A. hypogaea
is apparently quite promiscuous, nodulating readily
with strains of the cowpea miscellany, R. japonicum,
R. phaseoli, and R. lupini, and occasionally with
R. meliloti (160, 167, 168). Some Rhizobium strains
form nodules with no obvious infection threads on some
host species, e.g., Lupinus spp., and nodules with
prominent infection threads originating in a root hair
on others, e.g., Ornithopus sativus (54, 165). The
mode of infection is apparently a characteristic of
the plant and not the strain of Rhizobium.

The Arachis mode of infection has similarities
with early crown gall development where Agrobacterium
tumefaciens enters tissues through wounds and remains
largely intercellular.

8.3 NODULE DEVELOPMENT

8.3.1 Nodule Form

That nodule shape is a characteristic of the host
plant rather than the Rhizobium strain has long been
recognized (169-172). Nodule shape is determined by

the pattern of meristematic activity and may be broadly classified into round or oval, elongate to club-shaped, branched to coralloid, and collar nodules. As with cross-inoculation specificities, there seems to be no strict relationship between nodule shape and botanical classification, although Corby (173) found that for 400 wild Rhodesian species, nodule shapes within a tribe tended to be similar. However, within one genus nodule shape may vary — round for Vigna unguiculata and V. marina, and elongate for V. owahuensis (172). Annuals and trees may have similarly shaped nodules, e.g., Lupinus angustifolius and L. arboreus.

In the round or oval nodules such as on Glyine max or Vigna unguiculata, meristematic activity is initially spread through the nodule. After 10-20 days activity decreases and tends to be localized in pockets at the edge of the spherical bacteroid zone, and much of the increase in nodule size thereafter derives from the growth of already-infected cells (138, 174). Such nodules retain the capacity to increase in size even when they are about 60 days old. The bacteroid zone may be a single homogeneous unit or separated into sectors by uninvaded tissue continuous with the husk. Round nodules usually have a small attachment zone to the root, even though they grow quite large (up to 0.7 cm in diameter, as in Centrosema pubescens). The surface of many round nodules is often patterned by a series of ribs separating islands of protruding, lighter-colored cells that appear to be exfoliating and are possibly lenticullular in function (Plate 8.4e). If the bacteroid zone becomes divided into sectors the top of the nodule may become flattened as the meristematic foci are displaced towards the sides of the nodules and the nodule may develop secondary outgrowths while still retaining its basic oval form. Some round nodules such as Arachis hypogaea are not ribbed and may also have a broader attachment to the root with bacteroid tissue embedded in the root cortex so that the nodule appears hemispherical; the husk tissue may even grow around the subtending lateral root (Plate 8.14a).

The second major nodule shape is that found in Trifolium, Vicia, Pisum, and Medicago spp., for example, where the meristematic region forms a terminal cap functioning throughout the nodule's life, producing a young nodule that is at first hemispherical and then becomes basically cylindrical (Plate 8.4a). Such nodules continue to grow until they senesce and may reach

Plate 8.7. Trifolium subterraneum nodules. a. Young nodule showing the development of a vascular trace (vt) from the protoxylem point. The trace divides at the base of the nodule, and one of the bundles formed is cut transversely (vb). The nodule endodermis (ne) is continuous with the root endodermis (e), and merges with the nodule meristem. Bacteroid-filled cells have developed at the base of the nodule. U - Uninvaded cells in the bacteroid zone. b. Older nodule with clearly differentiated meristematic zone (m), differentiating zone (d) and bacteroid zone (b) with greatly enlarged cells, some uninvaded. ne - Nodule endodermis. c. Cells from mature bacteroid zone, packed with ovoid bacteroids enclosed singly in membrane envelopes. Some dense staining inclusions are present in the bacteroids. The nuclei are amoeboid in shape and condensed, staining densely with toluidene blue. Large intercellular spaces are present.

Plate 8.8. a. Bacteroids in a T. subterraneum nodule surrounded by a zone of low electron density and a membrane envelope (m). The bacteroids have a dispersed nucleoid and contain a few intracytoplasmic membrane vesicles (e.g., v) and polyribosomes, a crystalloid inclusion (c), and an aggregate of small circular profiles enlarged in 8.8b. The host cytoplasm is electron dense, with a few scattered ribosomes and endoplasmic reticulum fragments. (a) x21,375; (b) x30,000. c. Bacteroid crystalloid inclusion. (x24,375.) d. Residual infection vesicle containing unenlarged rhizobia, with poly-β-hydroxybutyrate inclusions. Note the very large increase in size of the rhizobia during bacteroid development. (x4875.) (M. R. Chandler and P. J. Dart, unpublished.)

Plate 8.9. a. Young bacteroids in a Pisum sativum nodule enclosed singly in membrane envelopes. The large nucleoid of the bacteroid is central at this stage of development, becoming more dispersed as the bacteroids age. (x17,390.) b. Light micrograph of P. sativum nodule showing older, ovoid bacteroids containing large inclusions of poly-β-hydroxybutyrate. These are the amylodextrin bodies described by Frank (262). A group of untransformed rhizobia (r) shows the great increase in size as bacteroids develop. u - Uninvaded cells. (M. R. Chandler and P.J. Dart, unpublished.)

Plate 8.10. a. Infection threads in a young Vigna unguiculata nodule. The rhizobia lie in single file and have little zooglaeal matrix surrounding them. The infection threads lie close to the host nuclei (n). Some rhizobia have been released and lie in the cytoplasm in a membrane-bound vacuole. v - Host cell vacuole. (x8140.) b. Cup-shaped plastid in a Trifolium pratense nodule with crystalline array of phytoferritin. s - Starch grain, c - host cytoplasm. (x37,740.) (M. R. Chandler and P.J. Dart, unpublished.)

Plate 8.11. a. Young Glycine max nodule with an infection vesicle adjacent to the lobed nucleus (n). Residual wall material is only present around part of the vesicle (arrow). Golgi bodies and endoplasmic reticulum lie close to the vesicle, and the host cytoplasm is packed with polyribosomes. (x10,150.) b. Newly invaded cell in a young G. max nodule with rhizobia dispersed singly or in pairs in large membrane-bound vacuoles. Whorls of rough endoplasmic reticulum, and Golgi bodies are also present. (x4900.) c, d. Rhizobia in young G. max nodule enclosed single in a membrane envelope with tubular profiles in the envelope space with one connection the envelope and bacterios in 8.11d (arrow). (c) x29,050; (d) x31,500. (M. R. Chandler and P. J. Dart, unpublished.)

Plate 8.12. a. Dividing Rhizobium in the host cytoplasm of a young Glycine max nodule. The Rhizobium has a condensed central nucleoid with associated polyphosphate granules and is packed with ribosomes. (x30,750.) b. Older rhizobia in a young G. max nodule, still enclosed singly in their membrane envelopes, and containing a large polyphosphate granule (p), some small poly-β-hygroxybutyrate granules (PHBA), and a few dispersed polyribosomes. (x25,500.) c. Mature bacteroid region of G. max nodule with several bacteroids per membrane envelope. Some fibrous material (f) also lies within the envelope space which is otherwise electron transparent. The bacteroids are packed with PHBA. Numerous plasmodesmata (arrow) connect the bacteroid-filled cell with the adjacent uninvaded cell. (x6000.) (M. R. Chandler and P. J. Dart, unpublished.)

Plate 8.13. a. Glycine max nodule bacteroid zone with uninvaded cells among the bacteroid filled cells. Uninvaded cells have smaller nuclei (n) and contain large starch granules (e.g., s) and little cytoplasm. Large intercellular spaces (i) are present between all cells. b. Mature G. max nodule cortex showing the nodule endodermis (ne) with greatly thickened walls containing simple pits. Crystalline inclusions (c) are present in several cells outside the endodermis. Startch grains are present in the inner cortical cells. The arrow indicates the start of a file of uninvaded cells that run radially through the nodule between the bacteroid-filled cells (b). c. Mature bacteroids in Phaseolus vulgaris nodule enclosed several per membrane envelope. The bacteroids are longer than in G. max, and contain much PHBA and some dense inclusions (arrows) which are possibly glycogen. The host cytoplasm outside the membrane envelope is much denser than in the adjacent uninvaded cell (u). i- Intercellular space. (x15,750.)

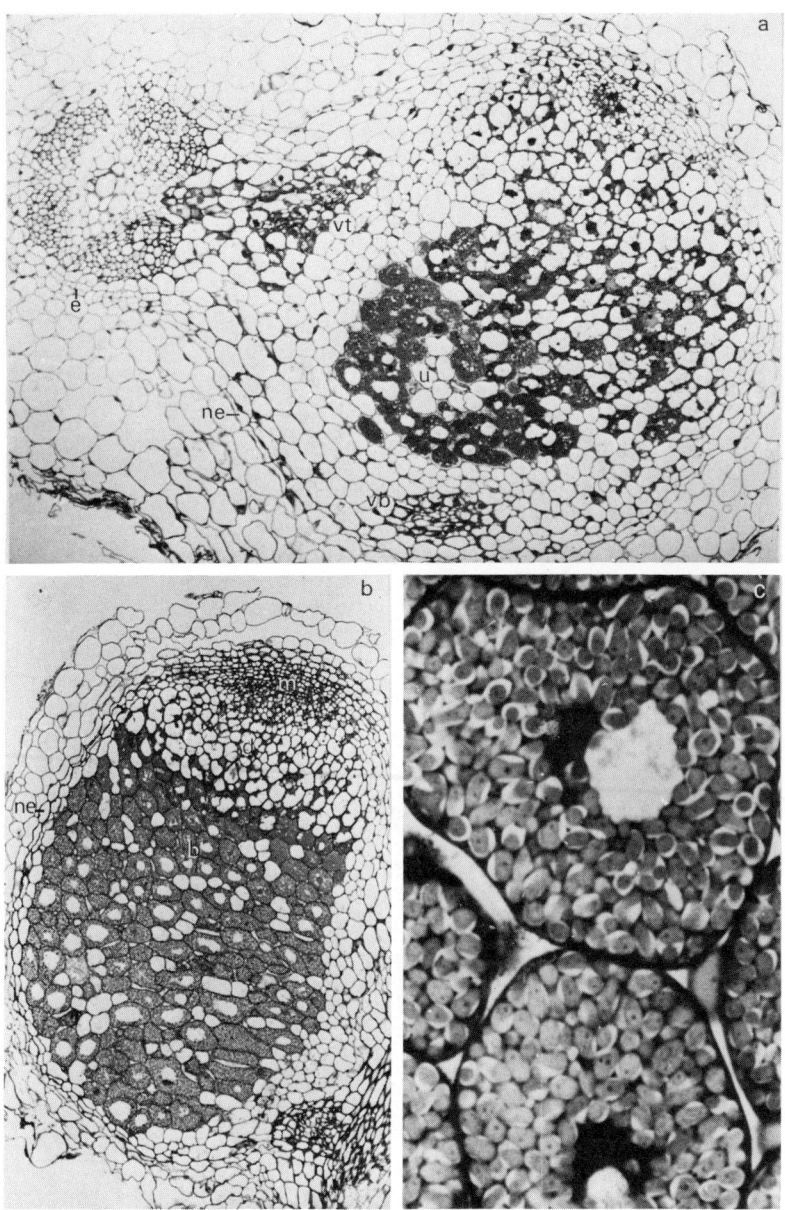

Plate 8.7

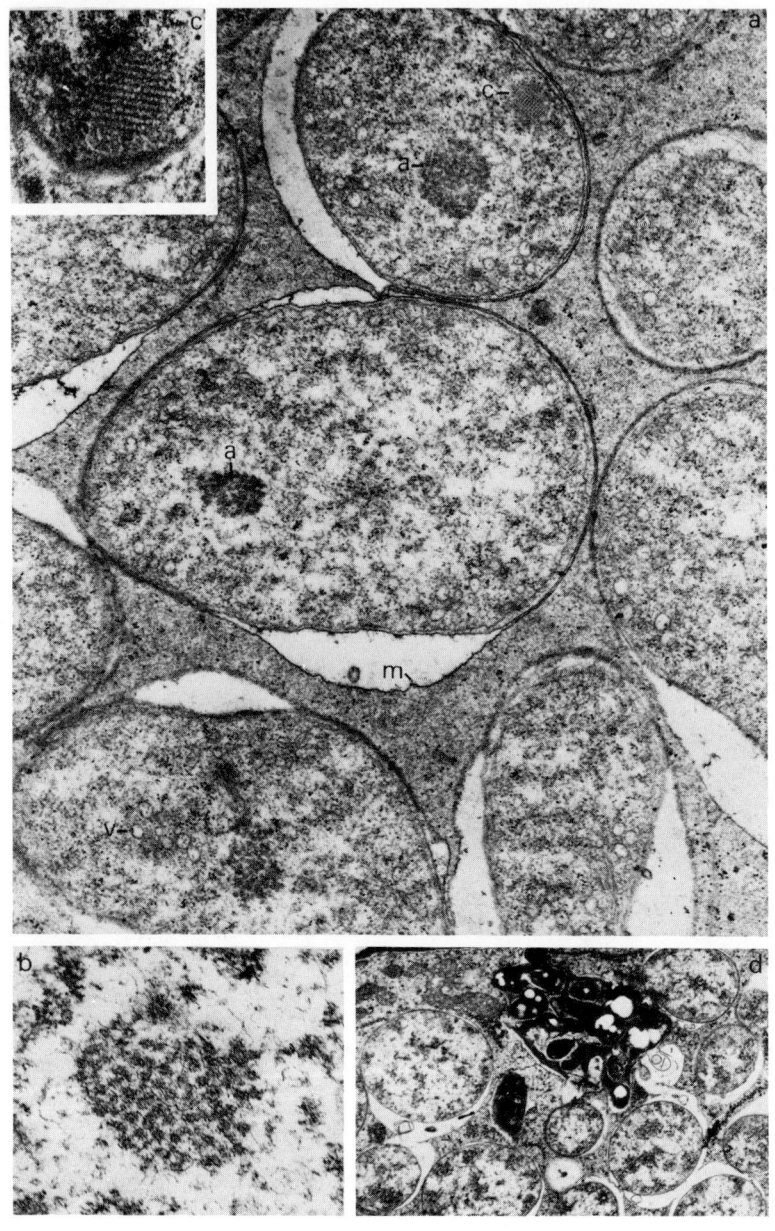

Plate 8.8

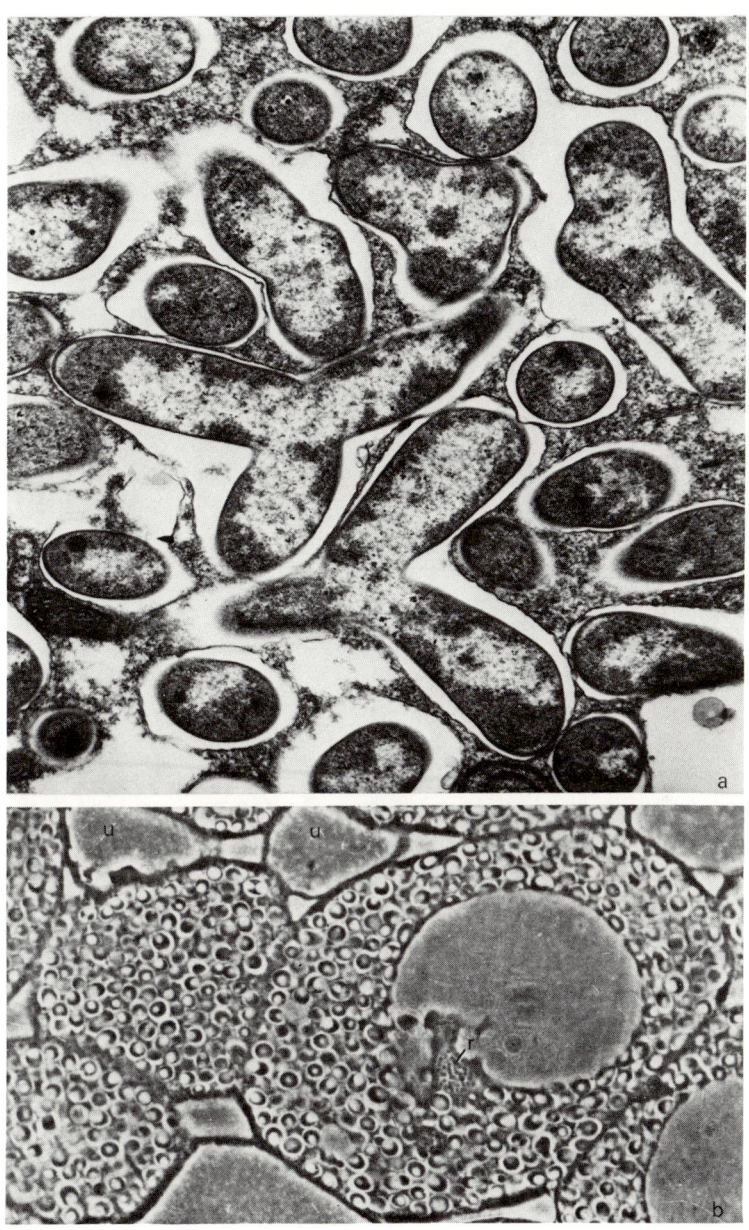

Plate 8.9

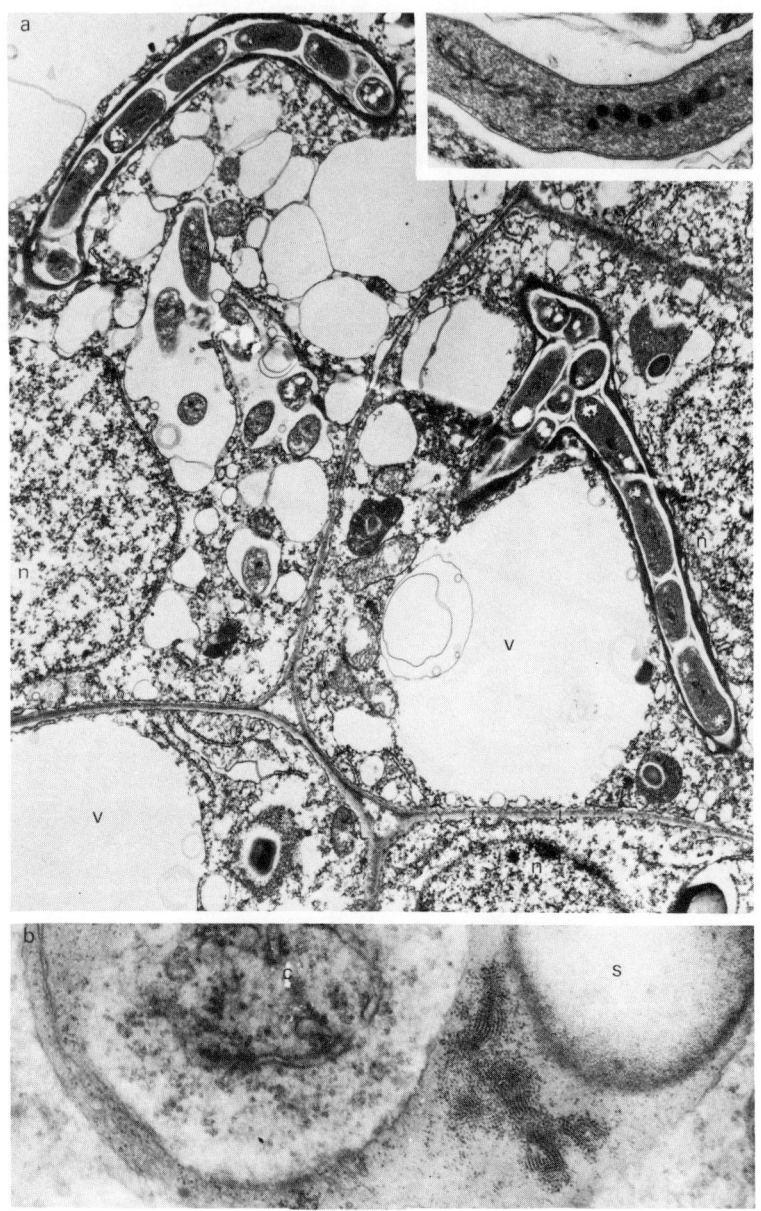

Plate 8.10

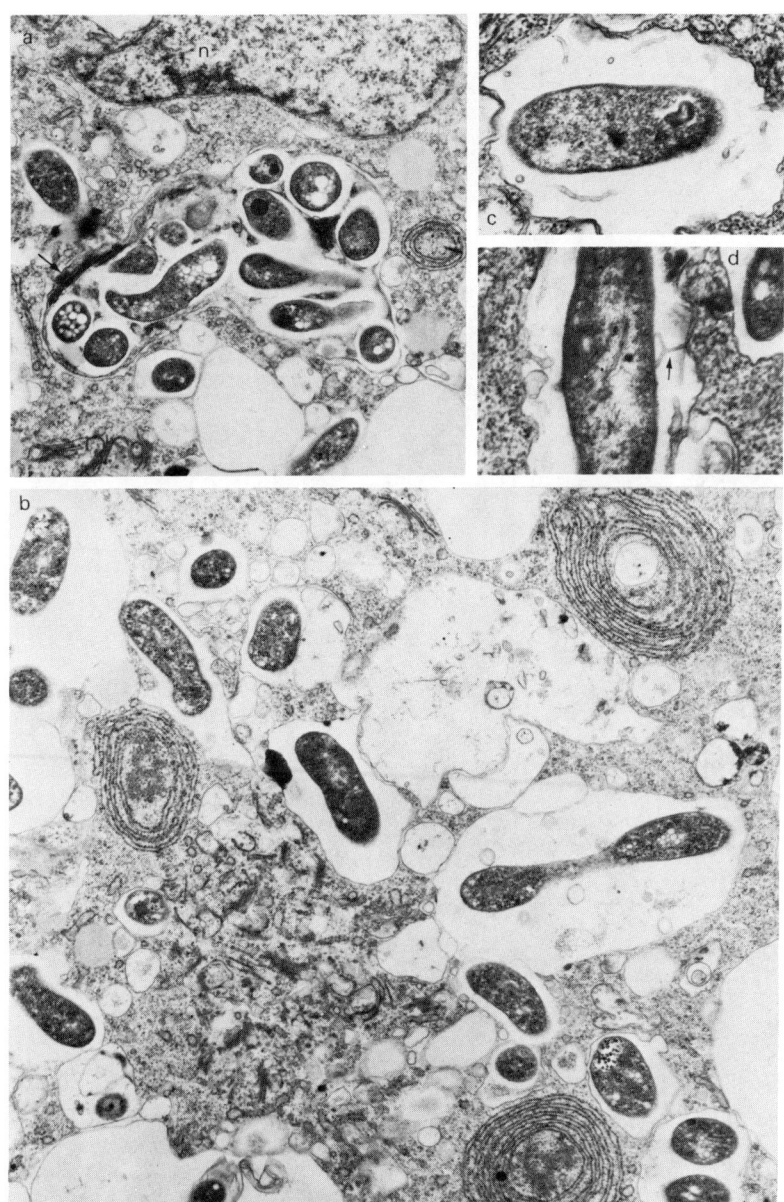

Plate 8.11

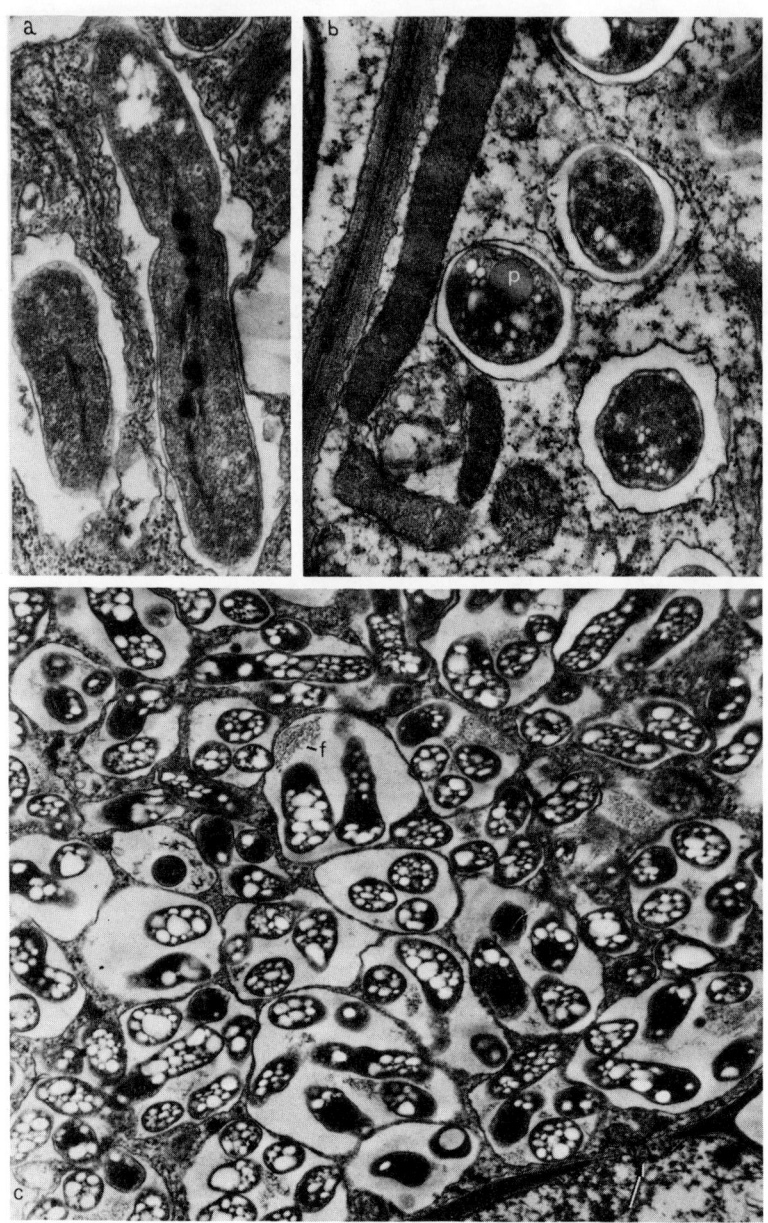

Plate 8.12

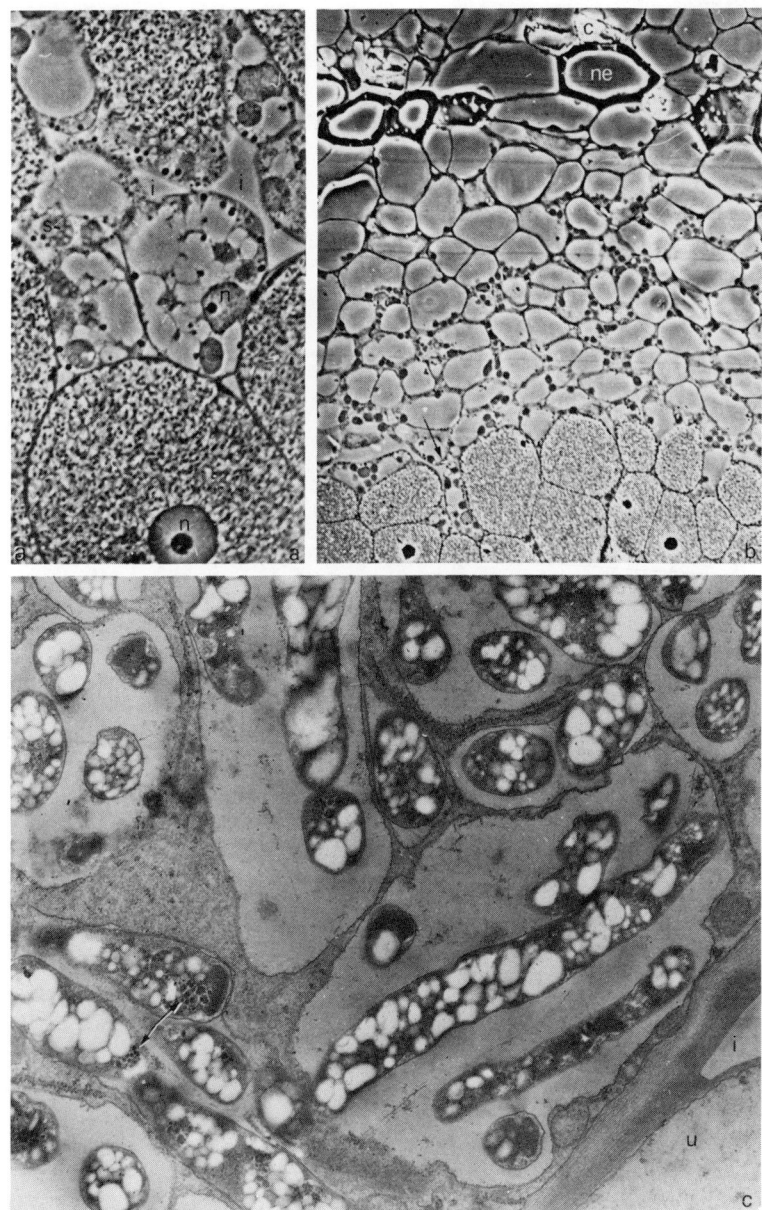

Plate 8.13

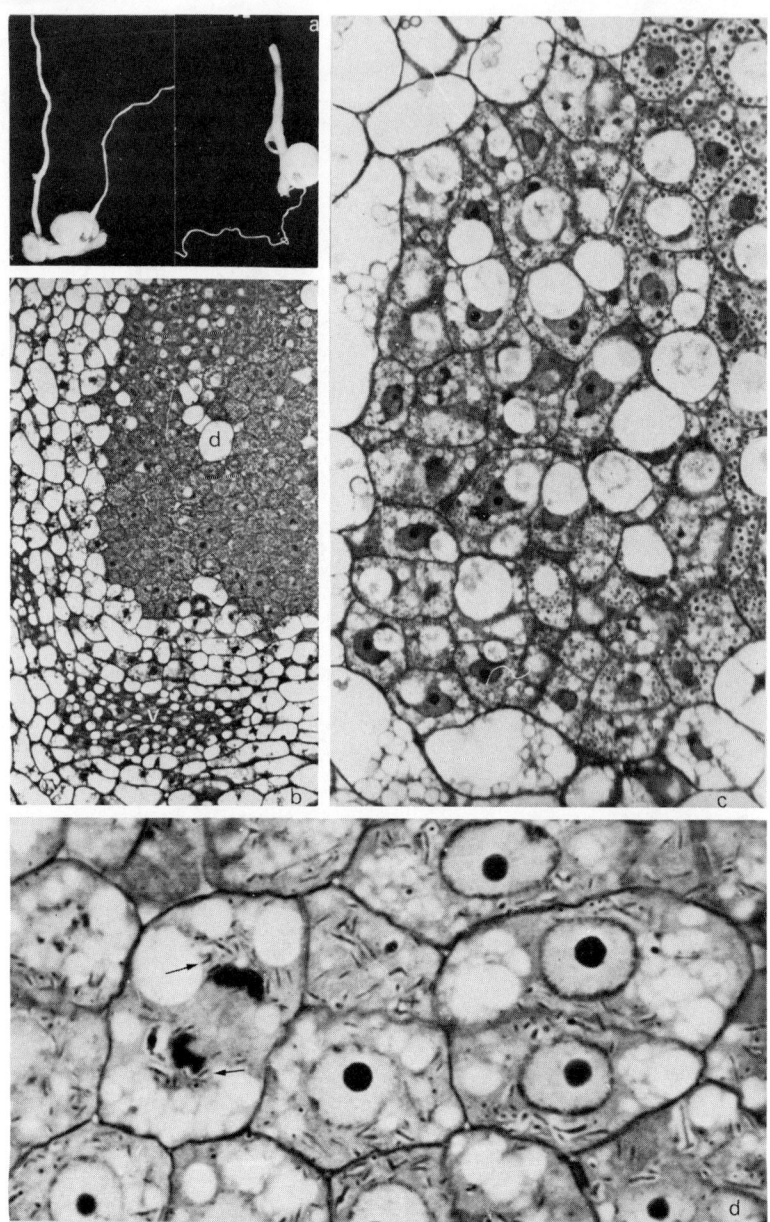

Plate 8.14. <u>a</u>. Round nodules of <u>Arachis hypogaea</u> showing the origin of the nodule in the axis of a lateral root. The left-hand nodule has grown around the root. <u>b</u>. Light micrograph of an <u>A. hypogaea</u> nodule showing the bacteroid zone with only a few large uninvaded duct cells (<u>d</u>). Starch granules are absent from the bacteroid-filled cells but prominent in the uninvaded cortex cells, clustered around the nuclei. <u>v</u> - Vascular trace.

up to 1 cm in length of Medicago sativa. As the nodule ages and grows in size this meristem may broaden so that the nodule becomes wider at its apex, and then divide into two or more parts, thus producing a branched nodule that may become corralloid with repeated branching. In some extreme cases a very large nodule cluster may be formed (e.g., Plate 8.4c). Branching is restricted to nodules that initially have a terminal meristem. High root temperature can increase meristematic activity often resulting in increased branching (III.8.3.5). Many species, such as Cajanus cajan, have nodules with a cylindrical to club-shaped form, which show less tendency to branch (Plate 8.4d). This varies within a genus: Acacia longifolia nodules are rarely branched and A. horrida nodules are often very corralloid (54, 172).

In the third type, as in Lupinus spp. (Plate 8.4f), the nodule forms a flattish collar around the root, particularly the tap root, and may even almost encircle it. The meristematic zone, initially a hemisphere at the apex of the nodule, becomes localized at the edge of the nodule so that it grows around the root. Collar nodules may also result from the merging together of several sites of infection that initially developed as discrete nodules (175). The point of attachment to the root tends to grow with the nodule. The collar form is less obvious on smaller lateral roots.

There appears to be no relationship between nodule size and the width of the attachment point to the root in other nodules, and for many this may be quite small as in Centrosema pubescens and Cajanus cajan. Nodule size, and the extent of branching, tends to have an inverse relationship with nodule number (142, 176, 177). Where a plant forms few nodules they are usually large. Some perennial, branched nodules such as those of Wisteria chinensis (178), Ulex europaeus, and Sarothamnus scoparius (179), grow to between 2 and 4 cm long, and even to 6 cm for Inga sp. (180). Nodules vary in robustness depending on the extent of sclerified or corky tissue in the husk with little in Trifolium spp. and much in Acacia nodules, for example.

Plate 8.14 (Cont'd). c. Meristematic region at the edge of the bacteroid zone of A. hypogaea nodule. All cells are invaded by Rhizobium, and there is a rapid development into large round bacteroids. Starch grains are again prominent in uninvaded cortex cells but absent in invaded cells. d. Young A. hypogaea nodule showing the rod-shaped rhizobia (e.g. arrows) in dividing cells. No infection threads are present.

Nodule number is determined by both plant and Rhizobium, and there may be only one to thousands on a plant.

The Rhizobium strain has little effect on the basic nodule shape, although the surface patterning of some round nodules (e.g., Glycine max) varies with the strain (54). The same Rhizobium strain forms nodules that are collar shaped as in Lupinus angustifolius or cylindrical in Ornithopus sativus (175). One strain forms round nodules on Sesbania spp. and coralloid ones on Mimosa spp. (181), round and rugged on Alysicarpus vaginalis, branched and slender nodules on A. longifolia (172).

Some roots produce pseudonodules, usually white, fluffy, callus-type growths that contain no rhizobia. These are particularly obvious on Medicago sativa (182, 376) and Arachis hypogaea (160) and occur also on uninoculated roots. In section they resemble callus tissue and may have a central vascular strand (182). Nematode galls may also have a superficial resemblance to nodules.

Nodule color varies, often pink to red-brown if active, due to the leghemoglobin in the bacteroid filled cells, but may be dark brown if the husk is corky or contains tannin, or white and fluffy if the husk is large with loosely attached cells as in nodules formed in water culture. Trifolium and Medicago spp. nodules exposed to the light may be green from chlorophyll in chloroplasts in the husk (54), and nodules of Lotus pedunculatus grown on agar slopes in tubes are reddish-purple from pigment in uninvaded cells in the cortex (183). On Vigna unguiculata, and V. mungo, Dolichos lablab, D. bifloris, Mimosa caesalpiniaefolia, M. bimucronata, M. pudica, M. invisa, Leucaena leucocephala, and Centrosema pubescens certain Rhizobium strains form nodules that are a deep purple to black due to a pigment in the bacteroid zone (172, 181, 184-186). For V. unguiculata — CB756 nodules several pigments are involved that are not anthocyanins or haem compounds, but possibly anthro-or napthoquinones (186).

8.3.2 Stimulation of Division in the Root Cortex

8.3.2.1 Location

It takes about 24 hours for the successful infection thread to grow from the root hair into the outer root cortex in Trifolium spp. (126). As it grows through the cortex cells, the thread usually passes close to the nucleus, which in Medicago sativa (187) and Pisum sativum may be enlarged due to DNA synthesis (188).

Often the cells through which the thread passes also enlarge (187-189), and cell division of adjacent cells occurs. In Ornithopus sativus penetration of the infection thread into the outer 2-3 cortical cells initiates several divisions in a radial plane of the cortex cells between it and the nearest protoxylem point (165). Although the thread may divide and more than one continue, thread growth is basically towards the stele along a radial axis (139, 187-190), presumably along a (hormone?) concentration gradient mediated from the stele. Incipient meristematic activity with cytoplasm synthesis and cell division occurs in advance of the infection thread(s) in Pisum sativum in the inner cortex adjacent to the endodermis (188, 191, 192) (Plate 8.3b). On reaching this zone (some 15 or so cells across) the thread(s) becomes more divided and ramifies through the cells at the center (Plate 8.2c).

Vesicles form on the threads, often at the end (Plate 8.6b and g). or a break in the thread wall occurs (Plate 8.6d-f), and ultimately a package of rhizobia embedded in zoogloeal matrix is budded off. The matrix disperses and the rhizobia are released into the cytoplasm, or they escape into the cytoplasm from the vesicle while it is still attached to the infection thread. The infection thread is enclosed by the plasmalemma of the invaded cell, and this in turn encloses the rhizobia in the latter form of release (130, 193, 194).

The rhizobia multiply and become dispersed through the host cells which enlarge and are then surrounded by dividing, uninvaded cells. Cell division becomes localized in a zone which caps the nodule initial on the side away from the stele (Plate 8.3d).

Much argument has centered around the location of the cell divisions that initiate the nodule, with many believing it was pericyclic and hence the nodule was analagous to a lateral root (139, 195-198). Others found that nodules developed from cortical cells, and this is the generally accepted view for most nodules so far examined (7, 8-11, 21, 138, 170, 171, 187, 188, 190, 192, 199-201). However, in Arachis hypogaea (160) and Aeschynomene indica (166) nodules arise in the axils of lateral roots apparently from pericyclic divisions. In nodules such as P. sativum arising in the inner cortex, some pericyclic divisions may also occur but rhizobia are released into the cortical cells. In others (e.g., Phaseolus vulgaris (138), Lupinus albus (165), Cyamopsis tetragonolobus (202), the nodule initial develops further away from the stele.

8.3.2.2 Ploidy in Nodule Cells and the Role of Hormones

Nodules are not randomly located on roots, and this suggests that specific foci need to be stimulated to divide and form the nodule, especially as infection threads can penetrate to the pericycle without initiating a nodule (190).

Nodules usually contain polyploid nuclei. Chromosome counts of Pisum sativum (190-192, 203-208), 3 Vicia spp. (190, 203-205), 10 Trifolium spp. (203-205, 207, 209, 210), 12 Melilotus spp. (204, 210), 6 Medicago spp. (204, 207, 209, 211), 4 Lathyrus spp. (190, 204, 206, 207, 212), Ornithopus sativus (209), Lotus corniculatus, Indigofera pseudo-tinctoria, Astragalus sinicus, Cassia mimosoides (207, 211), and Cicer arietinum (189) showed that dividing cells in the nodule meristem usually contained double the number of chromosomes found in the root tips. This occurred with both diploid and polyploid cultivars of several of the species (191, 204, 209, 210), for P. sativum ineffective nodules, and the few small nodules developing in the presence of calcium nitrate (204). Diploid divisions were sometimes found in the uninfected nodule cortex (203, 204, 207, 210).

Wipf and Cooper (190) proposed that the naturally occurring tetraploid cells were associated with nodule initiation, and that Rhizobium in the infection thread produced some secretion or hormone that stimulated such cells and their immediate neighbors to divide. Only the disomatic cells became infected. Oinuma (191), however, found only one disomatic cell in 38 mm of root of uninoculated tetraploid P. sativum and suggested that the increased ploidy associated with nodule initiation "was caused by a hormone-like substance secreted from the tip of the infection thread." Kodama (207) reported that most of the divisions in the spherical nodules of Desmodium fallax, Arachis hypogaea, Glycine max, Phaseolus angularis, and Vigna unguiculata subsp. cylindrica (syn. V. catiang Endl. var. sinensis) were diploid.

The bacteroid filled cells in many nodules have nuclei and usually also nucleoli that are larger than in uninvaded cells (138, 187, 190, 212-214) and photometric measurement after Feulgen staining showed that for P. sativum (213) and V. unguiculata (207) these nuclei contained 4c, 8c and even 16c levels of DNA. Uninvaded cells in the outer cortex of the nodule were 2c and in the inner cortex 4c for P. sativum, and the meristematic zone 4c and 8c. For V. unguiculata the husk tissue was 2c with both 2c and 4c nuclei in the meristematic regions.

Crown galls caused by Agrobacterium tumefaciens in many species contain up to 32-ploid cells (215, 216), although in the gall on Helianthus annuus most divisions contained diploid chromosomes (216).

Nutman (176) found that in Trifolium pratense, nodulated seedlings had fewer lateral roots than uninoculated plants, and that plants with fewer effective nodules developed more lateral roots than plants with many ineffective nodules and suggested that the lateral roots and nodules are physiologically homologous. Nodule and lateral roots on T. subterraneum tend to occupy mutually exclusive zones in the root (141), although nodules often arise in the axils of a lateral root on some species. Possibly, similar endogenous hormone levels are needed to initiate lateral roots and nodules, and the development of a lateral root changes these levels by its own hormone production or by attracting translocated hormones, so that nodulation in its vicinity is then either inhibited or promoted; auxin, for example, can both promote or inhibit some cell functions depending on its concentration.

Nodules of Pisum sativum (91, 197, 218), P. arvense and Ulex europaeus (219), Phaseolus vulgaris, Glycine max (218), and Lupinus luteus (94, 220, 221) contain auxin at much higher levels on a fresh weight basis than the rest of the root. Nodules of P. arvense and Ulex europaeus also contain two growth substances that inhibit Avena coleoptile growth (218). Within the nodule, the high auxin level may be due to an altered metabolism of the host cells rather than to production by the rhizobia themselves (94). While the IAA concentration of L. luteus nodules was high, indole carboxylic acid was present at much lower concentrations than in the roots (220). Thimann (197) proposed that auxin production by Rhizobium (see III.8.2.3.2) stimulated cell division of the pericycle producing a swelling in a manner analagous to lateral root development, which then becomes infected to form the nodule. Exogenous IAA promotes formation of root initials but inhibits their subsequent development and induces a radial expansion of the cortical cells.

Many legumes grown in soil have both nodules and an endotrophic mycorrhizal association. Nodule formation and N_2 fixation are stimulated by the presence of the mycorrhiza, particularly in soils with low available phosphate. The mycorrhiza colonizes virtually the whole root except the nodules which also suggests that the nodule hormone status differs from the rest of the root (222).

The way in which hormones may interact in nodule initiation has been shown in experiments with root pieces excised from P. sativum. If auxin was added to the basal culture medium only pericyclic cell division was induced, but if auxin and cytokinin were added, cortical cells divided. Auxin induced some hypertrophy in the inner cortex, but this was greatly increased if an extract of the stele tissue was also added. Although the stele extract also stimulated cell division alone and in conjunction with auxin, this was further increased by cytokinin addition. Extracts of cortex cells induced hypertrophy but not cell division (223). Wounding induces some polyploid divisions in the intact root which otherwise has only diploid divisions (224).

If only excised cortex cells of P. sativum are cultured, auxin alone induces an increase in cell size but no DNA synthesis or cell division. If 0.01 ppm kinetin (6-furfurylamino purine) is also added, two rounds of DNA synthesis occur beginning 24-32 hours after excision (225); the first doubles the chromosome complement, and the second is the normal process of DNA duplication preceding mitosis (226). Cell division begins by 48-60 hours (225), with 85% of the cells tetraploid or octaploid and 15% diploid when the explants are taken 5-6 mm from the root tips. Explants within 2 mm of the root tip contain 75% diploid mitoses and only 25% tetraploid (226).

Thus in nodule formation Libbenga et al. and Torrey (223, 226) proposed that cytokinin and auxin produced by Rhizobium in the infection thread (see III.8.2.3.2) induce endoreduplication and then, after a further round of DNA synthesis, tetraploid cell division in the nodule initial, and it is these cells which the infection thread invades and in which Rhizobium are released. In roots where a substantial amount of polyploidy exists naturally, i.e., without the intervention of Rhizobium, the same process of DNA synthesis may occur or tetraploid cell division may perhaps be initiated by the Rhizobium infection without preceding endoreduplication. There seems no reason to support the suggestion of Wipf and Cooper (190) that tetraploid cells are the preformed foci for the initiation of nodules; rather polyploidy in nodules is a consequence of the infection process.

Some factors necessary for infection are evidently supplied by the cotyledons as their removal from T. glomeratum and T. fragiferum seedlings up to 8 days after germination also reduces infection and prevents nodulation. The time of removal affects infection, and infections often form in a zone closer to the root tip (155). Cotyledons also supply some essential factor for nodulation of Glycine max. The transfer to the root occurs early because nodules form readily if cotyledon excision is delayed until 7 days after germination. An alcohol extract of the leaves (227) or cotyledons (228) added to these seedlings stimulated nodulation. Phaseolus vulgaris seedlings, on the other hand, nodulated even when the cotyledons were removed before germination (229).

Nodule abundance on Lotus corniculatus was increased by the addition to the root medium of indole, 2-phenyl-n-butyric acid, D- and L-leucine, barbituric acid, oxythiamine, and quercetin. A niacin antagonist α-picolinic acid prevented nodule formation, without apparent effect on plant or Rhizobium growth. Very few nodules formed in this experiment, however (97).

Sections of P. sativum roots showed that over 85% of the 400 nodules examined, and all lateral roots, originated opposite protoxylem points (191, 230); where nodules appeared to arise opposite the phloem, the initial cell divisions apparently occurred opposite the protoxylem (188). Removing a cotyledon before germination resulted in the formation of primary root nodules closer to the cotyledons, and doubled the proportion of nodules formed midway between xylem and phloem. There was a significant increase in the percentage of lateral roots bearing nodules for those lateral roots with most vascular connection to the excised cotyledon position, over lateral roots mainly connected to the remaining cotyledon. Phillips (230) suggested that an inhibitor, possibly abscisic acid, was translocated in the phloem from the cotyledons to the roots, which can inhibit nodule initiation subsequent to the infection process and also inhibit the polyploid divisions in pea root segments normally stimulated by cytokinin addition (116).

The results are difficult to reconcile with those of Libbenga et al. (223), who extracted a factor from the stele of P. sativum which promoted cell division of root explants. Possibly this fraction contains cytokinin produced in the root tip and carried to the root cortex via the xylem. Hence the higher endogenous level of cytokinin opposite the protoxylem points would promote cell division there, with the phloem factor restricting cell division in its vicinity.

8.3.3 Nodule Initiation

8.3.3.1 Meristematic Activity

The pattern of the meristematic activity is set early in nodule development and soon differentiates it from a lateral root initial. In the elongate nodules the meristem forms a hemispherical cap that produces files of cells mainly towards the bacteroid zone, with subsequent divisions along other axes. These meristematic cells have a relatively large nuclear volume, prominent nucleoli and little vacuolation. In the nodule initial the meristematic cells are roughly isodiametric, but in the more developed nodule they are often elongated in the transverse plane (Plates 8.6a, 8.7b, 8.14c). At the periphery of the meristem, activity may be continuous with the small groups of cells dividing and differentiating into vascular tissue in the nodule cortex. In some nodules there is a gap between these two zones. Infection threads rarely penetrate dividing cells after nodule initiation (54, 187, 188, 192).

In round nodules such as Glycine max and Phaseolus vulgaris meristematic activity is initially dispersed throughout the nodule with divisions in all planes. By about 18 days, cell division in the center of the nodule ceases and the meristematic activity is restricted to a cambium-like layer, 2-3 cells wide, surrounding the bacteroid zone, or localized groups of cells at the edge of this zone. Other meristematic regions occur in the cortex, giving rise to vascular bundles. During the initial meristematic phase, the rhizobia are released from infection threads, and are then thought to be disseminated by cell division (124, 138, 174). Uninvaded cells also remain in the bacteroid zone.

Some nodules apparently have few or no infection threads (Table 8.2b) and no uninvaded cells in the bacteroid zone. In such nodules, e.g., Arachis hypogaea and Lupinus spp., the rhizobia seem to be entirely spread by cell division (Plate 8.14c and d) with the invaded zone of the nodule initially meristematic. The rhizobia do not appear to be actively segregated during the host-cell division. In older nodules meristematic activity becomes localized at the edge of the bacteroid zone as in other round nodules (Plate 8.14c).

8.3.3.2 Cell Differentiation and Invasion by Rhizobium

Several tissues are found in nodules. The effective nodule is mainly filled with a central bacteroid-containing zone surrounded by an uninvaded cortex or husk in which lie vascular traces and sometimes sclerenchyma to the outside (Plates 8.2 and 8.7)

In the elongate nodules differentiation is rapid. Cell expansion is accompanied by vacuolation and the cytoplasm often forms strands supporting the nucleus in the center of the cell. Infection threads occasionally penetrate as far as the meristem itself which may be about 10 cells wide in Trifolium spp. Infection threads increase as the cells differentiate. The rate of host cell differentiation and development of the released rhizobia is dependent on root temperature, and to a limited extent on Rhizobium strain. This differentiating zone is much wider at cold temperatures (231).

In round nodules, infection is also accompanied by cell expansion. The pattern of infection thread penetration is similar in both types, although the size of the infection thread varies with the species — being quite small in Trifolium spp. (130, 232) and Medicago spp. (233, 234), Vigna unguiculata (Plate 8.10a), and relatively large in Cicer arietinum, for example. Before the infection penetrates the cell, there is usually an increase in the intercellular zoogloea and accompanying rhizobia. Material with similar electron density can also be found in intercellular spaces without rhizobia, suggesting that it is largely of host cell origin. Infection threads may run between cells for as much as two cell lengths and may cross cells before releasing rhizobia (Plate 8.6b). The infection thread is a roughly circular invagination of the host cell wall, quite thin at this stage, and is often lined by endoplasmic reticulum presumably involved in wall synthesis. The outline of the thread wall is often irregular unlike the rest of the host wall, and may be relatively more electron dense. Protrusions of the wall often contain an electron-dense matrix, suggesting the fusion of vesicular material with the wall. These are occasionally associated with microtubules although microtubules are not prominently associated with thread development. The thread wall contains some fibrous material oriented along the long axis. Both thread wall and zoogloeal matrix have strong peroxidase activity (235-237). The thread grows towards the nucleus, which may invaginate and surround the thread (130). _Rhizobium_ release usually begins with the formation of a vesicle at the end of the thread. The vesicle may be budded off or may remain attached to the thread. The wall layer either dissolves or is not very firm, as rhizobia penetrate the vesicle wall and are themselves "budded" off singly or in small groups with the plasma membrane that surrounds the thread enclosing the released bacteria and sometimes also a small amount of zoogloeal matrix.

Marked changes occur in the host cells just prior to invasion and subsequently. There is an increase in rough endoplasmic reticulum, which is often organized into parallel sheets, as well as free polyribosomes (130, 194, 238). Such a response also occurs when tissue is wounded (239). Golgi bodies become more active with peroxidase activity, (130, 240), there is an increase in membrane-bound vesicles from 0.05 to 0.5 μm across in the cytoplasm, and often membrane material accumulates as myelin figures. Vesicles can be seen budding (or fusing?) from the endoplasmic reticulum. In spherical nodules, large numbers of small vacuoles spread through the cytoplasm as well as larger more central ones.

As _Rhizobium_ multiplication proceeds, the vacuoles coalesce into one or two large central ones. The vacuoles of Trifolium spp. (54, 130) and Pisum sativum (241) are usually larger and more central throughout the differentiation zone, and sometimes contain a fibrous material of moderate electron density not found in uninvaded cells. The initially very low electron density of the cytoplasmic matrix increases, possibly associated with leghemoglobin synthesis (54, 242). Large plastids are prominent and may contain small oval starch grains, and usually large semi-crystalline arrays of phytoferritin (54, 130, 238) (Plate 8.10b). These plastids are often lobed, and may even present a circular profile with a central invagination of cytoplasm. As the rhizobia occupy more of the cytoplasm plastids are displaced towards the cell periphery, particularly adjacent to the inter-cellular spaces at the junction of neighoring cells (Plate 8.6c). The plastids become filled with elongate (rather than oval) starch grains and the phytoferritin becomes dispersed through the plastid and decreases in amount, presumably because the iron is mobilized for leghemoglobin and nitrogenase synthesis. The plastids, in turn, are lined by mitochondria separating them from the bacteroids. The mitochondria may be very long, as in Medicago truncatula (130), and contain much more prominent cristae than in the meristematic cells. Golgi bodies remain dispersed throughout the cytoplasm.

As the rhizobia occupy more of the cell, the rough endoplasmic reticulum prominent in Trifolium and Medicago spp. and Pisum sativum nodules becomes dispersed into smaller units. Bacteroid-filled cells appear to contain much less endoplasmic reticulum spread throughout the cell.

The invaded host cell wall thickens differentially. Walls are often very thick adjacent to the intercellular spaces. Thinner areas remain between invaded cells and between invaded and uninvaded cells, and which contain many plasmodesmata. These resemble the sieve plates but seem free of callose deposition.

Not all cells are invaded by infection threads or contain rhizobia. These uninvaded cells link up to form channels from the cortex into the bacteroid tissue (54, 243) (Plates 8.7, 8.13a). The thin layer of cytoplasm with its few organelles is markedly less electron dense than that in adjacent cells containing bacteroids. Plastids in these cells sometimes contain starch, in large round grains, but rarely phytoferritin. The nucleus is often smaller than in adjacent invaded cells.

Intercellular spaces are prominent in the nodule between cells in the bacteroid zone and nodule cortex. These were air filled in Glycine max grown in sand culture, forming a continuous link to the atmosphere outside the nodule (243).

Electron microscope probe analysis, where the X-ray emissions excited from the specimen by the electron beam are quantitatively analyzed, shows that the dense cytoplasm of the bacteroid-filled cells of Phaseolus vulgaris and Arachis hypogaea contains iron, not detectable in uninvaded cells or intercellular spaces (244).

In round nodules such as Glycine max, uninvaded cells form radial files. In the elongate Acacia longifolia and Viminaria juncea nodules these cells are much more plentiful than in Medicago or Trifolium spp. and form obvious longitudinal files rather than more transverse as in the latter species (54). The Rhizobium-filled cells in Acacia and Viminaria are elongate in the longitudinal direction, and in Glycine max nodules, in the radial direction.

In nodules where rhizobia seem to be spread entirely by cell division, uninvaded cells are not usual in the bacteroid zone, except when a duct-like array of large elongate cells traverses the zone (Plate 8.14b). These sometimes contain starch granules.

The deposition of starch in nodules follows a fairly regular pattern. In elongate nodules such as in Trifolium spp. and Pisum sativum starch is prominent in invaded cells of effective nodules only in the zone of bacteroid development, a few cells wide at the nodule apex, and occasionally in a few cells at the base of the nodule just before the zone of obvious bacteroid degeneration. In T. subterraneum nodules formed at 7° starch deposits may remain in the invaded cells of the bacteroid zone (245). In most nodules starch is present in the uninvaded cells of the bacteroid zone and inner cortex, with smaller amounts in actively fixing nodules.

8.3.3.3 Vascular Connections and the Nodule Cortex

Woronin (5) noted that the cortex of Lupinus mutabilis nodules contained several vascular traces that joined to the root. The differentiation of vascular traces in the nodule cortex begins just after the rhizobia spread through the nodule initial. These traces are initiated by divisions of the cortical cells, adjacent to the middle of the nodule initial, in a direction mainly parallel to the root radius forming a procambial strand which develops towards the root stele. There, pericyclic divisions and differentiation of existing cells into scalariform tracheids opposite the protoxylem points initiate the branching of the root vascular system. (Plate 8.7a).

Differentiation of the cells in the nodule traces is usually acropetal beginning at the root vascular connection. The number of these latter vary with the species, with usually a single connection for Ornithopus sativus (165), Phaseolus vulgaris (138), Arachis hypogaea (160), Pisum sativum (230) (although two have been also observed: 192, 196), Sesbania grandiflora (246), Aeschynomene indica (166), Cyamopsis tetrogonolobus (202), one or two for Glycine max (124, 247), Caragana arborescens (248), two for Vicia faba (6, 200), Viminaria denudata (249), Stizolobium deeringhianum (247), Cajanus indicus (166), C. cajan (250), and three to five for Cicer arietinum (189), and several for Crotolaria juncea. These connections to the root may be to a single protoxylem point, or to more than one as in Caragana arborescens, Vicia faba, Viminaria denudata, and Glycine max. The vascular bundles dichotomously branch at the base of the nodule with further branching of these traces in the nodule cortex, so that varying numbers run in a longitudinal direction around the nodule, depending on the species, nodule size and the number of lobes, e.g., 6-8 traces in Aeschynomene indica, up to 20 in Cicer arietinum, 24 in Caragana arborescens, and 126 in a nodule on a year-old Sesbania grandiflora plant. The vascular traces may join up at the top of round nodules such as Glycine max and Vigna unguiculata (124), but not in nodules with an apical meristem.

Each bundle has its own endodermis with Casparian strip thickening similar in structure and joined to that of the root (18, 130, 169, 171, 192, 251, 252). The arrangement of xylem and other cells within the bundles varies with the position in the nodules, with a collateral or bicollateral arrangement usually found near the base, and collateral (with xylem on the side away from the bacteroid zone) or amphicribral in other regions. The xylem elements have spiral thickening developing into scalariform. Little is known of the structure of the phloem tissue except that typical sieve elements occur in Pisum sativum nodules (192, 252).

Surrounding the phloem and xylem are pericycle cells (from 1-5 layers in P. sativum) (252), containing dense cytoplasm and a prominent nucleus (Plate 8.5a). Of 71 Genera examined, nodules of 8 in the Papillionatae contained modified pericycle or transfer cells with elaborate cell wall ingrowths, with a branched and reticulate form (252, 253, 254). In Vicia, Pisum, Lathyrus, and Medicago spp. these cells were distributed throughout the pericycle, and the ingrowths evenly spread over the cell wall, but in Trifolium spp. and Lupinus spp. the transfer cells were most developed adjacent to the xylem vessels, and in a nodule from Ononis repens the wall ingrowths were almost restricted to the walls bordering the xylem vessels and to the corners of walls bordering intercellular spaces. The wall ingrowths have many associated mitochondria (252) (Plate 8.5c). Transfer cells are thought to be concerned with the selective movement in the symplasm, perhaps against a concentration gradient, of amino acids into the xylem from the bacteroid zone and sugars from the phloem to the latter (254). Differentiation into transfer cells begins in Trifolium repens nodules more distal to the development of bacteroids, and a little after xylem vessels and phloem sieve tubes are obvious (252). In many effective nodules, the pericycle cells have dense contents but not wall protruberances (e.g., Phaseolus, Thermopsis, Cytisus, Amorpha, Ulex, Genista, Spartium, Doryncium, Lotus, Lespedeza, Coronilla spp. (252), Arachis hypogaea (54), and Glycine max (255); they were also lacking in several ineffective nodules and nodules from waterlogged conditions that fixed N_2 poorly (252, 254).

The vascular bundles are separated from the bacteroid zone by three or more cortex cells with relatively little cytoplasm. Outside the bundles are more cortical cells of varying thickness, depending on the species and nodule shape. These are generally isodiametric or slightly elongate tangentially and about the same size as in the root cortex, with large intercellular spaces.

Beyond the nodule endodermis there may be an outer cortex of cells that appear to be progressively sloughed off (e.g., Trifolium spp. (251), Pisum sativum (192), sometimes leaving the nodule endodermis as the outermost layer. In nodules on plants with a cork layer in the roots, the outer nodule cortex also develops a periderm, usually with the nodule endodermis acting as the phellogen (160, 165, 169, 178, 202, 214, 247, 248, 257). When the nodules of Viminaria denudata develop in waterlogged conditions, the phellogen is very active producing almost an aerenchyma layer (249). Nodules of several plants grown in water culture or water logged conditions also develop a conspicuous outer cortex layer (54).

The "inner" cortex is usually bordered by a nodule endodermis which may join at the base to the root endodermis (Plates 8.2c, 8.7a and b). Its cells usually have thickened, suberized walls and sometimes Casparian strips (171, 187, 192, 200, 214, 249, 251, 256). In round and collar nodules such as Glycine max, Phaseolus coccineus, and Lupinus spp. this endodermis is a continuous, single-cell-wide sheet of cells, slightly elongated in the tangential plane, which surround the nodule (124, 165, 169, 251). In nodules with an apical meristem this endodermis stops short of the apex and often merges with the meristem (192, 249, 251).

In some nodules, e.g., Glycine max (124, 251), Sesbania grandiflora (246), Cajanus cajan (250), the "endodermis" is a layer of sclerenchyma cells, with greatly thickened walls, containing simple pits, which retain their cytoplasm and nuclei (Plate 8.13b). Occasional unthickened cells occur. In Phaseolus vulgaris the thickenings appear to be irregular (138). The sclerenchyma and outer cortex cells of many nodules also contain tannin (170, 246).

Starch is usually abundant in cortex cells, with the amount reflecting the physiological state of the nodule. In some nodules (Arachis hypogaea (54, 160), Cajanus cajan (250), Phaseolus multiflorus (169), P. coccineus (251) large crystals, possibly calcium oxalate, are present in cortex cells (Plate 8.13b), and A. hypogaea contains prominent protein bodies (160, 161).

Fungi sometimes penetrate the nodule cortex (171, 258) and nematodes also seem attracted to Vigna unguiculata (259). In Trifolium repens, Meloidogyne javanica and Heterodera trifolii preferentially attacked nodules rather than the rest of the root, often with several larvae attacking each nodule, forming galls with giant cell formation in the vascular traces at the base of the nodule. Nodule structure seemed otherwise little affected (260).

8.3.4 Bacteroid Development

"The morphology of the bacteria taken directly from the nodules varies with the species of legume, the conditions of infection and growth, the age and size of the nodule, and the portion of the nodule examined. These bacterial cells are so characteristic, so varied and so beautiful in form as to be pleasing objects of study." [Harrison and Barlow, 1906 (261)].

The variable appearance of Rhizobium in nodules greatly worried early investigators. Nodules such as Pisum sativum contained small coccoid rods to greatly enlarged pleomorphic forms which may or may not contain inclusion granules. This led Frank (262) to propose that there were two forms of nodules on P. sativum: one form containing an organism rich in "albuminous" substance and the other "amylo-dextrin" [inclusion granules of poly-β-hydroxybutyrate (PHBA)]. This was disputed by Moeller (263) who rightly showed that both forms could exist in the one nodule. The inclusions, and dense staining properties of Rhizobium in nodules had led Brunchorst (264) in 1885 to coin the term "bacteroid" for these storage units of plant protoplasm that occupied such a large part of the nodule. Even Beijerinck (14) believed that "the bacteroids are organized albuminoid bodies which the plant has formed out of Bacillus radicicola, for the purpose of local storage of albumen — therefore an organ of the plant protoplasm, developed from bacteria which have wandered in." It was early recognized that in some nodules Rhizobium did not become pleomorphic, only larger while retaining an overall rod shape (265).

Could a bacterium be so variable? When it became apparent that several species of Rhizobium existed, it was also obvious that the species may differ in their development within the nodule. Since changes occur in the biochemical, cultural, and morphological properties of Rhizobium in the nodule it is difficult to arrive at a precise definition of a bacteroid. Historically, the morphological changes have been stressed. A bacteroid, then, is the enlarged, often pleomorphic form of Rhizobium in nodules, which usually

Nodule Development

contains nitrogenase, and which does not grow when cultured on the media so far tried (266).

8.3.4.1 Cross-Inoculation Relationships

The host plant mainly determines the pattern of developmentin the bacteroid zone. This can be modified by the Rhizobium strain, particularly in ineffective associations. Bacteroids are greatly enlarged oval-, club-, or Y-shaped bodies in Pisum, Vicia, Trifolium, Medicago spp., for example and rod-shaped in Phaseolus vulgaris, Glycine max, Vigna unguiculata. Although whole bacteroids extracted from the nodule had been examined by electron microscopy (267, 268), it was not until thin sections of nodules were examined that it became evident that the organization of the bacteroids in the plant cytoplasm also differed. The bacteroids in Glycine max (174), Phaseolus vulgaris (269), Viminaria juncea, Acacia longifolia, Vigna unguiculata (270), V. radiata, V. mungo (161), Ornithopus sativus (175, 271), Lotus corniculatus and L. uliginosus, L. hispidus, Astragalus glycyphyllos (272), are organized into groups, as many as 20 in some species, each group being enclosed within a plant-derived membrane envelope. In Medicago truncatula (273), and M. sativa (234), Pisum sativum (240, 241, 274, 275), Trifolium subterraneum (233), T. pratense, T. parviflorum, T. repens, Vicia faba and V. hirsuta (232), V. sativa V. atropurpurea (54, 233), Lupinus luteus (175), L. angustifolius (270), and Arachis hypogaea (161), the bacteroids are enclosed singly within a membrane envelope.

8.3.4.2 Clover, Pea, Medic Groups

Within the infection thread, rhizobia are short rods, usually 1 μm or less long, and often contain prominent PHBA inclusions and sometimes electron-dense inclusions (glycogen?). As in in vitro culture for R. trifolii (54), there may be a prominent polar gap between cell wall and plasma membrane containing electron-dense material. A rigid layer underlies the cell wall membrane, and an electron transparent zone is present between the rhizobia and the zoogloeal matrix. Small membrane blebs and pieces derived from the rhizobia are sometimes prevalent in this zone. This gap is probably not an artifact of preparation, because a similar space can be seen in infection threads in living root hairs (126); it could represent a capsule that does not take up electron-dense material.

On release from the infection thread vesicle, the rhizobia initially lie singly or in groups in the cytoplasm, often with some associated zoogloeal matrix. They may be surrounded by a membrane at the obvious interface between infection-thread-derived material and plant cytoplasm, but this sometimes appears incomplete (130, 233). This may be a fixation artifact or the membrane may be incomplete initially. Occasionally microtubules are associated with these initially released rhizobia (54). The rhizobia then divide many times, along with the membrane envelope, retaining their rod shape and eventually are spread throughout the host cytoplasm, which has increased in volume. By now each bacterium is enclosed singly in a membrane envelope, having enlarged slightly and lost the PHBA granules (196). Many "empty" membrane-bound vesicles abound in the cytoplasm near the rhizobia, and sometimes appear to join the membrane envelopes. The membrane envelope has a dense-light-dense appearance in thin section with dimensions 9-10 nm (194, 233) similar to those of the tonoplast and plasma membrane (377, 378) and wider than the membranes of the endoplasmic reticulum.

Most of the rhizobia in a cell enlarge rapidly into bacteroids together with those in host cells of similar age keeping in sequence. The increase in volume may be as much as 10 to 20-fold (233, 237, 240, 274, 275) (Plates 8.8a and d, 8.9a and b). The amount of nucleoid and ribosome material appears to decrease, the nucleoid remaining dispersed through the cytoplasm. A crystalline inclusion may be found in some *Trifolium subterraneum* bacteroids (Plate 8.8a and c) as well as occasional, small, electron-dense polyphosphate granules. In older T. repens (237), and *Pisum sativum* bacteroids (54), one or two large PHBA granules may occur, their appearance coinciding with a change of the bacteroid to a more oval shape (Plate 8.9b). A narrow rigid layer may persist in the cell wall although wall thickness is much reduced (54, 276). Between each bacterium and the membrane envelope there is a relatively electron empty space in thin sections, with the rhizobia usually in the center of it. Similar electron-transparent zones are found around intracellular bacteria in some animal tissues (277,

278); for Mycobacterium lepraemurium in mice liver, the space contains fibrillar material that includes peptidoglycolipid, and may protect it from the host cell and its lytic activities (277).

There are occasional pieces of membrane, sometimes as vesicles, in this space. The membrane envelope, Rhizobium cell wall and plasma membranes have some peroxidase activity when incubated with diaminobenzidine and hydrogen peroxide (235, 237, 240). A discrete, electron-dense, reaction product in T. repens may bridge the gap between membrane envelope and the bacteroid when they are closely opposed. The plasma membrane of the bacteroid, difficult to resolve in most fixations for electron microscopy, in Trifolium spp. forms occasional intracytoplasmic membranes (232, 233, 237) which may be quite elaborate (279).

Acid phosphatase activity, visualized as Pb precipitation by the Gomori reaction, is prominent around the infection thread vesicle wall, and on the membranes of the enclosed rhizobia in P. sativum nodules (240), but is not associated with rhizobia newly released into the cytoplasm. As bacteroids develop the membrane envelope and bacteroid membranes give a positive reaction, and at the onset of senescence the cytoplasm of some bacteroids reacts. Some vesicles, thought to be derived from the endoplasmic reticulum ("phytolysosomes"), are also active, particularly in the senescent zone. Similar results are found for Medicago sativa nodules (280).

The bacteroids pack the cell so completely that the neighboring membrane envelopes sometimes fuse laterally, so that the two tripartite membranes form into a single, symmetrical tripartite membrane of similar dimension to each of the original membranes (281).

When the bacteroids come to occupy most of the host cell, the nucleus, usually central near the vacuole, often has an amoeboid appearance (Plate 8.7c). During bacteroid development, peripheral chromatin and a large nucleolus, sometimes with a core low in RNA are prominent. In some cells not all rhizobia develop into bacteroids and small membrane-bound groups, often still embedded in zoogloeal matrix, remain (233). In common with those in the residual infection threads, they often contain PHBA and glycogen-like inclusions absent from the adjacent bacteroids.

Bacteroid senescence starts at the base of the nodule, at first in individual cells. The bacteroids change from elongate to oval shape, stain less densely, and form clumps. The integrity of their membranes is then lost with the envelope finally disappearing. The host cytoplasmic matrix thins, the organelles break down, with the nucleus usually last. After bacteroid degeneration is well established, small coccoid rhizobia can be seen in increasing numbers in intracellular spaces and in the cell, often among the bacteroids. The intracellular spaces expand, as does the width of the cell wall, presumably because it is also breaking down. The bacteroids and host cell contents finally dissolve completely except for isolated membrane fragments, and the host walls collapse on each other or break. Fewer rod-shaped rhizobia seem present in the oldest degenerate region, suggesting that they may be lysed as well.

Nodules remain functioning with senescent tissue at the base. The tissues change from red to brown as the leghemoglobin (LHb) breaks down. Some senescent tissue may appear green, presumably due to conversion of the LHb to biliverdin-type pigments (II.8).

8.3.4.3 Soybean, Phaseolus, Cowpea-type

Early bacteroid development has some similarities with the clover type. Rod-shaped rhizobia usually lie in single rank within the infection thread, and may appear distorted as though under pressure (Plate 8.10a). Several rhizobia are released from the infection thread, often adjacent to the nucleus, in vesicles bounded by a thin cellulose wall and then a membrane (Plate 8.11a). The zoogloeal matrix is less electron dense than in the clover type.

Groups of 1-3 rhizobia become "budded off" where the wall material appears lacking, into membrane bound vesicles. An electron-transparent space surrounds the bacteria. Repeated Rhizobium and membrane envelope division accompany their spread throughout the host cytoplasm while still predominantly enclosed singly in membrane envelopes. During this stage the PHBA inclusions are small, the nucleoid condensed and central with prominent fibrils, and in Glycine max and Vigna unguiculata, V. mungo sometimes also containing small dense polyphosphate granules (54, 161, 194) (Plates 8.10a, 8.11b, and 8.12a and b). Similar condensed mucleoids occur in Escherichia coli when protein synthesis is inhibited (282). Early in nodule development host cell division occurs at the stage when the rhizobia are dispersed as single units.

When the rhizobia are well spread through the host cell, further multiplication within each membrane envelope results in four or more being enclosed per envelope. Vesicles containing moderately electron-dense material are found attached to the membrane envelope in Phaseolus vulgaris, suggesting movement of material from the host cytoplasm across the membrane (242). Within the space between envelope and bacteria, 20-30 nm diameter tubules can be found in G. max, P. vulgaris, and V. unguiculata (54) (Plate 8.1lc and d). Similar tubular profiles have been observed in rough endoplasmic reticulum in the zygote of Gossypium hirsutum (283). Some fibrous material may lie in the large, electron-empty space between bacteroid and envelope. Increasing numbers of PHBA granules accumulate in the developing bacteroids and may make up more than 30% of the dry weight of mature G. max bacteroids (194). Small electron-dense granules (glycogen?) also accumulate in P. vulgaris, and in G. max large polyphosphate granules are found with some Rhizobium strains (54).

The bacteroid cell wall membranes retain their rigid layers. The bacteroids in P. vulgaris, Vigna spp. Acacia longifolia, Viminaria juncea elongate, and occasional cub- or Y-shaped forms develop (Plates 8.12c, 8.13c, 8.15b).

Associated with bacteroid development, there is initially an increase in host cell polyribosomes, followed by an increase in density of the cytoplasmic matrix. Plastids and mitochondria become located at the cell periphery. The nucleus lies near the center of the cell and is not as variable in shape or as obviously enlarged as in the clover type of nodule. The central vacuole is usually small.

8.3.4.4 Bacteroids of Other Cross-Inoculation Groups

In Lupinus spp. the rhizobia enter the cell from intracellular spaces, presumably through a localized breach in the wall; infection threads have not been found (163-165, 171, 284). The bacteroids are contained mostly singly in membrane envelopes. The gross enlargement of the rhizobia, as in clover nodules, does

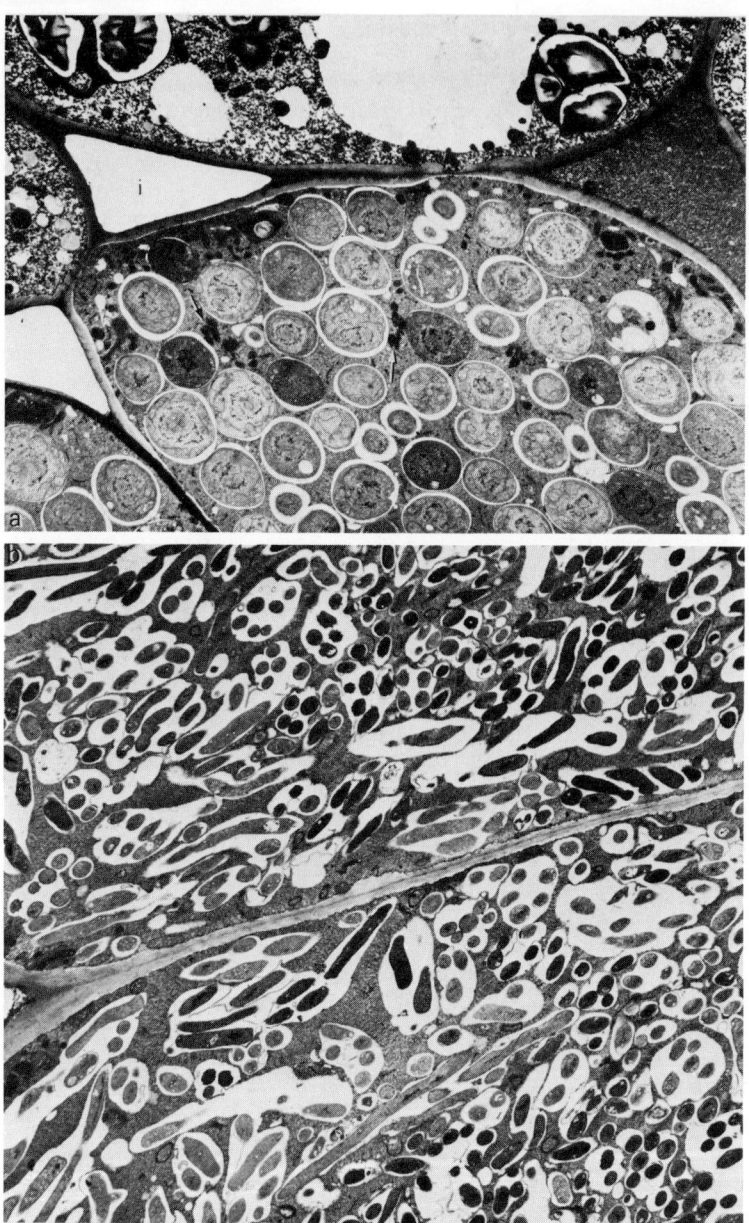

Plate 8.15. <u>a</u>. Edge of the mature bacteroid zone in an <u>Arachis hypogaea</u> nodule. The large oval bacteroids are enclosed singly in membrane envelopes. The host cytoplasm outside the envelope is much denser than in the adjacent uninvaded cells and contains crystalline inclusions (arrows), and a few plastids and associated mitochondria at the cell periphery bordering the large intercellular spaces (<u>i</u>). (x3431.) <u>b</u>. Bacteroid filled cells in a <u>Vigna mungo</u> nodule formed by the same <u>Rhizobium</u> strain in the A. <u>hypogaea</u> nodule of 8.15<u>a</u>. The bacteroids are rod shaped, elongate and enclosed severally in their membrane envelopes. (x3431.)
(M. R. Chandler and P. J. Dart, unpublished.)

not occur, although they may become long rods in
L. luteus (175) and occasionally club- and Y-shaped in
L. angustifolius (270). Bacteroids in L. luteus retain
their rigid layer (284). Lupinus bacteroids have a
central nucleoid zone with prominent fibrils early in
bacteroid development (175). An electron-empty space
lies between bacteroid and membrane envelope. Polyphosphate granules often lie in the nucleoid region
(175, 270, 284). There is a loss of nucleic acid as
bacteroids develop, so that in nodules on plants 6
weeks after emergence, DNA and RNA values per bacteroid
were only 40 and 13%, respectively, of those found in
the rhizobia of 1-week-old nodules (285).

In Lotus spp. rhizobia enter the cells through infection threads. They are then dispersed singly in
membrane envelopes through the cell. On vigorously
growing plants, further Rhizobium division results in
several bacteroids per membrane envelope (54). The
envelope space is again electron transparent with the
host cytoplasmic matrix greatly increased in electron
density. The cytoplasm of adjacent uninvaded cells is
not changed (Plate 8.16a). Lotus bacteroids are
basically rod shaped and not greatly increased in size.
They may contain several PHBA inclusions, a large
round polar body of moderate electron density
(unsaturated lipid?), polar glyocogen granules, and
polyphosphate bodies associated with the condensed,
central nucleoid and an occasional large crystalline
inclusion (286). Rhizobium strains differ in the
relative amounts of bacteroid inclusions and the
occurence of polyphosphate bodies is associated with
the production of acid in culture (287).

A cowpea Rhizobium strain is capable of effectively
nodulating the nonlegume Trema cannabina (family
Ulmaceae). The nodule structure differs from that of
legumes in that it contains a single central vascular
trace surrounded by invaded cells. In some of these,
several bacteroids are enclosed per membrane envelope
as in cowpea nodules, but otherwise the bacteria remain enclosed in infection thread like strands which
fill the whole host cell (288, 289).

8.3.4.5 Ineffective Nodules

Both plant and bacterial genomes interact to produce
ineffective nodulation (see III.7, III.11, III.12).
The same Rhizobium strain can form effective nodules

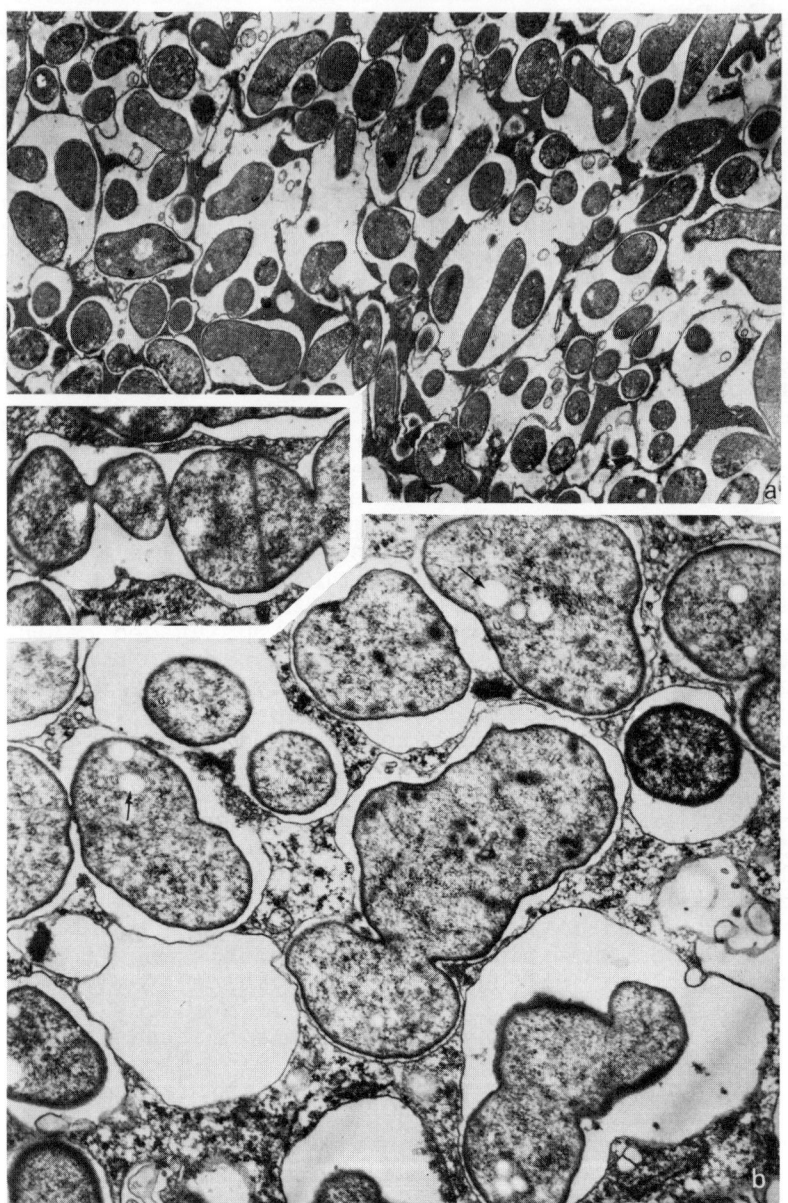

Plate 8.16. a. Bacteroids in a Lotus corniculatus nodule. Although larger than the rhizobia in the infection threads, they are basically rod shaped and enclosed several per membrane envelope. (x6391.) b. Bacteroids in Astragalus glycyphyllos nodule formed from a Lotus Rhizobium strain. The bacteroids are greatly enlarged, pleomorphic, have a dispersed nucleoid, a few polyribosomes and PHBA granules (arrow), and are enclosed singly in membrane envelopes. (x19,635.) The inset shows a large bacteroid with five lobes. (x16,940.) (M. R. Chandler and P. J. Dart, unpublished.)

on one species and ineffective nodules on another
closely related species from the same cross-inoculation
group. The symbiosis may "fail" at virtually any stage
in its development, with a consistent pattern of
development for any particular association. Generally
a plant forms more ineffective nodules than effective
and these are usually smaller and spread over the
whole root system.

Nobbe and Hiltner in 1893 (290) first described
ineffective nodulation on Pisum sativum, the large
nodules filled with rod-shaped bacteria rather than
bacteroids. Other ineffective P. sativum nodules are
small with a normal organization of tissues except
that starch is abundant throughout the nodule, and
Rhizobium rods rather than bacteroids fill the invaded
cells (291). In other associations bacteroids develop
but the degeneration of bacteroid and host tissue is
rapid (292). One P. sativum line produced completely
ineffective nodules containing leghemoglobin and
normal-appearing bacteroids (293). One R. leguminosarum mutant requiring adenine and thiamine for growth
was nonnodulating until small amounts of these were
added to the plant growth medium. The very small ineffective nodules formed in the presence of adenine
contained small invaded cells with dense cytoplasm,
tetraploid nuclei, and aborted infection threads. Nodules formed when thiamine was added contained enlarged
invaded cells in the center of the nodules in which
the rhizobia in their membrane envelopes enlarged
little. When thiamin and adenine were added together
the large nodules formed contained enlarged central
cells, but there was no sign of release of rhizobia
from the many infection threads (294).

A common pattern of development in T. pratense is
one where initial development of the nodule is similar
to that in effective nodules, but whereas degeneration
in effective nodules usually does not start much before they are 35 days old, and then proceeds slowly,
it is completed by 7-8 days in ineffective nodules.
Bacteroids do develop and fix some dinitrogen in this
brief period (292).

Nutman (295) has described ineffectiveness in
T. pratense determined by the genes ie, i_1, n, and d,
which were simply inherited and in which effectiveness
was dominant. No leghemoglobin was obvious, and starch
was abundant. The ineffective nodules formed on ie
homozygotes were as large and as numerous as effective

nodules on heterozygotes. The ineffective nodule developed relatively normally in the first 3 days with invasion of cells by infection threads. Subsequent development was abnormal. Large infection thread vesicles were formed with a prominent host cell wall preventing escape of the rhizobia. The vesicles, containing much zoogloea, detached from the infection thread, but then became the center for further deposition of fibrous wall material both around and through the vesicle with much associated membranous material. In 7-day-old nodules this stained as for lignin. The contained rhizobia and the host cytoplasm then degenerated leaving large masses of resistant wall and zoogloea in the cells surrounded by a few membrane fragments. Even when dispersed from the vesicle, rhizobia often retained associated zoogloea and could become surrounded by the wall material. In a few cells rhizobia developed into bacteroids that degenerated rapidly, along with the host cytoplasm. Host cells were less hypertrophied than in effective nodules and coupled with the aggregation of small uninvaded cells, this produced an irregular "tumorized" appearance. Plastids in invaded cells were large, containing starch and much phytoferritin in crystalline arrays (Plate 8.10b) (296, 297).

The ineffective i homozygotes formed smaller and more nodules than effective plants. Initial development was normal, but many infection thread vesicles were formed, with restricted release of rhizobia, so that large areas of the cytoplasm were occupied by zoogloeal matrix. Rhizobia were often released in groups and remained enclosed by a common membrane envelope. These divided little and rapidly lysed, seldom enlarging into bacteroids. There was much membrane synthesis in the host cytoplasm — either as vesicles or myelin-figure like whorls. By 12 days degeneration was virtually complete (296, 297). Degeneration of rhizobia occurs at the rod and at the bacteroid stage and is well advanced in 8-day-old nodules.

Infection threads and invaded cells are few in the small ineffective nodules formed by n homozygotes. Degeneration of rhizobia occurs at the rod and at the bacteroid stage and is well advanced in 8-day-old nodules.

In d ineffective nodules rhizobia or infection threads were rarely found, the whole nodules being formed of roughly isodiametric "parenchyma" cells.

Development of vascular traces was restricted. In a few nodules bacteria were released in a few cells, but these had degenerated by 4 days (297).

A riboflavin-requiring, auxotrophic strain of R. trifolii formed ineffective nodules on T. pratense with very low $N_2(C_2H_2)$ fixation. Release from infection threads and development in the invaded cells was slow. Bacteroids did not develop in many cells, the rhizobia contained PHBA granules, and large, membrane-bound masses of polysaccharide were formed. Some cells contained a few bacteroids alongside many small rods. Riboflavin addition to the plant growth medium induced normal nodule development (298). Antibiotic-resistant and antimetabolite-resistant mutants also often form ineffective nodules (III.12). The symbiosis broke down at various stages. Six strains formed tumor-like growths as for the d gene. With three other strains, rhizobia were not released from infection threads, and with others bacteroids did not develop. Two mutants formed nodules where some rhizobia in a host cell did not develop into bacteroids, remaining as rods, often several per membrane envelope. One mutant developed greatly enlarged, branched bacteroids (299).

Similar patterns of development have been described for ineffective T. subterraneum nodules (300). The large ineffective nodules formed on Medicago sativa by an adenine-requiring strain contained much starch, and many rod-shaped rhizobia among the bacteroids. With a uracil-requiring strain the rhizobia in the nodule remained mostly as rods that degenerated rapidly (301).

In effective Lotus pedunculatus nodules, pigment-containing cells were restricted to the husk. In an ineffective nodule these and a few, separate, rhizobia-containing cells intermingled but bacteroids were not formed. In an ineffective L. corniculatus association more cells were invaded; these varied in size but did not become filled with rhizobia. Starch was present in most cells. Degeneration was obvious in 16-day-old nodules (183).

One Glycine max ineffective association formed nodules with fewer cells invaded than for effective nodules, but these were packed with rhizobia. Degeneration of the nodules began on 4-week-old plants, but was not obvious for effective nodules until 17 weeks (292). In Phaseolus vulgaris (138) and Vigna unguiculata (54) ineffective nodules may have large volumes of uninvaded cortex cells surrounding a few, central, rhizobia-containing cells that rapidly collapse. In Cicer arietinum only 20% of the cells in the Rhizobium-containing zone were invaded compared with 60% in effective nodules, and these and the apical meristem soon degenerated. The invaded cells were smaller than uninvaded (189). In Cyamopsis tetragonolobus, although infection threads were numerous, few cells were invaded and these enlarged little (202).

On many plants, nodules that are formed late by normally effective strains are slow to develop and remain white and ineffective.

8.3.4.6 Role of the Plant in Bacteroid Development by the Same Rhizobium Strain

Effective nodules may be formed by the same Rhizobium strain in which bacteroid development varies considerably between host species. Lupinus luteus nodules formed by R. lupini strain D25 are typical collar nodules in which infection threads cannot be found. Bacteroids are rod shaped, sometimes elongate, and enclosed singly in membrane envelopes. Ornithopus sativus nodules formed by the same strain are cylindrical and sometimes branched and contain infection threads. In the mature bacteroid zone of the nodule, rod-shaped bacteroids, similar in size to those in Lupinus, are mostly enclosed several per membrane envelope and contain prominent PHBA inclusion granules (175).

Vigna unguiculata, V. radiata, and V. mungo nodules have very similar bacteroid development, with prominent infection threads, and several rod-shaped bacteroids, often containing PHBA inclusion granules, enclosed several per membrane envelope (III.8.3.3, Plate 8.15b). Effective nodules formed by the same strain on Arachis hypogaea seem to be disseminated entirely by host cell division (161), with rod-shaped rhizobia, often quite elongate, and enclosed singly in membrane envelopes dispersed with the other extranuclear organelles (Plate 8.14d). Rhizobia and envelope division continue in synchrony until the whole bacteroid zone is filled with cells containing such rhizobia. Together these then enlarge considerably, becoming oval and about 3 µm in diameter, but still enclosed singly in their envelopes (Plate 8.15a) (379).

In older nodules, meristematic foci are located at the edge of the bacteroid zone, where there is a rapid change of rhizobia from rod to oval form (Plate 8.14c). There are many fewer bacteroids per cell than in Vigna spp. The rod forms have a prominent central nucleoid often with associated polyphosphate. In the bacteroids the nucleoid is extremely condensed, often into a circular profile, staining so electron densely that individual nucleoid fibrils cannot be differentiated. The nucleoid volume is much reduced. These bacteroids often have elaborate, paired, intracytoplasmic membranes connected to the plasma membrane. Bacteroid development is accompanied by host cell hypertrophy, movement of plastids and their numerous associated small mitochondria to the cell periphery, and a striking increase in the electron density of the host cytoplasmic matrix. This often contains prominent crystalline aggregates (Plate 8.15a). Similar large round bacteroids have been observed in ineffective A. hypogaea nodules and effective nodules of A. erecta, A. nambyquarea, and A. villosulicarpa (302), Caragana tragacathoides (248) and Carmichaelia australis (171, 303).

In Lotus nodules the predominantly rod-shaped bacteroids in active nodules are usually enclosed several per membrane envelope (III.8.3.4.4, Plate 8.16a). In Astragalus glycyphyllos nodules formed by the same strains, bacteroids are enlarged and pleomorphic, and enclosed usually singly in membrane envelopes (Plate 8.16b). Intracytoplasmic membranes and inclusion granules are more prominent in Lotus. In both Lotus and Astragalus, tubules are present in the space between membrane envelope and bacteroid, sometimes apparently joining the two. Nodule senescence of Astragalus begins at the base as for clover nodules, while for Lotus senescence begins throughout the bacteroid zone (54).

8.3.4.7 Bacteroid-like Forms in Vitro

It is not known what induces bacteroid development in the nodule. Exchange of nucleic acid between host and symbiont, or a substance controlling the operation of Rhizobium genes, or changes in Rhizobium nutrition in the plant cell have been suggested. The pattern of cell wall synthesis obviously changes in the pleomorphic bacteroids, possibly due to changes in the rigid layer (304). During bacteroid development there is a progressive loss in ribosome number and the amount of DNA per Rhizobium, hence it is not surprising that bacteroids isolated from nodules do not divide (17, 266, 285). Enlarged, pleomorphic, bacteroid-like forms can be produced in both liquid and solid culture media by adding a variety of substances (24, 305), including caffeine, strychnine, and other alkaloids, plant extracts (but not usually with legume material) (306), blood (307), a variety of amino acids (308, 309), and yeast extract (310, 311) (Plate 8.17). Rhizobium species and strains vary in their propensity to form bacteroids in vitro (311). Pleomorphic bacteroid-like forms can be induced in strains that form few pleomorphic bacteroids in the nodule, e.g., R. phaseoli (306). As artificial bacteroid production increases, the viable count for the culture decreases. Increasing the available oxygen supply decreases the rate of bacteroid formation in liquid culture (310, 311).

Nucleic acid synthesis did not seem to be impaired in R. leguminosarum artificial bacteroids, but protein and cell wall synthesis was (310). Some mixtures of amino acids added to cultures induce even larger spherical bodies (5-7 µm in diameter) (309). R. trifolii strain TA1 needs only 0.5% yeast extract to induce bacteroids, but these contain glycogen, polyphosphate, and PHBA granules and aggregations of intracytoplasmic membranes absent from bacteroids in the nodule (311).

Many believed that induction of bacteroids in vitro was associated with dinitrogen fixation by the culture. Beijerinck (312) showed that any fixation was too small to be reliably demonstrated by Kjeldahl analysis.

Adding auxin (314) or kinetin (309) to the growth medium does not induce bacteroid formation in vitro. The induction by glycine suggests an analogy with L-forms of bacteria.

8.3.4.8 Bacteroid Zone Size

The volume of the bacteroid zone varies with nodule age and Rhizobium strain and is affected by both root temperature and the amount of light available to the plant. The proportion of bacteroid-filled cells to uninvaded cells within this zone also varies, with large species differences.

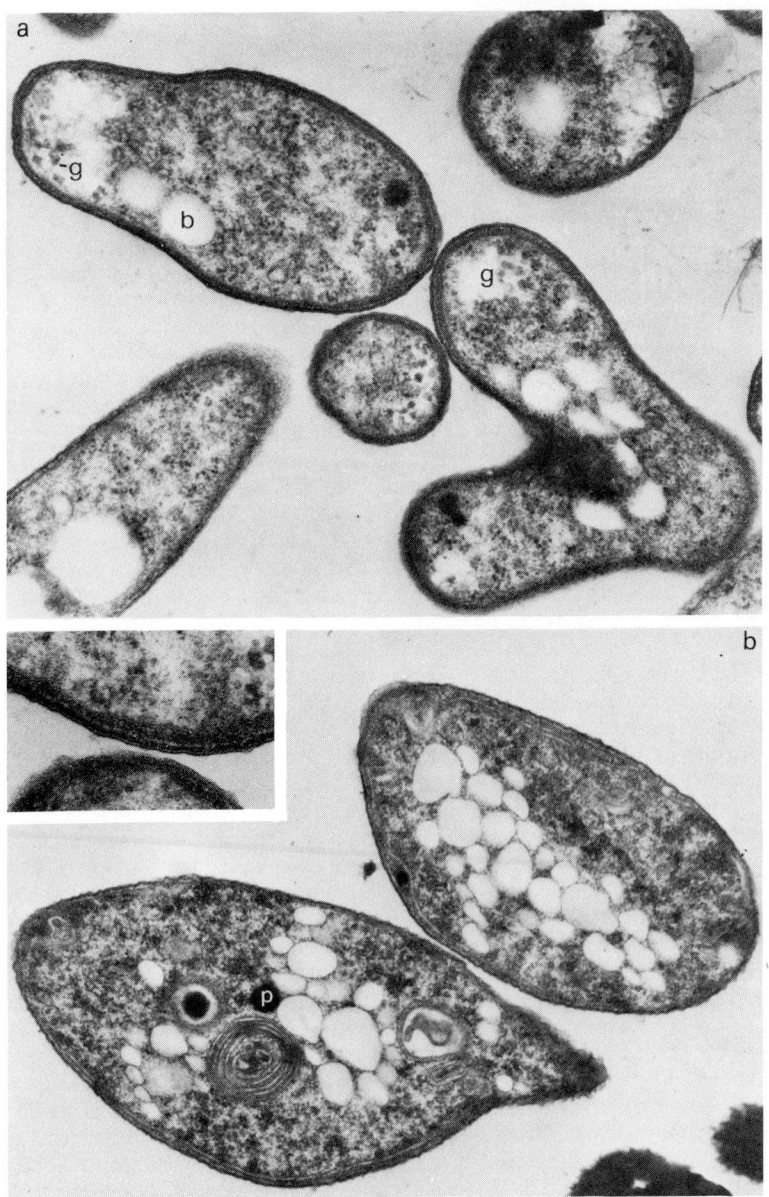

Plate 8.17. a. Artificial bacteroid forms of Rhizobium trifolii strain TA1 induced by 0.5% yeast extract in the growth medium. These rhizobia have glycogen deposits (g), PHBA granules (b), and a prominent rigid layer lining the cell wall membrane (inset). The space between the plasma membrane and cell wall membrane contains electron-dense material. x42,750; insert x81,000. b. Greatly enlarged R. trifolii artificial bacteroids formed in medium containing 0.5% casein hydrolysate. Many PHBA granules are present, some polyphosphate (p), and whorls of intracytoplasmic membranes. The cytoplasmic density is much greater than in mature bacteroids in the Trifolium nodules. (x27,750.) (M. R. Chandler, P. J. Dart, F. S. Skinner and R. J. Roughley, unpublished.)

Chen and Thornton (292) assessed the volume of bacteroid tissue in effective and ineffective Trifolium pratense and Glycine max nodules. For effective, tube-grown T. pratense nodules the overall volume of invaded tissue increased linearly with nodule length, with an increase in the organized bacteroid zone until nodules were about 30 days old, and then a decline over the next 25 days as the amount of senescent tissue at the base of the nodule increased. The volume of bacteroid tissue in ineffective nodules produced by the Coryn strain increased until nodules were about 7 days old and then declined to zero by 20 days.

For G. max the ineffective nodules formed by one strain were smaller than for effective ones and contained relatively less bacteria-filled tissue. Excluding the cortical tissue, about 83% of the nodule volume was occupied by bacteria in effective nodules, and only 43% in ineffective nodules. Calculations based on the volume of the organized bacteroid tissue, the time over which it lasted, and the amount of dinitrogen fixed suggested that rates of fixation per unit volume of active bacteroid tissue per day were similar for the effective and ineffective nodules in both species.

Nutman (142) showed that although nodule number of T. subterraneum cultivars and selections varied a great deal, similar amounts of dinitrogen were fixed by effectively nodulated plants. There was a hyperbolic relationship between average nodule length and nodule number per plant for all the many host selection-Rhizobium strain combinations tested. Nodule number and aggregate nodule length per plant were positively correlated, so that some ineffective associations produced a greater aggregate length of nodules per plant than the effective ones.

From median longitudinal sections cut with a freezing microtome, estimates were made of various nodule components. The total meristem and differentiating zone volume varied little with plant age and was less than one-twentieth of total nodule volume. The volume of bacteroid-containing tissue per plant was the same for abundantly and sparsely nodulating plants. Cortical and degenerate tissues increased with plant age. For ineffective associations the pattern was similar, although bacteroid containing tissue represented a much smaller component of nodule volume and the proportion was affected by the selection for nodule number.

Cold root temperatures affected this pattern. Roughley (231) cut wax sections of nodules grown at constant root temperatures of 7, 11, 15, and 19°C under artificial lighting. Only at 19° was the total production of new bacteroid tissue per plant balanced by degeneration, so that a constant amount of bacteroid tissue persisted over time. At lower root temperatures the amount of bacteroid tissue increased with time, most being formed at 11°, partly compensating for the poorer efficiencies in fixing N_2. The amount of degenerate tissue at the base of the nodule decreased as root temperature decreased, being absent at 7° on 40-day-old plants. At 19° only 70% as much nodule tissue was formed as at 11°C. Rhizobium strains TA1 and SU297 formed similar amounts of bacteroid tissue at 19 and 15°, but TA1 was the only strain to produce bacteroids in quantity at 7°, and also formed most at 11°. For both strains, nodule and bacteroid tissue volumes were greater on the abundantly nodulating selection than on the sparsely nodulating, at variance with Nutman's results for plants grown in a glasshouse. The maximum rate of dinitrogen fixation measured was 51 µg N/mm^3 bacteroid tissue·day.

In both Nutman's and Roughley's experiments plants were grown in cotton-wool-stoppered test tubes, where bacteroid degeneration starts much earlier than in nodules gown in open sand culture. When the ethylene produced in the enclosed system was removed by adsorption onto charcoal, or by circulating the air above the plants through mercuric perchlorate, plants nodulated earlier, nodules grew quicker and more dinitrogen was fixed, with nitrogenase activities over the first 25 days more than double those for control plants. In cotton-wool-stoppered flasks from which ethylene was continuously removed, nodule structure resembled that for pot-grown plants (103, 313).

For Lotus corniculatus and L. pedunculatus, the volume of the bacteroid zone in individual nodules increased for effective and partially effective associations as the nodules aged. The percentage number of uninvaded cells in this zone ranged from 11 to 27% (183).

The proportion of uninvaded tissue in the bacteroid zone varies between plant species and is also influenced by root temperature and Rhizobium strain. For one effective T. subterraneum association 42% of the volume of the bacteroid zone was occupied by uninvaded cells at 7°, and only 27% at 11° root temperature on 40-day-old plants (231). In Cicer arietinum about 40% of the cells in the bacteroid zone were estimated to be uninvaded (189), in Sesbania grandiflora up to 50% (246) and in Caragana arborescens as many as 80% (248).

8.3.5 Physiological Factors Influencing Nodule Initiation and Development

8.3.5.1 Temperature

Plants generally nodulate over a wide temperature range, the optimum varying with the species. The response to extreme temperatures can be much affected by the Rhizobium strain. Nodule structure is also dependent on root temperature. For Vigna unguiculata the few small nodules formed in a day temperature of 21° fixed little dinitrogen and contained little leghemoglobin. The few cells in the nodules containing rhizobia did not undergo hypertrophy, and were surrounded by uninvaded cells filled with starch. At 24° nodule structure was normal with virtually no starch in the bacteroid zone (315). Infection thread proliferation, cell invasion, and development of Rhizobium in Trifolium subterraneum nodules is much affected by cold root temperatures, the pattern modified by both host cultivar and Rhizobium strain. At 7° only strain TA1 formed many bacteroids. Bacteroid development was slower at 7 and 11° than at 19° occurring over a wider zone of cells. Senescence was also delayed. Low temperatures favored the distribution of starch throughout the nodule. Strain 0403 readily formed bacteroids at 19°, but at 15° was effective with an abundantly nodulating line but ineffective with the sparse line, with rapid degeneration of the host cells on invasion and no lasting development of rhizobia in the cells (231, 245).

At 30° R. trifolii strain TA1 formed effective nodules, but strain NA30 fixed little, and nodule structure was abnormal. Infection threads were much branched and distorted and bacteroid development slow. Many rhizobia did not develop into bacteroids, and were often severally enclosed in a membrane envelope among bacteroids enclosed singly; large masses of polysaccharide material resembling zoogloeal matrix

Nodule Development

often lay in host cytoplasm and contained occasional, very enlarged bacteria, Bacteroids and host cytoplasm rapidly degenerated. Nodules formed by strain TA1 at 30° appeared normal, apart from the occasional occurrence of rod-shaped rhizobia enclosed several per envelope. Transferring NA30 nodulated plants from 22 to 30° induced rapid breakdown of bacteroid tissue. Nodules formed by strain TA1 were less affected (316).

The morphology of effective nodules on T. subterraneum (especially cv Tallarook), T. pratense, T. patens, T. repens, and Medicago sativa formed at 20° changed greatly when transferred to a root temperature at 35°. Nitrogenase activity was completely lost by 2 days and bacteroids degenerated, but by 10 days the continued meristematic activity formed an enlarged apex of callus tissue. A few of the meristems divided to form nodules with up to 40 lobes. Other nodules, left at 35° or returned to 20°, differentiated roots from the tissue at the end of the vascular bundles (Plate 8.4b) (317). T. pratense occasionally forms roots from the end of nodules at lower temperatures (127). Older nodules on the perennial plants Sesbania grandiflora (246) and Caragana arborescens (248) also form nodule rootlets. In C. arborescens these usually develop apically on the branched nodules from the ends of the vascular traces and grow about 2 cm long. Sesbania forms nodules which are round initially although they may later develop more associated lobes. Up to 18 rootlets may emerge from these large multilobed nodules, developing from within the endodermis around the nodule vascular bundles, and emerging at right angles to them.

8.3.5.2 Combined Nitrogen

It has long been known that soluble nitrogenous compounds decrease or even prevent nodulation of legumes (ref. 23, 24, 318-320, and IV.10).

Combined nitrogen additions to nodules already fixing have drastic effects on them. The structure of M. sativa nodules was greatly altered when placed in 0.05% nitrate. Bacteroid tissue degenerated, and further development of bacteroids stopped. The meristem contained cells with reduced contents, thickened walls, and shrunken nuclei (321). When NH_4NO_3 was added to Medicago truncatula and Trifolium subterraneum grown in nitrogen-free conditions in sand, effects on nodule structure were obvious within 2 days although nodules were still pink. By 8 days the nodule fine structure had collapsed with bacteroids lysed and only rod-shaped Rhizobium cells intact and increasing in number. There was no effect on the rest of the plant growth (322). Adding NH_4NO_3 to nodulated roots of M. sativa and Lupinus spp. in sand had a similarly rapid and large effect on nodule fine structure (323).

Vigna unguiculata nodules collapse even more readily after ammonium nitrate additions with rapid structural changes; within 24 hours nodule color changes first to a deep purple-red and then brown (324). Adding NH_4NO_3 or urea at the equivalent of only 25 kg N/ha to nodulated plants with functioning nodules is sufficient to cause degeneration and loss of nitrogenase activity within 5 days. Foliar sprays of 1% urea have a similar effect (325).

Nodules can form and function on many plants in the presence of a continuous supply of combined nitrogen, e.g., Vigna unguiculata (54), Glycine max, Arachis hypogaea, Medicago sativa, Lespedeza stipulacea, Trifolium repens, and Lotus corniculatus (326), although they grow slowly in the presence of large amounts and are usually white and ineffective.

8.3.5.3 Other Environmental Factors

Light intensity and day length affect nodule formation, distribution, and structure (III.9). Few or no nodules develop on seedlings in the dark, and those that do on Medicago sativa degenerate within 2 weeks (327, 328). Placing nodulated plants of M. sativa and Trifolium pratense in the dark stops meristematic activity in the nodules within a few days, and induces bacteroid degeneration; infection threads and associated zoogloea increases (328). The few nodules that develop on Vigna unguiculata at low light intensities or in short days are usually ineffective with little leghemoglobin (145, 329).

Molybdenum, mangenese, copper, cobalt, boron, and phosphorus levels in the root medium all affect nodulation and nitrogen fixation (IV.10). Nodules formed in molybdenum-deficient conditions structurally resemble ineffective nodules with rapid bacteroid degeneration.

Waterlogging and soil drying both reduce nodule initiation and the longevity of already formed nodules (III.9 and IV.II). Salinity likewise affects nodulation (IV.10).

8.3.6 Nodulation of Excised Roots

Nodulation of excised roots does not occur readily, as cotyledons and photosynthesis supply the root with some factors necessary for nodulation. Raggio and Raggio (330) devised a technique for growing excised roots so that substances could be fed separately through an agar-containing vial attached to the cut end, or through the medium surrounding the rest of the root. Some root tips (2 cm long) excised 4 days after germination from young Glycine max or Phaseolus vulgaris seedlings, grew and nodulated if sucrose only was provided in the vial, and a nitrogen-free inorganic salts solution added to the roots. Nodulation of P. vulgaris was stimulated if the "organic" medium in the vial contained vitamins and yeast extract (331). Sucrose, glycine, thiamine, pyridoxine, nicotinic acid, and myoinositol seem to benefit nodulation (332). Not all roots nodulate. Nodulation is improved if a piece of hypocotyl is included in the explant (333, 334) and adventitious roots developing from a piece of hypocotyl excised after germination nodulate better than the main root (335, 336). Nodulation is also best if the roots develop in sand rather than on agar or in liquid (333, 335, 337). The nodules contain leghemoglobin and fix dinitrogen (102, 337, 338).

IAA and NAA added to the medium surrounding the root decreased nodulation while the antiauxins 1-napthoxyacetic acid and (p-chlorophenoxy)-isobutyric acid significantly stimulated nodulation of excised roots of P. vulgaris at $10^{-5}M$ and inhibited at $10^{-3}M$. There was a six-fold reduction in nodule formation with only 1 ppm GA in the external medium, while root growth was unaffected (339). Nodulation, but not root growth, was also inhibited by addition of the antimetabolites, benzene hexachloride, and methylglutamic acid, analogs of inositol and glutamine, via the root base. Pyridine-3-sulfonate, a niacin antimetabolite, stimulated nodulation (97). Ethylene at concentrations of 0.4 ppm and above, and 3% CO_2 also inhibited nodulation and dinitrogen fixation of P. vulgaris excised roots (102). Growing excised roots in the light inhibited nodulation, with blue light being more inhibitory than either white or red light (340); the optimum temperature for nodulation was 27°C (341).

Nitrate and urea supplied through the root base to P. vulgaris did not inhibit nodulation although they did even at 4.4×10^{-4} M, if added to the medium around the roots (337, 342). With 5 and 10% sucrose, nitrate via the root base even promoted nodulation. The inhibition by nitrate at 273 ppm N was overcome by increasing the sucrose concentration fed via the bases from 2 to 5 or 10% or by adding 0.2% sucrose, mannitol or L-arabinose to the external medium. In the latter case, the stimulation of Rhizobium growth may have rapidly reduced the external nitrate concentration. Suitable plant extracts supplied through the root base promoted nodulation of excised roots, Medicago sativa by extracts of M. sativa seeds (229), P. vulgaris by extracts of cotyledons, hypocotyls, and leaves of P. vulgaris, by horse chestnut fruits, and by factors in coconut milk (336, 340, 343). The coconut milk factor(s) may have cytokinin activity.

Trifolium repens and T. pratense roots developed from excised root tips also nodulate if fed the organic medium through the root base (344, 345). Of six sugars tested, sucrose (2% wt/vol) promoted most infections and nodules on T. repens, although root growth was similar. Kinetin, GA, and 2,4-D had little effect, except to decrease the numbers of roots infected as concentrations increased to 10^{-3} M. Infections but not nodulation were stimulated about 5-fold by 10^{-3} M IAA or NAA additions even though root growth was inhibited (344). Melilotus alba and Lotus corniculatus excised roots also nodulated (97), although Pisum sativum roots have so far proved resistant to nodulation despite numerous attempts.

8.3.7 Rhizobium-Host Interactions in Tissue Culture

An $N_2(C_2H_2)$-fixing symbiosis has been established in Glycine max callus tissue (346-350). Holsten et al. (346) grew root callus cells on a klinostat in liquid medium containing auxin, 2,4-dichlorophenoxyacetic acid (2,4-D), coconut milk and vitamins. Three to seven days after adding Rhizobium japonicum, the cells were transferred to a similar liquid medium containing no 2,4-D and coconut milk. During the next 7-20 days the rhizobia multiplied in a localized subsurface region of the callus tissue. A few callus cells were infected and the bacteria filled the whole cell and appeared to be enclosed predominantly singly by an electron-transparent zone occasionally bordered by a membrane.

PHBA inclusions in these bacteria increased with time. $N_2(C_2H_2)$ fixation was present with specific activity (per unit weight tissue) about 1% that of soybean nodule tissue.

The symbiosis can also be readily established with G. max callus tissue grown on agar. Suspension cultures of root callus cells were established in a defined medium containing minerals salts and myoinositol, thiamine, nicotinic acid, pyridoxine, and sucrose, the same organic constituents used when nodulating excised root cultures, but supplemented with 2,4-D. Some of this suspension culture is then pipetted onto the surface of an agar medium with the same constituents but lacking 2,4-D. Seven to fourteen days later this tissue was inoculated. Within a day the callus darkened due to the formation of lignin-like material (351). Callus growth continued with $N_2(C_2H_2)$ fixation detectable 5-7 days after inoculation. Autolysis frequently reduced activity after 28 days.

The symbiosis was most readily established with callus from the varieties Acme (also used by Holsten et al.), Norman and Mandarin, with N_2-fixing specific activities less than 1% of that for nodules at the time of their maximum activity. Rhizobium strains differed in the N_2-fixing activity measured. Reducing the level of nitrate and ammonium in the growth medium increased the N_2-fixing activity.

A cowpea Rhizobium strain is also able to form an association with G. max (348), Vicia hajastana, and Melilotus alba (352) callus cells. A small amount of acetylene reduction was also found for associations with R. trifolii, R. phaseoli, and R. meliloti strains with soybean callus. Acetylene reduction was not increased by oxygen tensions less than 0.2 atm and was inhibited by the presence of kinetin or 2,4-D in the medium (348). Succinic acid (349) and glutamine (350) at low concentrations both promote, sometimes synergistically, the specific activity of this association to levels about 10% of that found for nodules on roots of soybean.

Cowpea Rhizobium cultures have also been shown to produce comparable $N_2(C_2H_2)$ fixation in association with callus tissue of the nonlegumes Bromus inermis, Triticum monococcum, Brassica napus (352), and Nicotiana tabacum (353) as well as with soybean and cowpea tissue. The N. tabacum association also fixed $^{15}N_2$. The bacteria are mainly extracellular and only appear to invade moribund host cells. Bacteria growing free of the callus tissue on the agar also have nitrogenase activity, and it was suggested that the callus tissue provides some diffusible factor allowing the expression of the nif genes in the bacterial culture (352, 353).

8.3.8 Perennial Nodules

In many perennial plants, nodules apparently form and senesce in seasonal cycles (179, 354). On Wistaria sinensis, however, nodules may persist for up to 6 yr. The nodules grow apically every year with round portions representing the annual growth joined in a chain-like formation. Branching is rare. Each year's growth is in general larger than the previous one, reaching a maximum diameter of about 6mm. The older parts of the nodules are a dark brown and sometimes hollow. The nodules are ensheathed by a sclerenchyma layer with a periderm outside this and an inner cortex with persistent vascular strands. Rhizobia enter cells behind the apical meristem from infection threads and this is followed by the usual hypertrophy of the host cell and its nucleus and nucleolus. Uninvaded cells were spread among these. Some bacteria usually remain in the degenerate cells at the base of the nodule. During the winter the meristematic activity ceases, cells invaded by infection threads do not change, and rhizobia are not released. Starch, abundant in uninvaded cells and cortex during the growth period, mostly disappears (178).

Large, elongate Sophora and Robinia nodules have a similar pattern of annular constrictions suggesting seasonal growth (170). In Sesbania grandiflora large branched nodules were present on year-old plants believed to have been initiated in the first weeks of growth. Some of these nodules contained normal bacteroid-filled cells (the bacteroids were rod shaped and not enlarged) and a functioning apical meristem with infection thread invasion of the radially elongate, differentiating cells. The nodules had a sclerenchyma layer in the outer cortex, absent only around the meristem. Other old nodules had degenerated and were hollow with only the cortex persisting (246).

Perennial nodules, coralloid in form, on 3 to 6-year-old Ulex europaeus and Sarothamnus scoparius may reach 4 cm in length, with a fresh weight of up to 1 g. Constrictions occurred along the nodules. The cortex had a well-developed suberized layer, with continuous vascular strands, although the bacteroid zone at the base of some nodules had decayed and sometimes left a hollow. The nodules had an active apical meristem, and behind this a red (due to the leghemoglobin) bacteroid zone with adjacent older parts green (179).

Nodules on some annual and biennial legumes in Ireland and England may overwinter and resume activity in the spring. During the winter some species retain their leghemoglobin and may have some N_2-fixing activity if the soil temperature is 2° and above (355, 356). The survival depends on the severity of the winter. Overwintering nodules on annuals such as Vicia faba (54), Vicia hirsuta, V. angustifolia, V. lathyroides, Trifolium procumbens, T. dubium, T. arvense, Medicago arabica, and M. lupulina have an estimated maximum life of 6-8 months; on the perennial or biennial Anthyllis vulneraria, Melilotus officinalis, Onobrychis sativa, T. pratense, and T. hybridum, T. medium, Lathyrus montanus, L. pratensis, L. palustris, V. cracca, V. Sylvatica, V. sepium, and Ononis repens of 12-14 months (355). In colder conditions under snow, T. ambiguum and T. repens nodules shrivel but resume growth as the snow melts. The vascular connections at the base of the nodule remain intact, and the regenerated nodule lasts for about a month (357).

8.4 SELECTED METHODS

8.4.1 The Fåhraeus Slide Technique

This technique (42, 43) is used to study root infection by nodule bacteria. Each experimental plant is grown in a small open-sided glass cell that stands in a plugged boiling tube containing plant nutrient solution. The glass cell is made from a microscope slide and a coverslip separated by glass spacing pieces.

8.4.1.1 To Make Coverslip Spacing Pieces

Stick large coverslips onto gummed paper, and with a diamond rule a grid of two sets of lines at right angles to mark out 2-mm squares. Break glass along lines and place in water. The small glass squares float free and are then washed in chromic acid, water, and acetone and dried.

8.4.1.2 To Make Cell

Place a clean 3 x 1 in. slide on a sheet of paper on which are marked the positions of spacers and coverslip. With a fine glass rod place very small amounts of freshly made araldite mixture on the slide at the four spacer positions and at each position add a glass square and press flat. Put araldite on to the top of each square and add a second glass square at each position. The $1\frac{1}{2}$ x $\frac{7}{8}$ in. coverslip is similarly fixed to the spacer squares with araldite. The glass cells are stacked as they are made in sets of 20. A slide and 200g weight is placed on the top of each stack. The cells are baked for at least 1 hour at 100°C to harden the araldite. The completed cells are acid washed, thoroughly washed in water, and placed individually in petri dishes and sterilized.

The space between coverslip and slide can be varied by altering the number of individual thickness of the spacing squares. Two squares of No. 1 coverslip provide a suitable chamber for small seeded legumes such as Trifolium repens.

8.4.1.3 Seed Pretreatment and Sowing

Seed is first selected and surface sterilized by steeping in concentrated H_2SO_4 in small stoppered tubes for 20-30 min or by immersing briefly in 95% ethanol and then 0.2% (wt/vol) mercuric chloride acidified with 5 ml/l concentrated HCl for 3 min. The acid or mercuric chloride is drained off and the seed washed in many changes of sterile water. The seeds are spread on to a plate of Fåhraeus agar and placed in a closed room (about 2°C) for 24 hours and then incubated in an inverted position for about 36 hours at 18°C for temperate legumes, by which time the radicles are about 4 mm long.

The seedlings are placed on the freshly prepared agar of the Fåhraeus slide pushing the radicle gently underneath the coverslip, and each planted slide is placed into a 30 mm diameter boiling tube containing 25 ml of Fåhraeus' liquid medium (without agar). The plant culture tubes are held in a rack that shades the roots.

The Fåhraeus Medium

$CaCl_2 \cdot H_2O$	0.1 g
$MgSO_4 \cdot 7H_2O$	0.12 g
KH_2PO_4	0.1 g
$Na_2HPO_4 \cdot 12H_2O$	0.15 g
Fe citrate	0.005 g
Micronutrient solution	1 ml
Dist. H_2O	1000 ml
Agar	3.5 g

Micronutrient Stock Solution

$MnSO_4 \cdot 4H_2O$	2.23 g	or	Mn 0.05%
$CuSO_4 \cdot 5H_2O$	0.08 g	or	Cu 0.002%
$ZnSO_4 \cdot 7H_2O$	0.29 g	or	Zn 0.005%
H_3BO_3	1.86 g	or	B 0.05%
$Na_2MoO_4 \cdot 2H_2O$	0.121 g	or	Mo 0.005%
$CoSO_4 \cdot 6H_2O$	0.053 g		
Dist. H_2O	1 l		
	after Hewitt (358)		after Gibson (359)

Sterilize medium at 120°C for 15 min. Sterile media can be stored in stoppered bottles. Warm medium is run from a fine, sterile Pasteur pipette into the space between slide and coverslip, shortly before seeds are to be planted. For some work the slide cell can be filled with Fåhraeus' liquid medium without agar or with very finely sieved soil kept moist by capillarity.

8.4.2 Culture of Legumes for Nodulation Studies (360)

8.4.2.1 Axenic Culture in Test Tubes

Plants can be grown (i) totally enclosed in the tubes on agar slopes or deeps, or (ii) with only the roots enclosed.
For small seeded legumes, Trifolium, Medicago, Lotus spp., for example, 150 x 20 mm - 200 x 30 mm test tubes, loosely stoppered with cotton wool, containing 12-40 ml agar medium sloped to give ca. 7.5 cm depth of agar are convenient. Pregerminated seedlings, or surface-sterilized seed is placed ca. 1 cm below the top of the agar slope. One to two

milliliters of sterile ¼-strength nutrient solution without agar may be added and plants later watered with this solution. Inoculation at sowing is usual.

Agar Medium After Jensen (361)

CaHPO₄	1.0 g	or	100 g/l
K₂HPO₄	0.2 g		20 g/l
MgSO₄·7H₂O	0.2 g		20 g/l
NaCl	0.2 g		20 g/l
FeCl₃	0.1 g		10 g/l
Agar	12-15 g		Take 10 ml of each
Dist. H₂O	1 l		and make up to 1 liter, add sugar.

Adjust to pH 6.5-6.8

The agar is dissolved by heating, the media dispensed, and then autoclaved at 120°C for 15 min. An agar deep of the same composition (except for slightly less agar) may be made and covered with washed, sterilized sand, moistened with ¼-strength nutrient solution. Seeds or seedlings are planted in the sand. A deep of washed, neutralized vermiculite is also a useful growth medium for some legumes.

8.4.2.2 Root-Enclosed Assembly (359)

a. Use sufficient seedling agar per tube to provide a long slope reaching to the top of the tube (13 ml for 150 x 20 mm tube).
b. Cap the tubes with thin circles of aluminium foil (0.03 mm thickness, 44 mm diameter, cut out with a punch after the foil has been interleaved with paper to prevent sticking). Secure the cap with a strong rubber or plastic ring.
c. Make a small (5 mm) watering and ventilation hole in the cap near the side of the tube, and plug it with cotton wool.
d. Autoclave the tubes and set as slopes so that the plugged hole is uppermost and the agar reaches the aluminium cap.
e. Prepare seeds in the usual way, germinating them on water agar in inverted dishes so as to provide straight radicles 13-15 mm long.
f. Make a small hole in the aluminium cap opposite the plug and insert the radicle so that it lies

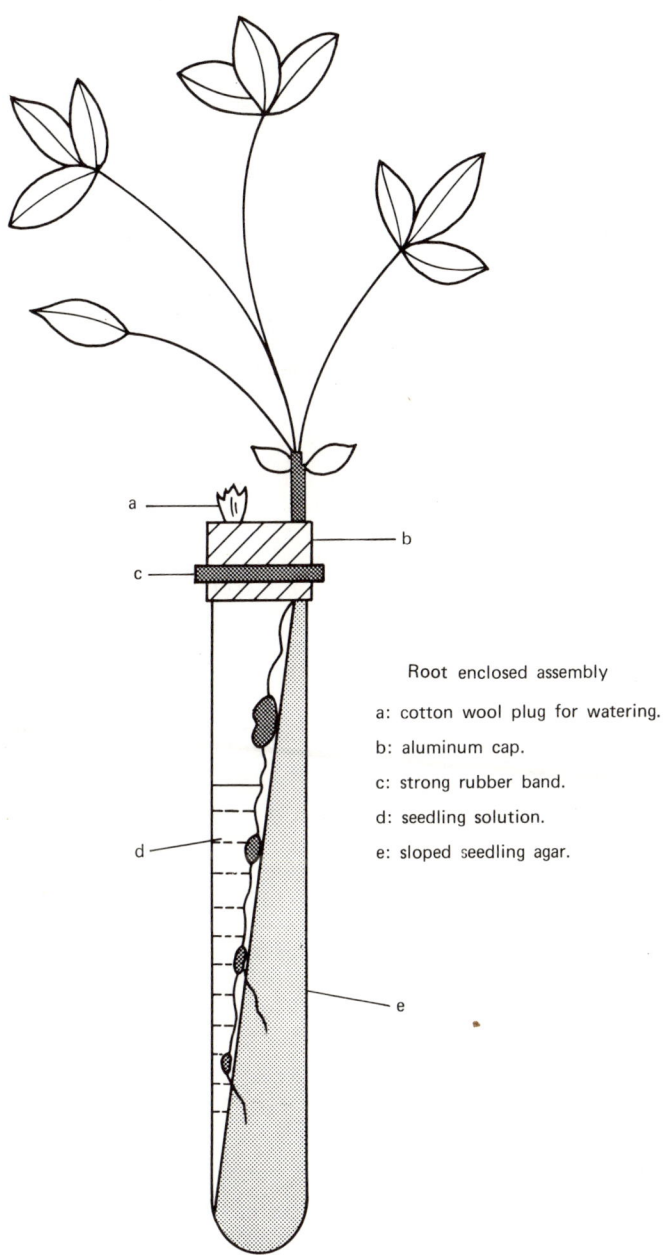

Root enclosed assembly

a: cotton wool plug for watering.
b: aluminum cap.
c: strong rubber band.
d: seedling solution.
e: sloped seedling agar.

along the surface of the slope. (In these operations the tips of the forceps and needle are kept in alcohol and flamed before use and cooled in sterile water before handling the seedling.)
g. Cover the planted tubes with specimen tubes containing a plug of moist cotton wool and resting on the rubber rings so as to prevent desiccation of the young seedling.
h. After 3 days most seedlings will have lost their seed coats, but in some cases this will need to be assisted with a sharp blade.
i. Add ¼-strength sterile seedling solution through the hole in the cap to within 12 mm of the top of each tube, preferably from a sterilizable automatic syringe, flaming the cannula every 4-5 tubes.
j. Add inoculum through the watering hole within a few days of setting up.
k. The tubes can be set up in an enclosed rack, with roots protected from the light.
l. Maintenance: It is generally not necessary to add more nutrient solution until the plants are 18-21 days old; further additions are made after another 4-5 days, and finally at 2-3 day intervals.

8.4.2.3 Sand Culture

Plants may be grown in coarse washed sand in modified Leonard bottle-jar assemblies (360) or in open pots. The sand can be heat or steam sterilized if required.

Nutrient Solutions:

Thornton (187)		Modified Knops (157)		Long Ashton (359)	
$Ca_3(PO_4)_2$	2.0 g	$CaSO_4 \cdot 2H_2O$	0.96 g	$CaCl_2 \cdot 6H_2O$	0.876 g
K_2HPO_4	0.5 g	K_2HPO_4	0.4 g	$Na_2HPO_4 \cdot 12H_2O$	0.478 g
$MgSO_4 \cdot 7H_2O$	0.2 g	$MgSO_4 \cdot 7H_2O$	0.2 g	$MgSO_4 \cdot 7H_2O$	0.368 g
NaCl	0.1 g			K_2SO_4	0.348 g
$FePO_4$	1.0 g				
		Water/liter			

Add 1 ml/l of trace element solution to each. Adjust to ca. pH 6.5

Fe EDTA (ferric ethylenediaminetetraacetic acid) may be used as the Fe source to provide ca. 5 ppm Fe in the nutrient solution.

(a) Dissolve 33.3 g of Na_2EDTA in 500 ml distilled H_2O containing 103 ml of 1 $\underline{N}$ NaOH. (b) Dissolve 17.8 g of $FeCl_2 \cdot 4H_2O$ in about 300 ml of hot (ca. 70°C) water containing 4 ml of 1N HCl. Mix solutions (a) and (b), add water to 950 ml, shake vigorously for 12 hours or overnight, make up the volume to 1 liter with H_2O (362).

Dispense in suitable container and autoclave. Add 1 ml of stock solution per liter of culture solution. FeEDTA can also be obtained already prepared as a dry powder.

8.5 CONCLUDING COMMENTS

Most legumes have still not been examined for nodulation (363). Nodules have been found on about 90% of the plants examined in the subfamilies Mimosoideae and Papilionaceae but on only about 23% of the Caesalpinoideae (180, 363-365). Most of the understanding of nodule development comes from a very few economically useful species. A much wider range of association needs to be examined to understand how the symbiosis with Rhizobium developed and particularly how infection occurs.

Nodules are complex structures; the symbiosis is finely controlled with differentiation following an ordered sequence, and this distinguishes it from callus or gall tissue (201). Specific host-Rhizobium interactions begin in the rhizosphere and operate during initial infection of the root, release of rhizobia into host cells, development into bacteroids, and the synthesis of nitrogenase. Nodule structure is remarkably homogenous within three or four different patterns of development. Nodule number and development are determined by interactions between host and Rhizobium genomes and further affected by competition between nodules on the plant for photosynthate and by the distribution of plant hormones, some possibly nodule produced, and by environmental factors such as temperature. The factors inducing nodule senescence are little understood; extending the period over which the nodule complement fixes N_2 efficiently should inrease yields a great deal.

8.6 GLOSSARY OF COMMON NAMES USED IN CHAPTER III.8

Anthyllus vulneraria	common kidney vetch
Arachis hypogaea	peanut; ground nut
Astragalus glycyphyllos	milk vetch
Cajanus cajan	pigeon pea, red gram
Canavalia gladiata	sword bean
Cicer arietinum	chick pea, gram
Crotolaria juncea	sunnhemp
Cyamopsis tetrogonoloba	quar, cluster bean
Dolichos lablab	hyacinth bean, lablab
Galega officinalis	goats' rue
Gossypium hirsutum	cotton
Helianthus annuus	sunflower
Lathyrus aphaca	yellow vetchling
L. odoratus	sweet pea
L. pratensis	meadow vetchling
L. sylvestris	narrow-leaved everlasting pea; flat pea
Lens esculenta	common lentil
Lotus corniculatus	birds-foot trefoil
L. hispidus	hairy trefoil
L. pedunculatus	greater trefoil
L. uliginosus	big trefoil
Lupinus albus	white lupin
L. angustifolius	blue lupin
L. arboreus	tree lupin
L. luteus	yellow lupin
Macroptilium atropurpureum	siratro
Medicago arborea	tree medick
M. denticulata	reticulated medick
M. lupulina	black medick
M. minima	little burr medic, small medic
M. sativa	alfalfa; lucerne
M. truncatula	barrel medick
Melilotus alba	white melilot, white sweet clover
M. indica	yellow sweet clover, senji
M. officinalis	yellow melilot

Glossary of Common Names Used in Chapter III.8

Onobrychis viciifolia	sainfoin
Ornithopus sativus	serradella
Phaseolus coccineus	runner bean
P. vulgaris	kidney, French, navy (baked) beans
Pisum arvense	field pea
P. sativum	garden pea
Pueraria phaseoloides	kudzu; puero
Robina pseudoacacia	black locust
Sarothamnus scoparius	broom
Stylosanthes guyanensis	oxley fine stem stylo
S. humilis	Townsville lucerne, stylo
Trifolium alexandrinum	berseem clover
T. ambiguum	Kura clover
T. angustifolium	narrow leaved clover
T. arvense	rabbits-foot clover or hares-foot trefoil
T. dubium	Shamrock, small hop clover
T. fragiferum	strawberry clover
T. glomeratum	cluster clover
T. hybridum	alsike clover
T. incarnatum	crimson clover
T. nigrescens	ball clover
T. ornithopodioides	birdsfoot fenugreek
T. parviflorum	small flowered clover
T. pratense	red clover
T. procumbens	large hop clover
T. repens	white, Dutch clover
T. scabrum	rough clover
T. subterraneum	sub(terranean) clover
Trigonella foenum-graecum	fenugreek
Ulex europeaus	gorse
Vicia atropurpureus	purple vetch
V. faba	broad bean; field bean
V. hirsuta	hairy tare or vetch
V. sativa	golden tare; vetch
V. villosa	smooth vetch
Vigna mungo	urd, black gram
V. radiata	green gram, mung
V. unguiculata	cowpea

8.7 REFERENCES

1. Dalechamps, J., *Historia Generalis Plantarum*, Lyons, 1, 487 (1587).
2. Malpighi, M., *Anatome Plantarum*, J. Martyn, London, 99 pp., 1679.
3. de Candolle, A. P., *Memoires sur la famille des Legumineuses*, Paris (1825).
4. Trevinarus, L. C., *Bot. Ztg.*, 11, 393 (1851).
5. Woronin, M. S., *Mem. Acad. Imp. Sci.*, St. Petersburgh, Ser. 7, 10, No. 6, 1 (1866); and *Ann. Sci. Nat. Bot.*, Ser. 5, 7, 73 (1867).
6. Erickssen, J., *Acta Univ. Lund.*, Part II, 10, 1 (1873).
7. Woronin, M. S., *Jahrb. Wiss. Bot.*, 11, 548 (1878).
8. Prillieux, E., *Bull. Soc. Bot. Fr.*, 26, 98 (1879).
9. Frank, B., *Bot. Ztg.*, 37, 377, 393 (1879).
10. Ward, H. M., *Phil. Trans. Roy. Soc. (London)*, Ser. B., 178, 539 (1887).
11. Prazmowski, A., *Landw. Vers. Sta.*, 37, 161 (1890).
12. Hellriegel, H., *Deutcher Naturforscher and Aerzte in Berlin*, 18-24 Sept., 290 (1886).
13. Hellriegel, H., and H. Wilfarth, *Beilageheft zu der Ztschr. Ver. Rubenzucher-Industrie, Deutchen Reichs*, 234 (1888).
14. Beijerinck, M. W., *Bot. Ztg.*, 46, 726, 741, 757, 781, 797 (1888).
15. Frank, B., *Landw. Jahrb.*, 19, 523 (1890).
16. Dawson, M., *Phil. Trans. Roy. Soc. (London)*, Ser. B., 192, 1 (1900).
17. Dawson, M., *Phil. Trans. Roy. Soc. (London)*, Ser. B., 193, 51 (1900).
18. Vuillemin, P., *Ann. Sci. Agron.*, 1, 121 (1888).
19. McCoy, E., *Proc. Roy. Soc. (London)*, Ser. B., 110, 514 (1932).
20. Nobbe, F., E. Schmid, L. Hiltner, and E. Hotter, *Ladw. Vers. Sta.*, 39, 327 (1891).
21. Atkinson, G. F., *Bot. Gaz.*, 18, 157, 226, 257 (1893).
22. Kirchner, D., *Beitr. Biol. Pfl.*, 7, 213 (1895).
23. Hiltner, L., *Arb. Biol. Bund Anst. Land-u. Forstw.*, 1, 177 (1900).
24. Fred, E. B., I. L. Baldwin, and E. McCoy, *Root Nodule Bacteria and Leguminous Plants*, University Wisconsin Press, Madison, 1932.
25. Perkins, A. T., *J. Agr. Res.*, 30, 243 (1925).
26. Jansen van Rensburg, H., and B. W. Strijdom, *Phytophylactica*, 1, 201 (1969).

References

27. Bhaduri, S. N., Ann. Bot. (London), 15, 209 (1951).
28. Purchase, H. F., and P. S. Nutman, Ann. Bot. (London), 21, 439 (1957).
29. Lim, G., Ann. Bot. (London), 27, 55 (1963).
30. Dart, P. J., J. Exp. Bot., 22, 163 (1971).
31. Houwink, A. L., and P. A. Roelofson, Acta Bot. Neerl., 3, 385 (1954).
32. Newcomb, E. H., and H. T. Bonnett, J. Cell Biol., 27, 575 (1965).
33. Scott, F. M., K. C. Hamner, E. Baker, and E. Bowler, Am. J. Bot., 43, 313 (1956).
34. Roberts, E. A., Bot. Gaz., 62, 488 (1916).
35. Dart, P. J., and F. V. Mercer, Arch. Mikrobiol., 47, 344 (1964).
36. Bell, J. K., and M. E. McCully, Protoplasma, 70, 179 (1970).
37. Head, G. C., Ann. Botany (London), 28, 495 (1964).
38. Jenny, H., and K. Grossenbacher, Soil Sci. Soc. Am. Proc., 27, 273 (1963).
39. Greaves, M. P., and J. F. Darbyshire, Soil Biol. Biochem., 4, 443 (1972).
40. Van Egeraat, A. W. S. M., Meded. Landbouwhogeschool Wageningen, 72 (1972).
41. Thornton, H. G., Proc. Roy. Soc. (London), Ser. B, 119, 474 (1936).
42. Fåhraeus, G., J. Gen. Microbiol., 16, 374 (1957).
43. Nutman, P. S., J. Exp. Bot., 10, 250 (1959).
44. Darbyshire, J. F., A Study of the Initial Stages of Clovers by Nodule Bacteria, Ph. D. Thesis, University of London (1964).
45. Purchase, H., Aust. J. Biol. Sci., 11, 155 (1958).
46. Sahlman, K., and G. Fåhraeus, LantbrHogsk. Annlr., 28, 261 (1962).
47. Haack, A., Zentr. Bakteriol. Parasitenk., Abt. II, 117, 343 (1964).
48. Yao, P. Y., and J. M. Vincent, Aust. J. Biol. Sci., 22, 413 (1969).
49. Li, D., and D. H. Hubbell, Can. J. Microbiol., 15, 1133 (1969).
50. Hepper, C. M., Rep. Rothamsted Exp. Stat. 1972, Part 1, 82 (1973).
51. Kleczkowska, J., P. S. Nutman, and G. Bond, J. Bacteriol., 48, 673 (1944).
52. Kleczkowski, A., and H. G. Thornton, J. Bacteriol., 48, 661 (1944).
53. Hubbell, D. H., Bot. Gaz., 131, 337 (1970).
54. Dart, P. J. (unpublished).

55. Thornton, H. G., and H. Nicol, Nature, 137, 494 (1936).
56. Chen, H. K., Nature, 142, 753 (1938).
57. Ljunggren, H., Physiol. Plant. Suppl., 5 (1969).
58. Solheim, B., and J. Raa, J. Gen. Microbiol., 77, 241 (1973).
59. Yao, P. Y., and J. M. Vincent (personal communication).
60. Skrdleta, V., Folia Microbiol., 14, 32 (1969).
61. Skrdleta, V., Soil Biol. Biochem., 2, 167 (1970).
62. Lindemann, W. C., E. L. Schmidt, and G. E. Ham, Soil Sci., 118, 274 (1974).
63. Marques Pinto, C., P. Y. Yao, and J. M. Vincent, Aust. J. Agri. Res., 25, 317 (1974).
64. Johnston, A. W. B., and J. E. Berringer, J. Gen. Microbiol., 87, 343 (1975).
65. Vincent, J. M., Proc. Linn. Soc. N. S. Wales, 79, iv (1954).
66. Evans, H. J., N. E. R. Campbell, and S. Hill, Can. J. Microbiol., 18, 13 (1972).
67. Jansen van Rensburg, H., and B. W. Strijdom, Phytophylactica, 3, 125 (1971).
68. Jansen van Rensburg, H., and B. W. Strijdom, Phytophylactica, 4, 1 (1972).
69. Jansen van Rensburg, H., and B. W. Strijdom, Phytophylactica, 4, 73 (1972).
70. Nutman, P. S., Proc. Roy. Soc. (London), Ser. B, 156, 122 (1962).
71. Albersheim, P., T. M. Jones, and P. D. English, Ann. Rev. Phytopathol., 7, 171 (1969).
72. Toms, G. C., in Chemotaxonomy of the Leguminosae, J. B. Harborne, D. Boulter, and B. L. Turner, Eds., Academic Press, London/New York, Vol. 1, pp. 367-462, 1971.
73. Hamlin, J., and S. P. Kent, Nature, 245, 28 (1973).
74. Bohlool, B. B., and E. L. Schmidt, Science, 185, 269 (1974).
75. Fisher, M. L., A. J. Anderson, and P. Albersheim, Plant Physiol., 51, 489 (1973).
76. Bateman, D. F., and R. L. Millar, Ann. Rev. Phytopathol., 4, 119 (1966).
77. Smith, W. K., J. Gen. Microbiol., 18, 33 (1958).
78. Lillich, T. T., and G. H. Elkan, Can. J. Microbiol., 14, 617 (1968).
79. Subba Rao, N. S., and K. S. B. Sarma, Plant Soil, 28, 407 (1968).
80. Fåhraeus, G., and H. Ljunggren, Physiol. Plant., 12, 145 (1959).

81. Fåhraeus, G., and H. Ljunggren, in The Ecology of Soil Bacteria, T. R. G. Gray, and D. Parkinson, Eds., Liverpool University Press, pp. 396-421, 1968.
82. Ljunggren, H., and G. Fåhraeus, Nature, 184, 1578 (1959).
83. Ljunggren, H., and G. Fåhraeus, J. Gen. Microbiol., 26, 521 (1961).
84. Munns, D. N., Plant Soil, 30, 117 (1969).
85. MacMillan, J. D., and R. C. Cooke, Can. J. Microbiol., 15, 643 (1969).
86. Solheim, B., and J. Raa, Plant Soil, 35, 275 (1971).
87. Bonish, P. M., Plant Soil, 39, 319 (1973).
88. Bonish, P. M., Plant Soil, 38, 307 (1973).
89. Green, P. B., Ann. Rev. Plant Physiol., 20, 365 (1967).
90. Molliard, M., Compt. Rend., 155, 1531 (1912).
91. Thimann, K. V., Trans. 3rd Comm. Int. Soc. Soil Sci., A, 24 (1939).
92. Kefford, N. P., J. Brockwell, and J. A. Zwar, Aust. J. Biol. Sci., 13, 456 (1960).
93. Rigaud, J., Physiol. Plant, 23, 171 (1970).
94. Dullaart, J., Acta Bot. Neerl., 19, 573 (1970).
95. Gibson, A. H., Aust. J. Agr. Res., 19, 907 (1968).
96. Tagiev, V. D., Izv. Akad. Nauk. SSSR, Ser. Biol., 2, 291 (1965).
97. Molina, J. A. E., and M. Alexander, Can. J. Microbiol., 13, 819 (1967).
98. Sarma, K. S. B., M. Lakshmi-Kumari, and N. S. Subba-Rao, Curr. Sci., 41, 717 (1972).
99. Munns, D. N., Plant Soil, 29, 257 (1968).
100. Fletcher, W. W., P. B. Dickenson, and J. C. Raymond, Phyton (Buenos Aires), 7, 121 (1956).
101. Garcia, M. M., and D. C. Jordan, Plant Soil, 30, 317 (1969).
102. Grobbelaar, N., B. Clarke, and M. C. Hough, Plant Soil, Special Vol., 215 (1971).
103. Day, J. M., P. J. Dart, and R. Percy, Rep. Rothamsted Exp. Stat. 1971, Part 1, 99 (1972).
104. Drennan, D. S. H., and C. Norton, Plant Soil, 36, 53 (1973).
105. Nutman, P. S., in Ecology of Soil-borne Plant Pathogens, K. F. Baker, and W. C. Snyder, Eds., University of California Press, Berkeley/Los Angeles, pp. 231-246, 1965.
106. Katznelson, H., and S. E. Cole, Can. J. Microbiol., 11, 733 (1965).

107. Radley, M., Nature, 191, 684 (1961).
108. Dullaart, J., and L. I. Dubba, Acta. Bot. Neerl., 19, 877 (1970).
109. Nutman, P. S., Ann. Bot. (London), 14, 79 (1952).
110. Fletcher, W. W., J. W. S. Alcorn, and J. C. Raymond, Nature, 184, 1576 (1959).
111. Mes, M. G., Nature, 184, 2035 (1959).
112. Chailakhyan, M. K., A. A. Megrabyan, N. A. Karaptyan, and N. L. Kaladzhyan, Izv. Akad. Nauk. Arm. SSR, 14, 25 (1961).
113. Giannattasio, M., and S. Coppola, Giorn. Bot. Ital., 103, 11 (1969).
114. Phillips, D. A., and J. G. Torrey, Physiol. Plant., 23, 1057 (1970).
115. Phillips, D. A., and J. G. Torrey, Plant Physiol., 49, 11 (1972).
116. Phillips, D. A., Planta, 100, 181 (1971).
117. Bonnier, C., Bull. Inst. Agron. Stat. Rech. Gembloux, 22, 167 (1954).
118. Trolldenier, G., Arch. Mikrobiol., 32, 328 (1959).
119. Weir, J. B., Plant Soil, 14, 85 (1961).
120. Schiel, E., E. L. G. de Olivero, R. N. Dieguez, J. C. Pacheco, and E. Enokida, Ann. Inst. Pasteur, 105, 332 (1963).
121. Van der Starre van der Molen, L. G., G. A. H. Bossink, and A. Quispel, Plant Soil, 26, 397 (1967).
122. Lie, T. A., and S. Brotonegaro, Plant Soil, 30, 339 (1969).
123. Schaede, R., Planta, 31, 1 (1940).
124. Bieberdorf, F. W., J. Am. Soc. Agron., 30, 375 (1938).
125. Kumarasinghe, R., Rep. Rothamsted Exp. Stat. 1974, Part 1, pp. 248-249, 1975.
126. Nutman, P. S., C. C. Doncaster, and P. J. Dart, Infection of Clover by Root-nodule Bacteria, Black and white 16 mm, optical sound track film available from The British Film Institute, 81 Dean Street, London W1V 6AA (1973).
127. Nutman, P. S., Biol. Rev. Cambridge Phil Soc., 31, 109 (1956).
128. Sahlman, K., and G. Fåhraeus, J. Gen. Microbiol., 33, 425 (1963).
129. Higashi, S., J. Gen. Appl. Microbiol., 12, 147 (1966).
130. Dart, P. J., and F. V. Mercer, Arch. Mikrobiol., 49, 209 (1964).

131. Bewley, W. F., and H. B. Hutchinson, J. Agr. Sci., 10, 144 (1920).
132. Thornton, H. G., and N. Gangulee, Proc. Roy. Soc. (London) Ser. B, 99, 427 (1926).
133. MacGregor, A. N., and M. Alexander, Plant Soil, 36, 129 (1972).
134. Burgin-Wolff, A., Ber. Schweiz. Bot. Ges., 59, 75 (1959).
135. Deinema, M. H., and L. P. T. M. Zevenhuizen, Arch. Mikrobiol., 78, 42 (1971).
136. Menzel, G., H. Uhlig, and G. Weichsel, Zentr. Bakteriol. Parasitenk., Abt. II, 127, 348 (1972).
137. Marshall, K. C., and R. H. Cruikshank, Arch. Mikrobiol., 91, 29 (1973).
138. McCoy, E., Zentr. Bakteriol. Parasitenk., Abt. II, 79. 394 (1929).
139. Peirce, G. J., Calif. Acad. Sci. Proc. 3rd Ser. Bot., 2, 295 (1902).
140. Roughley, R. J., P. J. Dart, P. S. Nutman, and C. Rodriguez-Barrueco, Proc. 9th Int. Grasslands Congr., pp. 451-455, 1970.
141. Roughley, R. J., P. J. Dart, P. S. Nutman, and P. A. Clarke, J. Exp. Bot., 21, 186 (1970).
142. Nutman, P. S., Aust. J. Agr. Res., 18, 381 (1967).
143. Lim, G., Microbiological Factors Influencing Infection of Clover by Nodule Bacteria, Ph. D. Thesis, London University (1961).
144. Wilson, P. W., Soil Sci., 35, 145 (1933).
145. Day, J. M., and P. J. Dart, Rep. Rothamsted Exp. Stat. 1968, Part 1, 87 (1969).
146. Roughley, R. J., and P. J. Dart, Plant Soil, 32, 518 (1970).
147. Nutman, P. S., R. J. Roughley, P. J. Dart, and N. S. Subba-Rao, Plant Soil, 33, 257 (1970).
148. Munns, D. N., Plant Soil, 28, 246 (1968).
149. Munns, D. N., Plant Soil, 29, 33 (1968).
150. Valera, L., and M. Alexander, Nature, 206, 326 (1965).
151. Tanner, J. W., and J. C. Anderson, Nature, 198, 303 (1963).
152. Tanner, J. W., and J. C. Anderson, Plant Physiol., 39, 1039 (1964).
153. Yoshida, S., and M. Yatazawa, J. Sci. Soil Manure, Japan, 38, 383 (1967).
154. Darbyshire, J. F., Ann. Bot. (London), 30, 623 (1966).
155. Nutman, P. S., (personal communication).

156. Nutman, P. S., Ann. Bot. (London), 13, 261 (1949).
157. Dart, P. J., and J. S. Pate, Aust. J. Biol. Sci., 12, 427 (1959).
158. Skrdleta, V., Rost. Vyroba, 17, 309 (1971).
159. Burton, J. C., and O. N. Allen, Soil Sci. Soc. Am. Proc., 14, 191 (1950).
160. Allen, O. N., and E. K. Allen, Bot. Gaz., 102, 121 (1940).
161. Chandler, M. R., and P. J. Dart, Rep. Rothamsted Exp. Stat. 1972, Part 1, 85 (1973).
162. Davey, M. R., and E. C. Cocking, Nature, 239, 455 (1972).
163. Milovidov, P. F., Zentr. Bakteriol. Parasitenk., Abt. II, 68, 333 (1926).
164. Schaede, R., Zentr. Bakteriol. Parasitenk., Abt. II, 85, 416 (1931).
165. Haack, A., Zentr. Bakteriol. Parasitenk., Abt. II, 114, 577 (1961).
166. Arora, N., Phytomorphology, 4, 211 (1954).
167. Saric, Z., J. Sci. Agr. Res. (Beograd), 16, 60 (1963).
168. Saric, Z., J. Sci. Agr. Res. (Beograd), 16, 90 (1963).
169. Tschirch, A., Ber. Deut. Bot. Ges., 5, 59 (1887).
170. Spratt, E. R., Ann. Bot. (London), 130, 189 (1919).
171. Dangeard, P. A., Le Botaniste, 16, 1 (1926).
172. Allen, O. N., and E. K. Allen, Soil Sci., 42, 161 (1936).
173. Corby, H. D. L., Plant Soil, Special Vol., 305 (1971).
174. Bergersen, F. J., and M. J. Briggs, J. Gen. Microbiol., 19, 482 (1958).
175. Kidby, D. K., and D. J. Goodchild, J. Gen. Microbiol., 45, 147 (1966).
176. Nutman, P. S., Ann. Bot. (London), 12, 81 (1948).
177. Nutman, P. S., in Encyclopedia of Plant Physiology, W. Ruhland, Ed., Vol. XV/1, pp. 1357-1379, 1965.
178. Jimbo, T., Proc. Imp. Acad. Japan, 3, 164 (1927).
179. Pate, J. S., Nature, 192, 376 (1961).
180. Norris, D. C., Trop. Agr., London, 46, 145 (1969).
181. Campelo, A. B., and C. R. Campelo, Pesq. Agropec. Bras., 5, 333 (1970).
182. Bonnier, C., Bull. Inst. Agron. Stat. Rech. Gembloux, 30, 31 (1962).
183. Pankhurst, C. E., N. Z. J. Sci., 13, 519 (1970).
184. Cloonan, M. J., Aust. J. Sci., 26, 121 (1963).
185. Dobereiner, J., Soil Biol. Int. News. Bull., 2, 33 (1964).

186. Maskall, S. M., P. J. Dart, and J. Carpenter, Rep. Rothamsted Exp. Stat. 1971, Part 1, 98 (1972).
187. Thornton, H. G., Ann. Bot. (London), 44, 385 (1930).
188. Libbenga, K. R., and P. A. A. Harkes, Planta, 114, 17 (1973).
189. Arora, N., Phytomorphology, 6, 367 (1956).
190. Wipf, L., and D. C. Cooper, Am. J. Bot., 27, 821 (1940).
191. Oinuma, T., Seibutsu (Sapporo), 3, 155 (1948).
192. Bond, L., Bot. Gaz., 109, 411 (1948).
193. Dixon, R. O. D., Arch. Mikrobiol., 56, 156 (1967).
194. Goodchild, D. J., and F. J. Bergersen, J. Bacteriol., 92, 204 (1966).
195. Van Tieghem, Ph., and H. Douliot, Bull. Soc. Bot. Fr., 35, 105 (1888).
196. Terby, J., Acad. Roy. Belg. Classe Sci. Mem., 8, 1 (1925).
197. Thimann, K. V., Proc. Nat. Acad. Sci. U. S., 22, 511 (1936).
198. Beijerinck, M. W., Zentr. Bakteriol. Parisitenk., Abt. I, Orig., 15, 728 (1894).
199. Schneider, A., Am. Naturalist, 27, 782 (1893).
200. Brenchley, W. E., and H. G. Thornton, Proc. Roy. Soc. (London), Ser. B, 98, 373 (1925).
201. Allen, O. N., and E. K. Allen, Brookhaven Symp. Biol., (6), 209 (1954).
202. Narayana, H. S., J. Indian Bot. Soc., 42, 273 (1963).
203. Wipf, L., and D. C. Cooper, Proc. Nat. Acad. Sci. U. S., 24, 87 (1938).
204. Wipf, L., Bot. Gaz., 101, 51 (1939).
205. Fujita, T., and S. Mitsuishi, Proc. Crop Sci. Soc. Japan, 22, 97 (1953).
206. Tatuno, S., and A. Kodama, Bot. Mag. (Tokyo), 78, 503 (1965).
207. Kodama, A., J. Sci. Hiroshima Univ., Ser. B, 13, 223 (1970).
208. Barrios, S., and J. G. Torrey, Caryologia, 22, 47 (1969).
209. Funke, C., Naturwissenschaften, 44, 98 (1957).
210. Bhaskaran, S., and M. S. Swaminathan, Nucleus (Calcutta), 1, 75 (1958).
211. Kodama, A., Bot. Mag. (Tokyo), 80, 92 (1967).
212. Sen, R., and P. N. Bhaduri, Cytologia, 34, 202 (1969).

213. Mitchell, J. P., *Ann. Bot. (London)*, **29**, 371 (1965).
214. Wendel, E., *Beitr. Allg. Bot.*, **1**, 151 (1918).
215. Rasch, E. M., *Exp. Cell Res.*, **36**, 475 (1964).
216. Kodama, A., *J. Sci. Hiroshima Univ.*, Ser. B, **13**, 265 (1970).
217. Gibson, A. H., *Proc. 11th Int. Bot. Congr., Seattle*, 70 (1969).
218. Link, J. K. K., and V. Eggers, *Bot. Gaz.*, **101**, 650 (1940).
219. Pate, J. S., *Aust. J. Biol. Sci.*, **11**, 516 (1958).
220. Dullaart, J., *Acta. Bot. Neerl.*, **16**, 222 (1967).
221. Kretovich, V. L., F. J. Alekseeva, and N. Z. Tsivina, *Sov. Plant Physiol.*, **19**, 421 (1972).
222. Crush, J. R., *New Phytol.*, **73**, 743 (1974).
223. Libbenga, K. R., F. van Iren, R. J. Bogers, and M. F. Schraag-Lamers, *Planta*, **114**, 29 (1973).
224. Matthysse, A. G., and J. G. Torrey, *Bot. Gaz.*, **130**, 62 (1969).
225. Phillips, R., and J. G. Torrey, *Develop. Biol.*, **31**, 336 (1973).
226. Libbenga, K. R., and J. G. Torrey, *Am. J. Bot.*, **60**, 293 (1973).
227. Raggio, M., and N. Raggio, *Phyton (Buenos Aires)*, **7**, 103 (1956).
228. Yatazawa, M., and S. Yoshida, *Nippon Dogo-Hiryogaku Zasshi*, **36**, 263 (1965).
229. Valera, C. L., and M. Alexander, *J. Bacteriol.*, **89**, 1134 (1965).
230. Phillips, D. A., *Physiol. Plant.*, **25**, 482 (1971).
231. Roughley, R. J., *Ann. Bot. (London)*, **34**, 631 (1970).
232. Mosse, B., *J. Gen. Microbiol.*, **36**, 49 (1964).
233. Dart, P. J., and F. V. Mercer, *Arch. Mikrobiol.*, **46**, 382 (1963).
234. Jordan, D. C., J. Grinyer, and W. H. Coulter, *J. Bacteriol.*, **86**, 125 (1963).
235. Dart, P. J., *4th Eur. Reg. Conf. Electron Microsc.*, Rome, 69 (1968).
236. Truchet, G., and Ph. Coulomb, *Compt. Rend.*, **272D**, 1499 (1971).
237. Gourret, J. P., and H. Fernandez-Arias, *Can. J. Microbiol.*, **20**, 1169 (1974).
238. Dart, P. J., and F. V. Mercer, *Proc. Linn. Soc. N. S. Wales*, **90**, 252 (1965).
239. Whaley, W. G., M. Dauwalder, and S. E. Kephart, in *Origins and Continuity of Cell Organelles*, J. Reinert and H. Ursprung, Eds., Springer-Verlag, New York, 1971.

References

240. Truchet, G., and Ph. Coulomb, *J. Ultrastruct. Res.*, 43, 36 (1973).
241. Kijne, J. W., *Physiological Plant Pathol.*, 5, 75 (1975).
242. Dart, P. J., *Rep. Rothamsted Exp. Stat. 1968*, Part 1, 91 (1969).
243. Bergersen, F. J., and D. J. Goodchild, *Aust. J. Biol. Sci.*, 26, 729 (1973).
244. Dart, P. J., and M. R. Chandler, *Rep. Rothamsted Exp. Stat. 1971*, Part 1, 99 (1972).
245. Dart, P. J., J. M. Day, and R. J. Roughley, *Rep. Rothamsted Exp. Stat. 1970*, Part 1, 84 (1971).
246. Harris, J. C., E. K. Allen, and O. N. Allen, *Am. J. Bot.*, 36, 651 (1949).
247. de Rothschild, D. I., *Rev. Inst. Municipal Bot., Buenos Aires*, 3, 3 (1963).
248. Allen, E. K., K. F. Gregory, and O. N. Allen, *Can. J. Bot.*, 33, 139 (1955).
249. Fraser, L., *Proc. Linn. Soc. N. S. Wales*, 56, 391 (1931).
250. Kapil, R. N., and N. Kapil, *Phytomorphology*, 21, 192 (1971).
251. Fraser, H. L., *Proc. Roy. Soc. Edinburgh, Ser. B*, 61, 328 (1942).
252. Pate, J. S., B. E. S. Gunning, and L. G. Briarty, *Planta*, 85, 11 (1969).
253. Briarty, L. G., *J. Microscopy*, 94, 181 (1971).
254. Gunning, B. E. S., J. S. Pate, F. R. Minchin, and I. Marks, *Symp. Soc. Exp. Biol.*, 24, 87 (1974).
255. Sprent, J. I., *New Phytologist*, 71, 443 (1972).
256. Lechtova-Trnka, M., *Le Botaniste*, 23, 301 (1931).
257. Okajima, T., M. Goto, and N. Okabe, *Bot. Mag. (Tokyo)*, 257, 177 (1972).
258. Chonkar, P. J., and N. S. Subba-Rao, *Can. J. Microbiol.*, 12, 1253 (1966).
259. Robinson, P. E., *Nature*, 189, 506 (1961).
260. Taha, A. H. R., and D. J. Rashi, *J. Nematology*, 1, 201 (1969).
261. Harrison, F. C., and B. Barlow, *Trans. Roy. Soc. Can.*, 2nd Ser., 12, 157 (1906).
262. Frank, B., *Ber. Deut. Bot. Ges.*, 10, 170 (1892).
263. Moeller, H., *Ber. Deut. Bot. Ges.*, 10, 242 (1892).
264. Brunchorst, J., *Ber. Deut. Bot. Ges.*, 3, 241 (1885).

265. Schneider, A., *Bull. Torrey Bot. Club*, 19, 203 (1892).
266. Almon, L., *Zentr. Bakteriol. Parasitenk*, Abt. II, 87, 289 (1933).
267. Baylor, M. B., M. D. Appleman, O. H. Sears, and G. I. Clark, *J. Bacteriol.*, 50, 249 (1945).
268. Palacios de Borao, G., *Microbiol. Espan.*, 2, 51 (1949).
269. Grilli, M., *Giorn. Bot. Ital.*, 71, 62 (1964).
270. Dart, P. J., and F. V. Mercer, *J. Bacteriol.*, 91, 1314 (1966).
271. Dilworth, M. J., and D. K. Kidby, *Exp. Cell Res.*, 49, 148 (1968).
272. Dart, P. J., *Rep. Rothamsted Exp. Stat. 1968*, Part 1, 89 (1969).
273. Dart, P. J., and F. V. Mercer, *J. Bacteriol.*, 85, 951 (1963).
274. Grilli, M., *Caryologia*, 16, 561 (1963).
275. Dixon, R. O. D., *Arch. Mikrobiol.*, 48, 166 (1964).
276. Mackenzie, C. R., W. J. Vail, and D. C. Jordan, *J. Bacteriol.*, 113, 387 (1973).
277. Draper, P., and R. J. W. Rees, *J. Gen. Microbiol.*, 77, 79 (1973).
278. Korner, H. R., *Z. Parasitenk.*, 40, 203 (1972).
279. Dart, P. J., and F. V. Mercer, *Arch. Mikrobiol.*, 47, 1 (1963).
280. Shilnikova, V. K., N. I. Korkina, A. M. Amchenkova, and A. A. Avakyan, *Isv. Akad. Nauk. SSSR Ser. Biol.*, 56, 282 (1972).
281. Gunning, B. E. S., *J. Cell Sci.*, 7, 307 (1970).
282. Zusman, D. R., A. Carbonell, and J. Y. Haga, *J. Bacteriol.*, 115, 1167 (1973).
283. Jensen, W. A., *Planta*, 79, 346 (1968).
284. Jordan, D. C., and J. Grinyer, *Can. J. Microbiol.*, 11, 721 (1965).
285. Dilworth, M. J., and D. C. Williams, *J. Gen. Microbiol.*, 48, 31 (1967).
286. Craig, A. S., and K. I. Williamson, *Arch. Mikrobiol.*, 87, 165 (1972).
287. Craig, A. S., R. M. Greenwood, and K. I. Williamson, *Arch. Microbiol.*, 89, 23 (1972).
288. Trinick, M. J., *Nature*, 244, 459 (1973).
289. Trinick, M. J., in *International Symposium on N_2 Fixation*, W. E. Newton and C. J. Nyman, Eds., Washington State University Press, Washington, pp. 507–517, 1975.
290. Nobbe, F., and L. Hiltner, *Landw. Vers. Sta.*, 42, 459 (1893).

291. Lohnis, M. P., Zentr. Bakteriol. Parasitenk., Abt. II, 80, 342 (1930).
292. Chen, H. K., and H. G. Thornton, Proc. Roy. Soc. (London), Ser. B., 129, 208 (1940).
293. Holl, F. B., Plant Physiol., Suppl., 51, 35 (1973).
294. Pankhurst, C. E. and E. A. Schwinghamer, Arch. Mikrobiol., 100, 219 (1974).
295. Nutman, P. S., Proc. Roy. Soc. (London), Ser. B., 172, 417 (1969).
296. Bergersen, F. J., and P. S. Nutman, Heredity, 11, 175 (1957).
297. Chandler, M. R., Dart, P. J., and P. S. Nutman, Rep. Rothamsted Exp. Stat. 1973, Part 1, 83 (1974).
298. Pankhurst, C. E., E. A. Schwinghamer and F. J. Bergersen, J. Gen. Microbiol., 70, 161 (1972).
299. Pankhurst, C. E., J. Gen. Microbiol., 82, 405 (1974).
300. Bergersen, F. J., Aust. J. Biol. Sci., 10, 233 (1957).
301. Truchet, G. M., and J. Denarie, Compt. Rend., 277D, 841 (1973).
302. Staphorst, J. L., and B. W. Strijdom, Phytophylactica, 4, 87 (1972).
303. Milovidov, P. F., Zentr. Bakteriol. Parasitenk., Abt. II, 73, 58 (1928).
304. Van Brussel, A. A. N., The Cell Wall of Bacteroids of Rhizobium Leguminosarum Frank. Ph. D. Thesis, University of Leiden (1973).
305. Jordan, D. C., Bacteriol. Rev., 26, 119 (1962).
306. Buchanan, R. E., Zentr. Bakteriol. Parasitenk. Abt. II, 23, 59 (1909).
307. Heumann, W., Ber. Deut. Bot. Ges., 65, 229 (1952).
308. Strijdom, B. W., and O. N. Allen, Can. J. Microbiol., 12, 275 (1966).
309. Skinner, F. S. (personal communication, 1973).
310. Jordan, D. C., and W. H. Coulter, Can. J. Microbiol., 11, 709 (1965).
311. Skinner, F. S., P. J. Dart, and M. R. Chandler, Rep. Rothamsted Exp. Stat. 1972, Part 1, 82 (1973).
312. Beijerinck, M. W., Versl. Mededeel. Afd. Natuurk, Ser. 3, 8, 460 (1891).
313. Day, J. M., P. J. Dart, and M. R. Chandler, Rep. Rothamsted Exp. Stat. 1972, Part 1, 84 (1973).

314. Dullaart, J., C. A. Wijffelman, and J. Haveman, Antonie van Leeuwenhoek J. Microbiol. Serol., 37, 219 (1971).
315. Dart, P. J., and F. V. Mercer, Aust. J. Agr. Res., 16, 321 (1965).
316. Pankhurst, C. E., and A. H. Gibson, J. Gen. Microbiol., 74, 111 (1973).
317. Day, J. M., and P. J. Dart, Rep. Rothamsted Exp. Stat. 1970, Part 1, 85 (1971).
318. Rautenberg, J., and J. Kuhn, J. Landw., 12, 107 (1864).
319. Wilson, P. W., The Biochemistry of Symbiotic Nitrogen Fixation, University of Wisconsin Press, Madison, 1940.
320. Vines, S. H., Ann. Bot. (London), 2, 386 (1888).
321. Thornton, H. G., and J. E. Rudorf, Proc. Roy. Soc. (London), Ser. B., 120, 240 (1936).
322. Dart, P. J., and F. V. Mercer, Arch. Mikrobiol., 51, 233 (1965).
323. Shilnikova, V. K., O. D. Sidorenko, and N. I. Korkina, Sel'skokhoz Akad, Part 2, 120 (1972).
324. Day, J. M., and P. J. Dart, Rep. Rothamsted Exp. Stat. 1970, Part 1, 85 (1971).
325. Eaglesham, A. J. R., J. M. Day, and P. J. Dart, Rep. Rothamsted Exp. Stat. 1973, Part 1, 84 (1974).
326. Allos, H. F., and M. W. Bartholomew, Soil Sci., 87, 61 (1959).
327. Thornton, H. G., Proc. Roy. Soc. (London), Ser. B, 104, 481 (1929).
328. Thornton, H. G., Proc. Roy. Soc. (London), Ser. B, 106, 110 (1930).
329. Doku, E. V., Exp. Agr., 6, 13 (1970).
330. Raggio, M., and N. Raggio, Physiol. Plant., 9, 466 (1956).
331. Raggio, M., N. Raggio, and J. G. Torrey, Am. J. Bot., 44, 325 (1957).
332. Raggio, N., M. Raggio, and R. H. Burris, Science, 129, 211 (1959).
333. Bunting, A. H., and J. Horrocks, Ann. Bot. (London), 28, 229 (1964).
334. Hough, M. C., B. Clarke, and N. Grobbelaar, Phyton (Buenos Aires), 23, 15 (1966).
335. Barrios, S., and M. Raggio, Phyton (Buenos Aires), 21, 209 (1964).
336. Schaffer, A. G., and M. Alexander, Science, 152, 82 (1966).

References

337. Cartwright, P. M., Ann. Bot. (London), 31, 309 (1967).
338. Raggio, N., M. Raggio, and R. H. Burris, Biochim. Biophys. Acta, 32, 274 (1959).
339. Cartwright, P. M., Wiss. Zeit. Univ. Rostock, 4/5, 537 (1967).
340. Grobbelaar, N., B. Clarke, and M. C. Hough, Plant Soil, Special Vol., 203 (1971).
341. Barrios, S., N. Raggio, and M. Raggio, Plant Physiol., 38, 171 (1963).
342. Raggio, M., N. Raggio, and J. G. Torrey, Plant Physiol., 40, 601 (1965).
343. Schaffer, A. G., and M. Alexander, Plant Physiol., 42, 557, 563 (1967).
344. Higashi, S., M. Abe, and G. Yamane, Rep. Fac. Sci. Kagoshima Univ. (Earth Sci. Biol.), No. 4 (1971).
345. Hepper, C. M., Rep. Rothamsted Exp. Stat. 1972, Part 1, 84 (1973).
346. Holsten, R. D., R. C. Burns, R. W. F. Hardy, and R. R. Hebert, Nature, 232, 173 (1971).
347. Child, J. J., and T. A. LaRue, Plant Physiol., 53, 88 (1974).
348. Phillips, D. A., Plant Physiol., 53, 67 (1974).
349. Phillips, D. A., Plant Physiol., 54, 654 (1974).
350. Phillips, D. A., in International Symposium on N_2 Fixation, W. E. Newton and C. J. Nyman, Eds., Washington State Univ. Press, Washington, pp. 367-373, 1975.
351. Velicky, I., and T. A. LaRue, Naturwissenschaften, 54, 96 (1967).
352. Child, J. J., Nature, 253, 350 (1975).
353. Scrowcroft, W. R., and A. H. Gibson, Nature, 253, 351 (1975).
354. Lange, R. T., Antonie van Leeuwenhoek J. Microbiol. Serol., 25, 272 (1959).
355. Pate, J. S., Aust. J. Biol. Sci., 11, 496 (1958).
356. Masterson, C. L., and P. M. Murphy, in Symbiotic Nitrogen Fixation in Plants, P. S. Nutman, Ed., Cambridge University Press, pp. 299-316, 1975.
357. Bergersen, F. J., F. W. Hely, and A. B. Costin, Aust. J. Biol. Sci., 16, 920 (1963).
358. Hewitt, E. G., Sand and Water Culture Methods Used in the Study of Plant Nutrition, Commonwealth Agricultural Bureaux, Farnham Royal, England, 1966.

359. Gibson, A. H., Aust. J. Biol. Sci., 16, 28 (1963).
360. Vincent, J. M., A Manual for the Practical Study of Root-Nodule Bacteria, IBP Handbook No. 15, Blackwell Scientific Publications, Oxford and Edinburgh, 164 pp., 1970.
361. Jensen, H. L., Proc. Linn. Soc. N. S. Wales, 66, 98 (1942).
362. Steiner, A. A., and H. van Winden, Plant Physiol., 46, 862 (1970).
363. Allen, E. K., and O. N. Allen, in Recent Advances in Botany, University of Toronto Press, pp. 585-588 (1961).
364. Barrios, S., and V. Gonzalez, Plant Soil, 34, 707 (1971).
365. Corby, H. D. L., Kirkia, 9, 301 (1974).
366. Nutman, P. S., in Nutrition of the Legumes, E. G. Hallsworth, Ed., Butterworth Sci. Publications, London, pp. 87-107 (1958).
367. Mukhopadhyai, A. K., Morphogenesis and Cytology of Nodules of Some Common Legumes, Ph. D. Thesis Presidency College, Calcutta (1967).
368. Narayana, H. S., and B. D. Gothwal, Proc. Indian Acad. Sci., Sect. B, 59, 350 (1964).
369. Schaede, R., Beitr. Biol. Pflanzen, 27, 165 (1941).
370. Laurent, E., Ann. Inst. Pasteur, 5, 105 (1891).
371. Scheerlinck, H., Ann. Soc. Sci. Bruxelles, 56, 250 (1936).
372. Nemec, B., Bull. Int. Acad. Sci. Boheme, Imprime Special, 1 (1915).
373. Milovidov, P., Spisy Vydav. Prir. Fak. Karl. Univ., 49, 3 (1925).
374. Prasad, D. N., and N. D. Deepish, Microbios, 4, 13 (1971).
375. Thomazini, L. J., and K. Arens, Ciencia Cult. (Sao Paulo), 23, 323 (1971).
376. MacGregor, A. N. and M. Alexander, J. Bacteriol., 105, 728 (1971).
377. Tu, J. C., J. Bacteriol., 119, 986 (1974).
378. Tu, J. C., J. Bacteriol., 122, 710 (1975).
379. Aufeuvre, M.-A., Compt. Rend., 277, 921 (1973).
380. Allen, O. N., and E. K. Allen, Soil Sci., 47, 63 (1939).

CHAPTER 9

Functional Biology of Dinitrogen Fixation by Legumes

J. S. PATE

Botany Department
University of Western Australia
Nedlands, Australia

- 9.1. Introduction, 474
- 9.2. Physiological and Biochemical Relationships between Nodule and Host, 474
  - 9.2.1. Nitrogen Metabolism of Nodules, 474
  - 9.2.2. Transfer of Fixed Nitrogen from Nodule to Host, 478
  - 9.2.3. Excretion of Nitrogen to the Rooting Medium, 480
  - 9.2.4. Supply of Assimilates and Other Factors to the Nodules, 481
  - 9.2.5. Carbohydrate Requirements of Symbiosis, 483
  - 9.2.6. Integration of Nodular Fixation and Host Plant Function, 485
- 9.3. Nodulation, Dinitrogen Fixation, and the Symbiotic Cycle, 489
  - 9.3.1. Parameters of Symbiotic Functioning, 489
  - 9.3.2. Symbiotic Cycle of the Annual Legume, 491
  - 9.3.3. Symbiosis in Perennial Legumes, 493
- 9.4. Physiological Factors Influencing Dinitrogen Fixation, 495
  - 9.4.1. Essential Chemical Elements, 495
  - 9.4.2. Light, Grazing, Defoliation, Shading, and Darkness, 497

9.4.3. Water and Gases in the Rooting Medium, 499
9.4.4. Temperature, 501
9.4.5. Combined Nitrogen, 503
9.5. Epilogue, 506
9.6. References, 506

9.1 INTRODUCTION

In this chapter any information of a physiological nature that furthers understanding of the functioning of symbiosis in the nodulated legume will be discussed. Three interrelated sections are included. The first deals with general aspects of the integration of physiological and biochemical functions in symbiosis, containing as well, a detailed account of the exchange of specific sets of metabolites between host plant and nodules. The second section evaluates the methods by which the seasonal cycle of symbiosis can be characterized and presents an outline of the biological forces shaping its course of development in annual and perennial legumes. The final section analyzes the effects of various external factors on symbiotic activity, particularly those potentially hostile to nodulation and dinitrogen fixation.

9.2 PHYSIOLOGICAL AND BIOCHEMICAL RELATIONSHIPS BETWEEN NODULE AND HOST

9.2.1 Nitrogen Metabolism of Nodules

It has become apparent from $^{15}N_2$ feeding studies on a wide range of dinitrogen-fixing systems, including detached legume nodules (1, 2) that ammonia is the first stable product of fixation and the starting material for incorporation of fixed nitrogen into organic compounds. In certain organisms, e.g., Nostoc (3, 4) and Alnus (5, 6), citrulline is a major product of fixation. Since the carbamyl nitrogen as opposed to the ornithine nitrogen of this compound becomes most heavily labeled in pulse feeding studies using $^{15}N_2$, the primary reaction of assimilation is assumed to occur via carbamyl phosphate and thence by reaction with ornithine to citrulline:

$\overset{*}{NH_3} + CO_2 + 1(?) \text{ ATP} \xrightarrow[\text{synthetase}]{\text{carbamyl phosphate}} H_2N-\overset{*}{C}\overset{O}{\underset{O}{\diagdown}}P + 1(?2) \text{ ADP} + Pi$

carbamyl phosphate

$$P-O-\overset{O}{\underset{\|}{C}}-\overset{*}{NH_2}$$

carbamyl phosphate

$+$

$NH_2-(CH_2)_3-\underset{\underset{NH_2}{|}}{CH}-COOH$

ornithine

$\xrightarrow{\text{ornithine transcarbamylase}}$

$\overset{O}{\underset{\|}{C}}-\overset{*}{NH_2}$
$|$
$NH-(CH_2)_3-\underset{\underset{NH_2}{|}}{CH}-COOH + Pi$

citrulline

The relationship of this system to the ornithine cycle is not clear, nor has it been suggested how ornithine acceptor molecules are replenished as the fixed nitrogen is channeled into citrulline.

In the majority of other diazotrophs, however, glutamic acid is the first organic compound to be heavily labeled in pulse experiments using $^{15}N_2$, and it is generally concluded to be the primary amino donor for synthesis of other amino compounds. Its synthesis takes place by reductive amination of α-ketoglutaric acid by the enzyme glutamic acid dehydrogenase, the system also held responsible for assimilation of ammonia and other combined inorganic forms of nitrogen by organisms heterotrophic for nitrogen (7, 8).

The two legumes studied to date, Glycine max (9-11) and Ornithopus sativus (12), and the nonlegume Myrica gale (13) appear to conform to the above pattern of assimilation. The time course of labeling of glutamic acid in $^{15}N_2$ feeding studies is fully consistent with its roles as primary product and amino donor (12), while isocitric acid dehydrogenase and glutamic acid dehydrogenase, enzymes responsible respectively for synthesis and amination of α-ketoglutaric acid, are present in nodules both within and outside the bacteroids (14, 15). As might be expected, the exchangeable pools of ammonia and glutamic acid in the nodule are small, while those for the secondary products are much larger. It is these derived compounds, especially aspartic acid and amides, which are the principal products exported from the nodule.

Amino transferases involving glutamic acid, aspartic acid, and alanine have been detected in nodules or nodule bacteroids (16, 17), suggesting that transamination reactions might be important in formation of new amino acids for nodule growth. At the same time the actively growing and N_2-fixing nodule may be dependent on amino acids synthesized elsewhere in the plant and translocated to it along with sugar and other assimilates via the phloem.

Multiple forms of aspartate amino transferase have been recorded in nodules of <u>Glycine max</u>, and the localization of these interpreted as demonstrating either a shuttle system for oxaloacetate between host cytoplasm and bacteroids, or the existence of several sites of assimilation of ammonia in the nodule (18).

Legume nodules and the roots that bear them usually generate a far greater exportable surplus of aspartyl than glutamyl compounds (19, 20), so that metabolic mechanisms additional to the tricarboxylic acid cycle are likely to be involved in generating oxaloacetate. An obvious possibility would be a phosphoenolpyruvic acid carboxylase system, already known to be widely distributed in plants (8) and active in the roots of legumes (21, 22):

$$\begin{array}{c}CH_2 \\ \| \\ C\text{-}O\text{-}PO_3H_2 \\ | \\ COOH\end{array} + {}^*CO_2 + H_2O \xrightarrow{Mg^2} \begin{array}{c}{}^*COOH \\ \backslash \\ CH_2 \\ | \\ C=O \\ | \\ COOH\end{array} + H_3PO_4$$

phosphoenol pyruvic acid   phosphoenol-pyruvic acid carboxylase   oxaloacetic acid

The activity of this system in nodules might make significant improvements in their carbon economy by allowing respired carbon dioxide to be reutilized, and even some carbon dioxide to be absorbed from the root medium. Reports of stimuli to legume growth by carbon dioxide supplied to the root environment (23, 24) might well be explained on the basis of such activity by roots and nodules.

Much interest centers on amide synthesis in nodules as these molecules usually serve as recipients of most of the fixed nitrogen. In pulse-labeling studies using $^{15}N_2$, the amido groups of amides achieve earlier and heavier labeling than the corresponding amino nitrogen (12, 13), implying that this grouping might receive nitrogen directly from the fixation process. The demonstration of glutamine synthetase in nodules of a legume (25) suggests that the usual pathway involving ammonia and glutamic acid might operate for this amide. The situation regarding asparagine is less clear, although the recent demonstration in Glycine max seedlings of a system catalyzing asparagine synthesis from aspartic acid and the amino group of glutamine (26) might indicate a similar reaction in nodules. In this system glutamine exhibits an affinity for aspartic acid 25 times greater than that exhibited by ammonia (26).

Apart from the primary reactions of assimilation mentioned above, very little is known regarding the metabolism of nitrogen in nodules. In amino acid composition they usually resemble closely the roots that bear them (27, 28), inferring a conventional pattern of metabolism, but these free amino acids usually constitute only a relatively small proportion of the nodule's ethanol-soluble nitrogen. Of the remainder much can be designated only as "unknown, ninhydrin-negative compounds" (29), although in certain legume nodules some of the nitrogen has been classified as "bound amino acid," detectable only after strong acid hydrolysis of the ethanol-soluble fraction. In nodules of Trifolium spp., γ-amino butyric acid is "bound" in this manner (30, 31); in nodules of Pisum sativum, β-alanine, homoserine, and γ-amino butyric acid (32). Since it is not possible to detect a similar bound fraction in these species in tissues other than nodules, special functions for "bound" amino acids in nodules are indicated. One possibility is that the "bound" fraction might function as a temporary store of fixed nitrogen, increasing when fixation rates exceed rates of export, diminishing when the reverse applies. Studies of synthesis and turnover of these compounds are needed to test this suggestion.

9.2.2 Transfer of Fixed Nitrogen from Nodule to Host

It is now generally agreed that neither the solutes generated in fixation nor protein derived from them accumulate to any extent within functional nodules (39-42), and that the xylem is the principal avenue of export from nodule to host shoot (36, 43).

Analyses of xylem fluids obtained from detached nodules, root stumps, or stems reveal that low molecular weight solutes, rich in nitrogen, are involved in export, the range of compounds utilized varying with the species (Table 9.1). Nodulated plants fixing dinitrogen employ the same set of solutes for xylem transport as do plants heterotrophic for nitrogen (37, 44, 45), reinforcing the conclusion (see III.9.2.1) that similar metabolic pathways are involved in assimilation of the two sources of nitrogen.

Export from nodules appears to be both active and selective since certain amino compounds are much more concentrated in nodule xylem fluid than in donor bacterial tissues, while others occur in large amounts within the nodule but do not feature prominently among the exported products (38). For many amino acids downhill concentration gradients exist from central fixing tissues to outer layers where the vascular strands are located, and since appropriate plasmodesmatal connections have been detected along this pathway, symplastic transfer of fixed dinitrogen to the nodule bundles has been suggested (46). The pericycle cells of the nodule vascular bundle are sometimes modified as "Transfer Cells" (sensu 46) and are envisaged as secreting selected amino compounds into the extracellular space of the nodule bundle. Water, entering osmotically across the bundle endodermis, then flushes these compounds out through the xylem (46).

The high levels of amino nitrogen in nodule xylem fluids (3-4 mg N/ml) indicate considerable efficiency in utilization of water in transport. Indeed, Pisum sativum nodules are estimated to require for export only twice their own volume of water per day, and almost half of this requirement might come from mass flow of sugars into the nodule via the phloem (47). But it is still debatable whether surface absorption would be fast enough to supply the additional water required for the export process, particularly in the case of large nodules in relatively dry soils. The

TABLE 9.1. Nitrogenous Solutes Employed in Xylem Transport of Nitrogen from Legume Roots

Species	Principal compound(s) exported	References
Albizzia lophantha[a]	Citrulline, allantoin, allantoic acid	33
Arachis hypogaea[b]	γ-Methylene glutamine	34
Glycine max[c]	Asparagine, aspartic acid, glutamine, glutamic acid	20
Lupinus spp.	Asparagine	35
Phaseolus vulgaris	Allantoic acid	19
Pisum spp.	Asparagine, glutamine, homoserine, aspartic acid	36, 36, 37
Trifolium repens	Asparagine	31
Vicia faba[c]	Asparagine, aspartic acid, glutamine, glutamic acid	38

[a]Tracheal sap analysis (? nodulated plants).
[b]Root bleeding sap (nonnodulated plants).
[c]Bleeding sap detached nodules.
Unmarked species — root bleeding sap nodulated plants.

cylindrical-type nodules of Pisum sativum, when detached and placed with their apical (meristematic) ends in water, can export nitrogen at a rate of only one-quarter of that of attached nodules (47), and surface uptake in spherical-type nodules (e.g., of Phaseolus spp. and Glycine spp.) might be even slower than this since the surface layers of these nodules are poorly adapted structurally for water/solute exchange (48-50). However, it has been demonstrated using tritiated water that nodules of Glycine max can trap water from the ascending transpiration stream (50), presumably by osmotic attraction across the endodermis separating the nodule from its parent root. This might prove sufficient to meet the relatively modest water budget of the nodule.

9.2.3 Excretion of Nitrogen to the Rooting Medium

The pioneer studies carried out in Finland on the excretion of nitrogenous solutes from effectively nodulated roots of Pisum sativum gave rise to the hypothesis that benefit from nodule to parent legume or surrounding plants occurred by excretion of simple amino compounds from nodules (51-53). The Finnish experiments provided instances where up to 80% of the nitrogen fixed in sterile sand cultures was released into the rooting medium, and since this occurred early in vegetative growth before nodule degeneration was evident, and did not occur to a comparable extent from uninoculated plants receiving nitrate nitrogen, it was concluded that surface excretion from healthy functioning nodules was involved.

The subsequent failure of many workers to obtain excretion from nodulated legumes [e.g., (53-56)] and the inconsistent results of those few workers who were able to demonstrate the phenomenon [e.g., (57, 58)] led to the conclusion that excretion is the exception rather than the rule under natural conditions, but is symptomatic of physiological stress or unusual climatic conditions and promoted especially in the somewhat artificial conditions of enclosed or semienclosed culture.

Since these early studies much additional information has accumulated on the physiology of root excretion. Using sterile cultures of plants in which roots developed in sand or water culture, in a fog box, or between sheets of filter paper, it has been shown that a wide range of compounds can be excreted, including sugars, amino acids, peptides, enzymes, vitamins, organic acids, nucleotides, and numerous other substances (59). Moreover, excretion is by no means the prerogative of legume roots, nor is there any real evidence to suggest that they are consistently more active in this respect than are nonlegumes (60, 61). Amounts released from roots are usually small — less than 0.5% of the carbon or nitrogen of the plant (59) — although excretion may be somewhat enhanced if roots experience anaerobiosis (62) or desiccation folfowed by remoistening (63). The spectrum of liberated nitrogenous solutes appears to be species specific (59), and often resembles closely solute composition of the root (64). It is perhaps significant that the most massive excretion is recorded in isolated root cultures, for in such a situation no shoot is present to sequester root assimilates [see (65)].

Tracer studies with ^{14}C on several species have shown that recently produced metabolites can contribute significantly to root excretions (59, 66). The root cap (66), the zone of root close to the apex (67), and sites of lateral emergence (59) have all been implicated in excretion, as has leakage from abraded root hairs and sloughed off epidermal cells (68). Rough handling of roots can lead to spurious leakage (69).

Bearing in mind the difficulties in extrapolating from data for sterile culture systems to the highly complex situation in soil, [e.g., (70)] and the paucity of recent data on excretion from nodulated legumes in laboratory culture, the part played by nodular excretion in providing nitrogen to surrounding organisms has still to be properly evaluated. However, from what was said earlier concerning the impervious nature of nodule surfaces, and the adequacy of xylem transport in releasing fixed nitrogen to the host, direct surface losses from the symbiotic organs would seem unlikely. This conclusion is fully consistent with the agronomic literature, where it is stated that underground transference of nitrogen is more likely to result from decay of roots and nodules than from active excretions (71-74). Even when the legume has been shaded, cut, or grazed — stresses known to promote nitrogen transfer — loss and decay of root and nodule tissue are still considered to be the major sources of benefit to surrounding plants.

9.2.4 Supply of Assimilates and Other Factors to the Nodules

From the time when the free-living Rhizobium is first attracted to and multiplies within the rhizosphere through to the stages of initiation, development, and maintenance of fully functional nodules, the host must supply the symbiotic organs with a wide variety of nutrient and regulatory substances in amounts and proportions matching the ever-changing requirements of symbiosis. The situation regarding the early stages of nodule development is dealt with elsewhere (III.8.3): suffice to say here that specific stimulatory factors seem to be involved, probably originating in the cotyledons and sequestered by the seedling root almost immediately after germination (75-77). Once the root has developed beyond a certain stage it becomes considerably more autotrophic with respect to these factors, as witnessed by its ability to nodulate true to form, or to continue to develop nodules if the shoot of another species (78) or genus (79, 80) is grafted on to it. Even better evidence of autonomy is displayed in isolated root cultures of legumes in which satisfactory nodulation occurs, and healthy, functional nodules form when the root is supplied via its proximal (hypocotyl) end with a relatively unsophisticated organic diet, e.g., in the case of Phaseolus vulgaris, a medium containing sucrose, glycine, inositol, nicotinic acid, pyridoxine and thiamine (81-84), or, in Pisum sativum, 2-5% glucose or sucrose as the only organic nutrient (85). So, self-sufficiency for non-carbohydrate organic nutrients for nodulation can be acquired by the root at a very early stage.

Although studies using isolated root cultures are useful in prescribing the minimum organic requirements for short periods of enforced autonomy, they offer no guide as to the situation in the complete plant, where the presence of a donor shoot may encourage nutritional dependencies by both root and nodules. Translocation studies provide useful information here, for any substance demonstrated to be transported in large amounts from the host to nodules can be classed as nutrient and possible regulator of symbiosis. Sugars formed in photosynthesis are obvious candidates in this connection both as energy source and as carbon donor for the fixation process. Feeding studies with $^{14}CO_2$ have made it clear that nodules benefit measurably from current photosynthate (86, 87), compete with the parent root for photosynthate (88), and consume a proportion of the carbon which they receive for synthesis of the amino compounds generated in fixation (35, 89). Based on known patterns of translocation, annual legumes do not differ basically from non-legumes. Thus, young seedlings show a fully integrated nutrition in which underground organs are nourished by all photosynthetic surfaces, while in older plants nutrition becomes progressively stratified with upper and middle stories of leaves mainly feeding the shoot apex and reproductive structures, while lower leaves, mainly nourish underground organs (90-94).

Somewhat different translocatory arrangements are likely in perennial legumes, but these have not been fully explored. In Trifolium repens all sinks — stolen apices, expanding leaves, nodules, roots, and reproductive structures — appear to draw from a common pool of assimilates donated by the mature leaves (94). This fully integrated pattern of nutrition may have adaptive significance in a species regularly subjected to treading and grazing.

Foliar feeding studies using ^{15}N (43) and ^{32}P (95) have proved that a translocatory pathway to the nodules exists for these elements, and since both occur predominantly in organic form in the phloem (96), quite specific nutritional effects on symbiosis are likely. The high sink capacity of nodules may well attract significant amounts of many phloem-mobile compounds with the recent demonstration of abstraction of the radiocarbon of foliar-fed ^{14}C-abscisic acid (97) by nodules as an example.

Finally, mention must be made of the somewhat confused situation regarding growth factors. Many Rhizobium strains require biotin in culture, while the growth of others is stimulated by thiamine and pantothenic acid (98). Since biotin and thiamine are excreted by plant roots (99), and transport of thiamine from root to shoot is suspected (100), it seems highly possible that any or all of such growth factors might be regularly supplied to nodules. The recent study (101) of symbiotically ineffective autotrophs of Rhizobium trifolii requiring an external source of riboflavin for effective symbiosis may well provide a useful tool for investigations in this connection. In the absence of added riboflavin, different levels of effectiveness are attained by the mutant on different Trifolium hosts, suggesting differences in the ability of the latter to furnish the deficient growth factor to the microsymbiont (101) (see III.12.2.1.3).

9.2.5 Carbohydrate Requirements of Symbiosis

It is an important goal of investigations of legume biology to measure how much carbohydrate is consumed in symbiotic N_2 fixation by nodules. One experimental approach to the problem consists of measuring the extent to which dry matter production by nodulated plants relying on dinitrogen falls short of that of plants receiving an equivalent amount of nitrogen in combined form and to assume that this difference represents the extra cost to the plant in using its nodules as a source of nitrogen. Using this method, values of 5.6 mg C (as carbohydrate) per mg N_2 fixed have been obtained for Glycine max nodules (102), and 1.2-4 mg C/mg N_2 fixed and 0.32 mg C/mg N_2 fixed for establishment and maintenance, respectively, of nodules of Trifolium subterraneum (103). Unfortunately the method fails to take into account carbohydrate expenditure in assimilation of the combined form of nitrogen, and, to give meaningful results, requires that there is absolute physiological comparability between nodulated and non-nodulated plants. This supposition has been disputed by several workers (104-106).

A second experimental approach is to measure directly the respiratory activity of detached nodules and to relate this to N_2 fixation. The intensity of nodular respiration does not differ appreciably from that of root tissues (107, 108), and is not noticeably different in ineffective and effective nodules (109). Respiration of nodules exhibits an RQ in excess of one, and generally above that of the parent root (110, 111). In comparisons of rates of detached or attached nodule respiration with rates of attached nodule fixation, the requirements of oxygen and dinitrogen were 5 vol O_2:1 vol N_2 for Glycine max (112) and 3:1 for Pisum sativum (47). These assessments refer only to carbon expenditure in respiration, and problems due to loss of activity may be encountered with detached nodules (113, 114).

Other studies have attempted to examine nodular requirements in relation to the functioning of the whole plant, by combining data for dry weight and nitrogen gains in plant parts with information on respiration of detached nodules, attached nodulated roots, and comparable roots without nodules receiving combined nitrogen. In one study of Glycine max (108), it was estimated that nodules respire 19 mg carbohydrate (7.6 mg C) per mg N_2 fixed, and that this corresponds to an expenditure of 16% of the total carbohydrate synthesized by the plant. In another, restricted to a 21-day period during the functional life of mature nodules of Pisum sativum (47, the total expenditure of the nodules was 10.3 mg carbohydrate (4.1 mg C)/mg N_2 fixed.

The estimated consumption of 0.8-19.0 mg carbohydrate/mg N_2 fixed is somewhat below the range of 20-30 mg carbohydrate/mg N_2 fixed by free-living diazotrophs (114-117). This would imply a somewhat higher general efficiency for symbiotic fixation, but it must be remembered that most values derived for the legume exclude the cost of establishing the symbiotic apparatus, and of maintaining the transport and processes of the nodules. If these costs were included the difference between free-living and symbiotic diazotrophs might be much reduced.

9.2.6 Integration of Nodular Fixation and Host Plant Function

The obligatory relationship between photosynthetic and N_2-fixing organs, necessitating a regular exchange of complementary sets of essential materials between root and shoot, constitutes a most valuable system for examining regulatory phenomena in higher plants. Studies with Pisum spp. provide a suitable example (Fig. 9.1). In the vegetative growth of this legume it has been estimated that the root and nodules command, respectively, 42 and 32% of the net photosynthetic gain of carbon by the shoot. The root utilizes 83% of its share in respiration, the rest in growth; the nodules' carbon budget allocates 16% to growth, 37% to respiration, and 47% to export as amino compounds. It can be seen that the shoot derives just over one-third of its requirements for carbon as materials cycled via the nodules (47). The translocatory system for nitrogen (Fig. 9.1) is dominated by xylem export from the nodules. Consequently, distal regions of the root receive only small amounts of nitrogen directly from the nodules, and in a medium deficient in combined nitrogen, young roots must receive the bulk of their requirements from nitrogen which has cycled through the shoot and back to the root via the phloem (43). In this manner as well as in its role as provider of carbohydrate, the shoot exercises direct nutritional control over growth of the root, and hence over the production of further sites for nodule initiation.

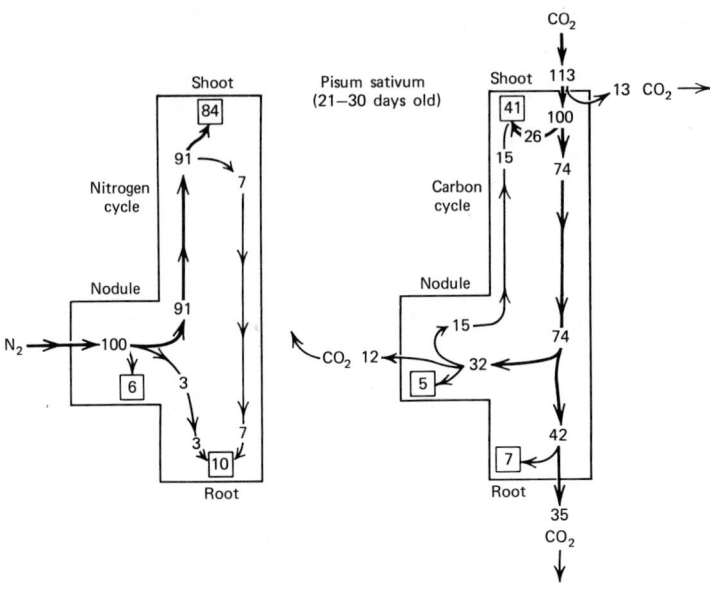

Fig. 9.1. Flow pattern for nitrogen and carbon in plants of garden pea (Pisum sativum var. Meteor) during the period 21-30 days after sowing. The figures given for nitrogen relate to the fate of 100 units of nitrogen fixed by the nodules, those for carbon to 100 units of carbon gained photosynthetically from the atmosphere. Arrows in an upward direction refer to xylem flow, those in a downward direction to translocation in the phloem. Figures in boxes refer to consumption in growth, those alongside arrows to amounts transported. Note: In the normal functioning of the plant 13 parts by weight of carbon flow within the carbon cycle (right) for every one of nitrogen flowing in the nitrogen cycle (left).

Integration of shoot and root metabolism is also manifest at the level of specific nitrogenous solutes. As seen earlier (III.9.2.2), nodules export a somewhat limited range of compounds, and the surplus amino groups attached to these provide nitrogen for temporary storage, or for synthesis of new amino compounds in the shoot, especially those derived from photosynthesis (118, 119). Analysis of phloem sap shows that root-derived and shoot-derived amino compounds accompany sugars, inorganic ions, and other assimilates in the translocation stream, and, as a result, solutes are dispensed to apical regions and reproductive structures in amounts and proportions closely related to their requirements in growth (119).

Considerable evidence indicates that carbohydrate supply to the nodule is the natural regulator of N_2 fixation. Thus, acetylene reduction assays have shown that a rapid decline in nitrogenase activity occurs following darkening, defoliation, or decapitation of host plants (20), and that greater activity is manifest in nodules during the photoperiod than at night (120), even when parent plants have been growing under constant temperature (121). Acetylene reduction activity has also been found to be greater in attached nodulated roots than in an equivalent mass of detached nodules (122), and since this difference is greater during the day than at night (123), marked dependence by nodules on current translocate is indicated. A rather similar situation exists in the nonlegume Alnus where diurnal fluctuations in nitrogenase activity have been shown to be closely correlated with the influx of photosynthate, but not with the current level of sugars in the nodules (124).

It has been shown that effective nodules of an annual species such as Pisum sativum require in translocate the equivalent of 30% of the net photosynthate of the plant (see Fig. 9.1), yet it is difficult to conceive how such a relatively small proportion of the plant's total weight can possibly maintain such high sink activity. The high auxin content of nodules (125, 126) may well elicit some form of directed transport to the fixing tissues, akin to that already suggested for translocation of assimilates to developing fruits and shoot apices (127, 128).

Several workers have explored the possibility that nitrogenase performance might be regulated by feedback control through accumulation in nodules of the products of fixation. Ammonia is a repressor of nitrogenase synthesis in free-living diazotrophs, and an inhibition of this nature has been suggested as a cause of lowered fixation activity under conditions where shortage of appropriate carbon acceptors in the nodule might allow free ammonia to accumulate (20). Nevertheless, it has been shown that both ammonia and a range of amino acids, at concentrations of up to 10 mM, fail, in the short term, to inhibit in vitro activity of nitrogenase in Lupinus (129). Unfortunately effects on enzyme synthesis and turnover in vivo have not been fully investigated.

Mention must be made of the strong diurnal rhythm in export of nitrogenous solutes from nodulated roots (130). This apparently possesses the same endogenous qualities displayed by other biological rhythms (89), and since the fluctuations in export activity can be out of phase with those for nitrogenase activity, considerable diurnal changes can occur in the concentration of free amino acids in the nodule (121). It has still to be ascertained whether these transient discrepancies between rates of synthesis and export from nodules have any significance in the regulation of the normal day's cycle of events in the nodule.

Since nodular fixation exhibits such a marked dependence on current photosynthesis, it is not surprising to find that poly-β-hydroxybutyrate, the major reserve polymer of legume bacteroids (131, 132), is not capable of maintaining high levels of nitrogenase activity in nodules at times when the normal supply of sugars to the root has become interrupted (20). Even if this polymer comprised 50% of the nodule's dry weight and was an immediately available carbon source, it would probably support only a day or two of normal nodule activity in an annual legume like Pisum sativum. It may be of advantage that the relatively meager reserves of polymer remain inviolate during temporary carbohydrate stress (20), but do become available slowly; and, on a long-term basis, as a means of supporting basal respiration during periods of starvation, e.g., during overwintering of nodules. Metabolic arrangements for utilization of poly-β-hydroxybutyrate would appear to favor such a supposition.

Just as it may be desirable, in the adaptive sense, that nodular fixation parallel current photosynthesis, so it may be undesirable for fixation of the nodules to be irrevocably linked with current growth of the host plant. At certain stages in the life cycle, photosynthetic activity may exceed current needs of growth, and, if this surplus is channeled into dinitrogen fixation, substantial reserves of soluble nitrogen may be accumulated. In the annual legume, nitrogen reserves established before flowering assume importance later as sources for fruit nutrition; in perennials nitrogen reserves formed in one season can be used to foster an early resumption of growth after a period of dormancy. An almost complete uncoupling of fixation from host growth can be witnessed in experiments involving plants whose shoot meristems have been removed, in rooted detached leaves (133), or in annual legumes maintained artificially in vegetative condition by removal of flower primordia (134, 135). Despite the fact that demands for nitrogen in shoot growth and reproduction have been decreased, or abolished within such systems, fixed soluble nitrogen still accumulates, and may eventually reach levels many times that encountered in the intact plant. There seems to be no upper limit to this accumulation and certainly no evidence whatever of feedback inhibition on nodular activity by the products of fixation.

9.3 NODULATION, DINITROGEN FIXATION, AND THE SYMBIOTIC CYCLE

9.3.1 Parameters of Symbiotic Functioning

Virtually every investigation of legume symbiosis requires a qualitative or quantitative diagnosis to be made of the relative success or failure of an association in forming nodules effective in providing the host legume with fixed nitrogen. This section lists some of the parameters that are used or measured and searches for those of greatest value in comparative studies of nodule functioning.

Although to some extent genetically determined, and consequently of some use in selection and breeding studies (136), the size, shape, distribution, and life span of nodules appear to be too variable within legumes as a whole to serve as general criteria of symbiotic functioning. Thus, there is no consistent relationship between large size and effectiveness (137, 138), high nodule number and ineffectiveness (139), or early "crown" nodulation and overall effectiveness (140-143). For obvious reasons nodule mass per plant (number times size) is usually better correlated with effectiveness (137, 144). The apparently infallible relationship that exists between the presence of leghemoglobin and ability to fix dinitrogen has suggested that number or mass of red nodules, or the leghemoglobin content of nodules or nodulated roots are even better measures of fixation potential

(145-150). The ultimate refinement in this direction is to measure, by anatomical investigation, the aggregate volume of healthy, pigmented bacterial tissue and to express fixation potential in terms of amount and duration of this quantity (151, 152). Though tedious in execution, this is a particularly instructive measurement since studies have shown that equal masses of nodules may contain quite different amounts of differentiated bacterial tissue (153).

There are many methods available for assessing the amount of dinitrogen fixed by an association (IV.12). Where only relative measurements are required, as for example in comparisons between different bacterial strains in laboratory or glasshouse studies, valuable information can be obtained from visual ratings or from comparisons involving plant and nodule dry weights (154-156), but where absolute measurements of N_2 fixation are required total plant nitrogen, percentage nitrogen in dry weight, and ratio of soluble to insoluble nitrogen will obviously provide more useful types of information. If no combined nitrogen has been available, dinitrogen fixation is assessed simply as total plant nitrogen minus that furnished by cotyledons, while if a known amount of nitrogen has been provided it must be subtracted (157). Other methods must be employed if the association has developed in a medium supplying an unknown amount of combined nitrogen, e.g., in soil. Here, the ^{15}N dilution technique is generally valuable (158), as are, if carefully calibrated, acetylene reduction assays (122, 159, 160). If a suitably virulent ineffective strain is available, a series of plants nodulated with it can be used to assess uptake of mineral nitrogen from the soil. A measure of dinitrogen fixation is then obtained by subtracting the nitrogen value obtained from this series from that accumulated by a comparable series of effectively nodulated plants (161). A somewhat similar comparative method, applicable so far only to Glycine max, utilizes as a control non-nodulating isolines of the host (162). These comparative methods assume that effectively nodulated plants and non-fixing controls perform equally well in scavenging combined nitrogen from the soil. This is a debatable point, and it must also be remembered that none of the techniques mentioned test whether any fixed nitrogen has been excreted or lost to the root surroundings during the course of the experiment.

From measurements of nodulation and dinitrogen fixation several forms of expression of symbiotic performance may be derived, each of potential value in comparisons between different associations and growth conditions. Nodule fixation efficiency per se, measured as N_2 fixed in unit time, may be expressed on the basis of nodule fresh weight (42), nodule dry weight (39, 163), aggregate bacterial tissue volume (153), or nodule hematin (145). However, quantities such as percent of nitrogen transferred from nodule to host (39), nodule:plant fresh weight ratio (42), relative nitrogen assimilation rate (expressed as mg plant N/mg nodule N·day) (164), and other parameters derived from growth curves for symbiotic fixation (159) are probably more useful when relating nodule activity to the growth of the host plant.

9.3.2 Symbiotic Cycle of the Annual Legume

It is of obvious advantage to a seedling developing in a nitrogen-deficient soil if nodules form promptly and commence to fix nitrogen before seed reserves become depleted. In small-seeded species (e.g., Trifolium spp., Medicago spp.) the cotyledonary reserves support the development of only a few nodules and a relatively small fraction of the total root system, but in larger-seeded species (e.g., Pisum spp., Vicia spp., Glycine spp.), much larger proportions of the root and nodule complement may be established before the seedling becomes fully autotrophic for carbon and nitrogen.

As soon as hemoglobin can be detected in the first nodules, and some time before nitrogen analyses provide reliable proof of fixation, it may be possible to detect benefit from effective symbiosis as a regreening of cotyledons and first leaves (156), development of red color in leaves of anthocyanin-producing species (165), and appearance in xylem sap of amide-nitrogen released from nodules (36, 44).

The growth curve for symbiotic dinitrogen fixation in annual legumes is essentially sigmoidal (39, 40, 159, 160, 166, 167). The first stage, equivalent to the "lag phase" of other growth processes, witnesses a large-scale retention of fixed nitrogen in the root and nodules, and, accordingly, a rather slow rate of shoot growth. Then the association enters an exponential phase of fixation, during which high rates of transfer from nodule to host are recorded, and the rate of growth in shoots becomes progressively faster than that of root and nodules. Finally a phase ensues exhibiting a progressive retardation of fixation rate and widescale degenerative changes in the nodule population. This may coincide with fruiting of the host plant.

Bearing in mind its exponential character, some useful parameters have been suggested for comparisons between different symbiotic cycles (159): (a) <u>initiation time</u> — the time after sowing when fixation first commences, (b) <u>doubling time</u> — the time taken during the exponential phase to double the amount of fixed nitrogen, and (c) <u>termination time</u> — the time of completion of the exponential phase of fixation. These three quantities vary greatly with species of legume (159), but appear to be sufficiently constant to enable acetylene reduction assays of performance by plants to predict a cultivar's yield or fertilizer requirements.

Despite apparent uniformity in respect of the general form of growth curves of symbiosis, species can differ widely in patterns and history of nodulation. In some associations, e.g., <u>Lupinus luteus</u> (168), winter-sown <u>Glycine max</u> (169), the top half of the primary root can carry the complete nodule complement and these nodules may persist in active state until the end of the growth cycle. In others, e.g., <u>Pisum</u> spp. (39), nodule initiation continues until all parts of the root have expanded, and rates of nodule turnover are such that only a fraction of the nodule population remains by the time of fruiting. In winter annuals, e.g., <u>Vicia sativa</u>, <u>V</u>. angustifolia, <u>Trifolium dubium</u> (134), an interrupted pattern of nodulation is observed, one set of nodules formed in autumn contributing overwintering reserves of nitrogen, another set forming after growth has commenced the following spring. Whatever the pattern of nodulation, growth of existing nodules, initiation of new nodules, and elimination of senescent nodules are usually so adjusted that nodule mass per root becomes a highly predictable quantity for an association at a particular stage in its development.

In all species examined so far, the maximum in nodule mass per plant is reached well before the corresponding maxima in nitrogen and dry weight, implying that there must be a considerable increase in fixation efficiency as the life cycle progresses. This is largely a reflection of increases in nodule size with large nodules usually possessing higher proportions of bacterial tissue than small ones. But the dramatic increase in fixation activity that is often observable at the onset of reproduction can also involve a greatly increased efficiency in output per unit volume of bacterial tissue (39). Experiments studying the effects of pod removal in Glycine max (120) indicate that this high level of nodular activity at reproduction may be prompted by the sink activity of the fruits. It is important that studies be made of the physiological characteristics of this "climacteric" in nodule performance, and of the degenerative changes in nodules which invariably follow it. This is particularly the case for legumes such as Glycine max, in which 75-90% of the yield of fixed dinitrogen in certain cultivars comes from nodular activity taking place after the time of flower initiation (170).

9.3.3 Symbiosis in Perennial Legumes

The large size, long life, and deep-rooting habit of most perennial legumes make them somewhat difficult subjects for nodulation studies, so that much of the information concerning them is of a somewhat fragmentary nature. However, this is not true for certain perennial pasture legumes. In the tropical species Centrosema pubescens (171) and the temperate species Trifolium repens (172) studies using soil cores have shown that nodule mass per root, and presumed fixation activity, follow a well-defined seasonal cycle, synchronized closely with the growth and net assimilation rate of the host. In Centrosema the seasonal decline in nodulation coincides with a time of decreasing net assimilation rate but occurs some 30-60 days before flowering. Just as with annual species a progressive elimination of smaller nodules occurs through the season in Centrosema, but nodule losses from roots are related more to checks in vegetative growth than to flowering and fruiting. The same holds true for three species of another tropical genus, Desmodium (173, 174), and here seasonal peaks in nodulation occur out of phase with flowering, there being no catastrophic loss of nodules from roots as plants commence to mature their fruits. In Desmodium uncinatum (173) the decline in nodule weight and number prior to peak vegetative weight coincides with senescence of the lower leaves, but some nodules survive fruiting to overwinter on the leafless plants.

Overwintering of nodules on herbaceous perennials in temperate climates appears to be quite common (134), even in mountain regions where plants are buried in snow during winter (175). Resuming growth and activity for a few weeks the following spring, these nodules may benefit host growth at a most critical time just before the new season's set of nodules become functional.

Truly perennial nodules, structurally (see III.8.3.8) and physiologically adapted to persist for several seasons of activity, have been described for several woody species, including Sophora, Robinia (176), Wistaria (177, 178), Sesbania (179), Ulex, and Sarothanmus (180). In Lupinus both herbaceous (181) and woody species (Plate 9.1) bear perennial nodules, the "mushroom type" organization of growth by lateral meristems being particularly conducive to forming nodules of great weight and girth — up to 50 g/nodule (possibly a record for a legume) in the case of Lupinus arboreus (Plate 9.1). Perennial nodules showing cylindrical construction and an apical pattern of growth are generally of much smaller dimensions, tending to increase their mass by repeated forking of the meristem (179).

Unless the history of individually tagged nodules is known, the precise age of a perennial nodule is very difficult to determine, since there is no distinct relationship between absolute age and size, lobe number, or extent of branching (182). A life span of 6 years has been claimed for nodules of Wistaria (177), and judging from the age of the root on which they are borne, nodules on Ulex europaeus and Lupinus arboreus may well remain active for almost that length of time.

Perennial nodules must all experience special problems relating to gas exchange, water supply, and the retention of adequate transport connections with the parent root, and since nonactive, supporting, and protective tissues accumulate as nodule mass increases, it is unlikely that a "giant" perennial nodule is as active per unit of mass as a smaller, ephemeral counterpart. Nevertheless, the bearing of perennial nodules may still be of adaptive significance since symbiotic potential will be retained on the main

Physiological Factors

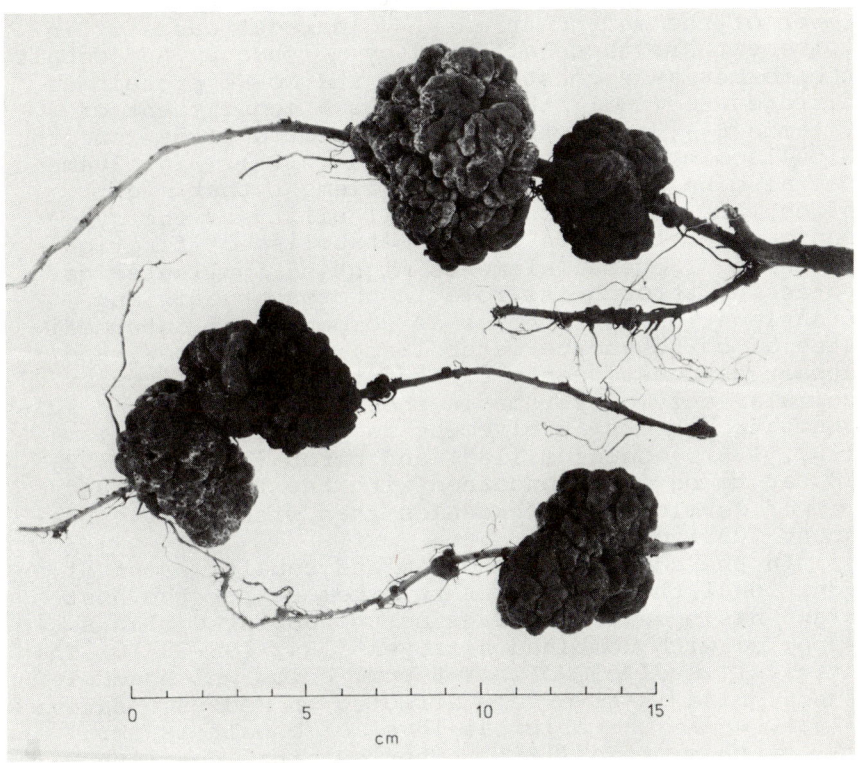

Plate 9.1. Perennial nodules from a 7-year-old tree of <u>Lupinus arboreus</u>. The nodules, weighing 20-50 g each, were healthy and still apparently in active growth when excavated.

framework of the root and, thus, permit fixation to continue even when conditions are unfavorable for growth of new roots and nodules.

9.4 PHYSIOLOGICAL FACTORS INFLUENCING DINITROGEN FIXATION

9.4.1 Essential Chemical Elements

Every essential plant nutrient must be classed as being of potential significance to symbiosis in view of its role in host plant functioning, yet specific functions in nodulation and dinitrogen fixation have been ascribed to only a few micronutrients and even fewer of the macronutrients. Clear-cut cases of involvement are those afforded by molybdenum and cobalt, the former as a constituent of the Mo-Fe protein of nitrogenase (183), the latter as a constituent of vitamin B_{12}, a growth factor suspected to be concerned with leghemoglobin synthesis (184, 185). Involvement can also be suspected for any element that might affect the production and availability of energy sources for fixation or the metabolism of fixation products. Several elements might be implicated here, potassium through its role in photosynthesis and translocation, phosphorus as component of intermediates of carbohydrate metabolism, iron, copper, and cobalt as constituents of essential metabolites or cellular enzyme systems in the nodule (186), and sulphur through its involvement in protein metabolism (187, 188). Calcium (189) and boron (133, 190-192) appear to be more concerned with the formation and proper development of nodules than with their subsequent functioning.

In the case of molybdenum and cobalt, proof of an additional though smaller requirement for the host plant has been obtained using uninoculated plants supplied with combined nitrogen (184, 193-195). The critical level of Mo is 4-8 ppm, below which nodule functioning is adversely affected in Medicago sativa (163), while the critical level of cobalt is about $1/300$ of that of Mo (196). When external supplies become deficient, molybdenum and cobalt are concentrated in nodules as opposed to host tissues (192, 197).

Compensatory interactions for number, size and efficiency of nodules operate characteristically in specific element deficiencies. In the case of calcium, where the major influence is on nodule initiation, a concentration range of 300-700 µM Ca in the external medium scarcely affects plant yield or nodule mass, but nodule numbers are decreased threefold by decreasing the calcium concentration from the top to the bottom of this range (189). Conversely, molybdenum and cobalt deficiency lead to abnormally high nodule numbers, but small size and low levels of leghemoglobin appear to impede functioning (198, 199, 184, 185). More extensive discussion of the effects of mineral elements on nodulation and dinitrogen fixation are to be found in IV.10.

Large differences between legumes may exist in the size of their requirements for a specific nutrient. For example, Medicago spp. appears to require more molybdenum than Trifolium spp. (198, 200), and Trifolium subterraneum more boron than Trifolium repens (201), but it is not clear whether these differences reflect demands in plant tissues or efficiency in absorption from the medium. In the case of calcium, the poor tolerance to deficiency of European legumes has been attributed to sluggish uptake from the soil (202), and the same applies for iron absorption by cultivars of Glycine max differing in susceptibility to chlorosis (203). But in Phaseolus vulgaris ability to grow at abnormally low tissue levels of potassium is said to confer advantage on certain varieties when supplies of this element are limiting (203).

The reaction of the medium can affect symbiosis both indirectly or directly. The indirect effect is essentially one on element availability; molybdenum and phosphate, for example, become more available (204) whereas boron and copper are less available (205) as the pH is raised. Acid soils often contain high soluble levels of manganese, iron, and aluminum, which is particularly hostile to symbiosis when combined with low levels of calcium (206, 207, 208). Direct effects of pH on nodulation are dealt with in IV.10.

9.4.2 Light, Grazing, Defoliation, Shading, and Darkness

The justification for grouping these factors together derives from the common influences that they may have on the supply of assimilates to root nodules.

Conditions of light intensity optimal for nodulation and symbiosis have been defined for several legume species (209, 210) and have been shown to vary with temperature and level of combined nitrogen (211). Since symbiotic injury arising from low-intensity lighting can be alleviated by supplying sugar or extra carbon dioxide, and since added combined nitrogen can offset the effect of too-high light intensity, the overall effect of light has been interpreted as operating through the carbohydrate:nitrogen balance of the plant (209, 212), but little progress has been made towards expression of these effects in more concrete terms. The "nitrogen hunger" condition commonly induced by excessive solar radiation still awaits evaluation in physiological terms, despite the demonstration that indirect effects due to elevated root temperatures may be involved (213).

In some annual species, e.g., Pisum sativum (210), Medicago tribuloides, and Vicia atropurpurea (214), light exercises far less effect on nodule initiation than on nodule growth and functioning, probably because a large proportion of the nodule complement is initiated while cotyledons are still furnishing carbohydrate (42). Certain legumes, e.g., Pisum sativum, will nodulate when grown in total darkness (215), although nodules never reach a large size or develop leghemoglobin.

The recorded effect of day length on nodulation of Glycine max (216) and Vigna sinensis (217) is that long days promote development of more and larger nodules than do short days, and since this effect occurs irrespective of photoperiodic effects on flowering of the host cultivar, a direct photosynthetic effect on symbiosis is suggested.

Nonphotosynthetic effects of light on nodulation are also in evidence. In Pisum and Vicia, exposure of root or shoot to far red light (730 nm) severely restricts nodule formation, and since red light (660 nm) reverses this inhibition, the phytochrome system has been implicated in the response (133). In isolated roots of Phaseolus vulgaris a more complex effect of white light is observed, exposure before inoculation stimulating nodulation, illumination after inoculation suppressing nodule development (218).

Prolonged darkness can have dramatic effects on already nodulated plants, presumably by restricting carbohydrate supply to the nodules. In Pisum sativum, 24 hours of darkness are sufficient to destroy leghemoglobin (53), reduce fixation almost to zero (219), and break down bacteroids (220). But, in other species, e.g., Glycine max, several days of darkness are required before leghemoglobin is destroyed (212) and reserve carbohydrates become depleted in the nodules (20).

The literature suggests that the effects of defoliation, grazing, and shading on symbiosis are similar, though not necessarily identical. The principal short-term effect of all these factors is a reduction in $N_2(C_2H_2)$-fixing ability of nodules (120), while, in the long term, nodule shedding may be induced and followed, perhaps, by the initiation of new nodules. Damage varies with the species and with the extent and frequency of removal or shading of photosynthetic surfaces (221, 222, 171, 223, 174). Legumes with a determinate pattern of root growth are usually more severely affected since replacement of nodules is not readily accomplished, particularly in older plants. In perennial species with large reserves of carbohydrate, nodule shedding may not be noticeable until several weeks after defoliation, and even then nodule mass per root may not become acutely depressed as initiation of a replacement set of nodules may have taken place (223, 174). Using legumes with roots grown in glass-sided boxes it has been observed that shading and defoliation have the effect of accelerating turnover of root and nodule tissues (72). A particularly high rate of turnover occurs in Trifolium repens, a stoloniferous species resistant to grazing.

9.4.3 Water and Gases in the Rooting Medium

Legumes are generally intolerant of either very arid or waterlogged environments, and there are good grounds for suggesting that the symbiotic organs are ultrasensitive in this respect. Legumes adapted to very dry soils develop nodules only in the deeper, moister strata and infection by Rhizobium may be confined to favorable seasons (224). In legumes of aquatic or marshy habitats nodulation occurs close to the water surface (225). In nonadapted species drought can induce nodule shedding, even after short periods of stress (226, 227, 217), while waterlogging can seriously restrict nodule growth and functioning (228, 53).

A recent exhaustive study in Glycine max has shown that water stresses applied atmospherically or osmotically to detached nodules or nodulated roots promote losses of $N_2(C_2H_2)$-fixing ability and that these losses are, within limits, proportional to the degree of desiccation (229). Effects on fixation are found to be reversible, provided that water loss from the nodule does not amount to more than 20% of its maximum fresh weight. Above this level, fixation is totally abolished, and irreversible structural damage is manifest as a disruption of plasmodesmatal connections and a collapse of the highly vacuolate cells of the nodule cortex (230). The implications of this are obvious in regard to the transport of assimilates to and from the bacterial tissues.

Nodulated plants of Vicia faba exhibit a high degree of correlation between $N_2(C_2H_2)$-fixing activity and soil water content. Activity is maximal at field capacity and is acutely suppressed when the soil dries out and flagging of lower leaves commences (50). Since these lower leaves are of special importance in nourishing the nodules (see III.9.2) assimilate starvation might be a direct cause of lowered fixation activity, but, there is also the possibility of effects of water shortage on the nodule's export mechanism (38). Aside from these effects, drought may generate high levels of salts around the root, thereby inhibiting fixation (231).

The situation in waterlogged soils comprises effects due to excess water per se and those due to buildup or depletion of dissolved and free gases. However, the single factor generally regarded as being especially detrimental to root development and functioning is low availability of oxygen (232, 233, 116). Tolerance limits for nodulation in Trifolium subterraneum have been given as a diffusion rate of 8.2×10^{-2} g $O_2/cm^2 \cdot m$ at which nodulation is confined to the more aerated pockets of the soil profile. A soil with a diffusion rate 3 times this level permitted unrestricted nodulation of this species (234). Unfortunately measurements of oxygen gradients on a micro scale still present technical difficulties, yet information at this level is essential for understanding effects in the immediate environs of the functioning nodule. Even very thin films of excess water on the nodule surface can lower C_2H_2-reducing activity of detached nodules, presumably by restricting oxygen diffusion (235, 236): Such films must form and disperse frequently at the soil-nodule interface, yet we are totally ignorant of their long-term effects on nodule functioning.

The deleterious effects of low pO_2 on symbiosis have been studied largely through use of water cultures. In Pisum sativum and Trifolium pratense fixation activity of whole nodulated plants is maximal when the culture solution is flushed with 20-21% O_2 (237, 238). Fixation is greatly suppressed by either sub- or supra-optimal levels of oxygen, and in these conditions assimilation of combined nitrogen is considerably less affected (237). In _Trifolium pratense_ most nodules form at 5% O_2, but these remain small and fix little N_2 (238). Tolerance of nodule initiation processes to oxygen deficiency may be of adaptive value in enabling a species to form nodules early in wet soils in spring. However, experiments on detached mature nodules of _Glycine max_ show that the efficiency of carbohydrate consumption in fixation is highest at near-atmospheric oxygen tensions and decreases dramatically with lowering of oxygen tension. This is regarded as being of major significance to a plant's economy when growing under waterlogged conditions (111).

It must not be forgotten that oxygen deficiency may derange growth and activity of parent roots as well as nodules. Flood-sensitive species such as _Pisum sativum_ show biochemical disturbance of root metabolism, through use of abnormal respiratory pathways (239), and in _Arachis hypogea_ leakage of sugars from roots is an important element in the syndrome of effects caused by anaerobiosis (240). Waterlogged roots of _Pisum_ and _Lupinus_ are apparently less active than normally aerated roots in biosynthesis and export of cytokinins and, as a result, the shoot of the waterlogged plant quickly suffers (241). Effects of this nature will have obvious repercussions on symbiotic activity.

Oxygen deficiency is not the only damaging factor in the stagnant root environment. Buildup of carbon dioxide, even to only 1% in the gaseous phase, has been shown to arrest growth of roots of _Pisum sativum_ (21), and 3% CO_2 or 0.4 ppm C_2H_4 are known to be sufficient to restrict nodulation of isolated roots of _Phaseolus vulgaris_ (242). The relevance of such influences under field conditions must be assessed.

9.4.4 Temperature

The extensive literature on the influence of temperature on symbiosis allows not only an overall picture to be obtained of response patterns in legumes as a whole, but also presents several accounts in depth of the effects on single species of variables such as host cultivar, _Rhizobium_ strain, and root/shoot temperature regime. Only general principles relating to these effects will be discussed here.

The optimum constant temperature promoting maximum yield of fixed nitrogen appears to be highly specific for a given association and usually lies within the range 24-30°C and close to or coinciding with the optimum for host plant growth (243, 244, 214). This optimum is clearly a compromise since it is usually below the optimum for rate of infection, final nodule number, and fixation efficiency per unit mass of nodules, but above the optimum for production of nodule tissue (245, 246, 153). In Pisum sativum var. "Iran," the paradoxical situation is found in which roots are resistant to most Rhizobium strains at 20°C, yet this temperature is optimal for N_2 fixation of already-nodulated plants. Nodulation of this cultivar proceeds normally at 26°C (247). Other cultivars of P. sativum appear to nodulate well over a wide range of temperatures (248, 249).

By comparing growth and nitrogen accumulation in nodulated plants with that possible in plants relying on an unrestricted supply of combined nitrogen, it is possible to measure the inadequacy of symbiosis under different temperature conditions. In temperate legumes the extent of such inadequacy is greatest at supraoptimal temperatures (243, 250), while in tropical species it is greatest at the lower end of the temperature range (250-252). Inordinately low levels of nitrogen in dry matter suggest that effects specific to N_2 fixation occur under these stress conditions.

At the lower end of the temperature range nodule initiation may be much restricted or even totally suppressed. The critical night temperature for nodulation of tropical species appears to lie near 18°C (250), and for temperate legumes is much lower, e.g., 7°C for Trifolium subterraneum (153), and 2°-6°C for cold-tolerant species such as Pisum spp. and Vicia faba (248, 253). Temperature transfer studies have shown that the root hair invasion process, occuring within the first few days after inoculation, has the most exacting temperature requirements (254, 247).

There is still uncertainty as to which specific physiological processes become limiting under temperature stress. In Trifolium subterraneum elevated shoot temperatures are less damaging to symbiosis than are comparably high temperatures in the root surroundings, so that root-located effects seem to be the more critical (213). A thermoperiodic effect is ruled out for this species since fixation is suppressed to a similar extent when elevated temperatures coincide with the photoperiod or dark period (250). In Pisum sativum, however, a specific inhibitory effect of high night temperatures is definitely involved (249). Elevated root temperatures in Trifolium subterraneum lead to rapid degeneration of bacterial tissues (250), while low-temperature stress slows differentiation of bacterial tissue and the associated synthesis of leghemoglobin (153). In either case, low levels of nodule efficiency are recorded. There is evidence of retention of nitrogen in root and nodule at low root temperatures (255), although this might be an effect rather than a cause of nodule malfunctioning.

It is also conceivable that N_2 fixation might be affected in a purely direct manner through temperature limiting the rate of nitrogenase activity. However, using plants raised under optimum temperature conditions, and transferring these just before C_2H_2 assay to a range of constantly maintained temperatures, it has been found that N_2-fixing activity exhibits a very wide temperature tolerance, and shows a broadly based optimum in activity spanning some 15-20°C of the temperature range (236, 250, 170). Cold lability is manifest, although usually only at near-freezing temperatures (256). In temperate species the lower limit for N_2 fixation is about 2°C, in tropical species, 10°C. In one tropical species, Vigna sinensis, the unusually high optimum of 40°C is recorded for N_2-fixing activity (236). It remains to be seen whether these apparently large differences between species really reflect different properties of their respective nitrogenases or differences in non-nitrogenase components of the fixing system. Whatever the case, the highly adaptable performance of nitrogenase in vivo is of obvious advantage to legumes experiencing wide daily and seasonal fluctuations in temperature, although it must be stated that activity in short-term assay need not necessarily denote comparable performance under prolonged stress in the field.

9.4.5 Combined Nitrogen

The inhibitory effect of high levels of combined nitrogen on nodulation has been known since very early studies on legumes (257-259), but progress has been made only recently towards elucidating the mechanisms underlying the response.

Observations on young seedlings suggest that root hair curling (260) and the subsequent formation of infection threads (261) are more susceptible to injury than are the later stages of nodulation and that nitrate and nitrite are more potent in this respect than ammoniacal forms of nitrogen (262). Nitrate applied to isolated root cultures via the rhizosphere inhibits nodulation more effectively than if fed to the base of the hypocotyl (263, 84), suggesting that external influences may be more important, but since nodulation can be stimulated or inhibited by foliar sprays of nitrogen (264, 265, 247), internal influences can obviously condition the response. Studies using divided root cultures have provided somewhat contradictory results suggesting in some cases a highly localized effect of nitrate (266, 158), and in another a more widespread influence (267). It has been proposed that nitrate acts externally through the catalytic action of its reduction product nitrite on destruction of indoleacetic acid, the presumed agent of root hair curling (268, III.8.2.5.2). Although this hypothesis is supported by experimental evidence (269, 261), it is still difficult to picture how events operate at the root hair surface, particularly in view of recent doubts regarding the specificity of indoleacetic acid in root hair curling (270).

After some nodules have formed and commenced fixation, it is difficult to distinguish inhibitory effects due to their presence from effects due to the continued presence of the added nitrogen. Nevertheless, it is agreed that effects on nodulation are, within limits, proportional to the nitrogen supply and to the frequency of its application (137, 271, 158); that certain species (272, 265), specific varieties of a species (273), and even specific bacterial partnerships with a host (157) are more tolerant of combined nitrogen than others; and that derangement of symbiosis is least likely to occur under conditions optimal for host plant growth (104, 157, 211).

The nodulation of several species has been found to benefit from small supplements of fertilizer nitrogen, especially if given at sowing or as the first nodules are forming (31, 253, 273-276). Even when early nodulation has been suppressed by combined nitrogen, the stimulus given to root growth may benefit later nodulation of minor roots (211).

It is possible to measure precisely how fixation has been affected by treatments of combined nitrogen by using the nitrogen balance technique (157, 265), the ^{15}N dilution technique (158, 277-280), acetylene reduction assays (281, 259), or a combination of these techniques (122). As a general rule, returns of fixed nitrogen become progressively diminished as the level of fertilizer nitrogen is raised, but in certain situations low levels of added nitrogen can actually benefit fixation (157). Consequently considerable finesse is required in both time and rate of nitrogen fertilization.

Physiological behavior of nodule-fed plants may be consistently different from that of plants heterotrophic for nitrogen. Generally, legumes grown on combined nitrogen yield better, have larger roots (104, 106), larger leaf area (105), and a higher proportion of soluble nitrogen (31, 106), but they may be slower to mature and produce seed of inferior quality than comparable symbiotic plants (104). Lower leaf areas, compounded with unusually high rates of dark respiration, have been held responsible for the poorer growth of Trifolium subterraneum plants relying on dinitrogen (105).

Studies of legumes deriving nitrogen partly from the medium and partly from nodules have provided clues as to the forms of physiological disturbance detrimental to symbiosis. With increasing level of nitrogen in the rooting medium of nodulated seedlings of Pisum and Trifolium, the competition between nodules and rootlets for photosynthate becomes tipped in favor of the rootlets (88). In Pisum arvense levels of combined nitrogen that severely suppress symbiotic fixation reduce downward translocation of photosynthetically fixed carbon, lower the efficiency of fixation per unit mass of nodules, and generate a pattern of nitrate reduction in which the greater share of assimilation takes place in the shoot (106, 122). Collectively, these effects suggest that consumption of electron donors and carbon skeletons in shoot nitrate reduction leads to shortage of assimilates for underground parts of the plant, and that under these conditions it is the nodules, not the roots, that are most severely affected.

9.5 EPILOGUE

Although it may be depressing to find at this point in time so many serious gaps in our understanding of the functional biology of nodulated legumes, there is no doubt that many notable advances have been made in this field within the past decade, and that future prospects in certain areas are extremely promising. Particularly great opportunities have now become possible with the advent of the acetylene reduction assay, dividends already being evident from several quarters, especially in the studies of the daily changes in activity of nitrogenase in living nodules, in examinations of symbiotic response to stress conditions, and in the monitoring of symbiotic activity during the life cycle of legumes under agricultural conditions. It has been seen that much is now known concerning the interchange of carbohydrate and nitrogen-containing materials between nodule and host, and, though still largely of a descriptive nature, this information should prove especially useful when searching for the systems that act as the natural regulators of symbiosis. Another area of potential interest is that relating to symbiotic performance under physiological or environmental stress, for much of practical and academic interest to the legume breeder can be gained from knowledge of the genetic and functional factors that affect the successful matching of host and bacterium. Finally, it is particularly gratifying to note the increasing number of contributions dealing with legumes native to tropical regions, particularly those of perennial habit. For too long our knowledge has rested on studies of a mere handful of agriculturally important species, mainly of temperate, European origin, and the justification for extending investigations to other new species is all too obvious every time that their functional biology turns out to be somewhat different from that of the more familiar "laboratory" examples.

9.6 REFERENCES

1. Bergersen, F. J., Aust. J. Biol. Sci., 18, 1 (1965).
2. Kennedy, I. R., Biochim. Biophys. Acta, 130, 285 (1966).
3. Linko, P., O. Holm-Hansen, J. A. Bassham, and M. Calvin, J. Exp. Bot., 8, 147 (1957).

References

4. Stewart, W. D. P., *Nitrogen Fixation in Plants*, Athlone Press, London, 1966.
5. Leaf, G., I. C. Gardner, and G. Bond, *J. Exp. Bot.*, **9**, 320 (1958).
6. Leaf, G., *Adv. Sci.*, **60**, 386 (1959).
7. McKee, H. S., *Nitrogen Metabolism in Plants*, Clarendon Press, Oxford, 1962.
8. Davies, D. D., J. Giovanelli, and T. Ap Rees, *Plant Biochemistry*, Blackwell, Oxford, 1964.
9. Zelitch, I., P. W. Wilson, and R. H. Burris, *Plant Physiol.*, **27**, 1 (1952).
10. Aprison, M. H., W. E. Magee, and R. H. Burris, *J. Biol. Chem.*, **208**, 29 (1954).
11. Fottrell, P. F., *Sci. Prog. Oxf.*, **56**, 541 (1968).
12. Kennedy, I. R., *Biochim. Biophys. Acta*, **130**, 295 (1966).
13. Leaf, G., I. C. Gardner, and G. Bond, *Biochem. J.*, **72**, 662 (1959).
14. Kennedy, J. R., C. A. Parker, and D. K. Kidby, *Biochim. Biophys. Acta*, **130**, 517 (1966).
15. Grimes, H., and P. F. Fottrell, *Nature*, **212**, 295 (1966).
16. Jordan, D. C., *Bacteriol. Rev.*, **26**, 119 (1962).
17. Grimes, H., and S. Turner, *Plant Soil*, **35**, 269 (1971).
18. Ryan, E., F. Bodley, and P. F. Fottrell, *Phytochemistry*, **11**, 957 (1972).
19. Pate, J. S., in *Nitrogen-15 in Soil-Plant Studies*, International Atomic Energy Agency, Vienna, pp. 165-187, 1971.
20. Wong, P. P., and H. J. Evans, *Plant Physiol.*, **47**, 750 (1971).
21. Stolwijk, J. A. J., and K. V. Thimann, *Plant Physiol.*, **32**, 513 (1957).
22. Jackson, W. A., and N. T. Coleman, *Plant Soil*, **11**, 1 (1959).
23. Mulder, E. G., and W. L. van Veen, *Plant Soil*, **13**, 265 (1960).
24. Berquist, N. O., *Bot. Not.*, **117**, 249 (1964).
25. Loomis, W. D., *Plant Physiol.*, **34**, 541 (1959).
26. Streeter, J. G., *Plant Physiol.*, **46** (Supl.), 44 (1970).
27. Hunt, E., *Am. J. Bot.*, **38**, 452 (1951).
28. Sen, A. P., and D. P. Burma, *Bot. Gaz.*, **115**, 185 (1954).
29. Miettinen, J. K., *Ann. Acad. Sci. Fenn. AII*, **60**, 520 (1955).

30. Butler, G. W., and N. O. Bathurst, Aust. J. Biol. Sci., 11, 529 (1958).
31. Copeland, R., and J. S. Pate, Occasional Symposium, British Grassland Society, 6, 71 (1970).
32. Virtanen, A. I., and J. K. Miettinen, Biochim. Biophys. Acta, 12, 181 (1953).
33. Bollard, E. G., Aust. J. Biol. Sci., 10, 292 (1957).
34. Fowden, L., Ann. Bot., 18, 417 (1954).
35. Pate, J. S., Plant Soil, 17, 333 (1962).
36. Wieringa, K. T., and J. A. Bakhuis, Plant Soil, 8, 254 (1957).
37. Virtanen, A. I., and J. K. Miettenen, in Plant Physiology, Academic Press, London, Vol. 3, pp. 565-668, 1963.
38. Pate J. S., B. E. S. Gunning, and L. G. Briarty, Planta (Ber.), 85, 11 (1969).
39. Bond, G., Ann. Bot., 50, 559 (1936).
40. Wilson, P. W., and W. W. Umbreit, Zent. Bakt II, 96, 402 (1937).
41. Tsujimura, K., J. Sci. Soil Manure (Japan), 21, 181 (1951).
42. Pate, J. S., Aust. J. Biol. Sci., 11, 366 (1958).
43. Oghoghorie, G. C. O., and J. S. Pate, Planta (Ber.), 104, 35 (1972).
44. Pate, J. S., and W. Wallace, Ann. Bot., 28, 83 (1964).
45. Wallace, W., and J. S. Pate, Ann. Bot., 31, 213 (1967).
46. Pate, J. S., and B. E. S. Gunning, Ann. Rev. Plant Physiol., 23, 173 (1972).
47. Minchin, F. R., and J. S. Pate, J. Exp. Bot., 24, 259 (1973).
48. Frazer, H. L., Proc. Roy. Soc. (Edinb.), Ser. B, 61, 328 (1942).
49. Weichsel, G., Flora, 151, 535 (1961).
50. Sprent, J. I., New Phytol., 71, 603 (1972).
51. Virtanen, A. I., and T. Laine, Biochem. J., 33, 412 (1939).
52. Virtanen, A. I., and H. Linkola, Nature, 158, 515 (1946).
53. Virtanen, A. I., Biol. Rev., 22, 239 (1947).
54. Ludwig, C. A., and F. E. Allison, Bot. Gaz., 98, 680 (1937).
55. Bond, G., and J. Boyes, Ann. Bot., 3, 901 (1939).
56. Myers, H. G., J. Am. Soc. Agron., 37, 81 (1945).

References

57. Wilson, P. W., and J. C. Burton, J. Agr. Sci., 28, 307 (1938).
58. Wyss, O., and P. W. Wilson, Soil Sci., 52, 15 (1941).
59. Rovira, A. D., Biol. Rev., 35, 35 (1969).
60. Rovira, A. D., Plant Soil, 7, 178 (1956).
61. Martin, J. K., Aust. J. Biol. Sci., 24, 1143 (1971).
62. Ivanov, V. P., G. A. Yacobsen, and B. S. Fomenko, Fiziol. Rast., 11, 630 (1964).
63. Katznelson, J., J. W. Rouatt, and T. M. B. Payne, Nature, 174, 1110 (1954).
64. Boulter, P., J. J. Jeremy, and M. Wilding, Plant Soil, 24, 121 (1966).
65. Butcher, D., and H. E. Street, Bot. Rev., 30, 513 (1964).
66. McDougall, B. M., and A. D. Rovira, New Phytol., 69, 999 (1970).
67. Pearson, R., and D. Parkinson, Plant Soil, 13, 391 (1961).
68. Ayers, W. A., and R. H. Thornton, Plant Soil, 28, 193 (1968).
69. Clayton, M. F., and J. A. Lamberton, Aust. J. Biol. Sci., 17, 855 (1964).
70. Martin, J. K., Aust. J. Biol. Sci., 24, 1131 (1971).
71. Walker, T. W., H. D. Orchiston, and A. F. R. Adams, J. Brit. Grassl. Soc., 9, 249 (1954).
72. Butler, G. W., R. M. Greenwood, and K. Soper, N. Z. J. Agr. Res., 2, 415 (1959).
73. Dilz, K., and E. G. Mulder, Plant Soil, 16, 229 (1962).
74. Williams, W., Occasional Symposium, British Grassland Society, 6 (Presidential Address), 1970.
75. Thornton, H. G., Proc. Roy. Soc., Ser. B, 104, 481 (1929).
76. Tanner, J. W., and I. C. Anderson, Nature, 198, 303 (1963).
77. Schaffer, A. G., and M. Alexander, Plant Physiol., 42, 557 (1967).
78. Richmond, T. E., Bot. Gaz., 82, 438 (1926).
79. Wagenbreth, D., Flora, 144, 84 (1956).
80. Rudin, B. A., H. G. Popapov, and V. F. Germanova, C. R. Acad. Sci., USSR, 88, 1063 (1953).
81. Raggio, M., and N. Raggio, Physiol. Plant., 9, 466 (1956).
82. Raggio, N., M. Raggio, and R. H. Burris, Science, 129, 211 (1959).

83. Bunting, A. H., and J. Horrocks, Ann. Bot., 28, 229 (1964).
84. Cartwright, P. M., Ann. Bot., 31, 309 (1967).
85. Roponen, I. E., Suomen Kemistilehti, B, 43, 54 (1970).
86. Bach, M. K., W. E. Magee, and R. H. Burris, Plant Physiol., 33, 118 (1958).
87. Hoshino, M., S. Nishimura, and T. Okubo, Proc. Crop. Sci. Soc. Japan, 33, 130 (1964).
88. Small, J. G. C., and O. A. Leonard, Am. J. Bot., 56, 187 (1969).
89. Pate, J. S., and J. M. Greig, Plant Soil, 21, 163 (1964).
90. Kursanov, A. L., Adv. Bot. Res., 1, 209 (1963).
91. Thrower, S. L., Symp. Soc. Exp. Biol., 21, 483 (1967).
92. Carr, D. J., and J. S. Pate, Symp. Soc. Exp. Biol., 21, 559 (1967).
93. Flinn, A. M., and J. S. Pate, J. Exp. Bot., 21, 71 (1970).
94. Harvey, H. J., Occasional Symposium, British Grassland Soc., 6 (1970).
95. Moustafa, E., M. Boland, and R. M. Greenwood, Plant Soil, 35, 651 (1971).
96. Lauchli, A., Ann. Rev. Plant Physiol., 23, 197 (1972).
97. Hocking, T. J., J. R. Hillman, and M. B. Wilkins, Nature, 235, 124 (1972).
98. Graham, P. J., J. Gen. Microbiol., 30, 245 (1963).
99. West, P. M., Nature, 144, 1050 (1939).
100. Bonner, J., Am. J. Bot., 29, 136 (1942).
101. Schwinghamer, E. A., Aust. J. Biol. Sci., 23, 1187 (1970).
102. Allan, F., Z. Pflanzenernähr, Düng., A, 20, 270 (1931).
103. Gibson, A. H., Aust. J. Biol. Sci., 19, 499 (1966).
104. Lyons, J. C., and E. B. Early, Proc. Soil Sci. Soc. Am., 16, 259 (1952).
105. Bouma, D., Ann. Bot., 34, 1143 (1970).
106. Oghoghorie, C. G. O., Ph. D. Thesis, University of Belfast, 1971.
107. Allison, F. E., C. A. Ludwig, F. W. Minor, and S. R. Hoover, Bot. Gaz., 101, 534 (1940).
108. Bond, G., Ann. Bot., 5, 313 (1941).
109. Virtanen, A. I., and A. Tietäväinen, Suomen Kemistilehti, B, 2, 1 (1953).

References

110. Allison, F. E., C. A. Ludwig, S. R. Hoover, and F. W. Minor, Bot. Gaz., 101, 513 (1940).
111. Bergersen, F. J., Ann. Rev. Plant Physiol., 22, 121 (1971).
112. Tjepkema, J. D., Ph. D. Thesis, University of Michigan, 1971.
113. Aprison, M. H., and R. H. Burris, Science, 115, 264 (1952).
114. Jensen, H. L., Proc. Soc. Appl. Bacteriol., 14, 89 (1951).
115. Burris, R. H., Proc. Roy. Soc., Ser. B., 172, 339 (1969).
116. Becking, J. H., in Nitrogen-15 in Soil-Plant Studies, International Atomic Energy Agency, Vienna, pp. 189-222, 1971.
117. Postgate, J., Plant Soil, Special Vol., 551 (1971).
118. Pate, J. S., in Recent Aspects of Nitrogen Metabolism in Plants, Academic Press, London, pp. 219-240, 1968.
119. Lewis, O. A. M., and J. S. Pate, J. Exp. Bot., 596 (1973).
120. Hardy, R. W. F., R. D. Holsten, E. K. Jackson, and R. C. Burns, Plant Physiol., 43, 1185 (1968).
121. Minchin, F. R., and J. S. Pate (unpublished date).
122. Oghoghorie, C. G. O., and J. S. Pate, Plant Soil, Special Vol., 185 (1971).
123. Bergersen, F. J., Aust. J. Biol. Sci., 23, 1015 (1970).
124. Wheeler, C. T., New Phytol., 70, 487 (1971).
125. Pate, J. S., Aust. J. Biol. Sci., 11, 516 (1958).
126. Dullaart, J., Acta Bot. Neerl., 19, 573 (1970).
127. Seth, A. K., and P. F. Wareing, J. Exp. Bot., 18, 65 (1967).
128. Bowen, M. R., and P. F. Wareing, Planta (Ber.), 99, 120 (1971).
129. Kennedy, I. R., Proc. Aust. Biochem. Soc., 3, 11 (1970).
130. Greig, J. M., J. S. Pate, and W. Wallace, Life Sci., 12, 745 (1962).
131. Forsyth, W. G. C., A. C. Hayward, and J. B. Roberts, Nature, 182, 800 (1958).
132. Schlegel, H. G., Flora, 152, 236 (1962).
133. Lie, T. A., Ph. D. Thesis, University of Wageningen, 1964.
134. Pate, J. S., Aust. J. Biol. Sci., 11, 496 (1958).

135. Roponen, I. E., and A. I. Virtanen, *Physiol. Plant.*, 21, 655 (1968).
136. Nutman, P. S., *Proc. Roy. Soc., Ser. B*, 172, 417 (1969).
137. Nutman, P. S., *Biol. Rev.*, 31, 109 (1956).
138. Holding, A. J., and J. King, *Plant Soil*, 18, 191 (1963).
139. Sloger, C., *Plant Physiol.*, 44, 1666 (1969).
140. Baird, K. J., *Phyton*, 7, 46 (1956).
141. Kamata, E., *Proc. Crop Sci. Soc. Japan*, 27, 245 (1958).
142. Lange, R. T., and C. A. Parker, *Nature*, 186, 178 (1960).
143. Bhaduri, P. N., and R. Sen, *Ind. J. Gen. Plant Breeding*, 28, 287 (1968).
144. Masterson, C. L., and M. T. Sherwood, *Occasional Symposium, Brit. Grassland Society*, 6 (1970).
145. Virtanen, A. I., J. Erkama, and H. Linkola, *Acta Chem. Scand.*, 1, 861 (1947).
146. Smith, J. P., *Biochem. J.*, 44, 585 (1949).
147. Jordan, D. C., and E. H. Garrard, *Can. J. Bot.*, 29, 360 (1951).
148. Ishizawa, S., and H. Toyoda, *Soil Plant Food*, 1, 47 (1955).
149. Graham, P. H., and C. A. Parker, *Aust. J. Sci.*, 23, 231 (1961).
150. Schwinghamer, E. A., H. J. Evans, and M. D. Dawson, *Plant Soil*, 23, 192 (1970).
151. Chen, K. H., and H. G. Thornton, *Proc. Roy. Soc., Ser. B*, 129, 208 (1940).
152. Nutman, P. S., *Aust. J. Agr. Res.*, 18, 381 (1967).
153. Roughly, R. J., *Ann. Bot.*, 34, 631 (1970).
154. Erdman, L. W., and U. M. Means, *Soil Sci.*, 73, 231 (1952).
155. Döbereiner, J., *Nature*, 210, 850 (1966).
156. Vincent, J. M., *A Manual for the Practical Study of the Root-nodule Bacteria, I.B.P. Handbook 15*, Blackwell, Oxford, 1970.
157. Pate, J. S., and P. J. Dart, *Plant Soil*, 15, 329 (1961).
158. Mishustin, E. H., and V. K. Shil'Mkova, *Biological Fixation of Atmospheric Nitrogen* (English Transl.), Macmillan, London, 1971.
159. Hardy, R. W. F., R. C. Burns, R. R. Herbert, R. D. Holsten, and E. K. Jackson, *Plant Soil*, Special Vol., 561 (1971).

160. Weber, D. F., B. E. Caldwell, C. Sloger, and H. G. Vest, Plant Soil, Special Vol., 293 (1971).
161. Virtanen, A. I., and A. Holmberg, Suomen Kemistilehti, B, 31, 98 (1958).
162. Weber, C. R., Agron. J., 58, 46 (1966).
163. Jensen, H. L., Proc. Linn. Soc. N. S. W., 71, 265 (1947).
164. Gibson, A. H., Aust. J. Biol. Sci., 18, 295 (1965).
165. Brockwell, J., J. Aust. Inst. Agr. Sci., 22, 260 (1956).
166. Erdman, L. W., J. Am. Soc. Agron., 21, 361 (1929).
167. Virtanen, A. I., J. Jorma, J. Erkama, and A. Linnasalmi, Acta Chem. Scand., 1, 90 (1947).
168. Dilworth, M. J., and D. C. Williams, J. Gen. Microbiol., 48, 31 (1967).
169. Bergersen, F. J., J. Gen Microbiol., 19, 312 (1958).
170. Hardy, R. W. F., R. C. Burns, and R. D. Holsten, Soil Biochem. Biol., 5, 47 (1973).
171. Bowen, G. D., Qld. J. Agr. Sci., 16, 267 (1959).
172. Young, D. J. B., J. Brit. Grassld. Soc., 13, 106 (1958).
173. Whiteman, P. C., Aust. J. Agr. Res., 21, 215 (1970).
174. Whiteman, P. C., and A. Lulham, Aust. J. Agr. Res., 21, 195 (1970).
175. Bergersen, F. J., F. W. Hely, and A. B. Costin, Aust. J. Biol. Sci., 16, 920 (1963).
176. Spratt, E. R., Ann. Bot., 33, 189 (1919).
177. Molisch, H., Pflanzenbiologie Japan (Jena), 227 (1926).
178. Jimbo, T., Proc. Imp. Acad. (Tokyo), 3, 164 (1927).
179. Harris, J. O., E. K. Allen, and O. N. Allen, Am. J. Bot., 36, 651 (1949).
180. Pate, J. S., Nature, 192, 376 (1961).
181. Viermann, J., Bot. Arch., 25, 45 (1929).
182. Allen, E. K., K. F. Gregory, and O. N. Allen, Can. J. Bot., 33, 139 (1955).
183. Hardy, R. W. F., and E. Knight, Jr., in Progress In Phytochemistry, Wiley, London, 407, 1968.
184. Ahmed, S., and H. J. Evans, Biochem. Biophys. Res. Commun., 1, 271 (1959).
185. Hallsworth, E. G., S. B. Wilson, and E. A. Greenwood, Nature, 187, 79 (1960).

186. Kliewer, N., and H. J. Evans, Plant Physiol., 38, 99 (1963).
187. Spencer, K., Aust. J. Agr. Res., 10, 500 (1959).
188. Jones, R. K., P. J. Robinson, K. P. Haydock, and R. G. Megarrity, Aust. J. Agr. Res., 22, 885 (1971).
189. Lowther, W. L., and J. F. Lonergan, Plant Physiol., 43, 1362 (1968).
190. Brenchley, W. E., and H. G. Thornton, Proc. Roy. Soc., Ser. B, 98, 373 (1925).
191. Loustalot, A. H., and E. A. Telford, J. Am. Soc. Agron., 40, 503 (1948).
192. Mulder, E. G., Plant Soil, 1, 94 (1948).
193. Wilson, S. B., and E. G. Hallsworth, Plant Soil, 22, 260 (1965).
194. Meagher, W. R., C. M. Johnson, and P. R. Stout, Plant Physiol., 27, 223 (1952).
195. Hewitt, E. J., Biol. Rev., 34, 333 (1959).
196. Wilson, D. O., and H. M. Reisenauer, Anal. Biochem., 6, 27 (1963).
197. Hallsworth, E. G., S. B. Wilson, and W. A. Adams, Nature, 205, 307 (1965).
198. Mulder, E. G., Plant Soil, 5, 368 (1954).
199. Hewitt, E. J., in Nutrition of the Legumes, Butterworths, London, pp. 15-42, 1958.
200. Andrew, W. D., and R. T. Milligan, J. Aust. Inst. Agr. Res., 20, 123 (1954).
201. Parle, J., in Nutrition of the Legumes, Butterworths, London, pp. 280-283, 1958.
202. Andrew, C. S., and D. O. Norris, Aust. J. Agr. Res., 12, 40 (1961).
203. Epstein, E., Mineral Nutrition of Plants: Principles & Perspectives, Wiley, New York, 1972.
204. Robson, A. D., D. G. Edwards, and J. F. Loneragan, Aust. J. Agr. Res., 21, 601 (1970).
205. Jackman, R. H., N. Z. J. Agr. Res., 4, 361 (1961).
206. Vose, P. B., and D. G. Jones, Plant Soil, 18, 372 (1963).
207. Munns, D. N., Aust. J. Agr. Res., 16, 743 (1965).
208. Masterson, C. L., Trans. 9th Int. Congr. Soil Sci., 2, 95 (1968).
209. Wilson, P. W., The Biochemistry of Symbiotic Nitrogen Fixation, University of Wisconsin Press, Madison, 1940.
210. Diener, T., Phytopath. Z., 16, 129 (1950).
211. Dart, P. J., and F. V. Mercer, Aust. J. Agr. Res., 16, 321 (1965).

212. van Schreven, D. A., Plant Soil, 11, 93 (1959).
213. Possingham, J. V., D. V. Moye, and A. J. Anderson, Plant Physiol., 39, 561 (1964).
214. Pate, J. S., Phyton, 18, 65 (1962).
215. McGonagle, M. P., Proc. Roy. Soc. Edinb., Ser. B, 63, 219 (1949).
216. Sironval, G., C. H. Bonnier, and J. P. Verlinden, Physiol. Plant., 10, 697 (1957).
217. Doku, E. B., Exp. Agr., 6, 13 (1970).
218. Grobbelaar, N., B. Clarke, and M. C. Hough, Plant Soil, Special Vol., 203 (1971).
219. Virtanen, A. I., T. Mosio, and R. H. Burris, Acta Chem. Scand., 9, 184 (1955).
220. Roponen, I. E., Physiol. Plant., 23, 452 (1970).
221. Leonard, L. T., J. Am. Soc. Agron., 18, 1012 (1926).
222. Eaton, S. V., Bot. Gaz., 91, 113 (1931).
223. Jones, R. J., J. G. Davies, and R. B. Waite, Aust. J. Exp. Agr. Anim. Husb., 7, 57 (1967).
224. Hannon, N., Ph. D. Thesis, University of Sydney, 1949.
225. Schaede, R., Planta (Ber.), 31, 1 (1940).
226. Wilson, J. K., J. Am. Aco. Agron., 23, 670 (1931).
227. Masefield, G. B., in Nutrition of the Legumes, Butterworths, London, pp. 202-215, 1958.
228. Virtanen, A. I., and S. von Hausen, J. Agr. Sci., 26, 281 (1936).
229. Sprent, J. I., Plant Soil, Special Vol., 225 (1971).
230. Sprent, J. I., New Phytol., 71, 441 (1972).
231. Sprent, J. I., New Phytol., 71, 449 (1972).
232. Russell, E. J., and E. W. Russell, Soil Conditions and Plant Growth, Longmans, Green and Co., London, 1958.
233. Carr, D. J., Enc. Plant Physiol., 16, 737 (1961).
234. Loveday, J., Aust. J. Sci., 26, 90 (1963).
235. Sprent, J. I., Planta (Ber.), 88, 372 (1969).
236. Dart, P. J., and J. M. Day, Plant Soil, Special Vol., 167 (1971).
237. Bond, G., Zeit Mikrobiol., 12, 93 (1961).
238. Ferguson, T. P., and G. Bond, Ann. Bot., 18, 385 (1954).
239. McManmon, M., and R. M. M. Crawford, New Phytol., 70, 299 (1971).
240. Rittenhouse, R. L., and M. G. Hale, Plant Soil, 35, 311 (1971).

241. Burrows, W. J., and D. J. Carr, Physiol. Plant., 22, 1105 (1969).
242. Grobbelaar, N., B. Clarke, and M. C. Hough, Plant Soil, Special Vol., 215 (1971).
243. Meyer, D. R., and A. J. Anderson, Nature, 183, 61 (1959).
244. Mes, M. G., Nature, 184, 2032 (1959).
245. Gibson, A. H., Aust. J. Biol. Sci., 20, 1087 (1967).
246. Small, J. G. C., M. C. Hough, B. Clarke, and N. Grobbelaar, S. Afr. J. Sci., 64, 218 (1968).
247. Lie, T. A., Plant Soil, 34, 751 (1971).
248. Stalder, L., Phytopathol. Z., 18, 376 (1952).
249. Roponen, I. E., E. Valle, and T. Ettala, Physiol. Plant., 23, 1198 (1970).
250. Gibson, A. H., Plant Soil, Special Vol., 139 (1971).
251. Mes, M. G., S. Afr. J. Sci., 55, 35 (1959).
252. Joffe, A., F. Weyer, and S. Saubert, S. Afr. J. Sci., 57, 278 (1961).
253. Korovin, A. I., and V. A. Vorob'ev, Fiziol. Rast, 14, 117 (1966).
254. Barrios, S., N. Raggio, and M. Raggio, Plant Physiol., 38, 171 (1963).
255. Gibson, A. H., Aust. J. Biol. Sci., 22, 829 (1969).
256. Moustafa, E., Phytochemistry, 8, 993 (1969).
257. Vines, S. H., Ann. Bot., 2, 386 (1888).
258. Fred. E. B., I. L. Baldwin, and E. McCoy, Root Nodule Bacteria and Leguminous Plants, University of Wisconsin Press, Madison, 1932.
259. Orcutt, F. S., and P. W. Wilson, Soil Sci., 39, 289 (1935).
260. Thornton, H. G., Proc. Roy. Soc., Ser. B, 119, 474 (1936).
261. Munns, D. A., Plant Soil, 29, 257 (1968).
262. Darbyshire, J. F., Ann. Bot., 30, 623 (1966).
263. Raggio, M., N. Raggio, and J. G. Torrey, Plant Physiol., 40, 601 (1965).
264. Cartwright, P. M., and D. Snow, Ann. Bot., 26, 251 (1962).
265. Dart, P. J., and D. C. Wildon, Aust. J. Agr. Res., 21, 45 (1970).
266. Gäumann, E., O. Jaag, and S. Roth, Ber. Schweiz. Bot. Ges., 55, 270 (1945).
267. Chailakhian, M. K., and A. A. Megrobian, C. R. Acad. Sci. USSR, 48, 138 (1945).

268. Tanner, J. W., and I. C. Anderson, Plant Physiol., 39, 1039 (1964).
269. Valera, C. L., and M. Alexander, Nature, 206, 326 (1965).
270. Yao, P. Y., and J. M. Vincent, Aust. J. Biol. Sci., 22, 413 (1969).
271. Allen, E. K., and O. N. Allen, Encyclopedia of Plant Physiology, W. Ruhland, Ed., Springer, Berlin, Vol. 8, pp. 48-118, 1958.
272. Allos, H. F., and W. V. Bartholomew, Soil Sci., 87, 61 (1959).
273. Richardson, D. A., D. C. Jordan, and E. H. Garrard, Can. J. Plant Sci., 37, 205 (1957).
274. Giobel, G., N. J. Exp. Stat. Bull., 436, 1 (1926).
275. Schmidt, G., Kühn Arch., 69, 165 (1955).
276. Ezedinma, F. O. C., Trop. Agr., 41, 243 (1964).
277. Norman, A. G., and L. O. Krampitz, Soil Sci. Soc. Am. Proc., 9, 191 (1945).
278. Thornton, G. D., Iowa State Coll. J. Sci., 22, 84 (1948).
279. Allos, H. F., and W. V. Bartholomew, Soil Sci. Soc. Am. Proc., 19, 182 (1955).
280. Loginov, Y. M., Agrokhimiya, 11, 21 (1966).
281. Moustafa, E., R. Ball, and T. R. O. Field, N. Z. J. Agr. Res., 12, 691 (1969).

CHAPTER 10

Physiological Chemistry of Dinitrogen Fixation by Legumes

F. J. BERGERSEN

Division of Plant Industry
CSIRO
Canberra, Australia

10.1. Introduction, 520
10.2. N_2-fixing Properties of Intact Nodules, 521
    10.2.1. Inorganic Nutritional Requirements of Nodulated Legumes, 522
    10.2.2. The Requirement for Photosynthesis by the Host, 524
    10.2.3. The Requirement for O_2 and the Role of Leghemoglobin, 526
    10.2.4. H_2 Metabolism in Nodules, 530
    10.2.5. The Nodule N Pool, 531
10.3. N_2 Fixation by Bacteroid Suspensions, 533
    10.3.1. The Search for the Activity Site in Nodules, 533
    10.3.2. Sensitivity to O_2 of the Bacteroid N_2-fixing System, 534
    10.3.3. The O_2 Requirement of Bacteroids, 536
    10.3.4. Substrates Supporting Bacteroid Respiration and Nitrogenase Activity, 538
    10.3.5. Bacteroid Electron Transport, 540
    10.3.6. Bacteroid Nitrogenase, 542
    10.3.7. NH_3 Assimilation, 543
    10.3.8. Induction of Nitrogenase, 545
10.4. Conclusion, 546
10.5. References, 548

10.1 INTRODUCTION

All legume nodules contain the same structural features, although there are many variations in detail. Consideration of the structure of these N_2-fixing organs (see III.8 for additional detail) can be helpful in understanding the functions that they perform. The active, N_2-fixing tissue is a central core of enlarged host parenchyma cells, which are packed with the endosymbiotic bacteria. This central tissue is enclosed in a host cortical layer, within which lie numerous pro-vascular strands connecting the nodule to the root vascular system. The soluble materials required in the nodule are supplied and the soluble products of nodule activity are removed through these strands and specialized structural features that are often found in association with them (1).

The cells of the central tissue usually form a compact mass, although in many nodules, especially the larger ones, uninfected interstitial cells, smaller than the bacteroid-filled cells, are distributed through it. These interstitial cells are usually more frequent towards the center of nodules and in the electron microscope, structures are seen that suggest that they are very active cells. They may be prominent in metabolite transfer (2) and may have functions related to those of nodule transport cells (1). The infected host cells of the nodule central tissue are usually densely packed with bacteria [termed bacteroids in mature nodule cells (3)] which are often grossly enlarged. These bacteria lie within membrane-bounded "infection vacuoles" of "membrane envelopes" (4). The infected host cells also contain the characteristic pigment leghemoglobin (5) and many mitochondria and amyloplasts, which are particularly numerous in the periphery of the cells adjacent to the intercellular spaces. The bacteria are, thus, located within the innermost of a number of enclosing structures, each of which has its own metabolic functions.

It is now clear, as will be seen in the course of this chapter, that the bacteroids contain all of the enzyme systems necessary for the reduction of N_2 to NH_3 and some of the enzymes for the initial steps of the assimilation of this NH_3 into amino acids. However, in order for these reactions to proceed, carbon substrates from the host plant must pass into the nodule, probably undergoing modifications en route;

O_2 must enter the tissues in quantities sufficient to support the activities of host cells and bacteroids without inactivating the O_2-sensitive N_2-fixing enzyme system, and the products of the reactions must be modified and translocated from the central nodule tissue to the sites of protein synthesis in the host plant. Furthermore, the range of conditions in which the bacteroids can develop and function is quite limited. Failure to maintain these conditions may limit N_2 fixation rates or even cause irreversible changes in the tissue which lead to premature breakdown of the symbiotic system.

Thus, the functions of the host component of nodules are indispensible to the functioning of the nodule as a whole, although the key processes of N_2 fixation are located wholly in the bacterial component of this symbiotic system. It is difficult to study host functions, and little substantial knowledge of their role in N_2 fixation in legume root nodules has been established. In contrast, N_2-fixing bacteroid suspensions can be prepared free of host components, and several of their properties are now well defined.

This chapter will, therefore, deal substantially with bacteroid functions, but an attempt will be made to present a total view whenever this is possible. However, the treatment cannot be exhaustive in a chapter of this length and the reader is referred to other chapters in this treatise for further details. Sufficient examples will be given to establish the various points being considered.

10.2 N_2-FIXING PROPERTIES OF INTACT NODULES

This section deals with requirements for N_2 fixation by intact nodulated plants and with information derived from the study of intact nodules still attached to the roots or freshly separated from them. Until about 1952, studies with intact nodulated plants were the only possible way in which N_2 fixation by legumes could be investigated. These studies yielded some important information upon which subsequent physiological and biochemical work was based. Some of this information was summarized by Wilson (6). Further important information was derived from agronomic and plant nutrition studies with legumes. In 1952 Aprison and Burris (7) reported, for the first time, N_2

fixation by nodules detached (excised) from soybean roots, by demonstrating incorporation of $^{15}N_2$ into nodule nonprotein nitrogen. Previous efforts to obtain fixation in detached nodules had been unsuccessful mainly because excessive precautions had been taken to guard against the activities of contaminating N_2-fixing bacteria from the soil (8). Studies with intact detached nodules using $^{15}N_2$ were then the main source of information until 1965, when success with nodule homogenates (9) and the later advent of C_2H_2 reduction as a tool, led firstly to the demonstration that bacteroids were the site of N_2-fixing activity (10, 11) and then to the isolation and purification of bacteroid nitrogenase (12-15). Detached soybean nodules were used for the first demonstration of an alternative substrate for nitrogenase. Hoch et al. (16) showed that N_2O was not only a specific competitive inhibitor of N_2 fixation, but also that it was reduced to N_2. Later, detached nodules provided an early example of the use of C_2H_2 as another alternative substrate in nitrogenase reactions (17).

10.2.1 Inorganic Nutritional Requirements of Nodulated Legumes

The mineral requirements of legumes growing with combined nitrogen are qualitatively similar to those of nodulated plants relying on fixed N_2. However, nodulated plants have increased requirements for certain elements and require at least one element not required at all by most non-nodulated species of legume. These increased requirements have been found to be related to (or appear to be logically related to) the synthesis and activity of nitrogenase, to certain activities of the symbiotic bacteria, or to nodule development. These matters are fully discussed by Munns (IV.10) and by Pate (III.9), but are briefly stated here as an introduction to an account of the physiology of nodules.

A specific requirement for Mo in N_2-fixing systems was first recognized by Bortels (18) in Europe. Anderson (19) found that very small dressings of molybdic oxide applied to acid soils in South Australia dramatically increased yields of subterranean clover and lucerne. Subsequently, Mo responses were reported for legumes in many parts of the world.

These results were soon found to be due to effects upon the N_2-fixing systems of these plants. Anderson found that the extent of the Mo responses depended upon the nitrogen status of the soils and on acid soils, no responses were obtained unless lime was used to promote nodulation (20). Other workers found that the symptoms of Mo deficiency in legumes resembled those of N deficiency (21). These studies thus pointed to the involvement of Mo in N_2 fixation in legume root nodules long before any biochemical evidence was obtained; it was not until 1968 (13) that soybean bacteroid nitrogenase was shown to contain Mo, although this element was found to be present in the nitrogenase of free-living bacteria some years previously (22, 23). Nitrate reductase, which also contains Mo, is present in the bacteroids of some nodules (24) and probably contributes to the high Mo requirement of nodulated legumes.

The author is not aware of any definitive evidence of a special requirement for Fe by nodulated legumes. However, this element is a constituent of the various cytochromes that are present in nodule bacteroids in relatively large amounts (25) and is present in both proteins of nitrogenase (13, 14), in bacteroid ferredoxin (26) and in the leghemoglobin that is present in nodules in quite high amounts (up to 20 nmoles/nodule in soybean, e.g., ref. 27). It seems probable, therefore, that the growth of nodulated legumes would require more Fe than that of nonnodulated plants that have none of these Fe-containing systems.

With the exception of the hemoproteins, all of the special Fe-containing proteins of nodules also contain acid-labile sulphur (13, 26). Nodulated legumes, therefore, have a high requirement for S, and S deficiency has been reported to produce symptoms of N deficiency in nodulated legumes. However, application of fertilizer-N does not always remove these symptoms because deficiency of S also restricts protein synthesis in many circumstances (28).

It has been reported that nodulated legumes may have an increased requirement for phosphate (29). It is pertinent to speculate that this may be related to the high requirement for ATP in N_2-fixing systems (30); increased cellular content of phosphate may favor more rapid phosphorylation of ADP to ATP in the bacteroids. However, there seems to be no such effect in other N_2-fixing systems (137).

Cobalt has been shown to be uniquely required by nodulated legumes (31-33). Evans and co-workers have shown that this requirement is attributable to the vitamin B_{12} coenzymes that are involved in nucleotide reductase and methylmalonyl mutase activities in the bacteria during nodule development and during N_2 fixation (34).

Two other elements, Ca and Cu, have been shown to be required in increased amounts by nodulated legumes. Ca seems to be specifically required during the early stages of infection (35) and may be concerned in pectolytic activity associated with the deformation and penetration of root hairs by the nodule bacteria (34). The role of Cu is obscure, but there is some evidence for its involvement in the synthesis of γ-amino-butyric acid in <u>Trifolium subterraneum</u> nodules (36). This amino acid has been found to be a prominent constituent of the nodules of some clovers (37), but its role in nodule metabolism is not clear.

10.2.2 The Requirement for Photosynthesis by the Host

The essential relationship between photosynthesis and N_2 fixation by nodulated legumes has been known and studied for many years. Experiments such as those of Wilson et al. (ref. 38, in which it was shown that increased CO_2 concentration in the atmosphere supplied to nodulated alfalfa or clover plants caused increased growth and N_2 fixation) led to an active period of investigation of "carbohydrate nitrogen relationships" (6). This concept is of little interest today, and many of the effects attributed to light may have been confounded with temperature changes during periods of high light intensity. However, much of the information recorded in these papers is of interest in reassessing the physiological relationships of the two processes. Effects of light intensity and quality are discussed by Gibson (IV.11).

Virtanen et al. (39) studied $^{15}N_2$ fixation by nodules detached from pea plants after various periods of darkness following a period of very active photosynthesis. The results showed that there was progressive exhaustion of substrates, causing declining N_2 fixation, during darkness. Plants returned to daylight after 24 hours in the dark quickly recovered their N_2-fixing activity. Bach et al. (40) studied the distribution of ^{14}C-labeled photosynthate in nodulated soybean plants. They found that a high proportion was translocated to the nodules where it was distributed among organic acids, amino acids, and carbohydrates. It was also shown that the addition of sucrose, glucose, and fructose to nodule slices enhanced N_2 fixation. These compounds were prominent among photosynthetic products found in nodules.

Pate (ref. 41; III.9) described experiments with nodulated pea plants, growing in N-free nutrient solution, in which $^{14}CO_2$ was fed to the shoots for 1-1½ hours. The shoots were then cut off and the sap exuding from the roots was analyzed. Asparagine, aspartic acid, and glutamine were all heavily labeled and these amino compounds had apparently been produced in the nodules. More recently, Minchin and Pate (42) have reported the results of a study of the carbon balance of nodulated pea plants during a 9-day period of active photosynthesis and N_2 fixation. It was found that 32% of the total photosynthate was translocated to the nodules, where 5% was utilized in nodule growth, 12% was consumed in respiration and 15% was returned to the shoots via the xylem as amino compounds generated from N_2 fixation.

The use of the C_2H_2 reduction assay has permitted the study of short-term fluctuations in N_2 fixation in response to changes in the illumination of the host plant. Hardy et al. (43) showed that nodules of field-grown soybeans were more active during the day than at night. It has been the consistent experience of workers in the author's laboratory, that the illumination received by soybean plants before measurement affects the magnitude of the C_2H_2-reducing activity of the nodules. For example, with 22 uniformly aged samples of detached nodules, those taken on days that had been bright and sunny for 2 hours before harvest had a mean activity of 6.7 ± 1.7 µmoles C_2H_4/hr·g. Those collected on days with more than 5/10 cloud during 2 hours prior to harvest, had a mean activity of 3.0 ± 0.5 µmoles C_2H_4/hr·g. In no case was the activity of a cloudy day sample as great as any bright day sample (44).

Some legume nodules continue to reduce C_2H_2 during the night, which suggests that substantial reserves of photosynthetic products may be present in the nodule tissue. We have studied C_2H_2 reduction by detached nodules and by nodulated tap-roots of glasshouse-grown soybeans in relation to light intensity during one complete photoperiod. The activity of detached nodules was much less variable and was always smaller than that of nodules remaining attached to the roots. The activity of nodulated roots fell during darkness to a level approximately the same as that of detached nodules (44). These results suggested the following conclusions: (a) Soybean nodules contain reserves of substrates that can sustain a fairly steady rate of nitrogenase activity (about 1/3 of maximum rates) during periods of low light intensity and during the night; and (b) the tap roots contain readily available additional substrates during the day, the amounts of which fluctuate in relation to light intensity. The possible nature of these photosynthetic and endogenous nodule substrates will be discussed later in this chapter (III.10.3.4).

It is clear from experiments such as those that have been described, that N_2 fixation in nodules is dependent upon a supply of photosynthetic compounds for use as a source of energy and reducing power and for the production of translocatable N compounds synthesized from the products of N_2 fixation.

10.2.3 The Requirement for O_2 and the Role of Leghemoglobin

Fred et al. (45) state that it was recognized as early as 1864 that air was needed for the development and functioning of root nodules on legumes. Several workers have reported deleterious effects of immersing the roots of nodulated pea plants in water (46-48). It was shown by aerating the water culture medium in which the plants were growing, that the effect was due to limitation of O_2 supply (47, 48). Similarly, Loewig (49) described beneficial effects on N_2 fixation by aerating soils in which soybeans were growing. In contrast to these results, growth of clover plants in controlled atmospheres (6) failed to show any difference between plants growing on symbiotically fixed N_2 or nitrate, in terms of response to O_2 pressure. In our laboratory [(50), Gibson, unpublished], clover plants are regularly grown on N_2 in unaerated cultures with the nodules completely immersed in culture solution, while other small seeded legumes must be grown in well-aerated media for good fixation of N_2 (51). Sprent (52) reported that wetting the surface of detached soybean nodules produced a drop in C_2H_2-reducing activities and Gibson (unpublished) has observed similar effects with nodulated lupin roots;

nodulated clover roots were not as susceptible to this treatment. These differences between legume species are possibly due to differences between nodule aeration pathways, since it will be shown (III.10.3.3) that the production of ATP by aerobic pathways is essential for N_2 fixation by bacteroids from all legume nodules that have been studied.

Allison et al. (53) concluded from studies of the effects of variations in O_2 pressure on soybean nodule respiration, that O_2 concentration within the tissues was very low. This conclusion is in agreement with the low redox potential measured by Ebertova (54) in nodule tissues (cf. ref. 136). Ferguson and Bond (55) found that increased O_2 concentration in the root medium favored N_2 fixation by red clover plants and Burris et al. (56) and Bergersen (57) found that O_2 pressures up to 0.5 atm progressively stimulated N_2 fixation by intact, detached soybean nodules, although higher O_2 pressures were inhibitory. Increasing N_2 fixation in response to increased pO_2 was associated with increased respiration (57). These experiments indicated that O_2 is required for N_2 fixation by legume root nodules and that at ordinary atmospheric concentrations, O_2 is a limiting factor.

This conclusion is confirmed by the examination of the state of oxygenation of nodule leghemoglobin. This pigment is present in the bacteroid-containing cells of nodule tissues and has an extremely high affinity for O_2 (II.8). However, in soybean nodules in air, it is only partially oxygenated (58, 59). This indicates that the O_2 concentration in nodule cells is very low. Increasing external pO_2 causes increased oxygenation (58) (Fig. 10.1a).

The energetic efficiency of N_2 fixation by soybean nodules is also affected by pO_2. At a pO_2 of 59 mm Hg, the R. Q. of the nodules was 1.3 and the ratio $CO_2:NH_3$ was 323. At pO_2 of 185 mm Hg, the R. Q. was 1.05 and the ratio fell to 22.6 (60).

The inhibitory effects of high pO_2 on N_2 fixation by detached soybean nodules (56) resembled competitive kinetics (57), but recent work with nitrogenase in cell-free systems (61) has shown that O_2 is an uncompetitive inhibitor. Reexamination of the earlier data shows that the inhibitory effect could be overcome by the use of higher N_2 concentrations, giving much increased fixation rates (Fig. 10.1b). This is usually impractical at atmospheric pressures with N_2 but can be achieved with C_2H_2 because of the greater

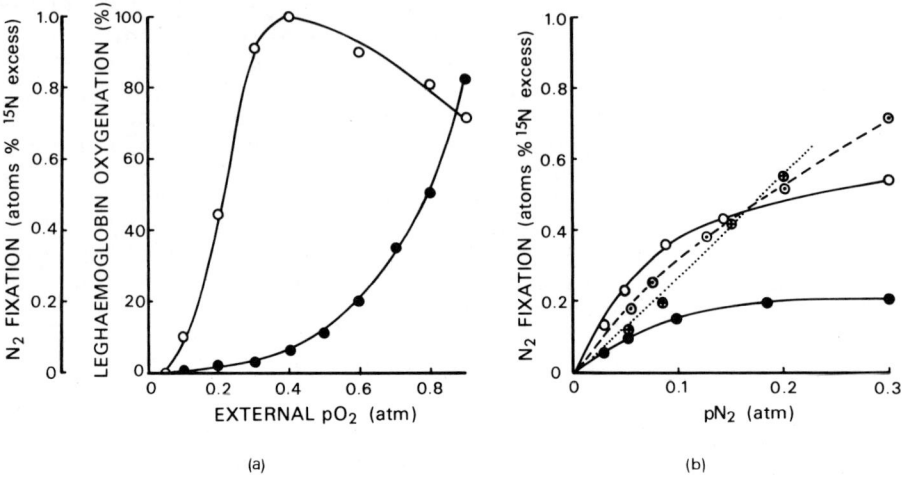

Fig. 10.1. Effects of pO_2 on intact, detached soybean nodules. (a) Increasing pO_2 causes increased oxygenation of leghemoglobin (●) and increased N_2-fixation (O). High pO_2 inhibits N_2-fixation. (pN_2 = 0.1 atm; nodules aged 35-36 days. Drawn from data of refs. 57, 58.) (b) High pO_2 depresses N_2 fixation at low pN_2 but increases fixation if N_2 is not limiting. ●, 0.2 atm O_2; O, 0.4 atm O_2; ⊙, 0.6 atm O_2; ⊕, 0.8 atm O_2. (Drawn from data used in ref. 57.)

solubility of this substrate (44). The reasons for these effects of O_2 appear to be related to kinetic effects of ATP concentration (15, 62). Production of ATP presumably increases in response to increasing pO_2 in the O_2-limited system that functions in nodules. Recent results with model experiments using bacteroid suspensions may lead to a modification of these conclusions (146).

Tjepkema (63) considered that the O_2-limitation in the N_2-fixing tissue of soybean nodules resulted from the high O_2 demand existing there and the relative impermeability to O_2 of the nodule cortex. We have recently studied the distribution of intercellular spaces in soybean nodule tissues (64). It was found that the entire nodule was traversed by interconnected, gas-filled intercellular spaces, and although there were fewer of these in the cortex than in the central tissue, they appeared to be adequate to conduct the observed rates of O_2 uptake with the development of only very small concentration differences between the outside atmosphere and the inner surface of the cortical zone. These gas-filled intercellular spaces appeared to be continuous from the outside of the nodule to the center, and in the bacteroid zone their distribution was related to the distribution of the uninfected interstitial cells. Electron microscopy of the intercellular spaces showed that the cytoplasm immediately adjacent to the spaces was invariably enriched in mitochondria and amyloplasts. It therefore seems likely that, if the oxidases of these organelles are less than saturated with O_2, they would constitute an efficient screen against the penetration of O_2 to the bacteroids further within these host cells. The continuum of O_2 sinks represented by the cortex mitochondria, the mitochondria of the interstitial cells (which appeared to be very active metabolically from the appearance of the various structures seen in their cytoplasm), and the mitochondria in the peripheries of the bacteroid-containing cells would, thus, be equivalent to the cortex in the theoretical considerations of O_2 penetration given by Tjepkema (63). Tjepkema and Yocum (135, 136) have recently presented further observations and electrical measurements supporting their contention that the nodule cortex is the main site of the restriction of the supply of O_2 to the central tissue.

The occurrence, synthesis, and properties of leghemoglobin in legume root nodules are discussed in II.8, but consideration of the O_2 relationships of nodule tissue would be incomplete without reference to the physiological role of this unique hemoprotein, whose biological functions seem to be so closely correlated with N_2 fixation. Early workers (65, 66) considered that leghemoglobin, being similar in many respects to mammalian hemoglobin and myoglobin, functioned in O_2 transport in nodules. Later work seemed to make this unlikely (5, 67) and many other roles were proposed (34, 60). More recently, the O_2 transport role has again been favored (34, 60, 68). Experimental evidence for an O_2-binding role has recently become available. Tjepkema (63) showed that the respiration of soybean nodule slices was depressed at low O_2 concentrations in the presence of CO which blocked O_2-binding by leghemoglobin. Bergersen, Turner, and

Appleby (69) showed that small concentrations of CO, which did not affect the H_2-evolving activity of nitrogenase when N_2 or C_2H_2 were absent, inhibited the evolution of H_2 by intact detached soybean nodules. This implicated a CO-sensitive step, present in intact tissue but absent from washed bacteroids or nitrogenase preparations. It was concluded that this step involved leghemoglobin. Experiments in which purified oxyleghemoglobin was added to suspensions of washed bacteroids, stimulating O_2-uptake, H_2-evolution, and C_2H_2-reduction at low pO_2, confirmed this conclusion (see 10.3.3).

Cytochemical study has confirmed that leghemoglobin is contained within the membrane envelopes that surround the bacteroids and is, thus, in contact with the bacteroid surfaces (27, 72, 138), although this had been a controversial matter previously (70, 71). The leghemoglobin is, thus, ideally located to perform a function related to O_2 uptake by the bacteroids. Many aspects of the mechanism of this function have been elucidated recently (see 10.3.3).

10.2.4 H_2 Metabolism in Nodules

Molecular hydrogen has been known to be intimately concerned with N_2 fixation since the observations of Wilson and his collaborators, who grew clover plants in controlled gaseous environments using H_2 as diluent for N_2 and O_2 (6). They found that H_2 was a specific competitive inhibitor of N_2 fixation, a finding that has been sustained through experiments with detached nodules (73) and enzyme preparations from soybean nodules (12) and with nitrogenases prepared from other sources (74, 75). This work led to an interest in the relationship between the presence of N_2-fixing systems and hydrogenases in various organisms. The sometimes erratic findings (76) were not clarified until it was found that many, but not all, N_2-fixing organisms have hydrogenases, while all nitrogenases evolve H_2 in the presence of ATP and reductant and the absence of reducible substrates (77, 78).

Evolution of H_2 was first described by Hoch et al. (16), with detached soybean nodules and the N_2-dependent exchange between D_2 and endogenous H donors was also described. These workers developed a diagram explaining their results which is still applicable to all N_2-fixing systems (79). This work was confirmed by the author (73), who also showed that H_2 was evolved by nodules even when N_2 was not limiting and that N_2 was a competitive inhibitor of H_2 evolution. It was also found that H_2 evolution increased with increased pO_2 and was only slightly inhibited by CO (due to the CO effect on leghemoglobin). The exchange reaction also responded to pO_2 but was very susceptible to CO inhibition. All of the effects described in this paragraph can be attributed to the properties of bacteroid nitrogenase (III.10.3.6).

The presence of a hydrogenase has been reported in bacteroids prepared from pea nodules (80), and it was considered that this enzyme might be concerned in the uptake of H_2 that had been evolved from the nitrogenase. Dixon (81) has confirmed this effect with nodules from pea and vetch plants. Some strains of R. leguminosarum produced nodules with hydrogenase in the bacteroids, while it was absent from nodules produced by other strains. Nodules with hydrogenase appeared to be more efficient in N_2 fixation with respect to O_2 consumption (and also presumably, with respect to consumption of energy-yielding substrates). It also appeared that the synthesis of hydrogenase in bacteroids was to a certain extent under host control since one strain of the rhizobia (ONA 311) produced nodules containing hydrogenase on Pisum sativum and Vicia bengalensis but not on V. faba.

Sprent observed that the declining rates of nitrogenase activity that are usually encountered in assays with detached nodules after 1-2 hours could be restored to rates similar to those initially attained, if the gas phase was replaced. It was suggested that this effect could be due in part to removal of accumulated inhibitory H_2 in the assay containers (52). However, it now appears that other volatile matabolites are involved (139).

10.2.5 The Nodule N Pool

The distribution of newly fixed N has been studied in several experiments using detached nodules and $^{15}N_2$. Aprison et al. (82) exposed soybean nodules to the isotope for 1-2 hours and then studied its distribution among the components of the nodule nonprotein N. The specific enrichment of glutamic acid was the highest and the pattern of distribution among other N compounds suggested that N_2 was fixed to NH_3 which then entered assimilatory pathways. Other experiments (83, 84) showed that the ^{15}N accumulated in the buffer-soluble fraction of soybean nodules and that the bacteroids were virtually unlabeled. Very short labeling experiments later showed clearly that the first free product of N_2 fixation was NH_3 that was rapidly incorporated into α-amino-N (85). Thus ended the prolonged controversy about the "key intermediate" of N_2 fixation in which the competitive merits of NH_3 and hydroxylamine had been supported by the Wisconsin group and the Helsinki group, respectively, since the 1930s (86). The matter had already been resolved for free-living bacteria by the work of the DuPont group in 1960, when cell-free extracts of Clostridium pasteurianum were found to incorporate N_2 into NH_3 (87).

Similar results were reported by Kennedy (88), who used pulse-labeling of serradella nodules with $^{15}N_2$. Incorporation of NH_3 into glutamic acid and glutamine was indicated, other amino compounds being the products of transamination reactions. It was noted in both soybean and serradella nodules (85, 88) that the newly fixed NH_3 was not in equilibrium with the bulk of the nodule soluble NH_3. Later results (89) indicated that serradella bacteroids contained most of the labeled glutamic acid; isocitric dehydrogenase and glutamic dehydrogenase were also present. Since the bacteroids were, thus, capable of synthesizing glutamic acid from NH_3 it seemed logical that only the NH_3 pool of the bacteroids was in equilibrium with the newly fixed NH_3.

The details of the distribution of fixed N from NH_3-glutamine-glutamic acid to the components of the N pool of the bacteroid-containing host cells, thence to the uninfected interstitial cells, to the nodule transport cells and vascular system have not been completely elucidated. However, the involvement of bacteroid and host cytoplasmic enzyme systems have been indicated (140, 141). Pate and his co-workers have studied the distribution of compounds containing fixed N from the nodules to the rest of the growing plant in Pisum sativum. Upward transport from the nodules via the xylem distributes fixed N to the shoots. This stream contains asparagine and glutamine amide-N, allantoic acid, aspartic acid, and homoserine (90).

The N nutrition of the rest of the plant is accomplished by redistribution from the shoots via the phloem (III.9).

10.3 N_2 FIXATION BY BACTEROID SUSPENSIONS

10.3.1 The Search for the Activity Site in Nodules

In III.10.2.5 above, experiments leading to the establishment of the nature of the products of N_2 fixation by legume root nodules were outlined. However, the location of the primary reactions was not established in about 10 years of research using detached soybean nodules and $^{15}N_2$, because the products of these reactions were freely soluble (83-85). Prior to 1965, attempts to obtain active fractions from disrupted nodules were unsuccessful (91). The work of Kennedy et al. (89) with serradella nodules strongly suggested that the bacteroids were the site of the primary reactions because glutamic acid, isolated from this fraction, contained the highest ^{15}N enrichment. In these experiments, nodules were fractionated by centrifuging filtered, undiluted homogenates, after they had been exposed to $^{15}N_2$ for various times. It was suggested that failure to observe labeling of bacteroids in similar work with soybean nodules was a consequence of the use of dilute buffers and the relative ease with which labeled material was lost from bacteroids during washing.

At about this time, work in the author's laboratory, which was quickly confirmed by others, had shown that if nodules were crushed under strictly anaerobic conditions, in a medium containing phosphate buffer, sucrose, and Mg^{2+}, a filtered brei could be prepared that would incorporate $^{15}N_2$ into NH_3 for up to 30 minutes, when supplied with up to 8% of O_2 in the gas phase (9, 84, 92). Similar observations were made independently by Sloger and Silver in their studies of N_2 fixation by nodular homogenates of the nonlegume, Myrica cerifera (160, 161). Higher pO_2 shortened the period of N_2 fixation, although the initial velocity of the reaction was increased. Evidence suggesting that oxidative ATP production was involved in N_2 fixation was obtained using these preparations, and the shortened time course with increasing pO_2 was considered to be a consequence of the sensitivity of the N_2-fixing system to O_2.

Within a few months of these reports, the preparation of N_2-fixing bacteroid suspensions was reported (10, 11) and active N_2 fixation by bacteroid cell-free extracts supplied with ATP and reductant was obtained (11). Bacteroid suspensions prepared and assayed in buffered sucrose medium fixed $^{15}N_2$ into NH_3 that was recovered from solution; only 5-6% of the total N_2 fixed was recovered from the bacteroid nonprotein N (10), an amount considered to be no more than that due to contamination of the bacteroid cells. These findings suggest that the original results with intact detached nodules (83-85) were not due to leakage of fixed N from the bacteroids as suggested by Kennedy et al. (89), but rather that soybean and serradella bacteroids differ in the extent to which they retain fixed N (60).

Koch et al. (11) found that improved bacteroid activity resulted from the use of polyvinylpolypyrrolidone in the presence of ascorbate, during maceration of the nodules. This treatment apparently removed harmful phenolic compounds that are released from the macerated plant tissue (34). The procedure is now almost universally used for the preparation of N_2-fixing bacteroid suspensions which has now been reported for several legumes, including lupins (15, 93) and peas (80).

Thus, conclusive evidence was at last obtained showing that the N_2-fixing system of legume root nodules is located in the bacteroids. The many earlier correlations (94) relating bacteroid tissue to N_2 fixation were explained, and interim theories (95), proposed to explain results in terms of a truly integrated, symbiotic metabolism involving host and bacteria (96), were abandoned.

10.3.2 Sensitivity to O_2 of the Bacteroid N_2-fixing System

The N_2-fixing activity of soybean nodule bacteroid suspensions is destroyed upon exposure to air for 15 minutes (97). This is apparently due to the extreme sensitivity of nitrogenase to O_2 (34). Lupin bacteroid nitrogenase appears to be less sensitive to O_2 (93), and the nitrogenase of bacteroids from some strains of R. japonicum seems to be more sensitive than others (60) (Fig. 10.2). The sharp difference

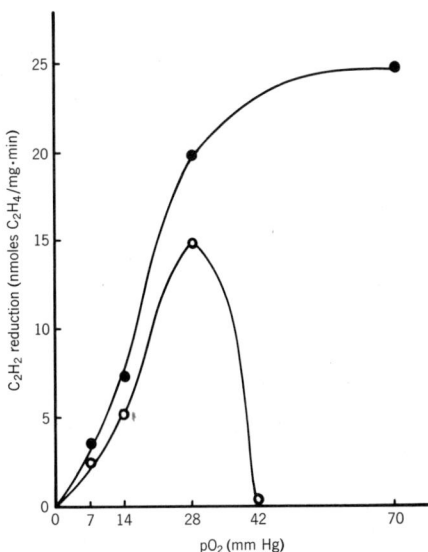

Fig. 10.2. Differences between strains of R. japonicum in the sensitivity to O_2 of bacteroids isolated from soy bean nodules. Assays contained 20 mg (dry wt) of bacteroids in 2.0 ml, shaken at 150 cycles/min at 30° for 10 min. O, strain CC705; ●, strain CB1809. (Data kindly supplied by J. Rigaud.)

between nitrogenase activity at 5% O_2 and 6% O_2 with strain CC705 suggests that bacteroids of this strain may have a "switch-off" mechanism, which is activated by pO_2, similar to that described in Azotobacter species (98).

The near anaerobic conditions which are found in nodule tissues are thus well suited to the preservation of bacteroid nitrogenase. This requirement in nodules for low concentrations of free O_2 also requires that a mechanism is provided that enables an adequate flux of O_2 for the provision of the ATP required by the bacteroids for N_2 fixation. Recent studies have shown that leghemoglobin is probably concerned in such a mechanism (refs. 63, 69, Sections III.10.2.3, II.8), and to this extent this hemoprotein could be considered to have a role in protecting bacteroid nitrogenase. However, statements to the effect that the leghemoglobin "traps" or "scavenges" O_2 (99)

are misleading. The main factors keeping pO_2 low in nodule tissue are the geometry of nodule structures, which restrict the rate at which O_2 can enter the bacteroid-containing tissues, and the O_2 demand of the large numbers of bacteroids that tightly pack the cytoplasm of the infected cells of this tissue. This explanation is illustrated by the following results. In the absence of leghemoglobin, dense suspensions of bacteroids remain active in C_2H_2 reduction for many hours with 5% O_2 in the gas phase and slow shaking (69). That is, the O_2 demand in the suspensions maintains a low O_2 concentration, which protects the nitrogenase, but the nitrogenase activity is low because there is insufficient ATP production. The addition of leghemoglobin to such suspensions allows increased nitrogenase activity, coupled with increased O_2 uptake by the bacteroids (69).

10.3.3 The O_2 Requirement of Bacteroids

The O_2 requirement of intact N_2-fixing nodules has been described (III.10.2.3) and some of the physiological conflicts between O_2 sensitivity and O_2 requirement have been outlined above (III.10.3.2). It seems to be fairly clear that the requirement is a consequence of the oxidative pathway of respiration in which ATP is produced in the bacteroids. However no studies of the details of phosphorylation coupled to the electron transport chain have been made. The effects of ATP concentration upon nitrogenase activity have been studied in cell-free systems and found to be complex (15, 62). These results suggest that the sigmoidal response to increasing pO_2 in N_2 fixation by bacteroid suspensions (97) may be a consequence of a linear increase in ATP concentration. In suspensions of soybean bacteroids, shaken with a gas phase containing O_2, nitrogenase activity was a linear function of ATP/ADP ratios (142).

Several recent papers have clarified many aspects of the role of leghemoglobin in stimulating nitrogenase activity in bacteroids (69) and its relation to conditions of O_2 supply. Wittenberg et al. (143) established that the effect was due to the O_2-binding properties of leghemoglobin and that special mechanisms involving superoxide or peroxide were not involved. Other O_2-carrying proteins had similar effects, although they were quantitatively inferior, thus eliminating the suggestion that a specific reaction between a terminal oxidase in the bacteroids and oxyleghemoglobin might be involved (69). Subsequently, several reports have been published (142-146, 153; see II.8 for a more detailed account) which suggest the following over-all statement of current understanding of the role of leghemoglobin, as studied in soybean root nodules:

1. In bacteroids, formation of ATP from ADP is more efficient in an oxidative pathway which operates best when the free, dissolved O_2 concentration in the bulk of the surrounding solution is between 0.01 and 0.1 µ$\underline{M}$, than in pathways operating at higher O_2 concentrations. Consequently, nitrogenase activity is greatest in this low range of O_2 concentration.
2. In the absence of leghemoglobin, bacteroid respiration is very restricted because of limited diffusion of free O_2 to active bacteroids in this range of low O_2 concentration.
3. Addition of leghemoglobin to such diffusion-limited systems removes the restriction because of macromolecular-carrier-facilitated diffusion of O_2. High bacteroid ATP concentration results and nitrogenase activity increases.
4. The equilibrium constant for oxygenation of leghemoglobin (0.04 x $10^{-6}\underline{M}$) ensures that bacteroids suspended in a solution of partially oxygenated leghemoglobin are in an environment in which the free O_2 concentration is poised in the range for optimum ATP production and nitrogen fixation. In nodule tissue, the leghemoglobin concentration is of the order of 1 m$\underline{M}$, (27), or 1 x 10^5 times greater than the O_2 concentration for maximum activity. The O_2 "buffering" effect would therefore be very great, minimizing effects of fluctuating O_2 or substrate supplied.

Recently the obligatory aerobic metabolism of rhizobia has been modified by the growth of $\underline{R.\ japonicum}$ under anaerobic conditions with nitrate substituting for O_2 as terminal electron acceptor (101). Soybean nodule bacteroids have nitrate reductase activity (102), and it has been possible to obtain anaerobic N_2 fixation and C_2H_2 reduction by soybean bacteroids in the presence of nitrate (103). The nitrite produced is inhibitory to nitrogenase activity, and only low rates of reaction were obtained. However, study of the initial rates of C_2H_2 reduction and comparison with rates obtained with O_2 as terminal electron acceptor showed that the O_2 pathway was 2.5 times as efficient as the nitrate pathway, in terms of electrons transferred to the terminal acceptor. Identification of the pathway to nitrate may provide valuable data for the establishment of the ATP-producing steps in the complete electron transport pathway.

The hemoproteins and flavoproteins involved in the respiratory chain will be discussed in III.10.3.5.

10.3.4 Substrates Supporting Bacteroid Respiration and Nitrogenase Activity

Energy-yielding substrates in nodules are derived from photosynthetic products (III.10.2.2). Organic acids are less abundant in nodules than sugars (40), but they are generally the best substrates for the support of O_2 respiration by bacteroid suspensions that have been prepared in dilute buffers from crushed nodules (104-107). High rates of N_2 fixation by soybean bacteroids prepared in buffer containing 0.3 $\underline{M}$ sucrose, are supported by endogenous substrates. Washing the bacteroids in buffers without sucrose diminished this endogenous activity and the rates were restored by the addition of succinate to the assays; fumarate and pyruvate were less effective (10), perhaps for reasons involving bacteroid permeability. Succinate was prominent among the organic acids leached from bacteroids in media of low osmotic strength following initial preparation and washing in sucrose medium (108). Sucrose at substrate concentrations did not stimulate N_2 fixation by bacteroids (10). However, the host cells have active invertase activity (109). It seems likely that hexoses may be the main substrates for bacteroids within the nodule cells, since several of the enzymes for hexose utilization have been detected in bacteroids from soybean, lupin, and serradella nodules (110, 121). Glucose is usually a poor substrate for bacteroids (103, 104). Kidby (110) proposed that this may be due to damage sustained by the bacteroid glucose transport mechanism during their preparation. When lupin bacteroids were kept at room temperature for a few hours, ability to utilize glucose was restored, and the restoration was sensitive to chloramphenicol, suggesting that synthesis of a permease may have been involved.

Pathways of glucose catabolism have not been investigated in bacteroids, but they are unlikely to be different from those of cultured rhizobia. Keele et al. (111) identified the Entner-Doudoroff pathway in R. japonicum cultures where the citric acid cycle seemed to be the main pathway for pyruvate utilization (111, 112).

It is the author's experience that bacteroids from several species of legume are notable for being able to sustain relatively high rates of respiration from endogenous reserves, for very long periods. Substantial reserves of carbon compounds exist in bacteroids. For example, soybean bacteroids contain up to 50% of their dry weight as poly-β-hydroxybutyrate, a polyester of D(-)-β-hydroxybutyrate (113). In bacteroids of other species of rhizobia, glycogen appears to be an important reserve material (114). The large quantities of poly-β-hydroxybutyrate present in the bacteroids is a striking feature of electron micrographs of sections of soybean nodules (4). Synthesis of this material is most rapid during the first week following nodule appearance, and the amount present in the bacteroids (as a percentage of dry weight) remains fairly constant throughout the N_2-fixing life of the nodules (115). It is possible that the synthesis is stimulated by the onset of O_2-limited conditions in the nodule tissue, since Senior et al. (116) found that the content of this polymer in Azotobacter beijerinckii increased from about 3% of the dry weight to 20-40% when O_2-limited conditions were imposed in chemostat studies.

It seems logical to propose that these large deposits of bacteroid polymers might be available for metabolism during darkness or when the supply of photosynthate was otherwise restricted. They may also be the source of reducing power that is responsible for the prolonged endogenous respiration of bacteroid suspensions. Wong and Evans (115) looked for diminished concentrations of poly-β-hydroxybutyrate in bacteroids from soybean plants that had been darkened for up to 16 days and from nodules at various intervals after detachment from the roots. In both experiments, no significant reduction in concentration of the polymer occurred during the period in which nitrogenase activity was declining to zero. The decline in activity was, thus, not due to exhaustion of polymeric reserves, and they concluded that there was no evidence supporting a direct role for poly-β-hydroxybutyrate as an energy source for the maintenance of nitrogenase activity. However, active poly-β-hydroxybutyrate depolymerase and β-hydroxybutyrate dehydrogenase were present in cell-free extracts prepared from the bacteroids. Failure to detect utilization of poly-β-hydroxybutyrate could have been due to the low nitrogenase activity of the nodules used (initially only 2 μmoles C_2H_4/g·hr. Compare ca. 10 μmoles/g·hr in refs. 17, 44 for detached nodules). It would also be difficult to measure relatively small depletions of the polymer in the presence of such large quantities in the bacteroids. The role of this polymer in soybean nodules should be investigated further, because of the energetic disadvantage that the symbiotic system suffers if these large reserves fulfill no useful function.

If β-hydroxybutyrate produced by depolymerization is utilized as a substrate, it is probably oxidized to acetoacetate by the action of β-hydroxybutyrate dehydrogenase and metabolized through the citric acid cycle, thus possibly giving rise to the succinate present in bacteroids that have high endogenous rates of respiration (see above). The glyoxylate pathway is not present in bacteroids during active N_2 fixation (112, 115). If it were present this pathway would tend to deprive the bacteroids of α-ketoglutarate required for the primary assimilation of fixed NH_3. Klucas and Evans (113) were able to use β-hydroxybutyrate as a substrate for the production of NADH which served as a weak electron donor to bacteroid nitrogenase in cell-free extracts when benzylviologen was used as an intermediate.

10.3.5 Bacteroid Electron Transport

Appleby and Bergersen (117) found that the cytochrome absorption spectra of centrifuged pellets of cultured R. japonicum and of bacteroids from soybean nodules were remarkably different. A full account of these differences has been given (25, 118, II.8). Cultured cells contained cytochomes a and a_3, cytochrome o, and a hemoglobin, which were not present in bacteroids. Cytochromes c 552, P-450 and P-420 were uniquely present in bacteroids. Cytochromes c 550 and b were present in similar amounts in both types of cell. Schemes of terminal electron flow, based on these differences in hemoproteins and in differences in respiratory behavior and CO and CN^- inhibition, were proposed. In bacteroids several possible terminal oxidases were suggested, but the main pathways were not elucidated. In cultured cells the main oxidases seemed to be cytochromes a_3 and o. Recent reports indicate the cytochrome P-450 is a component of an oxidase system which has optimum activity at free O_2 concentrations near 0.1 µM and which is responsible for the high bacteroid ATP concentrations necessary for nitrogenase activity (142, 144, 146; II.8).

Differences between the flavin coenzymes and flavoproteins of vegetative rhizobia and bacteroids also seem likely. Pankhurst et al. (119) studied a riboflavin-requiring auxotroph of R. trifolii that produced nodules of very low N_2-fixing capacity on red clover. The parent strain and prototrophic revertants from the auxotroph were fully effective. Electron and light microscopy showed that bacteroid development was arrested in nodules produced by the auxotroph, but when riboflavin was supplied to the root medium in which the plants were growing, bacteroid development resumed and nitrogenase activity ensued. Apparently the plants contained sufficient riboflavin to support infection and vegetative growth of the rhizobia during infection, but bacteroid formation had a higher requirement that the host could not meet unless exogenous riboflavin was supplied. Subsequent work (120) showed that bacteroids contained much more riboflavin + FMN and more FAD than vegetative cells (III.12).

Koch et al. (26) reported the presence of a flavoprotein that resembled a flavodoxin and the purification and properties of a ferredoxin present in soybean nodule bacteroids. Both of these components appear to function in electron transport to nitrogenase. Four types of pathways have been studied: (i) β-hydroxybutyrate→NAD→viologen, FMN or FAD→nitrogenase; (ii) NADPH-generating system→(ferredoxin + flavodoxin) or FMN, or FAD→nitrogenase; (iii) chloroplast photosystem

I→ferredoxin→nitrogenase; (iv) (<u>Clostridium</u> <u>pasteuri-</u>
<u>anum</u> hydrogenase + ferredoxin + H_2)→nitrogenase (34,
121). However, no conclusive results indicating the
significance or sequence of these pathways in bacteroids have yet been obtained. The high levels of FMN
and FAD and the presence of the flavodoxin and ferredoxin found in bacteroids seem to indicate that these
factors are involved in the natural electron flow to
nitrogenase (II.6).

Some of the changes in electron transport systems
that accompany bacteroid formation may be due to adaptation to low pO_2 in the nodule tissue. Rhizobia grow
very poorly at low pO_2, but such cultures contain
decreased contents of cytochromes $\underline{a}$-$\underline{a}_3$ and increased
cytochrome $\underline{c}$ (117, 122). Daniel and Appleby (101)
grew R. <u>japonicum</u> anaerobically with nitrate as terminal electron acceptor in place of O_2. Anaerobic
growth produced bacteria with a cytochrome pattern
resembling that of bacteroids from soybean nodules.
However, some details and in particular the properties
of P-450, were distinguishable from those of bacteroid
hemoproteins. Such cultures also synthesized other
factors that were active in electron transport to
nitrogenase in bacteroid extracts (123).

10.3.6 Bacteroid Nitrogenase

Nitrogenase from soybean bacteroids has been studied
more than nitrogenases from bacteroids of other legumes. Although no complete study, such as that of
<u>Klebsiella</u> <u>pneumoniae</u> nitrogenase (124), has been
done, enough is known to indicate that legume bacteroid nitrogenases closely resemble nitrogenases from
free-living N_2-fixing bacteria (II.2, II.3). This
resemblance will be seen from the following properties:

a. Bacteroid nitrogenase is soluble and released from
the cells upon breakage under anaerobic conditions
(11, 12, 15, 97).
b. For activity, there is an absolute requirement for
ATP and a low potential reductant (12).
c. The enzyme consists of two acidic proteins. The
Mo-Fe protein has now been highly purified (147).
It is eluted from DEAE cellulose at low ionic
strength (e.g., 35 mM $MgCl_2$), contains Mo, Fe, and
acid-labile sulfur in the ratio 1.3:29:26 per

molecule, and has a molecular weight of 200,000. The Fe protein is eluted from DEAE cellulose at higher ionic strength (e.g., 100 m$\underline{M}$ $MgCl_2$), contains Fe and acid-labile sulfur, and has an apparent molecular weight of 50,000-55,000. Both proteins are necessary for activity (13, 14, 34, 60).
d. Bacteroid nitrogenases catalyze the following reactions: reductant-dependent ATP hydrolysis, ATP- and reductant-dependent reduction of N_2 to NH_3, C_2H_2 to C_2H_4, and of CN^-. In the absence of these substrates, H^+ is reduced to H_2 and there is an exchange between D_2 and H^+ in the presence of ATP, reductant, and N_2 (11, 12, 15, 125).

Both nitrogenase proteins are inactivated in the presence of O_2 but the Fe protein seems to be more sensitive than the Mo-Fe protein. The Fe protein may also be cold-labile (60, 126), although this is not accepted by some workers (34). Under some circumstances, inactive Fe protein can be reactivated in the presence of Fe^{2+}, dithiothreitol, and reductant (62).

The kinetics of reactions mediated by bacteroid nitrogenases have also been studied. Kennedy (15) found that ATP reacted at several interacting sites on lupin bacteroid nitrogenase, and these sites interacted with ADP also. These results suggested that ATP and ADP produced conformational changes in the enzyme that may have a regulatory function in nature. Berbersen and Turner (62) studied the effects of the concentrations of soybean bacteroid nitrogenase components, ATP and reductant upon the apparent $K_m(N_2)$ and $K_m(C_2H_2)$. The results suggested that reactions mediated by this nitrogenase consisted of sequential steps and that the N_2 and C_2H_2 catalytic sites were located on the Mo-Fe protein. The Fe protein appeared to be a catalytic effector reacting at interacting sites on the Mo-Fe protein. The apparent K_m values were affected by many factors in cell-free assays, and the effects of ATP concentration were similar to effects found with intact nodules and bacteroid suspensions, where the apparent K_m increased with increased pO_2 (57, 97).

10.3.7 NH_3 Assimilation

Bacteroids may play a part in the early steps of the assimilation of NH_3 into amino compounds that are later translocated from the nodules to the shoot of the host plant. Kennedy (88) found that the primary assimilation step in serradella nodules was the incorporation of NH_3 into glutamic acid in agreement with earlier work with soybean (82). Bacteroids contain isocitric and glutamic dehydrogenases that would enable the formation of glutamic acid to proceed by diverting α-ketoglutarate from the citric acid cycle (88, 127). The presence in bacteroids of highly labeled glutamic acid in ^{15}N experiments with serradella nodules indicates that this pathway functions in this intact system (89). Kretovich et al. (128) have studied ketoacids in nodules and have established the existence of several enzymes concerned in amination and amino transfer reactions. In some of these experiments, it was clear that some of these reactions occurred in the bacteroids and some were catalyzed by plant enzymes. These workers stress the importance of ketoacids in these processes. Mulder and van Veen (129) showed that CO_2 stimulated nodulation and N_2 fixation in red clover grown in solution culture. Lowe and Evans (130) showed that CO_2 was required for the growth of Rhizobium spp., and that extracts of bacteroids from soybean nodules incorporated $^{14}CO_2$ into propionyl-CoA and phosphoenol pyruvate. These observations, indicating carboxylase activity in bacteroids, may be related to the importance of the synthesis of ketoacids as acceptors for fixed NH_3. Studies of aspartate aminotransferases from soybean nodules (131) have shown that the enzyme from bacteroids and from host cytosol had different apparent K_m values for glutamate and aspartate as substrates, whereas the K_m values for ketoacids were similar. These results suggested that the main function of the bacteroid enzyme could be aspartate synthesis, while that of the cytosol could be concerned mainly with transamination reactions. More recent findings have been reviewed by Dilworth (141). Failure to observe assimilation of NH_3 in N_2 fixation experiments with bacteroid suspensions has been attributed to the conditions imposed in the assays in which active oxidation of carbon substrates may prevent utilization of α-ketoglutarate from the citric acid cycle (60).

10.3.8 Induction of Nitrogenase

Rhizobia grown in ordinary culture media do not fix N_2, although convincing evidence has been obtained that the N_2-fixing system is located within the symbiotic, bacteroid forms of these bacteria (III.10.3.1). No completely in vitro induction or derepression of nitrogenase synthesis in rhizobia has been reported (34, 60), but recently it has been found that tissue cultures of root cells of some legumes and nonlegumes, when infected with certain strains of rhizobia, show evidence of nitrogenase activity (148-152). Apparently a diffusible factor(s) from the plant cells induces synthesis of nitrogenase in rhizobia growing on agar near, but not necessarily in contact with the plant cells (151, 152). Anaerobic growth induces modification of cytochromes in R. japonicum (101) and synthesis of other factors concerned in electron transfer to nitrogenase (123). These changes mimic changes found as a result of bacteroid formation in vivo, but no N_2 or C_2H_2 is reduced by such cultures, and no active nitrogenase components have been detected in them (3). These observations suggest that a specific mechanism inducing nitrogenase synthesis functions in the host tissues. Recent observations have detected nitrogenase activity in developing nodules at a stage in which active bacterial multiplication is occurring (27). The typical nongrowing property of bacteroids (132, 133) is thus not mandatory for N_2 fixation.

Dilworth and Parker (134) considered the induction of nitrogenase in nodule bacteroids. They proposed that the then available data supported the following mechanism. Genetic information for synthesis of the nitrogenase system has passed to the host during evolution. Early in nodule development, transmission of mRNA or of molecules of a protein or subunit from the host cell to the bacteria occurs. Synthesis or assembly of nitrogenase then occurs in the bacteria. Other possible mechanisms involving DNA transfer from host to bacteria were considered to be less probable. Although recent results seem to make such a mechanism unlikely (151, 152), it seems clear that the host is involved in the control of nitrogenase synthesis. The participation of factors from nonlegumes indicates that the control may not be as specific as previously thought. A possible role for plasmids in the control of symbiotic properties of rhizobia has been suggested

(154). Plasmid DNA has been detected in bacteroids (155). Recently, transfer of an R-factor between rhizobia and a non-N_2-fixing strain of Klebsiella has been used to support the proposition that rhizobia carry nif genes (156). These results, together with more direct evidence (157, 158) lead to the conclusion that the genetic information specifying nitrogenase is located in the bacteria but that special mechanisms may be involved in controlling its synthesis and activity.

Evans and Russell (34) proposed that the relationships which have been observed between the presence of nitrate reductase and nitrogenase in bacteroids (24, 102, 103) may be due to a constitutive subunit, consisting of a Mo-containing protein, being common to both enzymes. Evidence for this proposal comes from the observation that the constitutive Mo-containing subunit of Neurospora crassa nitrate reductase could be replaced by a Mo-containing subunit prepared from soybean bacteroid nitrogenase (34). It is also supported by recent genetic evidence (159).

10.4 CONCLUSION

The fixation of atmospheric N_2 by legume root nodules can be seen from this chapter to be an integrated metabolic sequence that involves both the host and the symbiotic bacteria, the so-called "bacteroids." The key process in which N_2 is reduced to NH_3 is located wholly within the bacteroids. This system depends on the host for the following essential supporting functions: (a) the supply of photosynthetic products as substrates for the generation of reducing power and ATP in the bacteroids; (b) the development of suitable structures producing near-anaerobic sites in the tissue in which the O_2-sensitive nitrogenase of the bacteroids is protected, while allowing sufficient

---

Fig. 10.3. A diagrammatic synthesis of the physiological chemistry of N_2 fixation in legume bacteroids discussed in this chapter. In some respects it is speculative, but all processes shown are supported by some experimental evidence or may reasonably be inferred from it. Pathways of electron flow are shown by dashed lines and are mainly derived from experiments with bacteroid suspensions. PHB, poly-β-hydroxybutyrate; Lb, leghemoglobin.

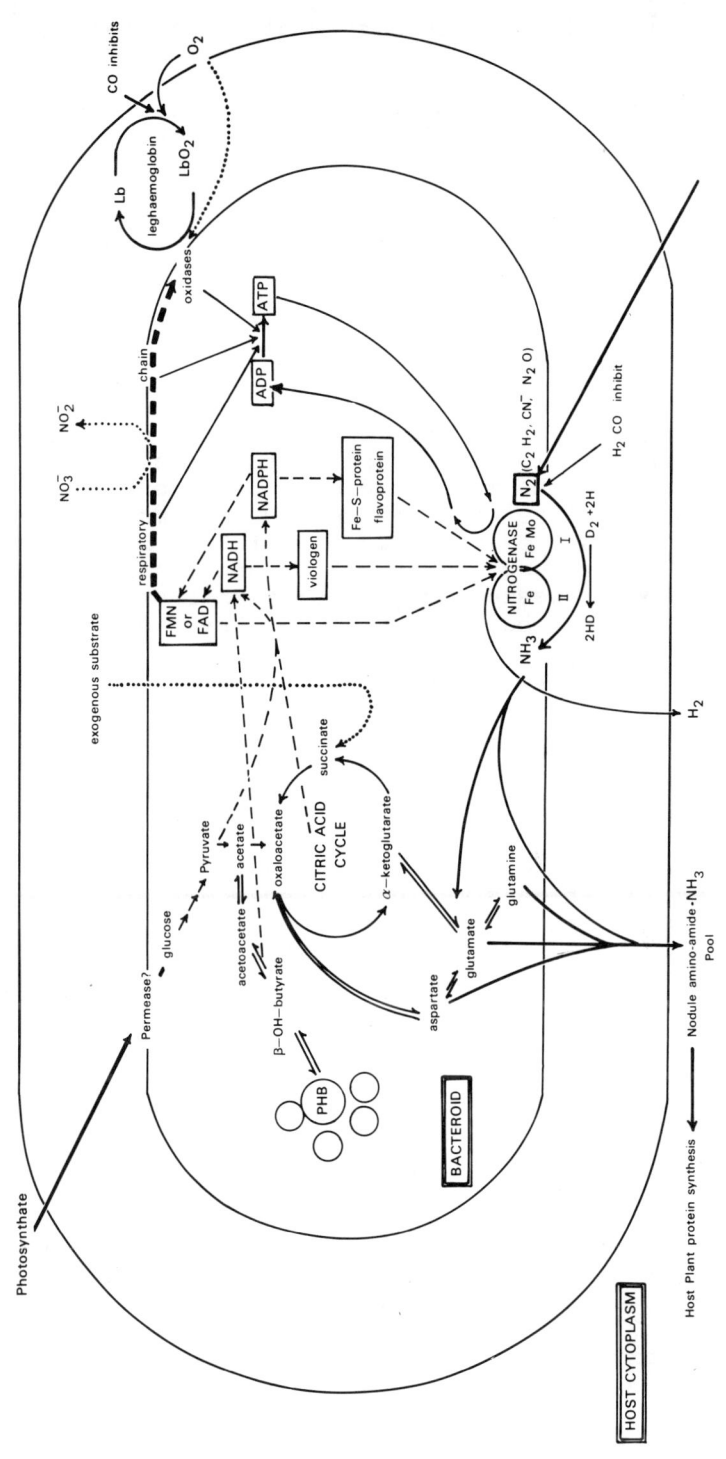

O_2 flux to allow an aerobic ATP-producing pathway to function with the aid of leghemoglobin; (c) rapid amination and amino-transfer reactions in the host supplementing those of the bacteria, and the rapid transfer of N_2 fixation products to the shoot preventing the accumulation of NH_4^+ in the bacteroids, thus minimizing possible repression of nitrogenase.

The aspects of the physiological chemistry of legume N_2 fixation that have been discussed in this chapter are summarized in Fig. 10.3 (appearing on page 547). Some aspects of this diagram are speculative but all are supported by experimental evidence or may be reasonably deduced from experiment or observation. The diagram particularly applies to the soybean system, which has been the most studied. Differences will be found in detail as other bacteroid systems are studied. For example, no reference is made to the mechanism in pea root nodule bacteroids, which apparently conserves reducing power wasted in H_2 evolution from nitrogenase (80, 81). Other bacteroids do not have poly-β-hydroxybutyrate deposits, and other polymer utilization mechanisms would function in these. For simplicity, these possibilities are omitted from the diagram. The reader is invited to compare Fig. 10.3 with a diagram prepared in 1968, which summarized the physiological chemistry of nodule N_2 fixation as it was understood by the author at that time (108). It will be seen that many of the details have been filled in in the intervening years, as the number of contributing scientists has increased and wider problems have been investigated in greater depth. It is probably inevitable that the figure presented in this chapter will be outdated with equal rapidity. That is one of the pleasures of working in a rapidly developing field of science.

10.5 REFERENCES

1. Pate, J. S., B. E. S. Gunning, and L. G. Briarty, Planta (Ber.), 85, 11 (1969).
2. Sprent, J. I., New Phytol., 70, 9 (1971).
3. Bergersen, F. J., in The Biology of Nitrogen Fixation, A. Quispel, Ed., North Holland Publishing, Amsterdam, 473, 1974.
4. Goodchild, D. J, and F. J. Bergersen, J. Bacteriol., 92, 204 (1966).

5. Smith, J. D., Biochem. J., 44, 585, 591 (1949).
6. Wilson, P. W., The Biochemistry of Symbiotic Nitrogen Fixation, University of Wisconsin Press, Madison, 1940.
7. Aprison, M. H., and R. H. Burris, Science, 115, 264 (1952).
8. Machata, H. A., R. H. Burris, and P. W. Wilson, J. Biol. Chem., 171, 605 (1947).
9. Bergersen, F. J., Biochim. Biophys. Acta, 115, 247 (1966).
10. Bergersen, F. J., and G. L. Turner, Biochim. Biophys. Acta, 141, 507 (1967).
11. Koch, B., H. J. Evans, and S. Russell, Plant Physiol., 42, 466 (1967).
12. Koch, B., H. J. Evans, and S. Russell, Proc. Nat. Acad. Sci. U. S., 58, 1343 (1967).
13. Klucas, R. V., B. Koch, S. Russell, and H. J. Evans, Plant Physiol., 43, 1906 (1968).
14. Bergersen, F. J., and G. L. Turner, Biochim. Biophys. Acta, 214, 28 (1970).
15. Kennedy, I. R., Biochim. Biophys. Acta, 222, 135 (1970).
16. Hoch, G., K. C. Schneider, and R. H. Burris, Biochim. Biophys. Acta, 37, 273 (1960).
17. Koch, B., and H. J. Evans, Plant Physiol., 41, 1748 (1966).
18. Bortels, H., Arch. Mikrobiol., 1, 333 (1930); ibid., 8, 13 (1937).
19. Anderson, A. J., J. Aust. Inst. Agr. Sci., 8, 73 (1942).
20. Anderson, A. J., and D. V. Moye, Aust. J. Agr. Res., 3, 95 (1952).
21. Evans, H. J., E. R. Purvis, and F. E. Bear, Soil Sci., 71, 117 (1951).
22. Mortenson, L. E., Biochim. Biophys. Acta, 127, 18 (1966).
23. Bulen, W. A., and J. R. LeComte, Proc. Nat. Acad. Sci. U. S., 56, 979 (1966).
24. Cheniae, G., and H. J. Evans, Plant Physiol., 35, 454 (1960).
25. Appleby, C. A., Biochim. Biophys. Acta, 172, 71 (1969).
26. Koch, B., P. Wong, S. A. Russell, R. Howard, and H. J. Evans, Biochem. J., 118, 773 (1970).
27. Bergersen, F. J., and D. J. Goodchild, Aust. J. Biol. Sci., 26, 741 (1973).
28. Anderson, A. J., and D. Spencer, Aust. J. Sci. Res., Ser. B., Biol. Sci., 3, 431 (1950).

29. McLachlan, K. D., and B. W. Norman, *J. Aust. Inst. Agr. Sci.*, 27, 244 (1961).
30. Burris, R. H., in *Chemistry and Biochemistry of Nitrogen Fixation*, J. R. Postgate, Ed., Plenum Press, London/New York, pp. 106-160, 1971.
31. Ahmed, S., and H. J. Evans, *Proc. Nat. Acad. Sci. U. S.*, 47, 24 (1961).
32. Hallsworth, E. G., S. B. Wilson, and E. A. N. Greenwood, *Nature (London)*, 187, 79 (1960).
33. Delwiche, C. C., C. M. Johnson, and H. M. Reisenauer, *Plant Physiol.*, 36, 73 (1961).
34. Evans, H. J., and S. A. Russell, in *Chemistry and Biochemistry of Nitrogen Fixation*, J. R. Postgate, Ed., Plenum Press, London/New York, pp. 191-244, 1971.
35. Lowther, W. L., and J. F. Loneragan, *Plant Physiol.*, 43, 1362 (1968).
36. Yates, M. G., and E. G. Hallsworth, *Plant Soil*, 19, 265 (1963).
37. Butler, G. W., and N. O. Bathurst, *Aust. J. Biol. Sci.*, 11, 529 (1958).
38. Wilson, P. W., E. B. Fred, and M. R. Salmon, *Soil Sci.*, 35, 145 (1933).
39. Virtanen, I. R., T. Moisio, and R. H. Burris, *Acta Chem. Scand.*, 9, 184 (1955).
40. Bach, M. K., W. E. Magee, and R. H. Burris, *Plant Physiol.*, 33, 118 (1958).
41. Pate, J. S., *Plant Soil*, 18, 333 (1962).
42. Minchin, F. R., and J. S. Pate, *J. Exp. Bot.*, 24, 259 (1973).
43. Hardy, R. W. F., R. D. Holsten, E. K. Jackson, and R. C. Burns, *Plant Physiol.*, 43, 1185 (1968).
44. Bergersen, F. J., *Aust. J. Biol. Sci.*, 23, 1015 (1970).
45. Fred, E. B., I. L. Baldwin, and E. McCoy, *Root Nodule Bacteria and Leguminous Plants*, University of Wisconsin Studies in Science, No. 5, Madison, 1932.
46. Golding, J., *Zentrblt. Bakt. Parasitenkde.*, Abt. 2, 11, 1 (1903).
47. Virtanen, A. I., and S. von Hausen, *J. Agr. Sci.*, 25, 278 (1935).
48. Virtanen, A. I., and S. von Hausen, *J. Agr. Sci.*, 26, 281 (1936).
49. Loewhig, W. F., *Plant Physiol.*, 9, 567 (1934).
50. Gibson, A. H., *Aust. J. Biol. Sci.*, 16, 28 (1963).
51. Brockwell, J., and F. W. Hely, *Aust. J. Agr. Res.*, 17, 885 (1966).

52. Sprent, J. I., *Planta (Ber.)*, **88**, 372 (1969).
53. Allison, F. E., C. A. Ludwig, S. R. Hoover, and F. W. Minor, *Bot. Gaz.*, **101**, 513 (1940).
54. Ebertova, H., *Nature (London)*, **184**, 1046 (1959).
55. Ferguson, T. P., and G. Bond, *Ann. Bot. N. S.*, **18**, 385 (1954).
56. Burris, R. H., W. E. Magee, and M. K. Bach, *Ann. Sci. Fenn.*, **60**, 190 (1955).
57. Bergersen, F. J., *J. Gen. Microbiol.*, **29**, 113 (1962).
58. Bergersen, F. J., *Nature (London)*, **194**, 1059 (1962).
59. Appleby, C. A., *Biochim. Biophys. Acta*, **188**, 222 (1969).
60. Bergersen, F. J., *Ann. Rev. Plant Physiol.*, **22**, 121 (1971).
61. Wong, P. P., and R. H. Burris, *Proc. Nat. Acad. Sci. U. S.*, **69**, 672 (1972).
62. Bergersen, F. J., and G. L. Turner, *Biochem. J.*, **131**, 61 (1973).
63. Tjepkema, J. D., *Oxygen Transport in the Soybean Nodule and the Function of Leghaemoglobin*, Ph. D. Thesis, University of Michigan, Ann Arbor, 1971.
64. Bergersen, F. J., and D. J. Goodchild, *Aust. J. Biol. Sci.*, **26**, 729 (1973).
65. Kasugai, S., H. Kubo, and K. Tsujimura, *J. Agr. Chem. Soc. Japan*, **19**, 765 (1943).
66. Keilin, D., and Y. L. Wang, *Nature (London)*, **155**, 227 (1945).
67. Burris, R. H., and P. W. Wilson, *Biochem. J.*, **51**, 90 (1952).
68. Yocum, C. S., *Science*, **146**, 432 (1964).
69. Bergersen, F. J., G. L. Turner, and C. A. Appleby, *Biochim. Biophys. Acta*, **292**, 271 (1973).
70. Dilworth, M. J., and D. K. Kidby, *Exp. Cell Res.*, **49**, 148 (1968).
71. Dart, P. J., *Proc. 4th Eur. Reg. Conf. Electr. Micr.*, Rome, pp. 69-70 (1968).
72. Truchet, G., *C. R. Acad. Sci. Paris*, **274**, 1290 (1972).
73. Bergersen, F. J., *Aust. J. Biol. Sci.*, **16**, 669 (1963).
74. Hwang, J. C., and R. H. Burris, *Fed. Proc.*, **27**, 639 (1968).
75. Lockshin, A., and R. H. Burris, *Biochim. Biophys. Acta*, **111**, 1 (1965).
76. Wilson, P. W., R. H. Burris, and W. B. Coffee, *J. Biol. Chem.*, **147**, 475 (1943).

77. Burns, R. C., and W. A. Bulen, Biochim. Biophys. Acta, 105, 437 (1965).
78. Mortenson, L. E., Biochim. Biophys. Acta, 127, 18 (1966).
79. Burris, R. H., Proc. Roy. Soc. London, Ser. B, 172, 339 (1969).
80. Dixon, R. O. D., Arch. Mikrobiol., 62, 272 (1968).
81. Dixon, R. O. D., Arch. Mikrobiol., 85, 193 (1972).
82. Aprison, M. H., W. E. Magee, and R. H. Burris, J. Biol. Chem., 224, 351 (1954).
83. Bergersen, F. J., J. Gen. Microbiol., 22, 671 (1960).
84. Klucas, R. V., and R. H. Burris, Biochim. Biophys. Acta, 136, 399 (1966).
85. Bergersen, F. J., Aust. J. Biol. Sci., 18, 1 (1965).
86. McKee, H. S., Nitrogen Metabolism in Plants, Clarendon Press, Oxford, pp. 55-64, 1962.
87. Carnahan, J. E., L. E. Mortenson, H. F. Mower, and J. E. Castle, Biochim. Biophys. Acta, 44, 520 (1960).
88. Kennedy, I. R., Biochim. Biophys. Acta, 130, 285 (1966).
89. Kennedy, I. R., C. A. Parker, and D. K. Kidby, Biochim. Biophys. Acta, 130, 517 (1966).
90. Oghoghorie, C. G. O., and J. S. Pate, Planta (Ber.), 104, 35 (1972).
91. Delwiche, C. C., Science, 151, 1565 (1966).
92. Bergersen, F. J. Biochim. Biophys. Acta, 130, 304 (1966).
93. Manorik, A. V., and E. P. Starchenkov, Dokl. Acad. Nauk SSSR, 186, 975 (1969).
94. Virtanen, A. I., Biol. Rev., 22, 239 (1947).
95. Bergersen, F. J., Bacteriol. Rev., 24, 246 (1960).
96. Jordan, D. C., Bacteriol. Rev., 26, 119 (1962).
97. Bergersen, F. J., and G. L. Turner, J. Gen. Microbiol., 53, 205 (1968).
98. Drozd, J., and J. R. Postgate, J. Gen. Microbiol., 60, 427 (1970).
99. Burris, R. H., Ann. Rev. Plant Physiol., 17, 541 (1966).
100. Wittenberg, J. B., C. A. Appleby, and B. A. Wittenberg, J. Biol. Chem., 247, 527 (1972).
101. Daniel, R. M., and C. A. Appleby, Biochim. Biophys. Acta, 272, 347 (1972).
102. Cheniae, G. M., and H. J. Evans, Biochim. Biophys. Acta, 26, 654 (1957).

103. Rigaud, J., F. J. Bergersen, G. L. Turner, and R. M. Daniel, *J. Gen. Microbiol.*, **77**, 137 (1973).
104. Burris, R. H., and P. W. Wilson, *Cold Spring Harb. Symp. Quant. Biol.*, **7**, 349 (1939).
105. Tuzimura, K., and H. Meguro, *J. Biochem. (Japan)*, **47**, 391 (1960).
106. Tuzimura, K., *J. Soil. Sci. (Japan)*, **32**, 320 (1962).
107. Bergersen, F. J., *J. Gen. Microbiol.*, **19**, 312 (1958).
108. Bergersen, F. J., *Proc. Roy. Soc. London, Ser. B*, **172**, 401 (1969).
109. Kidby, D. K., *Plant Physiol.*, **41**, 1139 (1966).
110. Kidby, D. K., *Carbon Metabolism in Legume Root Nodules*, Ph. D. Thesis, University of Western Australia, 1967.
111. Keele, B. B., P. B. Hamilton, and G. H. Elkan, *J. Bacteriol.*, **97**, 1184 (1969).
112. Johnson, G. V., H. J. Evans, and T. M. Ching, *Plant Physiol.*, **41**, 1330 (1966).
113. Klucas, R. V., and H. J. Evans, *Plant Physiol.*, **43**, 1458 (1968).
114. Bergersen, F. J., *J. Gen Microbiol.*, **13**, 411 (1955).
115. Wong, P. P., and H. J. Evans, *Plant Physiol.*, **47**, 750 (1971).
116. Senior, P. J., G. A. Beech, G. A. F. Ritchie, and E. A. Dawes, *Biochem. J.*, **128**, 1193 (1972).
117. Appleby, C. A., and F. J. Bergersen, *Nature (London)*, **182**, 1174 (1958).
118. Appleby, C. A., *Biochim. Biophys. Acta*, **172**, 88 (1969).
119. Pankhurst, C. E., E. A. Schwinghamer, and F. J. Bergersen, *J. Gen. Microbiol.*, **70**, 161 (1972).
120. Pankhurst, C. E., E. A. Schwinghamer, S. W. Thorne, and F. J. Bergersen, *Plant Physiol.*, **53**, 198 (1974).
121. Wong, P. P., H. J. Evans, R. V. Klucas, and S. Russell, *Plant Soil, Special Vol.*, 525 (1971).
122. Tuzimura, K., and I. Watanabe, *Plant Cell Physiol.*, **5**, 157 (1964).
123. Phillips, D. A., R. M. Daniel, C. A. Appleby, and H. J. Evans, *Plant Physiol.*, **51**, 136 (1973).
124. Eady, R. R., B. E. Smith, K. A. Cook, and J. R. Postgate, *Biochem. J.*, **128**, 655 (1972).
125. Bergersen, F. J., and G. L. Turner, *Biochem. J.*, **115**, 529 (1969).

126. Moustafa, E., Phytochemistry, 8, 993 (1969).
127. Mooney, P., and P. F. Fottrell, Biochem. J., 110, 17 (1968).
128. Kretovich, V. L., Z. G. Evstigneeva, K. B. Aseeva, O. N. Zagarian, and E. M. Martynova, Dokl. Akad. Nauk SSSR, 185, 942, 186, 1198 (1969); Bull. Akad. Nauk SSSR, Biol. Ser., (2), 208-214 (1969).
129. Mulder, E. G., and W. L. VanVeen, Plant Soil, 13, 265 (1960).
130. Lowe, R. H., and H. J. Evans, Soil Sci., 94, 351 (1962).
131. Ryan, E., F. Bodley, and P. F. Fottrell, Plant Soil, Special Vol., 545 (1971).
132. Almon, L., Zentrbl. Bakt. Parasitenkde, Abt. II, 87, 289 (1933).
133. Bergersen, F. J., IX Int. Cong. Soil Sci. Trans., Vol. II, pp. 49-63 (1968).
134. Dilworth, M. J., and C. A. Parker, J. Theor. Biol., 25, 208 (1969).
135. Tjepkema, J. D., and C. S. Yocum, Planta (Ber.), 115, 59 (1973).
136. Tjepkema, J. D., and C. S. Yocum, Planta (Ber.), 119, 351 (1974).
137. Bergersen, F. J., J. Gen. Microbiol., 84, 412 (1974).
138. Gourret, J-P., and H. Fernandez-Arias, Can. J. Microbiol., 20, 1169 (1974).
139. Van Straten, J., and E. L. Schmidt, Soil Biol. Biochem., 6, 231 and 347 (1975).
140. Dunn, S. D., and R. V. Klucas, Can. J. Microbiol., 19, 1493 (1973).
141. Dilworth, M. J., Ann. Rev. Plant Physiol., 25, 81 (1974).
142. Appleby, C. A., G. L. Turner, and P. K. Macnicol, Biochim. Biophys. Acta, 387, 461 (1975).
143. Wittenberg, J. B., F. J. Bergersen, C. A. Appleby, and G. L. Turner, J. Biol. Chem., 249, 4057 (1974).
144. Appleby, C. A., F. J. Bergersen, P. K. Macnicol, G. L. Turner, B. A. Wittenberg, and J. B. Wittenberg, in Proceedings of an International Symposium on Nitrogen Fixation: Interdisciplinary Discussions, W. E. Newton, and C. J. Nyman, Eds., Washington State University, Pullman, pp. 274-292, 1976.
145. Bergersen, F. J., and G. L. Turner, J. Gen. Microbiol., 89, 31 (1975).

146. Bergersen, F. J., and G. L. Turner, *J. Gen. Microbiol.*, 91, 345 (1975).
147. Israel, D. W., R. L. Howard, H. J. Evans, and S. A. Russell, *J. Biol. Chem.*, 249, 500 (1974).
148. Holsten, R. D., R. C. Burns, R. W. F. Hardy, and R. R. Hebert, *Nature*, 232, 173 (1971).
149. Phillips, D. A., *Plant Physiol.*, 53, 67 (1974).
150. Child, J. J., and T. A. LaRue, *Plant Physiol.*, 53, 88 (1974).
151. Child, J. J., *Nature*, 253, 350 (1975).
152. Scowcroft, W. R., and A. H. Gibson, *Nature*, 253, 351 (1975).
153. Stokes, A. N., *J. Theor. Biol.*, 52, 285 (1975).
154. Dunican, L. K., and F. C. Cannon, *Plant Soil*, Special Vol., 73 (1970).
155. Sutton, W. D., *Biochim. Biophys. Acta*, 366, 1 (1974).
156. Dunican, L. K., and A. B. Tierney, *Biochem. Biophys. Res. Comm.*, 57, 62 (1974).
157. Phillips, D. A., R. L. Howard, and H. J. Evans, *Plant Physiol.*, 28, 248 (1973).
158. Bishop, P. E., H. J. Evans, R. M. Daniel, and R. O. Hampton, *Biochim. Biophys. Acta*, 381, 248 (1975).
159. Kondorosi, A., I. Barabas, S. Svab, L. Orosz, T. Sik, and R. D. Hotchkiss, *Nature*, 246, 153 (1973).
160. Sloger, C., and W. S. Silver in *None-Heme Iron Proteins: Role in Energy Transfer*, A. C. San Pietro, Ed., Antioch Press, Yellow Springs, pp. 299-302, 1965.
161. Sloger, C., and W. S. Silver, *IX International Cong. Microbiol.*, Abstracts of Papers, Moscow, 285, 1966.

CHAPTER 11

Genetic Aspects of Nodulation and Dinitrogen Fixation by Legumes: The Macrosymbiont

B. E. CALDWELL
Department of Crop Science
North Carolina State University
Raleigh, North Carolina, U.S.A.

H. G. VEST
Department of Horticulture
Michigan State University
East Lansing, Michigan, U.S.A.

11.1. Introduction, 558
11.2. Types of Nodulation Responses and General Nodulation Effects, 559
    11.2.1. Nonnodulation, 561
    11.2.2. Ineffective Nodulation, 562
    11.2.3. Effective Nodulation, 564
        11.2.3.1. Inefficient Type, 564
        11.2.3.2. Efficient Type, 565
    11.2.4. Rhizobium-induced Chlorosis, 567
    11.2.5. Grafting Studies, 568
11.3. Nodule Distribution as Determined by the Host, 569
    11.3.1. Strain Acceptance, 569
    11.3.2. Nodulation Patterns, 570
11.4. Plant Growth Stages and Nodulation, 571
11.5. Increasing Nodule Mass by Breeding in the Host, 572
11.6. New Concepts of Increasing N_2 Fixation through Host Genetics, 573
11.7. References, 574

11.1 INTRODUCTION

Genetic variability in plants responding to invasion of plant pathogens is well known. This variability has been utilized successfully as a basis for the development of plants resistant to certain of these pathogens. Similar variability is evident in legumes invaded by symbiotic bacteria of the genus Rhizobium, but this has not been exploited.

Nodulation responses of legumes inoculated with Rhizobium species range from differences associated with bacterial species to individual cultivar-Rhizobium strain interactions. The generally held concept is that a specie or species of Rhizobium will have a particular affinity for one or more legume species. This has given rise to the classification of these bacteria into cross-inoculation groups. The concept of genetic control of nodulation is implied when Rhizobium species are able to induce nodulation on some legume species, but fail to induce nodulation of others. However, it is the variable nodulation responses that are observed within legume species and cross-inoculation groups that have provided the material for genetic studies.

The first report of nodulation variability of a host to strains of Rhizobium was by Vorhees (1). He observed that soybean (Glycine max) cultivar 'Haberlandt' did not nodulate when planted in plots containing R. japonicum. Five other soybean cultivars planted in the same plots were well nodulated. Interplanting of Haberlandt seed and seed of nodulating cultivars did not result in the nodulation of Haberlandt. Therefore, Vorhees concluded that different cultivars of the same legume had different powers of resistance to association with the symbiotic bacteria.

Later, Wilson, Burton, and Bond (2) concluded that host genetics were involved in the control of dinitrogen fixation in the nodules of sweet clover (Melilotus alba). These conclusions were based on their studies of four species of sweet clover and several strains of R. meliloti. They observed that some sweet clover species inoculated with certain strains of R. meliloti always nodulated well and fixed dinitrogen. However, some strains fixed high amounts of dinitrogen with one cultivar but not with another. Additional research by Burton and Wilson (3) clearly indicated host-symbiont specificity between nine strains of R. meliloti, three cultivars of Medicago sativa (alfalfa), and four other Medicago species.

Aughtry (4) and Nutman (5) have provided additional support and verification for the existence of genetic variability for nodulation in the host. Nutman (6) extended his studies to show that host factors could control the amount of N_2 fixed. In similar studies with alfalfa, Erdman and Means (7) reported significant host-strain interactions between three alfalfa cultivars and 21 strains of R. meliloti. These early observations of host-strain interactions provided a firm basis for the investigation of the host genetics of nodulation and N_2 fixation.

11.2 TYPES OF NODULATION RESPONSES AND GENERAL NODULATION EFFECTS

Interactions between the legume host and strains of a Rhizobium species relate to nodulation and N_2 fixation. These observed interactions are the result of genetic variation in the host and the bacteria and the interaction of these two genetic systems produce a continuum of variation. Only the genetic variation of the host will be considered in this chapter, while the genetics of the bacteria will be considered in the following chapter (III.12).

Plant appearance, number, size, weight and interior color of nodules, and nitrogen content of various plant parts are characteristics that have been used to measure nodulation responses and N_2 fixation. Using these characteristics the investigators often classified nodulation responses as either "effective" or "ineffective." Allen and Baldwin (8) were the first to use this classification. Effective nodulation was associated with strains that produced large nodules around the tap root, while ineffective strains induced many small nodules scattered over the root system. Effectively nodulated plants were green and healthy, while the ineffectively nodulated plants were small and chlorotic as were the uninoculated controls.

However, these two terms are not sufficiently descriptive to distinguish all possible nodulation responses. Distinction is not always made between the production of nodules by the host and the ability of the bacteria to induce nodulation or to fix N_2. In our research we have observed specific host-strain combinations in which the plant produced normal-appearing nodules and also expressed symptoms indicating a lack of N_2 fixation. Because the two terms were found to be inadequate, a more extensive classification system was developed (9), and a description of it is presented in Table 11.1. A key feature of this system is the separation of infection of the plant and the nodule formation process of the plant from N_2 fixation by the bacteria. To date the system has been adequate to classify most of the observed host-strain interactions. In the following sections we will relate reported host-strain interactions to this system of classification.

TABLE 11.1. Classification and Description of Nodulation Responses

Classification	Description
I. Nonnodulated	No nodules, no external evidence on root that nodule formation has been initiated. Plant is small and chlorotic, typical of nitrogen deficiency.
II. Nodulated	Roots show that nodule formation has been initiated.
A. Ineffective	Roots bear cortical proliferations and/or small nodular-like structures with white, green, or pink interiors. Plant is small and chlorotic, typical of nitrogen deficiency.
B. Effective	Roots are nodulated
1. Inefficient	Nodules have green or white interior, sometimes pink. Plant is small to medium with degree of chlorosis depending on the degree of N_2-fixing efficiency.
2. Efficient	Nodules have pink-red interior. Plant is dark green and healthy in appearance.

[a]Descriptions are of plants grown on N-limited medium.

11.2.1 Nonnodulation

The most easily recognized host-strain interaction is the lack of nodulation of a plant exposed to bacteria of a compatible species. This is nonnodulation. Nutman (10) reported that the failure of a genotype of Trifolium pratense (red clover) to nodulate when inoculated with strains of R. trifolii was controlled in the host by a single recessive gene rr in conjunction with a cytoplasmic factor called σ. In this case, the inability of the bacteria to induce nodulation was apparently due to failure of the bacteria to penetrate the root hairs even though the root hairs did curl. Using another Trifolium species, Gibson (11) reported that many naturally occurring strains of R. trifolii would not induce nodulation on Trifolium subterraneum cultivar "Woogenellup."

Williams and Lynch (12) reported a nonnodulating condition in soybean that was controlled by a single recessive gene which they designated no no. This has since been changed to rj_1 rj_1 (13). Studies with rj_1 rj_1 soybean genotypes and near isogenic normal nodulating Rj_1 Rj_1 lines have shown that R. japonicum is plentiful in the rhizosphere of the nonnodulating plants (14), that there was no difference in tryptophan excretion between nodulating and nonnodulating genotypes, and that nonnodulation is controlled in the roots of the plants (15). Elkan (16) reported a study of near isogenic nodulating and nonnodulating soybean genotypes in which root excretions of the nonnodulating genotype inhibited the formation of nodules by the nodulating genotype. However, the growth of R. japonicum was not affected by the extract.

Hubbell and Elkan (17) compared R. japonicum strains that would and would not induce nodulation on rj_1 rj_1 soybean genotypes, and reported that strains that induced nodulation were physiologically different from strains that did not induce nodulation. The characteristics of the strains capable of inducing nodulation included capsule formation, failure to reduce triphenyl tetrazolium chloride, and inability to metabolize nitrate. The plant characteristics associated with the nonnodulating genotypes were also studied by comparing inoculated and uninoculated plants of nodulating and nonnodulating genotypes (18). During the time before visible nodulation, uninoculated Rj_1 Rj_1 roots had higher amounts of protein, reducing sugars, and smaller amounts of free amino acids than did the rj_1 rj_1 roots. In the inoculated roots of both genotypes the amount of these compounds decreased. The decrease of reducing sugars and free amino acids was more rapid in inoculated rj_1 rj_1 than in the Rj_1 Rj_1 roots. High C/N ratios were associated with the Rj_1 Rj_1 plants and low C/N ratios were associated with nonnodulating plants.

Although much work has been done, the mechanisms that determine the nonnodulating response are not understood. However, the near isogenic genotypes are a valuable research tool in investigating nitrogen nutrition of the soybean.

11.2.2 Ineffective Nodulation

Although the ineffective response appears to have been the most widely studied, it remains a challenge to the researcher for the process is poorly understood. The term *ineffective* has been used with respect to nodule induction, nodule formation, and N_2 fixation. In the classification system used here, ineffective nodulation means that infection occurs and nodulation is induced, but the sequence from induction to complete normal nodule formation by the host is interrupted.

Nutman (19) reported that most plant responses to inoculation with certain Rhizobium strains were genetic. In later studies (20), he reported that a single recessive gene (i_1 i_1) in red clover conditioned an ineffective nodulation response when inoculated with Rhizobium trifolii strain A. The plants were small and showed nitrogen deficiency symptoms. However, homozygous i_1 i_1 plants produced normal nodules when inoculated with other R. trifolii strains. He also found that i_1 i_1 plants could be restored to normal-effective nodulation by the presence of a recessive repressor which he called m. Thus, i_1 i_1 m_1 m_1 plants were effectively nodulated and i_1 i_1 M_1_ were ineffectively nodulated.

Nutman (21) later described another recessive gene, ie ie (with modifiers), in red clover that conferred an ineffective symbiosis on red clover plants inoculated with certain R. trifolii strains. The root system of seedling ie ie plants appeared normal and ineffective nodules formed readily, but the plants remained small and showed symptoms of nitrogen deficiency. Although, the plant tops of ie and i_1 plants are indistinguishable the two genes are not identical, and segregate independently. In addition the i_1 reaction is very strain specific, but the ie reaction is not. Also, plants homozygous for ie differ from those homozygous for i_1 in size and number of ineffective nodules formed and, the mechanism of suppression by other host factors.

By studying the effect of ie and i_1 on nodule structure Bergersen and Nutman (22) found different mechanisms involved in conditioning the ineffective ness controlled by these two genes. The i_1 ineffectiveness was found to be due to the failure of the bacteria in the host cells to produce bacteroids. The ie ineffectiveness was due to abnormal cell division where the bacteria were released from the infection threads, producing a tumor-like growth within the nodules, but no bacteroids. With specially selected strains of R. trifolii varying degrees of effectiveness were found, and the effectivenesses were associated with corresponding amounts of bacteroid formation. In some nodules tumorized growths and normal bacteroids were found side by side.

Three host genes controlling ineffective nodulation in soybeans have been reported. Caldwell (13) found that the cultivar "Hardee" was stunted and showed symptoms of nitrogen deficiency when inoculated with R. japonicum strains of the cl and 122 serogroups (R. japonicum collection at Beltsville, Md.). Small, white cortical protuberances were produced on the roots of the plants. This ineffectiveness is controlled by a single dominant gene, Rj_2 (13), and the control is in the roots of the plant (23). This gene, Rj_2, appears to be independent of the nonnodulating gene rj_1 rj_1, (Caldwell, unpublished).

The cultivar Hardee is ineffectively nodulated with R. japonicum strain 33. This ineffectiveness is controlled by a single dominant gene Rj_3, which is distinguishable from Rj_2 (24). There is some evidence that Rj_2 and Rj_3 are linked (Vest, unpublished). An interesting difference in the nodulation responses controlled by Rj_2 and Rj_3 is the biological relationship of strains that induce this ineffective response. One, Rj_2, conditions the ineffective response to most strains of serogroups cl and 122, suggesting the possibility that the serological (surface) properties of the strains may be involved in this response; while the other, Rj_3, conditions an ineffective response to a single strain within a serogroup and not to the other serologically related strains, suggesting that factors other than serological properties of the strain are involved in the ineffectiveness.

Another ineffective response in soybean is controlled by the dominant gene Rj_4 (25). The cultivar "Hill" (Rj_4 Rj_4), when inoculated with R. japonicum strain 61, produces ineffectivenodules. The source of this ineffectiveness is the cultivar "Dunfield", which was selected from a plant introduction, and which has been used extensively in breeding programs. As a result some commercial varieties which have Dunfield in their ancestry contain Rj_4.

11.2.3 Effective Nodulation

11.2.3.1 Inefficient Type

Normal nodulation of the legume root with a subsequent lack of N_2 fixation is defined as inefficient symbiosis. An obvious example of effective nodulation and inefficient dinitrogen fixation is that of the soybean cultivar "Peking" inoculated with R. japonicum strains of serogroup 123. Large normal-appearing nodules are formed on the roots of Peking, but the plants remain chlorotic and small, symptoms typical of nitrogen deficiency. The interior of the nodules are white, green, or slightly pink. A very small amount of N_2 is fixed only in the nodule with slightly pink interiors (Sloger, unpublished). The cultivar Peking forms an efficient symbiotic association with other strains of R. japonicum, and strains of serogroup 123 form efficient symbiotic associations with other soybean genotypes.

Although the host genetics of the Peking x serogroup 123 inefficiency has not yet been worked out, there is an obvious interaction between the host and bacteria. Work done several years ago (26, 27) classifying host-strain responses are examples of earlier reports of inefficient associations between soybean genotypes and certain R. japonicum strains. However, there was no attempt to study these interactions genetically.

Gibson (28) reported an inefficient nodulation response in the subterranean clover (Trifolium subterraneum) cultivar "Northan First Early" inoculated with R. trifolii strain NA30. Nodulation responses of plants from Northam First Early crossed with six other cultivars indicated that this inefficiency was controlled by a single major gene. However, a number of modifying genes did influence the expected response ratios, and heterozygotes were intermediate in efficiency. Both Northam First Early and strain NA30 were able to establish efficient symbioses in other host-strain associations.

11.2.3.2 Efficient Type

Within the proper cross-inoculation groups the majority of the legume-Rhizobium symbiotic associations fall into the effective nodulation, efficient N_2 fixation classification. However, within this classification some combinations are more efficient than others due to interactions between the plant and bacterial genotypes.

Gibson (29) studied the efficiency of N_2 fixation in 15 lucerne (Medicago sativa) cultivars inoculated with single strains of R. meliloti that differed widely in the source plants and locations from which they were isolated. He reported considerable differences between the strains and the cultivars as to the amount of N_2 fixed. Gibson (29) further concluded that some cultivars were more efficient in their symbioses than were others, that some bacterial strains were more efficient in their symbioses than others, and that there were cultivar x bacterial strain interactions affecting the efficiency of N_2 fixation. The efficiency of combinations as classified by Gibson ranged from inefficient through intermediate to highly efficient.

In a subsequent study Gibson and Brockwell (30) evaluated nodulation and N_2 fixation efficiency in genotypes of Trifolium subterraneum, subsp. subterraneum, subsp. yannimcum, and subsp. brachycalycinum, inoculated with strains of R. trifolii isolated from the habitat of each species. They found host-strain specificities in the degree of N_2 fixation efficiency, which did not follow any taxonomic grouping of the host genotypes, nor any natural grouping of the bacterial strains. Strains isolated from one subspecies would induce effective nodulation of the other subspecies.

However, the strains isolated from a particular subspecies were generally most efficient on that subspecies.

At present there is very little genetic information of host influence on the efficiency of these symbioses. However, the extensive use and acceptance of acetylene reduction (see IV.12) as a technique of measuring N_2 fixation should encourage work on the genetic aspects of these associations.

Another approach to measuring the efficiency of symbiosis has been used recently by Abel and Erdman (31) and Caldwell and Vest (32). Abel and Erdman inoculated the soybean cultivar "Lee" with many strains of R. japonicum and found significant differences in yield, weight of nodules per plant, leaf color, oil and protein percent, and fresh plant weight. These results were obtained in soil known to be free of R. japonicum prior to planting. These associations were not observed when the same material was planted in soil known to contain populations of R. japonicum.

Caldwell and Vest (32) inoculated five soybean cultivars with several strains of R. japonicum, planted them for three successive years in soil free of R. japonicum, and measured yield differences. The variety and strain performance was influenced by the environment each year, but in general, a strain that was associated with high or low yields other years. They also reported that the consistency of the relative yields across cultivars and environments associated with certain R. japonicum strains indicated the presence of a homeostatic type of response. In other studies (33) intercultivar grafts of roots and shoots were used to obtain a relationship between root and shoot effects and nodule activity. It was found that there were specific shoot effects independent of the roots, and specific root effects independent of the shoots associated with specific nodule activity. Specific N_2-fixing activity of the nodules measured by acetylene reduction was also correlated with shoot dry weight and photosynthesis of upper leaves. These results confirm the interaction between the plant and bacterial strains in soybeans and suggest that the interaction is complex.

Another example of a host-strain interaction in the efficient category is that of the soybean cultivar "Clark" inoculated with R. japonicum strain 35. The young plants form large normal appearing nodules, but remain small and chlorotic for approximately 6 weeks. Then the plants turn green and begin to grow, and N_2 fixation is similar to normal nodulated plants. The cause of this time lag in N_2 fixation is not understood.

In other legumes, Kunelius and Clark (34) studied the yield of two birdsfoot trefoil (Lotus corniculatus) genotypes inoculated with four strains of Lotus rhizobia. They reported that high plant dry weights and dry matter yields were associated with one of the strains, while low yields were associated with another.

11.2.4 Rhizobium-induced Chlorosis

Rhizobium-induced chlorosis was first reported by Erdman and co-workers (35). Although this condition was initially attributed to a minor element deficiency, further investigation showed that the chlorosis appeared when certain soybean genotypes were inoculated with certain strains of R. japonicum. The cultivar Lee and R. japonicum serogroups 76 and 94 have been used to study this host-strain interaction. In field studies the chlorosis appeared on second, third, or later trifoliates; however, plants often recovered and leaves formed later were a normal green. Normal leaves remained green, and moderately chlorotic leaves could become green; occasionally pods were chlorotic (36). Severely chlorotic leaves became dessicated and fell to the ground. In the greenhouse, the plants were severely chlorotic and often died. Nodules formed were normal in size, shape, and distribution and had the normal dark red interiors. Johnson and Clark (37) showed that the control of the expression of chlorosis was in the roots.

Biochemical studies by Owens and Wright (38) have shown the chlorosis to be caused by a toxin. The toxin was isolated from leaves, nodules, and the cultural medium on which the strains were grown (39). This toxin was synthesized in the nodules and was translocated to young developing leaves where it blocked chlorophyll synthesis (37). Owens et al. (40) named the toxin "rhizobitoxine" and reported that it was a sulfur containing amino acid similar to cystathionine, which blocks the cleavage of cystathionine by β-cystathionase to form homocystine in the bacterium, Salmonella typhimurium. β-Cystathionase affinity for rhizobitoxine is more than 10,000 times greater than for its normal substrate cystathionine (41). The structure of rhizobitoxine was identified as 2-amino-4-(2-amino-3-hydroxypropoxy)-trans-but-3-enoic acid (42).

Soybean genotypes vary in their susceptibility to the toxin (43). The cultivar "Otootan" was resistant, while Lee was very susceptible. We have studied progeny from the cross Lee and Otootan in an attempt to determine the mode of inheritance. The studies, although inconclusive, indicated a single genetic factor with modifiers.

11.2.5 Grafting Studies

Many reported host-strain interactions have been investigated genetically. However, it is also of interest to determine the site of control of the nodulation response. This has been accomplished in some studies by grafting. Hely et al. (44) showed that Trifolium ambiguum (kura clover) roots, which would not form nodules when inoculated with R. trifolii, did form nodules when grafted with a T. repens (white clover) scion. All nodules formed were normal and efficient. The best nodulation occurred using a simple graft where the T. repens scion was entirely dependent on the roots of T. ambiguum as compared to the approach-graft relationship. In a similar study, Evans and Jones (45) reported that T. ambiguum roots with T. hybridum (alsike clover) as a scion also formed nodules. When inoculated with R. trifolii, effective nodulation also was obtained on the sexual hybrid between these two species.

From the Trifolium studies one would conclude that the control of nodulation resides in the above ground portion of the plant. By contrast, the root appears to be the site of control for at least one ineffective trait in the soybean. Caldwell et al. (23) studied the effect of grafting on two soybean genotypes, one of which was ineffectively nodulated by certain strains of R. japonicum. They concluded that the control of the Rj_2 type of ineffectiveness was in the roots. Similar results have been obtained utilizing the nonnodulating trait. Regardless of the scion used, a root stock from a nonnodulating genotype was never nodulated. We have also observed a similar response with the Rj_3 ineffective trait. However, Fischer, Lawn, and Brun (33) reported that the efficiency of N_2 fixation involves both the scion and the root. It is apparent that the control site of all interactions is not located in the same portion of all plants, thus, indicating different mechanisms of action.

11.3 NODULE DISTRIBUTION AS DETERMINED BY THE HOST

Two mechanisms will be reviewed here — strain acceptance and nodule distribution whereby the macrosymbiont can influence nodulation and N_2 fixation.

11.3.1 Strain Acceptance

In addition to its influence in specific host-strain interactions, the genotype of the host plant can also act as a biological sieve when exposed to a population of N_2-fixing bacteria in the soil. The first and most obvious discrimination is among species, i.e., R. trifolii will not infect soybeans. However, discrimination also exists within a compatible species. Robinson (46) obtained isolates from nodules of red clover and subterranean clover, and tested them for compatibility with both plants. He found that each host species nodulated faster and more effectively when inoculated with an isolate obtained from the roots of that species. Since both legumes had originally been exposed to the same R. trifolii field population, Robinson concluded that the host exerted a selective influence in accepting infections by specific R. trifolii strains in the population.

In a similar study Vincent and co-workers (47, 48) showed that when strains of R. trifolii were mixed and placed on different hosts, the host exercised selection pressure and determined the strains' relative success in the induction of nodulation. They mixed two strains of R. trifolii and inoculated red and subterraneum clover. Strain 157 occurred most frequently on red clover while strain 204 was most common on subterraneum clover. They found a similar relationship for cultivars within a species.

Working with soybeans, Caldwell and Vest (49) showed the influence that genotypes exert in selecting

R. japonicum strains from a field population. Serological data obtained from nodules of 17 genotypes for three consecutive years showed that even though the genotypes were in adjacent rows, the nodules of each genotype did not contain equal numbers of the strains present in the soil. Some of the 16 genotypes differed significantly from the check cultivar Lee; but the cultivar "Pickett," developed using Lee as a recurrent parent, had the same distribution of R. japonicum strain as did Lee. No studies were made as to the efficiency of N_2 fixation of the various compatible relationships.

An additional example of the selective influence of the host is demonstrated by the soybean cultivar Peking—R. japonicum strain 110 relationship. When grown in a field where the R. japonicum population of the soil contained as much as 60% of strain 110, usually less than 1% of the nodules of Peking contained strain 110. However, when Peking plants were grown in sterile soil or sand in the greenhouse and inoculated with only strain 110, normal effective nodules were formed.

The investigation indicates the significance of the relationship of compatibility to competition for infection sites. Studies to ascertain the nature of competition and the mechanism of exclusion of a Rhizobium genotype by the host are needed.

11.3.2 Nodulation Patterns

The host genotype has also been shown to influence the distribution of nodules on the root system. Wright (50) using the nodulation response of plants classified R. japonicum strains into two categories, Types A and B; Type A strains induced nodules clustered around the tap root, while Type B strains induced nodules scattered on the lateral roots. Type A strains fixed 1.5 times as much N_2 as Type B strains on the soybean cultivar "Ito San" (51). Allen and Baldwin (8) observed similar variations in nodule distribution with large nodules clustered around the tap root for some host-strain combinations and small nodules scattered over the entire root system for others.

Lange and Parker (52) investigated host effects on nodule distribution. They inoculated the lupines, Lupinus angustifolis, L. digitatus, L. lutens, and L.

mutabilis, with nine Rhizobium strains known to form an effective symbiosis with all the species listed. They observed three distinct nodulation patterns: (1) crown nodulation with nodules located on the tap root above the first lateral root, (2) tap root nodulation with nodules located on the tap root below the first lateral root, and (3) lateral root nodulation with nodules prevalent on the lateral roots. The dominant nodulation pattern varied with the host species.

Studies on the genetic control of nodule distribution are limited to the report of Bhaduri and Sen (53), in which they reported three nodulation patterns for species of Phaseolus and soybeans — localized, diffuse and mixed. A single genetic factor controlled the nodule distribution of one of these patterns in Phaseolus species. When P. aureus var. "N.P. 28", which had a localized nodulation pattern, was crossed with P. trilobus, which had a diffuse nodulation pattern, the diffuse pattern was dominant.

An explanation for different distribution patterns was reported by Purchase and Nutman (54). In T. pratense, infection of the root hairs by nodulating bacteria does not take place at random, but is restricted to specific foci, which are available for infection over a limited period of time. This type of mechanism could be involved in determining the nodulation pattern of a genotype. In our soybean studies we have observed differences in nodule distribution. Small nodules distributed primarily on secondary roots were normally associated with an ineffective host-strain combination. Large normal nodules found mainly on the tap root or on lateral roots near the tap root were usually associated with an effective host-strain combination. However, effective host-strain nodulation combinations have been observed from normal size nodules scattered over the root system. This trait is difficult to characterize due to its sensitivity to environmental variation, inoculum density, root diseases, and other factors. More extensive investigations are needed in this area.

11.4 PLANT GROWTH STAGES AND NODULATION

The relationship of stages of plant development and nodulation is not well understood. Pate (55) reported the synchronization of nodulation and plant development in two cultivars of peas (Pisum aureus).

Cultivars differed as to the initiation and point of maximum nodule production, but in general both cultivars exhibited the same trends and patterns. The conclusion from 2 years of data was that nodule initiation was complete by the time the plants grew to the seven-leaf stage, and the number of nodules declined during the remainder of the life cycle of the plant. Dinitrogen fixation efficiency, however, increased in the remaining nodules sufficiently to offset the loss in the number of nodules.

Additional information on the nitrogen nutrition at various stages of plant development would be of value in determining the feasibility of using applied nitrogen to supplement fixed nitrogen. This coupled with the development of nonrepressable bacterial strains could have profound influence on plant yield. Also, knowledge of N_2 fixation by various strains could lead to the development of a population of Rhizobium in the soil that would complement one another throughout the life of the plant.

11.5 INCREASING NODULE MASS BY BREEDING IN THE HOST

There are only a few reports of increasing nodule mass by selection in the host. Jones and Burrows (56) made crosses between white clover (Trifolium repens) plants selected for high total nodule mass. Selection for this characteristic was made after inoculation with a known effective and efficient strain of R. trifolii. The selection of these plants was evidently successful in that both the F_1 and F_2 progeny of the crosses had significantly higher nodule scores than did the parental population. Nodule mass increased significantly in F_1 but there was only a slight increase from F_1 to F_2. They further observed that the efficiency of N_2 fixation on a nodule basis decreased as nodule size increased. As selection increased the total amount of nodule tissue, there was also a significant increase in nodule number in the F_1. These results (56) are in agreement with those of Chen and Thornton (57), who showed a correlation between effectiveness and total nodule tissue in red clover and similar results of Jones (58) with white clover. Bowen (59) reported a highly significant correlation between individual plant growth and total nodule weight in

Centrosema pubescens. The plant weight in this study was related to nodule weight and not nodule number.

One other report that indicates the complexity of the relationship of nodule tissue to plant growth is that of Nutman (6) who reported that the amount of nodular tissue formed was constant and independent of the number of nodules. He found that specific nodule volume was a property of the host and was independent of the bacteria. Yet plants inoculated with effective strains showed variation in nodule mass that was not correlated with yield. These data indicate that the number of nodules alone is not important in N_2 fixation, that the total nodule volume does not always indicate the amount of active tissue, and that tissues may vary in activity depending on the host-strain relationship. Physiological processes have also been shown to regulate dinitrogen fixation (III.8, III.10).

11.6 NEW CONCEPTS OF INCREASING N_2 FIXATION THROUGH HOST GENETICS

Geneticists and plant breeders have been effective in utilizing genetic variability to develop cultivars resistant to plant pathogens. They have also been effective in selecting cultivars that will respond to fertilizers, especially applied nitrogen. Very limited attention has been given to improving the nitrogen nutrition of legumes by breeding for more effective and efficient host-strain combinations. There are several approaches that could be taken to increase N_2 fixation.

One procedure would involve searching for a strain of Rhizobium that will fix the greatest amount of N_2 with a particular genotype. In current legume cultivar development, the cultivar is released based on performance in a given geographical region. Therefore, the average response to a general population of Rhizobium is a selection criterion. A similar approach would be to identify the inefficient strain in a region and develop a resistance to infection by these strains. In this approach, the susceptibility to the efficient strains would be preserved. Cultivars developed by this approach would have very specific adaptibility and a requirement that the compatible strain be present or applied with the inoculum.

A second procedure would be to select cultivars that will nodulate when high levels of N are present in the soil. If such a procedure were employed, host-strain combinations could be employed that would utilize applied N to complement the symbiotically fixed nitrogen.

Another area that needs further investigation relates to the N_2 fixed at various stages of plant growth. The rate of fixation at each stage of growth has been reported by Hardy et al. (60) and Weber et al. (61). For some legumes an increase in fixation during early development would be beneficial. For soybeans, an increase in fixation during flowering and pod set could contribute to increased yields. Also, prolonging period of maximum fixation would be of value.

Host-strain interactions investigated to date, and summarized in this chapter, indicate that genetic variability for nodulation and N_2 fixation does exist. However, this variability remains to be exploited.

11.7 REFERENCES

1. Vorhees, J. H., J. Am. Soc. Agron., 7, 139 (1915).
2. Wilson, P. W., J. C. Burton, and V. S. Bond, J. Agr. Res., 55, 619 (1937).
3. Burton, J. C., and P. W. Wilson, Soil Sci., 47, 293 (1939).
4. Aughtry, J. D., Cornell Univ. Agr. Exp. Stat. Memoir, 280, 18 pp., 1948.
5. Nutman, P. S., Nature, 157, 463 (1946).
6. Nutman, P. S., Symp. Soc. Exp. Biol., 13, 42 (1959).
7. Erdman, L. W., and U. M. Means, Agron. J., 45, 625 (1953).
8. Allen, O. N., and I. L. Baldwin, Wisc. Agr. Exp. Stat. Res. Bull., 106, 56 pp., 1931.
9. Vest, G., D. G. Weber, and C. Sloger, in Soybean Science, Am. Soc. Agron. Monograph, 1973.
10. Nutman, P. S., Heredity, 3, 263 (1949).
11. Gibson, A. H., Aust. J. Agr. Res., 19, 907 (1968).
12. Williams, L. F., and D. L. Lynch, Agron. J., 46, 28 (1954).
13. Caldwell, B. E., Crop Sci., 6, 427 (1966).
14. Clark, F. E., Can. J. Microbiol., 3, 113 (1957).
15. Tanner, J. W., and I. C. Anderson, Can. J. Plant Sci., 43, 542 (1963).

16. Elkan, G. H., Can. J. Microbiol., 7, 851 (1961).
17. Hubbell, D. H., and G. H. Elkan, Can. J. Microbiol., 13, 235 (1967).
18. Hubbell, D. H., and G. H. Elkan, Phytochemistry, 6, 321 (1967).
19. Nutman, P. S., Heredity, 8, 35 (1954).
20. Nutman, P. S., Heredity, 8, 47 (1954).
21. Nutman, P. S., Heredity, 11, 157 (1957).
22. Bergersen, F. J., and P. S. Nutman, Heredity, 11, 175 (1957).
23. Caldwell, B. E., K. Hinson, and H. W. Johnson, Crop Sci., 6, 495 (1966).
24. Vest, G., Crop Sci., 10, 34 (1970).
25. Vest, G., and B. E. Caldwell, Crop Sci., 12, 692 (1972).
26. Boyes, J., and G. Bond., Ann. Appl. Biol., 29, 103 (1942).
27. Ruf, E. W., and W. B. Sarles, J. Am. Soc. Agron., 29, 724 (1937).
28. Gibson, A. H., Aust. J. Agr. Res., 15, 37 (1964).
29. Gibson, A. H., Aust. J. Agr. Res., 13, 388 (1962).
30. Gibson, A. H., and J. Brockwell, Aust. J. Agr. Res., 19, 891 (1968).
31. Abel, G. H., and L. W. Erdman, Agron. J., 56, 423 (1964).
32. Caldwell, B. E., and G. Vest, Crop Sci., 10, 19 (1970).
33. Fischer, K. S., R. J. Lawn, and W. A. Brun, Am. Soc. Agron. Abstr., 33 (1972).
34. Kunelius, H. T., and K. W. Clark, Can. J. Plant Sci., 50, 717 (1970).
35. Erdman, L. W., H. W. Johnson, and F. E. Clark, Plant Dis. Rptr., 40, 646 (1956).
36. Johnson, H. W., U. M. Means, and F. E. Clark, Agron. J., 50, 571 (1958).
37. Johnson, H. W., and F. E. Clark, Soil Sci. Am. Proc., 22, 527 (1958).
38. Owens, L. D., and D. A. Wright, Plant Physiol., 40, 927 (1965).
39. Owens, L. D., and D. A. Wright, Plant Physiol., 40, 931 (1965).
40. Owens, L. D., S. Guggenheim, and J. L. Hilton, Biochim. Biophys. Acta, 158, 219 (1968).
41. Owens, L. D., Science, 165, 18 (1969).
42. Owens, L. D., J. F. Thompson, R. G. Pitcher, and T. Williams, J. C. S. Chem. Commun., 714 (1972).
43. Johnson, H. W., and U. M. Means, Agron. J., 52, 651 (1960).

44. Hely, F. W., C. Bonnier, and P. Manil, Nature, 181, 1272 (1953).
45. Evans, M., and D. G. Jones, Ann. Bot. (London), N. S., 28, 222 (1964).
46. Robinson, A. C., Aust. J. Agr. Res., 20, 1053 (1969).
47. Vincent, J. M., and L. M. Waters, J. Gen. Microbiol., 9, 357 (1953).
48. Vincent, J. M., and L. M. Waters, Aust. J. Agr. Res., 5, 61 (1954).
49. Caldwell, B. E., and G. Vest, Crop Sci., 8, 680 (1968).
50. Wright, W. H., Soil Sci., 20, 95 (1925).
51. Wright, W. H., Soil Sci., 20, 131 (1925).
52. Lange, R. T., and C. A. Parker, Nature, 186, 178 (1960).
53. Bhaduri, P. N., and R. Sen, Indian J. Gen. Plant Breeding, 28, 287 (1966).
54. Purchase, H. F., and P. S. Nutman, Ann. Bot. (London), N. S., 21, 439 (1957).
55. Pate, J. S., Aust. J. Biol. Sci., 11, 366 (1958).
56. Jones, D., and A. C. Burrows, J. Agr. Sci., CAMB, 71, 73 (1968).
57. Chen, H. K., and H. G. Thornton, Proc. Roy. Soc. London, Ser. B, 129, 208 (1940).
58. Jones, D. G., J. Sci. Food Agr., 13, 598 (1962).
59. Bowen, G. D., Queensland J. Agr. Sci., 16, 267 (1959).
60. Hardy, R. F. W., R. D. Holsten, E. K. Jackson, and R. C. Burns, Plant Physiol., 43, 1185 (1968).
61. Weber, D. F., B. E. Caldwell, C. Sloger, and H. G. Vest, Plant Soil, Special Vol., 293 (1971).

CHAPTER 12

Genetic Aspects of Nodulation and Dinitrogen Fixation by Legumes: The Microsymbiont

E. A. SCHWINGHAMER

Division of Plant Industry
CSIRO
Canberra, Australia

12.1. Introduction, 578
12.2. Mutation, 579
    12.2.1. Loss of Symbiotic Ability, 580
        12.2.1.1. Effect of Resistance to Some Adverse Factors in Artificial Culture or in the Field, 580
        12.2.1.2. Effect of Resistance to Antibiotics or Antimetabolites, 583
        12.2.1.3. Effect of Auxotrophy, 586
    12.2.2. Gain of Symbiotic Ability, 588
        12.2.2.1. Screening Effect of Plant Passage, 588
        12.2.2.2. Mutagenic Treatment and "Direct" Assay on Plants, 589
        12.2.2.3. Mutation to Prototrophy and Selection for Maximum Synthesis of Some Growth Factors, 590
    12.2.3. Modification of Strains for Increased Survival or Competitive Ability, 592
    12.2.4. Modification of Strains for Genetic or Ecological Research, 593

12.3.   Genetic Transfer within the Genus
        Rhizobium, 594
        12.3.1.   Transformation, 594
        12.3.2.   Transduction, 601
        12.3.3.   Conjugation, 602
        12.3.4.   Plasmid Transfer or Control, 604
12.4.   Genetic Transfer between Rhizobium and Related
        Genera, 605
12.5.   Other Genetic Aspects, 607
        12.5.1.   Replication Mapping, 607
        12.5.2.   Rhizobium-Phage Interactions, 607
                  12.5.2.1.   Lysogeny, 607
                  12.5.2.2.   Host-controlled
                              Modification, 608
                  12.5.2.3.   Mapping of the Phage
                              Genome, 608
                  12.5.2.4.   Transfection, 609
        12.5.3.   Bacteriocinogeny, 609
12.6.   Bacterium or Host Plant: Which Partner
        Controls the Key Steps in the Symbiosis? 610
        12.6.1.   Inferences from Studies on the Bio-
                  chemistry of Dinitrogen Fixation, 610
        12.6.2.   Dilworth-Parker Hypothesis: Sharing
                  of Genetic Information, 611
12.7.   Conclusions, 612
12.8.   References, 614

12.1 INTRODUCTION

Despite the considerable number of publications concerning genetics of Rhizobium which have appeared during the past two decades our knowledge of the inheritance of factors controlling symbiotic ability of Rhizobium is still quite limited. The primary obstacle, similar to that encountered in studies of inheritance of pathogenicity in plant pathogens (especially bacteria), has been the procedural problem of quantitatively assaying heterogeneous populations on the host plant. This obstacle should, hopefully, be at least partly overcome by indirect (in vitro) selection methods developed recently, and some progress in mapping of bacterial genes for symbiosis may be anticipated.

Research relevant to genetics of the legume nodule microsymbiont may be divided into two chronological periods. The period from the first isolation of the bacterium in 1888 to about 1953 was marked by observations or experiments concerning variability (spontaneous mutation or adaptation) of cultural and symbiotic properties through storage in culture media or soil and through plant passage. Studies on taxonomic groupings or on the validity of established cross-inoculation groups were also well represented. Research since 1953 has dealt with mutation (spontaneous and induced) and with mechanisms of genetic exchange occurring in Rhizobium. Studies of genetic exchange involved mainly transfer of cultural markers and examination of experimental variables, although some nonquantitative demonstrations of transfer of symbiotic properties have been described. While this chapter is intended as a comprehensive review of the genetics of Rhizobium, the volume of the relevant literature precludes a survey of most of the papers published during the first period. For these the reader is referred to several extensive reviews on biological properties of the rhizobia — notably those of Fred et al. (1), Wilson (2), and Allen and Allen (3, 4) — and to a brief survey of genetics of Rhizobium by Lorkiewicz (5).

The major emphasis in the present review centers on Rhizobium genetics research of the past decade, a period that has experienced rapid expansion of interest and publication in this field. Research on transformation in rhizobia prior to 1967 has been reviewed by Marečková et al. (6), and a brief survey of genetics literature appearing before 1969 was included as part of a review of genetics of the symbiosis (mainly the host plant) by Nutman (7). The genetics of the legume host plant, the macrosymbiont, is the subject of the preceding paper (III.11). Readers not fully conversant with the terminology and mechanisms of genetic transfer in bacteria are referred to the text of Hayes (8) for general information; several reviews concerning more detailed information on specific mechanisms are also indicated where appropriate in this chapter.

12.2 MUTATION

Research concerning the role of mutation in contributing to genetic variability in Rhizobium will be considered mainly with regard to effect on the two arbitrarily defined symbiotic properties commonly assigned to the microsymbiont: infectiveness (ability to induce nodule formation), and effectiveness [ability to fix dinitrogen (N_2) in the nodules]. Loss of either character, particularly effectiveness, has been observed much more frequently than gain, although it is recognized that the common use of effective strains in the laboratory limits observations on spontaneous gain in symbiotic ability. The use of mutagens and selective growth procedures has greatly facilitated study of change in both directions (Tables 12.1 and 12.2). Mutation involving cultural characters will be considered in this chapter only insofar as they relate to pleiotropic effects on symbiosis or to marker application. Most of the reports concerning attempts to correlate differences in symbiotic ability with various cultural or physiological properties are also omitted here because known mutants were not used (i.e., no common genetic background in the strains compared) and few positive correlations were encountered. For this subject the reader is referred to several reviews (ref. 1, 4; III.7) of the literature prior to 1958 and to a brief discussion of more recent papers by Damery and Alexander (9).

12.2.1 Loss of Symbiotic Ability

12.2.1.1 Effect of Resistance to Some Adverse Factors in Artificial Culture or in the Field

Partial or full loss of symbiotic ability in some strains of rhizobia during culture and storage on commonly used artificial media has been observed by many investigators (1, 3, 4). As an example, in one of the more extensive studies of longevity and effectiveness, Means and Erdman (10) found marked differences in the stability of strains within all cross-inoculation groups of rhizobia, although differences between groups were also observed. The authors suggested the feasibility of selecting inoculant strains for improved stability under different conditions of growth or storage. Storage in soil has also been reported to decrease or increase effectiveness in different strains (1, 3, 4, 11).

It is probable that growth of rhizobia on most artificial media will almost invariably result in some shift in the balance of symbiotically active and inactive cells in a culture. The presence in the medium of inhibitory levels of growth factors like glycine or some other amino acids (12-21) strongly favors the growth of resistant variants that also tend to be less effective than the wild-type cells. Most of the evidence for such "attenuation" of effectiveness stems from serial adaptation of mass cultures to increasing concentrations of some amino acids, but the mutational basis of the shift in symbiotic properties has been established with resistant clones (17, 19, 22). D-Amino acids, some of which are normal components of bacterial cell wall peptides, exert a particularly strong selective effect favoring survival of ineffective or noninfective cells (13, 14, 18-20, 22). It is likely that complex media containing such growth factors at slightly inhibitory levels have contributed to the many reported cases of strain instability in culture. This belief is supported by the observation (23) that yeast extract when present in a medium of 3.5 g/l (no obvious effect at 1 g/l) is inhibitory to R. leguminosarum and induces rapid formation of "artificial bacteroids." Loss of infectiveness or effectiveness was observed in two strains of R. meliloti following serial culture in a medium containing yeast extract at concentrations above 5 g/l (24). The authors emphasized the potential hazard involved in growth of commercial inoculants when media are nutritionally enriched to shorten the generation time. Commercial yeast extracts were found to vary considerably in their glycine content and R. meliloti strains were more tolerant to glycine than were strains of three other species (25). An interaction between glycine and monovalent cations was the main cause of yeast extract toxicity, and this toxicity could be prevented by adding Ca^{2+} or other divalent cations to the medium.

There is less information available concerning specific conditions in the field that would favor survival and growth of mutants with altered symbiotic properties. However, there is some indication that ineffective nodulation is associated with tolerance to adverse factors in some field soils. The prevalence of ineffective, clover-nodulating rhizobia in sandy acid soils (26-29) has been ascribed to a buildup of ineffective mutants that are tolerant to high concentrations of metal ions (26, 27, 29, 30) like Mn^{2+}, Cu^{2+}, or Al^{3+}. Similarly, rhizobia acquiring resistance to tannic acid or gallic acid (31) may contribute to predominance of ineffective bacteria in old-field prairie soils in which inhibitory levels of such phenolic compounds accumulate following growth of certain species of plants.

Bacteriophages exert a selective effect on variant types during growth of rhizobia in culture and a similar role has been implicated under some field conditions (ref. 3, 4; III.7). Variants differing from the parent strains in colonial morphology or nodulating properties were isolated (32, 33). These results were confirmed and extended in R. trifolii (34-37) where a high proportion of phage-resistant mutants of effective strains were ineffective, whereas resistant variants from ineffective parent strains only rarely showed a gain in effectiveness. Loss of effectiveness occurred independently of change in colony morphology. Some ineffective variants were stable in culture and in symbiosis, while others remained unstable even after several passages through the plant. Kleczkowska concluded (35) that the action of phage involved selection of prexisting ineffective mutants in a culture rather than "induction" of mutants as implied in some previous reports. She further considered it unlikely that phage could directly cuase failure of legume crop nodulation (as in "alfalfa sickness") by destruction of nodule bacteria (38), but rather that failure could result from an increase in the proportion of resistant, ineffective rhizobia in the soil.

Suppression of effective nodulation in some soils has also been attributed to microbial antagonism (ref. 3, 4, 39-41; III.7), although evidence is largely lacking concerning the relative importance of a selective genetic basis (selection or induction of ineffective resistant variants) versus a nonselective lethal or growth-inhibitory effect. However, it has been shown that inoculant strains differed in their ability to nodulate the host plant in soil preinoculated with some actinomycetes (42), and that continued growth of rhizobia in the presence of antagonistic actinomycetes (41) in culture can result in impaired symbiotic ability.

Aside from impairment of symbiotic ability by microorganisms that are lethal (phage) or inhibitory (actinomycetes, etc.) to rhizobia, microbial attack on the host legume plant apparently can also cause ineffective nodulation. Thus, in one report (43) a predomi-

nance of ineffective type nodules was observed on
white clover plants infected with clover phyllody
"virus" and inoculated with effective R. trifolii
(see ref. 44 for more recent evidence that the pathogen
is mycoplasma-like rather than viral). Bacteria from
such nodules were still effective in nodulation tests
on disease-free plants, indicating a direct, nongenetic
suppression of the symbiosis on diseased plants. However, in analogous experiments on white clover,
bacterial isolates obtained from ineffective type
nodules on diseased plants remained ineffective in
tests on healthy plants (45). The authors attributed
the ineffectiveness to an unknown modification in the
rhizobia induced either by the "virus" or by the
infected tissue and considered some possible genetic
mechanisms, including mutations and genetic transfer.
The ineffective rhizobia were also found to be highly
competitive for nodulation (46), even on diseased
plants, thereby having gained a possible ecological
advantage. The disagreement between the above results
may be partly due to differences in virulence of the
pathogen used, since the level of pathogenicity
appeared to influence the incidence or degree of loss
of effectiveness (45).

12.2.1.2 Effect of Resistance to Antibiotics or Antimetabolites

Mutants resistant to toxic agents like antibiotics or
antimetabolites (Table 12.1) have been used mainly for
marker purposes in genetic or ecological experiments
and for biochemical study of associated blocks in
symbiosis. Streptomycin resistance has been the most
frequently used marker character for genetic experiments because highly resistant, single-step mutants
are easily isolated and remain stable in culture or
through plant passage. Resistance to this antibiotic
occurs independently of changes in symbiotic ability
in many strains (34, 47-50), although loss of effectiveness has been found to occur readily (depending on
the level of resistance) in some strains of R. trifolii
(51, 52). Nonalteration of the symbiotic characteristics of the marked strains is a requisite for most
genetic or ecological experiments. However, in a
search for additional markers, Schwinghamer (53) found
spontaneous mutation for resistance to viomycin and
neomycin to be closely associated phenotypically with

TABLE 12.1. Symbiotic Properties of Rhizobium as Affected by Mutation for Resistance to Various Growth Inhibitory or Toxic Agents

Mutation for resistance to	Most commonly observed change in symbiotic properties, relative to inf^+ eff^+ wild-type strain
Antibiotics:	
Group I (str, chl, spi)	None, or partial loss of eff
Group II (D-cyc, nov, van, pen)	Partial loss of eff, in < 50% of mutants
Group III (vio, neo)	Loss of eff, in high proportion of mutants
Other toxic or inhibitory substances:	
Glycine	Loss, or partial loss of eff in some mutants
Some amino acid analogues	Loss of eff in ca. 50% of mutants (R. trifolii)
Some D-amino acids	Loss of inf in ca. 50% of mutants (R. leguminosarum)
Metal ions (Mn^{2+}, Cu^{2+}, Al^{3+})	Loss of eff
Phenolic compounds from plants	Loss of eff
Pesticides, seed-coat toxins, etc.	No change in inf or eff?
Other agents:	
Bacteriophage (virulent)	Loss of eff
Microbial antagonists	Loss of eff?

Abbreviations: str, streptomycin; chl, chloramphenicol; spi, spiramycin; D-cyc, D-cycloserine; nov, novobiocin; van, vancomycin; pen, penicillin; vio, viomycin; neo, neomycin; inf, infectiveness; eff, effectiveness.

loss of effectiveness in three cross-inoculation groups of rhizobia. Mutation for resistance to antibiotics (49) that inhibit cell membrane or wall synthesis (D-cycloserine, vancomycin, penicillin, etc.) in bacteria resulted in a less frequent and less drastic (partial loss of effectiveness) pleiotropic effect. Mutation for resistance to antibiotics known to inhibit protein synthesis (streptomycin, spiramycin, chloramphenicol) appeared least likely to affect symbiotic properties (see III.7).

Alteration of cell wall or membrane structure or function was postulated as contributing to loss of effectiveness in resistant mutants (22, 49). Subsequent investigations (54-60) on the biochemical nature of viomycin resistance in R. meliloti have established that resistance involves decreased uptake of the antibiotic as a result of lowered cell membrane permeability (57, 58). The latter condition was attributed to a higher content of phospholipids which link to viomycin-sensitive sites on the cell envelope by calcium bridging and complex the antibiotic (59, 60). This finding supports the suggestion that loss of cell wall or membrane integrity through any of a number of mutations affecting these cell structures may be an important factor in failure of rhizobia to develop into bacteroids in ineffective nodules.

Ineffective mutants of R. trifolii and R. meliloti (9) isolated on the basis of resistance to kanamycin produced more polysaccharide than the effective parent strains but did not differ consistently in vitamin B_{12} synthesis or in excretion of vitamins or amino acids (12).

Other studies on pleiotropic effects of mutation for resistance to toxic substances (Table 12.1) have involved antimetabolites, rather than antibiotics, because of their greater specificity of action in the cell. Almost half of the mutants selected from effective strains for resistance to D-amino acids (D-alanine, D-histidine, D-leucine, D-methionine) and several amino acid analogs (norleucine, norvaline, and p-fluorophenylalanine) were defective in symbiosis (22). Mutants were cross-resistant, and the substances readily induced spheroplast formation in nonmutant, sensitive cells. These results were considered to support the interpretation of cell wall/membrane defects as previously applied to antibiotic-resistant mutants. Unlike the latter, the block in symbiosis occurred at two different stages of nodulation, depending on the species. Most of the R. trifolii mutants were able to produce ineffective nodules on clover, while those of R. leguminosarum were usually nonnodulating on pea (22). Partial auxotrophy (nutritional dependence) was also observed in some of the resistant mutants.

12.2.1.3 Effect of Auxotrophy

Auxotrophy (Table 12.2), like resistance to antibiotics or antimetabolites, has been used in research on genetics and physiology of Rhizobium, both as a marker (Table 12.3 and 12.4) and as a tool for probing the nature of induced defects in symbiosis (61, 62). Unlike mutants resistant to antibiotics or other substances, auxotrophs offer the important advantage of easy detection of back mutants, which allow study of genetic or biochemical associations to be carried further. In particular, auxotrophs allow study of the possible role of specific bacterial metabolites at various stages of nodule formation. Loss of symbiotic ability in most auxotrophs is unrelated to the growth factor requirement (63, 64). Evidence for a possible valid genetic-biochemical relationship should meet at least two of the following three criteria: (1) a high frequency of loss in symbiotic ability among mutants with a specific nutritional dependency, (2) a high frequency of restoration (genetic) of symbiotic activity among prototropic back mutants, and (3) partial or full biochemical promotion of effective symbiosis by addition of the required growth factors to plants inoculated with the auxotroph.

A wide range of amino acid or vitamin requirements was noted (63) among auxotrophs isolated from an irradiated effective strain of R. trifolii, but all of the auxotrophs were noninfective, parasitic, or ineffective, and all prototrophic revertants examined remained defective. There was, thus, no evidence of close relationship between requirement and symbiotic activity. In view of the rather high dose (survival 10^{-4}) of UV radiation used by these workers, the high incidence of reduced symbiotic activity among such a diverse group of auxotrophs (see ref. 65 for similar results with nonauxotrophs of R. meliloti) might also be explained on the basis of possible plasmid inheritance and plasmid elimination. In contrast, a

TABLE 12.2. Symbiotic Properties of Rhizobium as Affected by Presence or Absence of Some Growth Factor Requirements[a]

Type of change observed in symbiotic properties of auxotrophs	Growth factors required by auxotrophs	Restoration of symbiotic activity[b]		
		Biochemical	Genetic Prototrophic mutants	Prototrophic recombinants
Gain of eff	gly (66, 67)			
None (no loss of eff)	pan, etc. (64) cys, met (66, 67)			
Partial loss of eff	thi, chol, inos, his, pur, pyr; partial, complex requirements (64)	−	+, −	
Loss of eff	amino acids, vitamins, pur (63) lys (65, 130, 131) pur, pyr (66, 67) rib (64, 62) leu (68)	− + +	− − + + +	− +
Loss of inf	ade + thi (49, 64) his (190) amino acids, vitamins, pur (63)	+(partial)	+ − −	

[a]Growth factors: gly, glycine; his, histidine; pan, pantothenate; cys, cysteine; met, methionine; thi, thiamine; chol, choline; inos, inositol; leu, leucine; rib, riboflavin; ade, adenine; pur, purines; pyr, pyrimidines. Reference numbers are in parentheses. Symbiotic properties: inf, infectiveness; eff, effectiveness.
[b]Biochemical "restoration" assay: required growth factors added to plants inoculated with the auxotroph. + or − denotes restoration or nonrestoration, respectively.

riboflavin-dependent mutant of R. trifolii (62, 64) showed full restoration of effectiveness in prototrophic revertants and a fully effective plant response following addition of riboflavin or its coenzymes to the inoculated plant. This auxotroph has been used in physiological studies concerning the nature of the riboflavin requirement during symbiosis on clover (61, 62). Another auxotroph, a R. leguminosarum mutant that had a double requirement (adenine + thiamine) and was noninfective on pea (64), also recovered the parental level of effective nodulation by back mutation to prototrophy, but addition of the growth factors to the plant allowed only a partial response; the nodules produced did not develop beyond the ineffective stage.

An in vitro-in vivo relationship has also been found in purine- and pyrimidine-requiring mutants of R. meliloti by Dénarié and co-workers (66, 67). Reversion to prototrophy was accompanied by restoration of effectiveness, but plants inoculated with the auxotrophs did not respond to purine or pyrimidine amendment. Amino acid auxotrophs showed no impairment in symbiotic ability, but glycine auxotrophs were more effective than the parent strain. A recently described leucine auxotroph (68) of R. meliloti appears to be very similar to the riboflavin mutant of R. trifolii in that effectiveness can be restored by back mutation and effective nodulation can be promoted (nongenetic change) by biochemical amendment. In addition, genetic restoration can be accomplished by transduction to prototrophy (see III.12.3.2).

While the changes associated with mutation to auxotrophy generally involve lowered symbiotic ability, there appear to be exceptions, depending on the growth factor required and on the initial level of effectiveness of the parent. The increased effectiveness of the glycine auxotrophs (66, 67) is of added interest in view of the commonly observed opposite effect associated with resistance to glycine.

12.2.2 Gain of Symbiotic Ability

12.2.2.1 Screening Effect of Plant Passage

The question of whether the symbiotic effectiveness of rhizobia is measurably modified by passage through the plant (i.e., in the nodule) has been the subject of much research (1, 3, 4, 11, 69). Earlier reports of significant changes in effectiveness following host passage could not be confirmed (11, 69), when pure lines (clones) of rhizobia were used in bacteriologically controlled nodulation experiments. Nutman discussed some of the practical difficulties of reliably demonstrating gain or loss, and differences in genetic stability or homogeneity of strains used by different workers were considered to be an important factor in the conflicting results. Similarly reported cases of gradual effective "adaptation" to various legumes by ineffective rhizobia in field soil probably involved selective multiplication of a few effective bacteria already present in the soil (7).

It appears plausible that plant passage could impose some selective pressure against pre-existing variants of lower or greater effectiveness, probably in the early stage of root infection. The possible role of the plant in such "recognition" of strains has more recently been examined (70), and clover plants were found to discriminate against ineffective strains in mixed inocula of R. trifolii. At the least, it can be assumed that plant passage would eliminate non-infective rhizobia or bacterial contaminants (including atypical colony types that may be mistaken as variants of Rhizobium) from highly heterogeneous or contaminated cultures, with a resultant apparent "gain" in effectiveness among clones isolated from effective nodules.

Aside from selective effects of the plant on infection by existing mutants it has also been noted that spontaneous variation may occur in the nodule (11, 71), as in soil or in culture, at an unknown but presumably very low frequency. In particular, Krassilnikov and Melkumova (71) found the diversity of forms to be much greater in the nodule than in culture media. In the light of present-day knowledge of chemical mutagenicity by numerous substances at various levels of toxicity, the possibility of increased mutation rates in the nodule, notably during the host cell disintegrative phase of ineffective or senescent nodules, should not be overlooked.

12.2.2.2 Mutagenic Treatment and "Direct" Assay on Plants

Use of mutagens to induce an increase in symbiotic ability of rhizobia has been limited largely by practical difficulties of detecting the desired mutants (72). Gain of infectiveness (i.e., extension of host range) can be reasonably attempted by direct inoculation of the nonpermissive host with mutagen-treated rhizobia, although quantitation is not feasible. Radiation-induced mutants of R. trifolii that nodulated pea seedlings ineffectively were detected in this manner (73). Irradiated bacteria in these experiments were grown in mass culture for phenotypic expression of mutation before plant inoculation, and nodule isolates were identified by presence of antibiotic resistance markers.

Direct screening of mutagen-treated bacteria for increased effectiveness on the plant is even less practicable since only a very limited number of infections (nodules) can be examined and only a major change (e.g., ineffective to effective) can be detected with reasonable reliability on a single-nodule basis. For detection of smaller increments of effectiveness, inoculation with individual clones is essential; this, in turn, is feasible only if the clones are preselected in vitro on the basis of some linked marker, as discussed later. An increase in effectiveness of clones grown from irradiated R. meliloti has been noted (74). Since the plant tests involved only a small number of clones selected as colony-type variants, the detection of three clones with altered effectiveness would appear to indicate association with the rough colony type, but the changes were considered to occur independently. Increased effectiveness was also reported for strains of R. lupini (75) and R. japonicum (76) following irradiation with gamma rays. Clones tested on plants were selected at random from colonies of surviving bacteria, but unfortunately (with regard to conclusions concerning a significant mutagenic effect) the amount of spontaneous variation present in the nonirradiated, wild-type bacteria was not similarly assayed on a clonal basis. Some streptomycin-resistant and neomycin-resistant (strain-dependent) mutants of R. meliloti were found (47) to be more effective than the parent strain.

12.2.2.3 Mutation to Prototrophy and Selection for Maximum Synthesis of Some Growth Factors

Restoration of effectiveness in some auxotrophs by reversion to prototrophy (Table 12.2), aside from the relevance to genetic or biochemical study of the symbiosis, offers a potential means of improving the symbiotic activity of a strain of Rhizobium. In view of accumulating evidence that ability of the bacterium to synthesize certain growth factors favors effective symbiosis, at least on some hosts, it was postulated (64) that some prototrophic back mutants might surpass the original, nonauxotrophic parent in N_2 fixation. A model has been proposed (72) for strain improvement by isolating suitable auxotrophs and testing on plants only the prototrophic clones that have been preselected on the basis of maximum growth or of growth factor synthesis in a minimal medium. Alternatively the selection pressure could be directly applied to the nonauxotrophic parent strain for isolation of variants with superior synthetic capacity. Plant nodulation tests could then be confined to a small number of clones in which the probability of mutation alteration of genes affecting symbiotic activity is very high, assuming that the growth factor requirement involved satisfies the criteria already discussed. Such procedures could facilitate attempts to induce small but significant gains in the effectiveness of partly effective inoculant strains that are exceptional strains in other characteristics (e.g., saprophytic competence), since such small increments of effectiveness can only be reliably detected on a clonal basis.

Evidence has already been presented for a slight gain of effectiveness in some prototrophic revertants from a purine auxotroph of R. meliloti (66, 67). Support for a relationship between maximum nutritional independence and maximum symbiotic capacity stems also from studies of nonmutant strains of rhizobia. Shemakhanova and co-workers (77, 78) compared the vitamin-synthesizing capacity, in culture, of strains having different levels of symbiotic activity and concluded that the more "active" strains were also superior in synthesis of riboflavin or pyridoxine in R. phaseoli, and of cobalamin or riboflavin in three other species. "Exceptional" strains of R. japonicum were observed to be more sensitive to asparagine than were strains of normal effectiveness, and this difference was suggested as a possible criterion for laboratory selection of superior strains (79, 80). Nodules produced by the exceptional strains also were high in molybdenum (a component of nitrogenase) content, but not in leghemoglobin; strain differences were tentatively ascribed to differences in the N_2-fixation system or in incorporation of ammonia into arginine relative to other amino acids.

Information gleaned from studies on nutritional requirements and on depression of symbiotic activity by excess glycine or some other metabolites in culture media points to the laboratory culture medium as a likely major factor affecting symbiotic capacity of inoculant strains. A possible depressant effect of growth on nitrogen-rich (gelatin) media was recognized as early as 1893 (1, 3), but complex media enriched by yeast extract or plant extracts have been commonly used for growth and maintenance of rhizobia. It would appear that use of synthetic media with a minimum content of added growth factors should tend both to eliminate many ineffective mutants (either auxotrophic or resistant to inhibitory levels of amino acids) and to favor the predominance of the most effective, nutritionally independent variants. A review (208) concerning effects of mutations on symbiotic properties of Rhizobium has recently appeared.

12.2.3 Modification of Strains for Increased Survival or Competitive Ability

Inoculant strains may differ in their ability to survive in the soil and to colonize the rhizosphere (refs. 29, 81-83; also IV.9), as well as to nodulate legume plants. The predominance of certain native rhizobia populations in different soil environments also indicates the adaptive ability of the bacteria under stress of adverse environment or microbial competition. It should, therefore, be possible to select saprophytically competent strains (82) from such populations or to modify inoculant strains by mutation or other genetic means. However, suitable experimental procedures for detection and isolation of variants with superior tolerance to most adverse factors (biological, chemical, physical) in the complex field soil environment are lacking. Furthermore, as mentioned before, mutation for resistance to factors like bacteriophage, antibiotics, and various toxic substances that can occur in the soil frequently is linked with depressed symbiotic ability. Nonetheless, mutants that were resistant to some pesticides but retained the symbiotic activity of the parent strain have been isolated (84).

Pigmented variants (85, 86) which were more resistant to heat, to UV light, and presumably also to sunlight, were ineffective; the variants recovered infectivity but lost their pigmentation character following irradiation or plant inoculation. Unfortunately, no other markers were used to identify the nonpigmented parent strains. A small proportion of cells of some strains of Rhizobium were also described as heat resistant (86, 87), although resistant clones were not isolated. Mutants of Rhizobium resistant to some herbicides have been isolated (88), but no information was given concerning change in symbiotic properties. The finding of a mutagenic effect (89) for several of the herbicides indicates the need for considering induction as well as selection of variants by pesticides. Adaptation of rhizobia to a fungicide (90) has also been described, but the effect on symbiotic ability was not indicated.

Lysogenization of inoculant strains with temperate phages having a wide range of lytic activity was suggested (72) as one means of increasing the ability of a strain to compete with native rhizobia in the rhizosphere. The possibility of increasing survival ability of rhizobia on inoculated seeds through use of mutants resistant to legume seed-coat toxins (91) also warrants further examination.

12.2.4 Modification of Strains for Genetic or Ecological Research

The potential of some of the marker systems already described has not yet been fully exploited in genetic recombination experiments in Rhizobium. Availability of markers (e.g., streptomycin or spectinomycin resistance) that in many strains are apparently not closely linked with symbiotic ability, and of other markers (e.g., viomycin resistance, purine, or riboflavin requirement) that are closely associated, provides the necessary selection systems for mapping of factors for symbiosis. Manipulation of the level of symbiotic ability within one strain (22, 47, 63, 64, 66, 67) allows considerable flexibility in the combinations of characters that can be analyzed. In particular, a series of mutants derived from one effective parent but defective at different stages of nodule formation provides the necessary material for intrastrain genetic analysis or for biochemical analysis.

Mutation for resistance to some antibiotics is potentially very useful also for marker purposes in ecological studies (47, 48, 50, 92), although identification of rhizobia reisolated from nodules has in the past been based almost exclusively on serological methods (III.7, IV.9). Mutants used for such studies should be resistant to high levels of an antibiotic, stable in culture as well as in plant passage and, in the case of double markers for resistance to two antibiotics, must not be cross-resistant. Unlike most mutants intended for use in genetics or biochemical experiments these mutants should be free of pleiotrophic effects involving symbiotic properties. Resistance to streptomycin (47, 48, 50, 92) and spectinomycin (50) appears to meet these requirements in most rhizobia, although change in symbiotic properties has been encountered in mutants of occasional strains (50-52). Strains doubly marked with resistance to streptomycin and kanamycin have been successfully used in field strain competition studies (48, 92).

Rhizobia that are readily identifiable by colony pigmentation or by formation of black nodules provide another potentially useful method for strain identification. Colony pigmentation and inability to nodulate legumes were linked in the bacteria studied by Gillberg (85), but Norris (93) described a red-pigmented Rhizobium that effectively nodulated Lotononis plants and suggested the use of such markers for ecological studies. Rhizobia of the "cowpea group" which produce black nodules on some legumes, can be identified even more directly (94). The ability of a strain to induce black nodule formation can be lost without alteration of effectiveness and can occur in ineffective or effective strains (95). Genetic transfer of this marker system to other groups of rhizobia would appear to be limited by the range of distribution of the corresponding host plant factors for nodule pigmentation, but this limitation would not apply to colony pigment markers.

12.3 GENETIC TRANSFER WITHIN THE GENUS RHIZOBIUM

12.3.1 Transformation

The major portion of the literature concerning genetic transfer within Rhizobium or between Rhizobium and related genera of bacteria has dealt with transformation (Table 12.3) by DNA, by other types of cell extracts, or by culture filtrates. Interest in this area of research with rhizobia was generated largely by Krassilnikov's reports (96-98) on interspecific transfer of nodulating ability by growth of nonnodulating bacteria in culture filtrates of "virulent" bacteria. However, the main stimulus came from the extensive exploratory work of R. Balassa and her coworkers on methods of preparation of transforming material, choice of markers or strains, and other experimental factors affecting transformation. The published (Table 12.3) and unpublished results of Balassa are summarized in a review (99). A considerable number of reports of transformation in Rhizobium (Table 12.3) that have appeared since have dealt largely with cultural characters, relative to study of competence or other experimental variables. Transfer of symbiotic properties has also been reported for most cross-inoculation groups, but, as discussed later, the validity of some of this work has been questioned.

With regard to cultural markers (mainly streptomycin resistance) a number of general observations can be drawn concerning experimental variables likely to affect the efficiency of transfer. Competence of recipient cells was generally found to be maximum during the early exponential phase of growth (100-108). Żelazna and co-workers found competence in R. trifolii to occur in cycles (107, 109) corresponding roughly to the cell generation time, but this pattern depended on factors like cell concentration (109), aeration (108, 110), amount of capsular polysaccharide produced (107), and the medium used (105, 106, 110). Similar results were obtained with R. meliloti (111). The saturation concentration of DNA varied from ca. 0.2-0.5 µg/ml in Balassa's experiments to 10-40 µg/ml in experiments of other workers. A growth period corresponding to 2-3 cell divisions was needed for full phenotypic expression of transformation (101, 102, 107, 112, 113). Balassa (101) found that nontransforming DNA depressed the efficiency of transforming DNA while others (111) observed a stimulatory effect, which was attributed to protection of transforming DNA against nuclease action. Evidence for the occurrence of two nucleases in cowpea Rhizobium led to the suggestion (106) that nucleases may be responsible for the fairly high DNA concentration requirement for transformation of rhizobia.

TABLE 12.3. Summary of Reports on Transformation Involving Rhizobium

Species[a] or genera	Characters transferred or examined[b]		References[d]
	Cultural Markers	Symbiotic (or tumorigenic) properties[c]	
	Intrageneric transformation:		
Rhizobium spp.		inf$^+$	96-98
Rhizobium spp.		inf$^+$	117
Rhizobium spp.		(inf$^+$)	118, 119
Rhizobium spp.	str^r, str^d, pen^r, cys$^+$, others	inf$^+$, eff$^+$	100, 101, 112, 113, 121-123, 191
R. lupini	str^r, str^d		102, 114
Rhizobium spp.		inf$^+$	125
R. trifolii		inf$^+$	124
R. meliloti	prophage		192
R. trifolii	str^r, chl^r, c.v.^{r}		193
R. trifolii, R. meliloti		(inf$^+$)	126
R. phaseoli, R. leguminosarum		inf$^+$	194
Rhizobium spp.		(inf$^+$)	127
R. lupini	str^r, str^s, str^d		115, 116
R. trifolii		eff$^-$, (eff$^+$)	36

R. phaseoli, cowpea Rhizobium	pen^r, fructose utilization		103
Rhizobium spp.	str^r		107
R. trifolii	str^r		108, 109
R. trifolii	str^r (linked with inf$^-$)	inf$^-$	51, 52
R. trifolii	str^r (linked with colony type)		110
R. meliloti	str^r		111
R. japonicum	str^r	inf$^+$	104, 195
R. japonicum, cowpea Rhizobium	ade$^+$, gelatinase activity		105, 106

Intergeneric transformation:

Rhizobium, Pseudomonas		inf$^+$	97, 98
Agrobacterium, Rhizobium		(tumors)	147
Agrobacterium, Rhizobium		tumors	146
Agrobacterium, Rhizobium		tumors	148–150
Agrobacterium, Rhizobium	str^r		151
Rhizobium, Agrobacterium	sulfa^r		151
Rhizobium, Azotobacter	str^r, c.v.^{r}, other properties		152

[a] Species names are omitted where more than two species were used (mainly interspecific transfer). Where two species were used the donor is listed first.
[b] Abbreviations: str, streptomycin; pen, penicillin; c.v., crystal violet; sulfa, sulfanilamide; chl, chloramphenicol; cys, cysteine; ade, adenine; inf, infectiveness; eff, effectiveness. Superscripts: r (resistant), s (sensitive), d (dependent), + or − (presence or absence of function).
[c] Parentheses and underlining indicate negative results (i.e., no transfer of the character in question).
[d] Papers listed in approximate order of publication.

Transformation frequencies reported for cultural markers ranged from a low level of about 10^{-7} to very high values approaching 1% (99, 103, 107, 111). In general the frequency paralleled the apparent degree of relationship between donor and recipient, being lowest for interspecific transfer and highest for intrastrain transfer, particularly when DNA from already-transformed donors was used. The amount of cellular slime (polysaccharide) produced by a strain also influenced the transformation frequency, which was higher with semirough or rough mutant strains than for mucoid wild-type strains (102, 107, 114). The level of streptomycin resistance transferred was similar to that of the donor. Streptomycin resistance was transferred independently of colony type (107) in R. trifolii, but streptomycin-resistant, mucoid transformants were obtained (110) when a rough-colony recipient was exposed to DNA from a semimucoid strain — an example of an intermediate effect resulting from a partial transfer of the donor genotype. In a study of the relationship between streptomycin resistance and dependence, it was concluded that the controlling factors were either allelic or closely linked within a complex locus (114-116).

The early attempts to demonstrate transfer (interspecific) of symbiotic properties between rhizobia were made with culture filtrates. Krassilnikov (96-98) reported that strains of rhizobia that were normally noninfective on lucerne became infective after being grown for several months in filtrates of lucerne bacteria. Similarly, growth of noninfective strains in filtrates of clover bacteria led to virulence on clover. The newly acquired infectiveness was readily lost following growth of the bacteria in culture but was retained and "reinforced" by plant passage. Morphological or physiological characteristics of the donors were not transferred. "Muff-shaped" pseudonodules were also observed on clover inoculated with R. trifolii culture filtrates in which other, noninfective rhizobia had been grown. Peterson (117) interpreted the infections to have been produced not by the intended recipient bacteria but by filter-passing forms of the R. trifolii donor bacteria. The recipient bacteria added to the filtrate merely served as "feeders" for regeneration of the filter-passing bacteria. However, negative results were obtained in similar experiments by Izrailsky (118, 119), who

asserted that the pseudonodules produced on lucerne by
R. meliloti filtrates were not nodules, but mostly
sterile growths induced by bacteria-free filtrates.
He concluded that the filtrate experiments lacked
adequate controls to detect a possible low level of
cross-infection (resembling pseudonodules) by the
recipient strains used and stressed the hazards of
interspecific cross-nodulation (2, 120) in transformation experiments. Balassa's experiments involved
interspecific transformation of infectiveness by an
"in vivo" procedure (121) and by culture filtrates
(122, 123), disrupted-cell supernatants (122, 123),
or DNA preparations (100, 112, 113). Positive results
were claimed for all three methods, but the possibility of donor cell carry-over was not unequivocally
excluded. This possibility applies particularly to
the in vivo transformation procedure in which penicillin was used to lyse the penicillin-sensitive donor
bacteria in nodules. In the DNA transformation
experiments the conclusions concerning joint transfer
of infectivity, colony type, and antigenic characteristics (100, 112) from R. meliloti to R. japonicum and
R. lupini imply a close relationship between these
three characters, contrary to accumulated evidence for
nonlinkage as based on independent mutation.

Interspecific transformation of infectiveness by
"polysaccharide" extracts from infective strains was
reported (124, 125), but suitable markers for positive
identification of the transformant isolates were again
lacking. Using similar procedures, Humphrey and
Vincent (126) were unable to demonstrate transformation
and found that some bacterial cells survived the extraction procedure. They stressed the obvious but
often-overlooked need for reliable markers to allow
positive demonstration of mixed genomes in genetic
transfer experiements involving symbiotic properties.
Wagenbreth (127) using DNA extracts, also failed to
provide confirmation of interspecific transfer between
strains from a number of species of Rhizobium. Small
pseudonodules of unknown origin appeared on some non-inoculated plants as well as on plants inoculated with
recipient strain bacteria. Some rhizobia were able to
survive the DNA extraction procedure or a 24-hour
"sterilization" treatment with alcohol, and formation
of occasional ineffective nodules by DNA-treated
recipient bacteria was ascribed to surviving, weakened
donor cells. The DNA-base composition of isolates

from such nodules was similar to that of the donor strain rather than the recipient strain. These results underscore the pitfalls of donor bacteria contamination, particularly in transformation experiments in which very long growth periods followed the addition of DNA. Viable phage was also observed in DNA extracts from a lysogenic R. meliloti donor strain (111) and the use of phenol extraction procedures or of prophage-cured donors was recommended to circumvent this problem.

Kleczkowska (36) transformed an effective strain of R. trifolii to ineffectiveness, but reverse transfer did not occur. Nodule isolates were serologically of the recipient strain type, although the very high frequency of transformants (ca. 8%) among the randomly selected clones could have resulted from selective enrichment of mutant or transformed ineffective cells during the long (4-21 days) growth period of cells in the presence of high levels of DNA. Exposure of the effective recipient cells to phage plus DNA from an ineffective donor gave a higher proportion of ineffective clones than did either phage or DNA alone. On the basis of the frequency of loss in effectiveness following treatment with phage or DNA, and of the much greater stability of the ineffective state (relative to the effective), it was suggested (36) that loss can result from change in any one of several bacterial (and plant) factors.

When a streptomycin-resistant, nonnodulating mutant was used as donor and the sensitive, effective parent used as recipient, all streptomycin-resistant transformants (52) were nonnodulating. Attempts to restore nodulating ability by transformation to streptomycin sensitivity failed, and the authors relate this failure to a complex change, possibly a cell wall alteration (aside from ribosome protein changes) in streptomycin-resistant mutants. The authors suggest that the resistance locus in R. trifolii may be complex, as in R. lupini (114-116), and that only some alleles of this locus may interfere with infectivity, since the probability of loss of infectiveness was dependent on the level of streptomycin resistance. The strains of R. trifolii used in the experiments (52) may be somewhat unique in the close linkage between streptomycin resistance and loss of symbiotic properties, since such loss occurred relatively infrequently among many other strains of Rhizobium (34, 47, 49, 50).

12.3.2 Transduction

Transduction in Rhizobium (Table 12.4) has thus far been shown to occur only in R. meliloti. Kowalski (128, 129) found that 12 of 21 temperate phages were able to transduce streptomycin resistance at a frequency of 10^{-5} to 10^{-7}. Lysates from an induced lysogenic donor or from an infected donor gave similar

TABLE 12.4. Summary of Reports on Transduction and Conjugation in Rhizobium

Species	Character transferred or examined[a]		References
	Cultural markers	Symbiotic properties	
Transduction:			
R. meliloti	str^r		128, 129
R. meliloti	str^r, lys^+	(eff^+)	65, 130, 131
R. meliloti	cys^+		132
R. meliloti	leu^+ (linked with eff^+)	eff^+	68
Conjugation:			
R. lupini	auxotrophy, pigmentation		133-135
E. coli, R. lupini	antibiotic resistance (R factor)		136, 137
R. trifolii, R. phaseoli		inf^+	139
R. leguminosarum	str^r, pen^r		138
R. trifolii	auxotrophy		63

[a] Abbreviations: lys, lysine; leu, leucine. Other symbols defined in Table 12.3.

results, and all transductants examined were lysogenic. The phenotypic lag period of 4-6 hours was similar to that observed for transformation in R. trifolii. A lysine-dependent (lys⁻) ineffective mutant could also be transduced to prototrophy (65, 130, 131) at similar frequencies by lysates from an effective lys⁺ donor. Transductant clones and prototrophic revertants tested on the plant remained ineffective, thus indicating no close linkage between lys⁻ and ineffectiveness. Subsequently, effectiveness and leucine independence in R. meliloti were co-transduced (68), thereby confirming the close linkage or single gene change implied from back mutation to prototrophy and from leucine-promoted effective symbiosis (see III.12.2.1.3). This work provides the first demonstration for the feasibility of mapping genetic factors for symbiotic ability by indirect selection of recombinants in vitro, a procedure for which some auxotrophic mutants are ideally suited.

Kowalski's results involved generalized, nonspecific transduction, but specialized transduction has also been reported. Sik and Orosz (132) were able to transduce at low frequency only cysteine nonrequirement among several auxotrophic markers studied in R. meliloti, and high-frequency-transducing lysates were obtained from the cysteine transductants. Thus both types of transduction occur in R. meliloti.

12.3.3 Conjugation

Evidence for the existence of conjugation in Rhizobium (Table 12.4) has, until very recently, hinged largely on the experiments of Heumann and co-workers. The initial work involved genetic mapping of markers for auxotrophy, antibiotic resistance, and colony pigmentation in several types of bacteria isolated from soil, including a star-forming type designated as a probable member of the family Rhizobiaceae (133). Similar studies were later extended to a strain isolated from a lupin nodule (but not tested for nodulating ability) and designated as R. lupini. The results indicated a circular linkage map with two main marker groups (134). Anomalous differences in recovery of parental markers and some recombinant types were later explained by the isosexual (capable of acting as donor or recipient) nature of the strain. The physiological state of the cells of a mating strain determined its function; it acted as a recipient during the log phase of growth and as a donor during the stationary phase (135, 136). Female strains were obtained through elimination of fertility factors during cold storage or by treatment with sodium dodecyl sulfate. One-way transfer experiments (135) revealed the presence of two fertility factors, which could be independently eliminated and which regulated transfer of two different chromosomal regions. A recent preliminary report (137) mentions the successful transfer of the resistance factor, RP_4 (confers resistance to ampicillin, kanamycin, and tetracycline), from Pseudomonas aeruginosa to R. lupini. In R. lupini it acts as a fertility inhibitor, whereas the same factor does not inhibit the fertility system in Escherichia coli. On the basis of the above interesting findings it would appear that conjugation may now be readily applied to genetic analysis of symbiotic properties in Rhizobium. Unfortunately no evidence was presented that the designated R. lupini strain, which apparently shows some relationship to Pseudomonas, is capable of nodulating lupin or other legumes; the full relevance of Heumann's detailed genetic mapping experiments to Rhizobium genetics thus awaits appropriate identification of the isolate used.

Conjugation between penicillin-resistant and streptomycin-resistant strains of R. leguminosarum has been reported (138), but the limited data presented and small cell populations examined do not adequately rule out spontaneous mutation. A recent report (63) on conjugation among five strains of R. trifolii indicates that conjugation may occur fairly commonly in this genus. The frequency of prototrophic recominants ranged from 10^{-7} to 1.8×10^{-3}, in crosses between singly auxotrophic mutants or between auxotroph and wild type. A period (several hours) of phenotypic expression was observed in crosses in which elimination of the donor cells by streptomycin was delayed for different time periods. Prototrophic back mutants from ineffective auxotrophs remained ineffective; prototrophic exconjugants were apparently not tested on plants.

12.3.4 Plasmid Transfer or Control

Several reports appearing since 1967 indicate that extrachromosomal inheritance, particularly inheritance of factors for symbiotic properties, also occurs in Rhizobium. Up to 12% of the clones from a strain of R. trifolii spontaneously lost infectiveness on clover (139) and treatment with acridine orange increased the frequency to as high as 72%. This very high frequency could have resulted from the method of determining infection, based on observations (microscopic) of infected root hairs rather than nodules. Evidence for episomally mediated transfer of infectiveness from R. trifolii to R. phaseoli was derived from similar observations on clover roots inoculated with a mixture of bacteria in which donor cells had been "eliminated" by a donor specific phage. Identification of the infecting bacteria by immunological and physiological criteria was based largely on one nodule isolate. A very high incidence of loss of infectivity in R. japonicum was also observed (140) following treatment of the bacteria with ultraviolet light or acriflavine. These results were, however, reinterpreted in subsequent work (141); instability of effectiveness was restricted to "fast-growing cultures" that contained fast-growing contaminants, and plasmid-eliminating treatments simply selected for the contaminant cells which lacked nodulating ability. Slow-growing strains remained stable to these treatments.

The case for plasmid inheritance of symbiotic properties is supported by the recent work of Dunican and co-workers (142-144) with R. trifolii. Viomycin resistance, known to be closely associated with loss of effectiveness (53, 55; see III.7.6.5.3) was used as a marker for detecting loss of effectiveness following treatment of effective, viomycin-sensitive bacteria with plasmid-eliminating agents. Treatment with acridine orange or ethidium bromide significantly increased the frequency of viomycin resistance whereas a prophage-inducing agent and a mutagen did not (144). Thus factors for both viomycin sensitivity and symbiotic efficiency were presumed to be located on plasmids, possibly on the same plasmid. The validity of this convenient assay depends, among other factors, on the absence of selective enrichment of spontaneous mutants by plasmid-curing agents. Absence of such an effect was shown for treatment with phenolic compounds (142) but not for acridine orange or ethidium bromide. The authors considered plasmid loss as a possible factor contributing to prevalence of ineffective strains in some field environments, e.g., under high concentrations of metals (in acid soils) or phenolic compounds. Preliminary evidence (143) for the presence of satellite DNA in cell lysates of R. trifolii was obtained by dye-buoyant-density-gradient analysis. Viomycin-resistant derivatives obtained by plasmid-curing treatment were not included in the analysis, and the relationship of the extrachromosomal DNA to viomycin sensitivity or symbiotic properties remains to be established.

No evidence for episomal control of viomycin resistance was found in mutants of R. meliloti (56, 57), since resistance was not eliminated by plasmid-curing treatments and did not involve production of antibiotic-inactivating enzymes, a characteristic of R-factor-mediated antibiotic resistance in other bacteria (145). The question of presence or absence of plasmids relative to sensitivity or resistance to viomycin in Rhizobium needs further clarification.

The reported occurrence of fertility factors in a strain of R. lupini (137) and the transfer of an antibiotic resistance factor to this strain from Escherichia coli was discussed previously (III.12.3.3).

12.4 GENETIC TRANSFER BETWEEN RHIZOBIUM AND RELATED GENERA

Transformation has been used in some attempts to bridge the taxonomic gap between Rhizobium and other apparently related genera, notably Agrobacterium and Pseudomonas (Table 12.3). Krassilnikov's (96-98) claims for transfer of virulence by growth of non-virulent bacteria in culture filtrates of infective bacteria included transfer from several rhizobia to soil bacteria tentatively identified as Pseudomonas. Two of the Pseudomonas cultures reportedly acquired the ability to produce nodule-like swellings on roots of clover following several cycles of prolonged growth in R. trifolii filtrates and a third culture became infective on lucerne following similar growth in R. meliloti filtrates. Klein and Klein (146) first reported DNA-mediated transfer of tumor-inducing ability from Agrobacterium tumefaciens to a strain of

R. leguminosarum. The transfer was accomplished with bacterins (suspensions of heat-killed donor bacteria), culture filtrates, or DNA extracts of A. tumefaciens, although less readily than between strains of Agrobacterium. The identity of isolates from tumors produced by the transformed culture was not confirmed by marker criteria. Manil (147), using DNA extracts from A. tumefaciens, was unable to transfer tumor-inducing ability on Helianthus plants to R. meliloti. However, Kern (148) also reported transfer of ability to form tumors (on stems of Solanum or other plants) from A. tumefaciens to R. leguminosarum. Isolates from the tumors retained this property on the new host but were no longer symbiotically effective on roots of pea; these isolates and the donor strain formed only swellings or pseudonodules on pea. One of the transformed isolates was compared with the donor and recipient strains in an extensive study of morphologic, physiologic, and biochemical properties (149, 150). In some properties it resembled one of the parent strains but in a number of criteria, including DNA homology, it was intermediate. Further evidence for transformation between these two genera was provided by reciprocal transfer of antibiotic resistance markers (151). Transformation rates of $4-7 \times 10^{-6}$ for streptomycin resistance and $13-16 \times 10^{-6}$ for sulfonamide resistance were observed. Antigenically the transformants resembled the recipient strain.

Intergeneric transformation between strains from three species of Rhizobium (donor) and two species of Azotobacter has also been reported (152). The observed frequency for transformation of low-level streptomycin resistance or crystal violet resistance ranged from 0.05 to 0.95%, and the transformants resembled the donor, rather than the recipient, in a number of unrelated morphological or biochemical properties. The high frequencies (for intergeneric transfer) and the range of apparent co-transfer of nonselected characters appear to warrant confirmation with more definitive mutant marker systems.

Although the experimental evidence for intergeneric genetic exchange involving Rhizobium is still limited relative to exchange within Rhizobium, the possibility of transcending some of the generic limits established by older taxonomic criteria is strengthened by available data on DNA base content and homology (153, III.7). Within the family Rhizobiaceae there is

Other Genetic Aspects

evidence of considerable DNA homology between Rhizobium (especially the fast-growing group of rhizobia) and Agrobacterium (154). A lesser but significant degree of genetic relatedness appears to exist between Rhizobium and Azotobacter, Pseudomonas, or Arthrobacter. One may reasonably anticipate some modification of the range of symbiotic N_2 fixation along these lines of homology (See III.12.7).

12.5 OTHER GENETIC ASPECTS

Although research on genetics of rhizobia has dealt largely with mutation and methods of genetic transfer, considerable effort has also been directed towards other, miscellaneous areas relating to Rhizobium genetics. Several of these subjects, especially those concerning rhizobiophage, are also reviewed by Vincent (III.7), and will be considered only briefly here.

12.5.1 Replication Mapping

Replication mapping (155), whereby chromosomal loci are mapped according to their sequential response to pulsed mutagenic treatment in synchronized cell populations, has recently been used in Rhizobium. N-methyl-N'-nitro-N-nitrosoguanidine treatment (156) was used to determine the relative location of factors for resistance to streptomycin and to chloramphenicol in a strain of R. trifolii. The chromosomal attachment site of a prophage in R. meliloti was located (157) by pulsed superinfection of synchronized bacteria (prophage previously heat-induced) with the homologous clear-plaque mutant phage and analysis of the clear/turbid plaque ratios. Replication mapping appears to be a potentially useful means of approximate mapping of factors for symbiotic ability if used with several markers that are phenotypically linked (or not linked) to infectiveness or effectiveness.

12.5.2 Rhizobium-Phage Interactions

12.5.2.1 Lysogeny

Lysogeny is known to occur among many strains from at least four cross-inoculation groups of Rhizobium

(158-167). Lysogenization by temperate phage arising from the virulent form by mutation (a rare event among bacteriophage, relative to loss of the temperate state) has also been reported in R. trifolii (37). Lysogenic rhizobia and the temperate phage they produce are of considerable interest for their application to both general and specific transduction, as already described. There is no evidence for alteration of symbiotic properties due to introduction of a prophage by lysogenization, although lysogenic conversion of a somatic antigen in R. trifolii following lysogenization has been observed (168). Prophage-controlled immunity to homologous phage or related phage, therefore, contrasts with mutation-acquired resistance to virulent phage, which depresses symbiotic ability in some rhizobia (33, 35, 36). Prophage-linked (specific) transduction (132) of an auxotrophic marker has, however, been encountered with a temperate phage of R. meliloti, and the possibility of similar association of factors for symbiosis with a prophage cannot be excluded.

12.5.2.2 Host-controlled Modification

Host-controlled or phenotypic modification (172) of phage DNA by some strains of bacteria and the recognition of modified DNA by bacteria of other "restricting" strains has been described for two temperate phages of R. leguminosarum (169, 170) and has also been observed in a phage of R. trifolii (171). Restriction in one strain of R. leguminosarum could be prevented by heating (sublethal temperature) or UV irradiation of the bacteria, or by infecting with phage at a high phage/bacterium input ratio (170). Modification may occur in virulent or temperate phage but is more relevant to temperate phage and transduction because it allows manipulation of the host range of such phages. It also points to the likely occurrence, in restricting strains, of nucleases that may inhibit transfer (172, 173) of bacterial, plasmid, or phage DNA, particularly in interspecific or intergeneric transfer attempts.

12.5.2.3 Mapping of the Phage Genome

The genetic mapping of a temperate phage of R. meliloti has recently been initiated (132, 157, 174). Crosses

Other Genetic Aspects   609

and in vivo and in vitro complementation tests involving a variety of mutant characters established a linear sequence of cistrons for regulator function, early or late function, tail formation, and lysozyme synthesis. The DNA from this phage resembled the bacterial DNA in base composition and was active in transfection experiments.

12.5.2.4 Transfection

Transfection (infection of bacteria with DNA from phage) (175) has been demonstrated recently in R. meliloti. A low frequency was noted (132) for transfection (10^{-8} per phage equivalent DNA) of intact cells by a temperate phage, but use of spheroplast cells and helper phages gave a 300-fold increase in frequency. In another report, no plaques were produced (176) by infecting cells of R. meliloti with DNA from a phage, but positive results were again obtained with spheroplast cells. Wide variation in frequency between experiments was attributed to differences in competence of spheroplasts.

12.5.3 Bacteriocinogeny

Genetic factors controlling the production of bacteriocins by bacteriocinogenic bacteria are generally considered to be located on extrachromosomal elements or plasmids (145), although defective prophages may be implicated in the production of some large, phage-like bacteriocins. Bacteriocinogeny appears to occur widely in all species of rhizobia examined (171, 177) but the nature of the controlling genetic element has not been determined. Lotz and Mayer (178) described a phage tail-like bacteriocin, apparently devoid of nucleic acid, produced by a lupin nodule isolate. Schwinghamer et al. (179) identified the bacteriocinogenic activity of one strain R. trifolii with short-tailed, phage-like particles that contained double-stranded DNA and closely resembled defective phage. Bacteriocins produced by strains of R. trifolii comprised several types on the basis of various criteria like specificity, inducibility by UV-irradiation, filterability, or sensitivity to enzymes (171).

12.6 BACTERIUM OR HOST PLANT: WHICH PARTNER CONTROLS THE KEY STEPS IN THE SYMBIOSIS?

12.6.1 Inferences from Studies on the Biochemistry of Dinitrogen Fixation

There is as yet little information (excepting that obtained from some symbiotically defective mutants with linked metabolic blocks) derived from genetic experiments that can aid in identifying key biochemical functions of either symbiont. Conversely, information accruing from research on biochemistry of dinitrogen fixation can provide at least a partial answer to the question of whether bacterium or plant genome controls the synthesis of certain substances that are essential for fixation in nodules. This applies particularly to nitrogenase and to leghemoglobin (II.8). In the case of nitrogenase the localization of the enzyme in the bacteroid establishes the bacteroid as the site of synthesis. The similarity of bacteroid nitrogenase to that of free-living N_2-fixing bacteria (II.2) also supports the contention that the bacterium carries much or all of the necessary genetic information, although some induction or regulatory message must be provided by the plant. In the case of leghemoglobin, experiments comparing this protein as produced on different host plants by one strain of Rhizobium and, conversely, produced on one host by different strains, have provided evidence that synthesis is controlled by the plant (180-182). The possibility that the bacterium contributes to synthesis of the heme component of leghemoglobin has also been examined (181, 183). Information concerning control of synthesis of some other nodule components (e.g., cytochromes, flavoproteins) may similarly be obtainable, depending on availability of suitable methods for assay in vitro.

Relatively little biochemical research effort has been directed towards an understanding of nodule development at stages other than the N_2-fixing phase of effective symbiosis, with the result that few key substances can be singled out for genetic study. Rather, genetic research may logically precede and implement biochemical study of the prebacteroid stages of the nodule by locating induced blocks in symbiosis and providing suitable bacterial genotypes (also host plant genotypes) for biochemical analysis.

12.6.2 Dilworth-Parker Hypothesis: Sharing of Genetic Information

Some apparent similarities between biological N_2-fixing systems have long challenged botanists and others concerned with the evolutionary development of such systems. The legume-nodule symbiosis was considered by Parker (184) to have evolved from free-living N_2-fixing bacteria, through a succession of stages ranging from a loose bacterial-plant root association to the existing symbiotic association in which neither partner alone can fix N_2. Alternatively (though considered less likely), fixation ability could have been acquired by a plant-pathogenic bacterium like A. tumefaciens which already had the ability to invade roots and establish tumorous growth. This hypothesis has served as a stimulus and guide for attempts at intergeneric transfer which, in turn, will test the validity of the two evolutionary alternatives. This is especially true for the more recent model of Dilworth and Parker (185), which also assumes (in view of current evidence for fixation by isolated bacteroids) that Rhizobium was once a free-living, N_2-fixing organism. Following establishment of an in vivo root association, retention of nitrogenase activity by the bacterium during its saprophytic phase of existence in the soil was considered an ecological disadvantage. Selection favored the bacterial variant in which nitrogenase activity was confined to the plant phase of its existence. The authors discounted simple repression of genes for nitrogenase synthesis as an explanation for the complete absence of N_2-fixing ability by rhizobia in culture, largely because constitutive mutants had not been observed. Instead, a part of the missing genetic information for this function and for other shared functions was envisaged as having been transferred to the plant and incorporated in the plant genome. Synthesis of nitrogenase, leghemoglobin, or other substances that are not produced by either partner alone thus could result from sharing of genetic information at the DNA level or, more likely, at the level of interchange of RNA or complementary proteins between bacterium and host cell. Complementation, in particular, could help explain differences in degree of effectiveness of different strain-cultivar combinations.

This interesting model, which has far-reaching implications in genetic, taxonomic, and biochemical aspects of the symbiosis, appears to be reinforced by the numerous recent reports of uptake and integration of bacterial DNA by roots or shoots of higher plants. Particularly relevant to the hypothesis for sharing of information in the nodule is the evidence that A. tumefaciens (186, 187) apparently transfers to the infected plant tissue a considerable part of its genome, including that portion necessary for induction of autonomous tumors. Unlike legume nodules, however, the tumors do not require the continued presence of the inducing bacteria, suggesting that the transfer of essential information was complete and one-way, from bacterium to plant. The bacterial DNA, although incorporated into the genome of the somatic tumor cells, apparently is not translocated to germinal tissue for permanent incorporation and transmission to the plant progeny. Such incorporation is implied for ancestral legumes (185). Despite obvious differences in the two bacterial-plant associations, the findings for the Agrobacterium-plant system point to means whereby Rhizobium may originally have "banked" a dispensable (for saprophytic existence) portion of its genome with a legume, as visualized by Dilworth and Parker.

12.7 CONCLUSIONS

Research on genetics of Rhizobium has evolved through a series of stages involving emphasis on spontaneous variation, induced mutation, search for mechanisms of genetic transfer, search for cultural characters phenotypically linked to differences in symbiotic ability, and study of experimental variables affecting efficiency of recombination. At the present stage, all of the main mechanisms (DNA transformation, transduction, conjugation) of genetic exchange in bacteria have been reported to occur in Rhizobium, although some of the claims are equivocal and warrant confirmation. Availability of these mechanisms plus use of linked marker systems suitable for indirect preselection (in culture) of recombinants should allow some mapping of factors for symbiotic ability to proceed. At this point, however, it should be recognized that "effectiveness in N_2 fixation in the root nodules is

Conclusions 613

determined by the interaction of bacterial strain and host plant factors which are independently subject to genetic variation," as concluded by Nutman (188) in a study of naturally occurring variation in the symbiosis of R. trifolli and Trifolium pratense. Progress towards an understanding of genetic interaction between the microsymbiont and macrosymbiont will require coordinated genetic analysis of both bacterium (subject to limitations imposed by partial genome transfer) and plant (III.11), analogous to studies of pathogenicity and host response to some fungal plant pathogens (189). Finally, such analyses should not overlook the possibility of shared genetic information.

Since the original literature search for this review was made, research activity has intensified in several areas of Rhizobium genetics to such an extent that the readers attention is directed to some of the more recent papers in order to indicate the current research trend. These fall within three related areas of interest:

(1) Intergeneric transfer (by conjugation or DNA transformation) of plasmids, involving mainly antibiotic resistance transfer factors (R-factors), from Escherichia coli or Pseudomonas to Rhizobium (194-200). Through this approach an F-like R-factor transferred (by DNA transformation) from Pseudomonas aeruginosa to R. trifolii has been used to mediate transfer of nif (N_2-fixing) genes from a strain of R. trifolii to a strain of Klebsiella aerogenes which was reportedly incapable of fixing dinitrogen (201).

(2) Search for plasmid systems occurring naturally in Rhizobium. Evidence for occurrence of resistance transfer factors in R. japonicum has been obtained from curing treatments with acridine orange and transfer of multiple antibiotic resistance from R. japonicum to Agrobacterium tumefaciens (202). A high frequency of loss of infectiveness following treatment with curing agents, suggesting plasmid-borne factors for symbiosis, has been reported for a strain of R. trifolii (203) and a strain of R. meliloti (204).

(3) Comparison of biophysical properties of DNA from cultured rhizobia, bacteroids, and the host plant. DNA from cultured cells of R. japonicum showed no measurable hybridization with soybean plant DNA (205). In fast-growing strains of Lotus rhizobia the bacteroid DNA content was lower than that of the cultured

bacteria but the DNA's were otherwise indistinguishable and a satellite DNA was present in both types of cells (206). Bacteroids of Cicer and Phaseolus rhizobia had a lower GC (guanine + cytosine) content (207) than the cultured cells.

Recent literature concerning plasmid-borne nif genes (potentially transferable to rhizobia) in non-symbiotic bacteria is reviewed in the chapter which follows.

12.8   REFERENCES

1. Fred, E. B., I. L. Baldwin, and E. McCoy, Root Nodule Bacteria and Leguminous Plants, University of Wisconsin Press, 1932.
2. Wilson, J. K., Cornell Univ. Agr. Exp. Stat. Mem., 221, 48 pp., 1939.
3. Allen, E. K., and O. N. Allen, Bacteriol. Rev., 14, 273 (1950).
4. Allen, E. K., and O. N. Allen, Handb. Pflanzenphysiol., VIII, 48 (1958).
5. Lorkiewicz, Z., Acta Microbiol. Polon., 8, 225, (1959).
6. Marečková, H., V. Škrdleta, and M. Plišková, Rost. Vyroba, 13, 325 (1967).
7. Nutman, P. S., Proc. Roy. Soc. London, Ser. B, 172, 417 (1969).
8. Hayes, W., The Genetics of Bacteria and their Viruses, 2nd ed., Blackwell Science Publications, Oxford, 1968.
9. Damery, J. T., and M. Alexander, Soil Sci., 108, 209 (1969).
10. Means, U. M., and L. W. Erdman, Proc. Soil Sci. Am., 27, 305 (1963).
11. Nutman, P. S., J. Bacteriol., 51, 411 (1946).
12. Gostkowska, K., Acta Microbiol. Polon., 12, 158 (1963).
13. Hamdi, Y. A., Arch. Mikrobiol., 63, 227 (1968).
14. Hamdi, Y. A., Acta Microbiol. Polon., Ser. B, 1, 85 (1969).
15. Holding, A. J., S. N. Tilo, and O. N. Allen, Trans. 7th Int. Congr. Soil Sci., Madison, Comm. III, 2, 608 (1960).
16. Longly, B. J., T. O. Berge, J. M. Van Lanen, and I. L. Baldwin, J. Bacteriol., 33, 29 (1937).

References

17. Staphorst, J. L., and B. W. Strijdom, Phytophylactica, 3, 131 (1971).
18. Strijdom, B. W., and O. N. Allen, Can. J. Microbiol., 12, 275 (1966).
19. Strijdom, B. W., and O. N. Allen, S. Afr. J. Agr. Sci., 10, 623 (1967).
20. Strijdom, B. W., and O. N. Allen, Phytophylactica, 1, 147 (1969).
21. Wolf, M., and I. L. Baldwin, J. Bacteriol., 39, 344 (1940).
22. Schwinghamer, E. A., Can. J. Microbiol., 14, 355 (1968).
23. Jordan, D. C., and W. H. Coulter, Can. J. Microbiol., 11, 709 (1965).
24. Staphorst, J. L., and B. W. Strijdom, Phytophylactica, 4, 29 (1972).
25. Sherwood, M. T., J. Gen Microbiol., 71, 351 (1972).
26. Holding, A. J., and J. King, Plant Soil, 18, 191 (1963).
27. Holding, A. J., and J. F. Lowe, Plant Soil, Special Vol., 153 (1971).
28. Jones, D. G., and C. Burrows, Soil Biol. Biochem., 1, 57 (1969).
29. Masterson, C. L., Proc. 9th Int. Congr., Soil Sci., Adelaide, 2, 95 (1968).
30. Singer, M., A. J. Holding, and J. King, Proc 8th Int. Congr. Soil Sci., Bucharest, 3, 1021 (1964).
31. Blum, U., and E. L. Rice, Bull. Torrey Bot. Club, 96, 531 (1969).
32. Almon, L., and I. L. Baldwin, J. Bacteriol., 26, 229 (1933).
33. Krassilnikov, N. A., Mikrobiologiya, 10, 396 (1941).
34. Gupta, B. M., and J. Kleczkowska, J. Gen. Microbiol., 27, 473 (1962).
35. Kleczkowska, J., J. Gen Microbiol., 4, 298 (1950).
36. Kleczkowska, J., J. Gen Microbiol., 40, 377 (1965).
37. Kleczkowska, J., Plant Soil, Special Vol., 47 (1971).
38. Demolon, A., and A. Dunez, C. R. Acad. Sci., Paris, Ser. D, 202, 1704 (1936).
39. Hely, F. W., F. J. Bergersen, and J. Brockwell, Aust. J. Agr. Res., 8, 24 (1957).
40. Holland, A. A., and C. A. Parker, Plant Soil, 25, 329 (1966).
41. van Schreven, D. A., Plant Soil, 21, 283 (1964).
42. Damirgi, S. M., and H. W. Johnson, Agron. J., 58, 223 (1966).

43. Vanderveken, J., Ann. Inst. Pasteur, Paris, 107, 143 (1964).
44. Maramorosch, K., R. R. Granados, and H. Hirumi, Advan. Virus Res., 16, 135 (1970).
45. Joshi, H. U., A. J. H. Carr, and D. G. Jones, J. Gen. Microbiol., 47, 139 (1967).
46. Joshi, H. U., and A. J. H. Carr, J. Gen Microbiol., 49, 385 (1967).
47. Imšenecki, A. A., A. N. Parijskaya, and L. Erraiz Lopez, Mikrobiologiya, 39, 343 (1970).
48. Obaton, M., C. R. Acad. Sci., Paris, Ser. D., 272, 2630 (1971).
49. Schwinghamer, E. A., Antonie van Leeuwenhoek, 33, 121 (1967).
50. Schwinghamer, E. A., and W. F. Dudman, J. Appl. Bacteriol., 36, 263 (1973).
51. Żelazna-Kowalska, I., Plant Soil, Special Vol., 67 (1971).
52. Żelazna-Kowalska, I., and Z. Lorkiewicz, Acta Microbiol. Polon., Ser. A, 3, 11 (1971).
53. Schwinghamer, E. A., Can. J. Microbiol., 10, 221 (1964).
54. Alexander, D. C., D. C. Jordan, and M. McKague, Can. J. Biochem., 47, 1092 (1969).
55. Hendry, G. S., and D. C. Jordan, Can. J. Microbiol., 15, 671 (1969).
56. Jordan, D. C., Proc. 6th Int. Congr. Chemother., University of Tokyo Press, 456 (1970).
57. Jordan, D. C., Y. Yamamura, and M. E. McKague, Can. J. Microbiol., 15, 1005 (1969).
58. MacKenzie, C. R., and D. C. Jordan, Biochem. Biophys. Res. Commun., 40, 1008 (1970).
59. MacKenzie, C. R., and D. C. Jordan, Can. J. Microbiol., 18, 1168 (1972).
60. Yu, K. K., and D. C. Jordan, Can. J. Microbiol., 17, 1283 (1971).
61. Pankhurst, C. E., E. A. Schwinghamer, and F. J. Bergersen, J. Gen. Microbiol., 70, 161 (1972).
62. Schwinghamer, E. A., Aust. J. Biol. Sci., 23, 1187 (1970).
63. Lorkiewicz, Z., W. Zurkowski, E. Kowalczuk, and A. Górska-Melke, Acta Microbiol. Polon., Ser. A, 3, 101 (1971).
64. Schwinghamer, E. A., Can. J. Microbiol., 15, 611 (1969).
65. Kowalski, M., Acta Microbiol. Polon., Ser. A, 2, 115 (1970).

66. Dénarié, J., C. R. Acad. Sci., Paris, Ser. D, 269, 2464 (1969).
67. Scherrer, A., and J. Dénarié, Plant Soil, Special Vol., 39 (1971).
68. Kowalski, M., and J. Dénarié, C. R. Acad. Sci., Paris, Ser. D, 275, 141 (1972).
69. Virtanen, A. I., Biol. Rev., 22, 239 (1947).
70. Robinson, A. C., Aust. J. Agric. Res., 20, 1053 (1969).
71. Krassilnikov, N. A., and T. A. Melkumova, Izv. Akad. Nauk, Ser. Biol., 5, 693 (1963).
72. Bergersen, F. J., J. Brockwell, A. H. Gibson, and E. A. Schwinghamer, Plant Soil, Special Vol., 3 (1971).
73. Schwinghamer, E. A., Am. J. Bot., 49, 269 (1962).
74. Jordan, D. C., Can. J. Bot., 30, 125 (1952).
75. Yemtsov, V. T., Izv Akad. Nauk SSSR, Ser. Biol., 732 (1962). (Cited by Hamatová, 1964.)
76. Hamatová, E., Acta Microbiol. Polon., 13, 247 (1964).
77. Shemakhanova, N. M., and I. P. Bunko, Mikrobiologiya, 37, 433 (1968).
78. Shemakhanova, N. M., and O. D. Sidorenko, Mikrobiologiya, 39, 1026 (1970).
79. Döbereiner, J., A. A. Franco, and I. Guzman, Pesq. Agropec. Bras., 5, 155 (1970).
80. Pedrosa, F. O., A. J. Nascimento, R. Alvahydo, and J. Döbereiner, Pesq. Agropec. Bras., 5, 373 (1970).
81. Brockwell, J., W. F. Dudman, A. H. Gibson, F. W. Hely, and A. C. Robinson, Proc. 9th Int. Congr. Soil Sci., Adelaide, 2, 103 (1968).
82. Chatel, D. L., R. M. Greenwood, and C. A. Parker, Proc. 9th Int. Congr. Soil Sci., Adelaide, 2, 65 (1968).
83. Vincent, J. M., in Soil Nitrogen, W. V. Bartholomew and F. E. Clark, Eds., Am. Soc. Agron., Madison, pp. 384-435, 1965.
84. Gillberg, B. O., Arch. Mikrobiol., 75, 203 (1971).
85. Gillberg, B. O., Arch. Mikrobiol., 69, 260 (1969).
86. Gillberg, B. O., Nature (London), 222, 574 (1969).
87. Gillberg, B. O., Arch. Mikrobiol., 62, 328 (1968).
88. Kaszubiak, H., Acta Microbiol. Polon., 17, 41 (1968).
89. Kaszubiak, H., Acta Microbiol. Polon., 17, 51 (1968).

90. Golebiowska, J., H. Kaszubiak, and M. Pajewska, Acta Microbiol. Polon., 16, 153 (1967).
91. Masterson, C. L., Ann. Inst. Pasteur, Paris, Supplement, 109, 216 (1965).
92. Obaton, M., Plant Soil, Special Vol., 273 (1971).
93. Norris, D. O., Aust. J. Agri. Res., 9, 629 (1958).
94. Cloonan, M. J., Aust. J. Sci., 26, 121 (1963).
95. Stanford, N. P., A. B. Campelo, and J. Döbereiner, IVth Reun. Latino-Am. Inoculantes Legumin., Porto Alegre, Brazil (1968).
96. Krassilnikov, N. A., C. R. Acad. Sci., USSR, 31, 75 (1941).
97. Krassilnikov, N. A., Mikrobiologiya, 14, 230 (1945).
98. Krassilnikov, N. A., Usp. Sovrem. Biol., 40, 179 (1955). (Cited by Izrailsky, 1957.)
99. Balassa, G., Bacteriol. Rev., 27, 228 (1963).
100. Balassa, R., Naturwissenschaften, 43, 133 (1956).
101. Balassa, R., Acta Microbiol. Acad. Sci. Hung., 4, 85 (1957).
102. Balassa, R., and M. Gábor, Mikrobiologiya, 30, 457 (1961).
103. Gadre, S. V., L. Mazumdar, V. V. Modi, and V. Parekh, Arch. Mikrobiol., 57, 388 (1967).
104. Marečková, H., Arch. Mikrobiol., 68, 113 (1969).
105. Raina, J. L., and V. V. Modi, J. Gen. Microbiol., 57, 125 (1969).
106. Raina, J. L., and V. V. Modi, J. Gen. Microbiol., 65, 161 (1971).
107. Żelazna, I., Acta Microbiol. Polon., 12, 166 (1963).
108. Żelazna, I., Acta Microbiol. Polon., 13, 283 (1964).
109. Żelazna, I., Acta Microbiol. Polon., 13, 291 (1964).
110. Żelazna-Kowalska, I., Z. Lorkiewicz, and M. Dziak-Hoffman, Acta Microbiol. Polon., Ser. A, 3, 3 (1971).
111. Żelazna-Kowalska, I., and Z. Lorkiewicz, Acta Microbiol. Polon., Ser. A, 3, 21 (1971).
112. Balassa, R., Acta Microbiol. Acad. Sci. Hung., 4, 77 (1957).
113. Balassa, R., Nature, 188, 246 (1960).
114. Balassa, R., and M. Gábor, Acta Microbiol. Acad. Sci. Hung., 11, 329 (1965).
115. Gábor, M., Genetics, 52, 905 (1965).
116. Gábor, M., Symp. Biol. Hung., 6, 75 (1965).

117. Peterson, N. V., Mikrobiologiya, 24, 275 (1955).
118. Izrailsky, W. P., Mikrobiologiya, 23, 22 (1954).
119. Izrailsky, W. P., Mikrobiologiya, 26, 541 (1957).
120. Kleczkowska, J., P. S. Nutman, and G. Bond, J. Bacteriol., 48, 673 (1944).
121. Balassa, R., Naturwissenschaften, 42, 422 (1955).
122. Balassa, R., Magy. Tud. Akad. Agrartud. Oszt. Kozlemen., 2, 307 (1953).
123. Balassa, R., Acta Microbiol. Acad. Sci. Hung., 2, 51 (1954).
124. Ljunggren, H., Nature (London), 191, 623 (1961).
125. Lange, R. T., and M. Alexander, Can. J. Microbiol., 7, 959 (1961).
126. Humphrey, B. A., and J. M. Vincent, Nature (London), 199, 927 (1963).
127. Wagenbreth, D., Arch. Mikrobiol., 52, 154 (1965).
128. Kowalski, M., Acta Microbiol. Polon., 16, 7 (1967).
129. Kowalski, M., Acta Microbiol. Polon., Ser. A, 2, 109 (1970).
130. Kowalski, M., Genet. Polon., 12, 201 (1971).
131. Kowalski, M., Plant Soil, Special Vol., 63 (1971).
132. Sik, T., and L. Orosz, Plant Soil, Special Vol., 57 (1971).
133. Heumann, W., Naturwissenschaften, 47, 330 (1960).
134. Heumann, W., Mol. Gen. Genet., 102, 132 (1968).
135. Heumann, W., A. Pühler, and E. Wagner, Mol. Gen. Genet., 113, 308 (1971).
136. Pühler, A., Dissertation, Naturwissenschaftliche Fakultät der Univ. Erlangen-Nürnberg, pp. 1-30 (1971). (Cited by Heumann et al., 1971.)
137. Pühler, A., H. J. Burkardt, and W. Heumann, Proc. Soc. Gen. Microbiol., xxvi (1972).
138. Bose, P., and G. S. Venkataraman, Experientia, 25, 772 (1969).
139. Higashi, S., J. Gen. Appl. Microbiol., 13, 391 (1967).
140. Jansen van Rensburg, H., B. W. Strijdom, and C. J. Rabie, S. Afr. J. Agri. Sci., 11, 623 (1968).
141. Jansen van Rensburg, H., and B. W. Strijdom, Phytophylactica, 3, 125 (1971).
142. Cannon, F. C., L. K. Dunican, and M. J. Farrell, Proc. Soc. Gen. Microbiol., 62nd Gen. Meeting, Dublin, ix (1971).
143. Cannon, F. C., L. K. Dunican, and F. O'Gara, Biochem. J., 125, 103P (1971).

144. Dunican, L. K., and F. C. Cannon, Plant Soil, Special Vol., 73 (1971).
145. Novick, R. P., Bacteriol. Rev., 33, 210 (1969).
146. Klein, T., and R. M. Klein, J. Bacteriol., 66, 220 (1953).
147. Manil, P., Bull. Inst. Agron. Stat. Rech., Gembloux, 28, 272 (1960).
148. Kern, H., Arch. Mikrobiol., 51, 140 (1965).
149. Kern, H., Arch. Mikrobiol., 52, 206 (1965).
150. Kern, H., Arch. Mikrobiol., 52, 325 (1965).
151. Kern, H., Arch. Mikrobiol., 66, 63 (1969).
152. Sen, M., T. K. Pal, and S. P. Sen, Antonie van Leeuwenhoek, 35, 533 (1969).
153. Dixon, R. O. D., Ann. Rev. Microbiol., 23, 137 (1969).
154. Gibbins, A. M., and K. F. Gregory, J. Bacteriol., 111, 129 (1972).
155. Ryan, F. J., and S. D. Cetrullo, Biochem. Biophys. Res. Commun., 12, 445 (1963).
156. Lester, L. P., Ph. D. Thesis, Purdue University, (1971).
157. Szende, K., Plant Soil, Special Vol., 81 (1971).
158. Davies, R. J., Jr., Bacteriol. Proc., 10 (1958).
159. Kowalski, M., Acta Microbiol. Polon., 15, 119 (1966).
160. Kowalski, M., R. Staniewski, and J. M. Ziemiecka, Ann. Inst. Pasteur, 105, 237 (1963).
161. Marshall, K. C., Nature (London), 177, 92 (1956).
162. Moskalenko, L. N., and Ya. I. Rautenstein, Mikrobiologiya, 38, 340 (1969).
163. Ördögh, F., and K. Szende, Acta Microbiol. Acad. Sci. Hung., 8, 65 (1961).
164. Rautenstein, Ya. I., and L. N. Moskalenko, Mikrobiologiya, 39, 507 (1970).
165. Schwinghamer, E. A., and D. J. Reinhardt, Aust. J. Biol. Sci., 16, 597 (1963).
166. Szende, K., and F. Ördögh, Naturwissenschaften, 47, 404 (1960).
167. Takahashi, I., and C. Quadling, Can. J. Microbiol., 7, 455 (1961).
168. Barnet, Y. M., and J. M. Vincent, J. Gen Microbiol., 61, 319 (1970).
169. Schwinghamer, E. A., Aust. J. Biol. Sci., 18, 333 (1965).
170. Schwinghamer, E. A., Can. J. Microbiol., 12, 395 (1966).

171. Schwinghamer, E. A., Soil Biol. Biochem., 3, 355 (1971).
172. Arber, W., and S. Linn, Ann. Rev. Biochem., 38, 467 (1969).
173. Holloway, B. W., Bacteriol. Rev., 33, 419 (1969).
174. Orosz, L., and T. Sik, Acta Microbiol. Acad. Sci. Hung., 17, 185 (1970).
175. Spizizen, J., B. E. Reilly, and A. H. Evans, Ann. Rev. Microbiol., 20 371 (1966).
176. Staniewski, R., Z. Lorkiewicz, and Z. Chomicka, Acta Microbiol. Polon., Ser. A, 3, 97 (1971).
177. Roslycky, E. B., Can. J. Microbiol., 13, 431 (1967).
178. Lotz, W., and F. Mayer, J. Virol., 9, 160 (1972).
179. Schwinghamer, E. A., C. E. Pankhurst, and P. R. Whitfeld, Can. J. Microbiol., 19, 359 (1973).
180. Broughton, W. J., and M. J. Dilworth, Biochem. J., 125, 1075 (1971).
181. Cutting, J. A., and H. M. Schulman, Biochim. Biophys. Acta, 192, 486 (1969).
182. Dilworth, M. J., Biochim. Biophys. Acta, 184, 432 (1969).
183. Godfrey, C. A., and M. J. Dilworth, J. Gen. Microbiol., 69, 385 (1971).
184. Parker, C. A., Nature (London), 179, 593 (1957).
185. Dilworth, M. J., and C. A. Parker, J. Theor. Biol., 25, 208 (1969).
186. Stroun, M., P. Anker, P. Gahan, A. Rossier, and H. Greppin, J. Bacteriol., 106, 634 (1971).
187. Yajko, D. M., and G. D. Hegeman, J. Bacteriol., 108, 973 (1971).
188. Nutman, P. S., Heredity, 8, 35 (1954).
189. Flor, H. H., Ann. Rev. Phytopathol., 9, 275 (1971).
190. Lorkiewicz, Z., and A. Melke, Acta Microbiol. Polon., Ser. A, 2, 75 (1970).
191. Balassa, R., Abstr. 7th Int. Congr. Microbiol., Stockholm, 49 (1958).
192. Szende, K., T. Sik, F. Ördögh, and B. Györffy, Biochim. Biophys. Acta, 47, 215 (1961).
193. Ellis, N. J., G. G. Kalz, and J. J. Doncaster, Can. J. Microbiol., 8, 835 (1962).
194. Yamane, G., and S. Higashi, Bot. Mag., Tokyo, 76, 149 (1963).
195. Marečková, H., Zbl. Bakt. II, 125, 594 (1970).
196. Datta, N., R. W. Hedges, E. J. Shaw, R. B. Sykes, and M. H. Richmond, J. Bacteriol., 108, 1244 (1971).

197. Datta, N., and R. W. Hedges, J. Gen. Microbiol., 70, 453 (1972).
198. Dunican, L. K., and A. B. Tierney, Mol. Gen. Genet., 126, 187 (1973).
199. O'Gara, F., and L. K. Dunican, J. Bacteriol., 116, 1177 (1973).
200. Beringer, J. E., J. Gen. Microbiol., 84, 188 (1974).
201. Dunican, L. K., and A. B. Tierney, Biochem. Biophys. Res. Commun., 57, 62 (1974).
202. Cole, M. A., and G. H. Elkan, Antimicrobial Agents and Chemotherapy, 4, 248 (1973).
203. Żurkowski, W., M. Hoffman, and Z. Lorkiewicz, Acta Microbiol. Polon., Ser. A, 5, 55 (1973).
204. Parijskaya, A. N., Mikrobiologiya, 42, 119 (1973).
205. Rake, A. V., Genetics, 71, 19 (1972).
206. Sutton, W. D., Biochim. Biophys. Acta, 366, 1 (1974).
207. Agarwal, A. K., and S. L. Mehta, Biochem. Biophys. Res. Commun., 60, 257 (1974).
208. Dénarié, J., G. Truchet, and B. Bergeron, in Symbiotic Nitrogen Fixation in Plants, P. S. Nutman, Ed., Cambridge University Press, 1975.

CHAPTER 13

The Genetic Basis of Dinitrogen Fixation in *Klebsiella pneumoniae*

STANLEY STREICHER

Department of Biology
Massachusetts Institute of Technology
Cambridge, Massachusetts, U.S.A.

R. C. VALENTINE

Plant Growth Laboratory and Agronomy and
 Range Science Department
University of California
Davis, California, U.S.A.

13.1. Introduction, 624
13.2. Dinitrogen-fixing Genes of Bacteria, 625
    13.2.1. Symbiotic Root Nodule Bacteria of Leguminous Plants, 625
    13.2.2. Free-living Organisms, 627
13.3. A Close Relative of Escherichia coli that Harbors the nif Genes, 629
    13.3.1. Klebsiella pneumoniae, 629
    13.3.2. Dinitrogen-fixing Mutants, 631
    13.3.3. Search for a Transducing Phage, 634
13.4. Transductional Analysis, 635
    13.4.1. Generalized Transduction, 635
    13.4.2. Transduction of nif Genes, 638
    13.4.3. Cotransduction of Most nif Genes with his, 638
    13.4.4. Rare nif$^-$ Loci Unlinked to his, 639
    13.4.5. Fine Structure Mapping of nif, 642
13.5. Deletions of nif, 642
13.6. Conjugational Analysis, 644
    13.6.1. Circular Linkage Map, 644
    13.6.2. Transfer of nif by a "Supermale," 644
13.7. Genetic Construction of N_2-fixing Escherichia coli, 646

13.8. Additional Evidence for a Major nif Operon Near
       his, 647
       13.8.1.   Biochemical Evidence, 647
       13.8.2.   Transcription and Translation of the
                 nif Operon(s) by Escherichia coli, 649
13.9. Regulation of Nitrogen Fixation in K.
       pneumoniae, 650
13.10. Future Outlook, 652
13.11. Acknowledgment, 653
13.12. References, 653

13.1 INTRODUCTION

Many of the current concepts of the chemical nature of biological dinitrogen (N_2) fixation have been formulated by biochemists. Now the possibility of studying the genetic basis of dinitrogen fixation provides a fresh approach, which may lead to a better understanding of the regulation and mechanism of this important reaction. This new subfield of dinitrogen fixation was recently ushered in by the discovery of both a system of transduction (1, 2) and conjugation (3, 4) for manipulating the N_2 fixation genes of Klebsiella, a free-living N_2 fixer related to Escherichia coli. Transducing phage P1 was found to mediate the transfer of "N_2 fixation DNA" from actively fixing donor strains of dinitrogen fixation minus (nif$^-$) recipients which subsequently acquired the N_2-fixation trait allowing them to utilize N_2 as sole source of nitrogen (1). This experiment, first performed in our laboratory a few years ago, has been followed up by transductional analysis of the genes essential for N_2 fixation as summarized below. Recently, Dixon and Postgate were able to transfer the N_2 fixation genes using the techniques of bacterial mating (3, 4). In this sexual process N_2 fixation genes were transferred from the nif$^+$ Klebsiella donor to nif$^-$ Klebsiella recipients as well as E. coli yielding recombinants capable of fixing dinitrogen. With these two techniques at hand it is possible to study the genes governing nitrogenase and the nitrogenase system. There is also considerable interest in the construction of specialized transducing phages such as λ (nif), a useful tool for biochemical and genetic studies designed to answer some of the following questions. Where are the nif genes located on the genetic linkage map? How many genes are involved

and are they clustered as a nitrogenase operon? How
does ammonium ion control the expression of nitrogenase?
Which nif⁻ lesions block activity and assembly
of the nitrogenase complex? Can nonfixing organisms be
converted to N_2 fixers after receiving nif DNA? These
are points of major concern and will be discussed in
the various sections below. In the following pages we
use the term "N_2 fixation (nif) genes" to refer to
those segments of DNA that code for and regulate
proteins essential for the overall process of dinitrogen
reduction in the cell.

13.2 DINITROGEN-FIXING GENES OF BACTERIA

An increasing number of microorganisms are known that
fix N_2 (5; III.1). Obviously these microbes all harbor the
essential genes for the nitrogenase system of proteins
in a segment(s) of their DNA. Past investigators have
observed the workings of the N_2-fixation genes mainly
through studies of spontaneous or induced mutations
that alter or even completely block the N_2 fixation
process.

13.2.1 Symbiotic Root Nodule Bacteria of Leguminous Plants

In a classic monograph on Rhizobium published in 1932,
Fred et al. (6) discussed at length the variable traits
of root nodule bacteria (Rhizobium) that may alter
patterns of host plant specificity, time of nodulation,
abundance of nodulation, and effectiveness of nodule
N_2 fixation. Occasionally these bacteria even lose
their capacity to form nodules on host plants. More
recently, noneffective strains have been readily produced
by various chemical mutagens (7). In addition
a wide variety of antibiotic-resistant and auxotrophic
mutants of Rhizobium have been isolated (for complete
reference list and detailed discussion of Rhizobium
genetics, see III.12). In some cases, nutritional
auxotrophy was found to be linked to defective symbiotic
properties (8). Also some prototrophic
revertant strains showed improved symbiotic characteristics
over the wild type (9). The nature of the bacterial
genes responsible for functional nodulation
involves not only the genes coding for the nitrogenase

enzyme system but also genes concerned with the infective process and maintenance of the symbiotic relationship. Plant genes are also involved in symbiotic N_2 fixation (1); for example, the protein moiety (globin) of leghemoglobin, a pigment necessary for symbiotic N_2 fixation, is coded for by the plant (ref. 11, II.8). There is still uncertainty about the plant versus bacterial origin of the nitrogenase genes of root nodules. The strongest argument favoring the bacterial origin is that the bulk of nitrogenase is localized within the mature bacteroids (12) presumably being synthesized there from the DNA template of the bacterium. Evidence adduced from cross inoculation experiments revealed that bacterial DNA coded for the Fe-Mo protein (68). The alternate explanation, that the plant genome supplies the genetic information for nitrogenase, as been hypothesized (13).

There are several reports in the literature of DNA transfer in Rhizobium. Balassa (14) has described genetic experiments involving transformation (DNA uptake and recombination) in which as much as 0.05% of the recipient cells were transformed for various nutritional and drug-resistance markers. This work was reproduced by Kleczkowska (7), who was able to obtain ineffective strains of Rhizobium by exposure of cultures to mutagens, virulent phage, or DNA from a spontaneously ineffective variant. Her attempts to construct effective strains of Rhizobium from ineffective strains were not successful. Transformation has also been used to transfer plant tumor-inducing ability into Rhizobium from the related crown-gall bacterium, Agrobacterium tumefaciens (15, 16). A system of conjugation has been reported for one species of Rhizobium (17). Levels of recombination were high, with as many as 10% of the recipient cells receiving DNA from donor cells, but N_2 fixation markers were not studied. Derivatives of this strain as presently maintained do not produce nodules on host plants (Heumann, personal communication); because of keen interest in this organism as a tool for studying the genetic basis of symbiotic N_2-fixation, a detailed classification of this organism as an authentic strain of Rhizobium is anxiously awaited.

Although these systems of genetic transfer have considerable potential for the study of symbiotic N_2 fixation genetics, there are experimental disadvantages to working with Rhizobium. The most serious drawback is that the N_2-fixation genes have at this time, been found to be expressed only in the symbiotic state, the free-living Rhizobium being unable to fix dinitrogen. The current hypothesis is that the essential genes for N_2 fixation are activated by compounds present in the microenvironment of the plant nodule (see III.10), a condition not readily achieved in vitro. In the past, this has necessitated working with whole plants, which requires relatively long periods of time, effort, and space. The recent development of a Rhizobium-plant symbiotic tissue culture system (18) may prove to be an important technical advance for studies of the genetic basis of symbiotic N_2 fixation.

13.2.2 Free-living Organisms

Many species of free-living bacteria and blue-green algae can fix dinitrogen, but little is known at present regarding the genes governing this in most of these organisms. Bacterial mutants that are no longer able to utilize dinitrogen as sole nitrogen source have been isolated by many workers. Karlson and Barker (19), the first to isolate N_2-fixation mutants, described a mutant of Azotobacter that had lost the ability to fix N_2 when growing on glucose as an energy source but could fix N_2 when glucose was replaced by ethanol. Other substrate conditional mutants have been isolated from Azotobacter. For example, a mutant strain of A. vinelandii blocked in a reaction association with N_2 fixation but not ammonia utilization was isolated (20). The mutant was able to fix N_2 in media containing lactate or pyruvate, but was unable to fix N_2 in media containing glucose, sucrose, or glycerol as sole carbon and energy source. In light of current knowledge it seems probable that this strain was blocked in a crucial ancillary step of N_2 fixation but not nitrogenase itself. The fact that growth of the mutant strain occurred on glucose or the other substrates in the presence of available ammonia does not rule out the possibility of a partial metabolic block preferentially affecting the N_2 reduction system.

Temperature-sensitive mutants of A. vinelandii that fix N_2 at 29° but not at 39°C (21) have recently been isolated. Most of the temperature-sensitive mutants isolated were not affected in their N_2-fixing activity when shifted from 29°C to 39°C. Dinitrogen fixation continued undiminished; however, no further increase activity occurred with growth after shift to the elevated temperature. The nitrogenase present was presumably "diluted out" on continued growth with no further replenishment from new synthesis. These mutants were classified as "biosynthetic" in reference to some step in biosynthesis, regulation, or assembly of the nitrogenase proteins and/or other ancillary reactions supporting N_2 fixation that were apparently blocked at the high temperature. A second class of mutants referred to as "enzymatic" mutants were blocked in N_2-fixing activity when shifted to 39°C. The nitrogenase complex of these strains were not inactivated, as such, on shift to high temperature; these lesions might be in other proteins specifically involved in N_2 fixation such as ferredoxin and flavodoxin. Mutants that fix N_2 at restricted ranges of pH have also been isolated (22). Mutant strains which apparently produce defective nitrogenase have been isolated from Azotobacter vinelandii (23, 24) and Clostridium pasteurianum (25).

This work provides a background of information concerning the biochemical classes of nif⁻ mutations. From analysis of A. vinelandii and Clostridium pasteurianum nif⁻ strains, three biochemical classes of nitrogenaseless mutants were found. All three classes were deficient in dithionite-dependent nitrogenase activity. Two classes contained either active Fe protein or active Mo-Fe protein; the third class was deficient in the activities of both nitrogenase components. Sorger and Trofimenkoff (24) were able to reconstruct nitrogenase activity by mixing extracts, each of which were deficient in one component. They were able to identify the Mo-Fe protein in one mutant by demonstrating the loss of the complementing activity in extracts prepared from cells grown on media deficient in Mo. The in vitro complementation activity of extracts of the other class was not affected by the Mo deficiency and was assumed to contain only the Fe protein. Extracts from this latter class also were able to stimulate the nitrogenase activity in wild-type extracts, which have often been found to contain limiting quantities of Fe protein. Brill and co-workers (23, 24) determined the specific deficiencies in their mutant strains by complementation with the separate components of wild-type nitrogenase. They found one C. pasteurianum nif⁻ strain and one

A. vinelandii nif⁻ strain each deficient in Mo-Fe protein activity; i.e., extracts regained nitrogenase activity by addition of purified Mo-Fe protein, whereas addition of Fe protein alone did not stimulate activity. Extracts prepared from another A. vinelandii mutant failed to complement with either component.

The absence of genetic analysis with the above organisms limits the scope of the investigation to finding in vitro complementation groups. A major drawback in interpreting this work is that more than one genotype could yield identical biochemical phenotypes that would be mistakenly classified together. Each nitrogenase component, for example, might require the functioning of several gene products, some specific for one component (e.g., enzymes of Mo metabolism), and others required for both components (e.g., enzymes which synthesize $S^=$) in addition to the structural genes for the proteins in each component. Regulatory lesions might also affect the synthesis of one or both components. Thus the need for genetic analysis is evident.

Unfortunately, no system of DNA transfer is yet known for these bacteria. The recent report of transformation in blue-green algae may allow the genetic study of N_2 fixation in these important organisms. Shestakov and Khyen (26) have recently presented evidence that the blue-green alga Anacystis nidulans can undergo genetic transformation. DNA from drug-resistant strains transformed wild-type cells to drug resistance at frequencies as high as 0.07%. Dinitrogen fixation markers were not tested. Other encouraging findings regarding future studies on the genetic basis of N_2 fixation in blue-green algae include the isolation of algal phages raising the possibility of transductional analysis of the nif genes (27). Also, Asato and Folsome (28) describe a procedure for genetic mapping of a blue-green alga genome, an important prerequisite for genetic analysis of N_2 fixation in these ecologically important organisms.

13.3 A CLOSE RELATIVE OF ESCHERICHIA COLI THAT HARBORS THE NIF GENES

13.3.1 Klebsiella pneumoniae

Klebsiella (29) seems to be an ideal organism for genetic studies of N_2 fixation. K. pneumoniae is

Dinitrogen Fixation in Klebsiella pneumoniae

related to the common colon bacterium, E. coli; in fact, it has a chromosomal map (Fig. 13.1) (30) very similar to that of E. coli (31) and Salmonella typhimurium (32). As discussed below, Klebsiella mates with E. coli and is sensitive to many coliphages. There are several distinguishing biochemical and physical features such as the higher G + C content of its DNA (60% compared to 50% for E. coli; (see ref. 33) and the larger number of nitrogen sources used by Klebsiella (34). DNA hybridization tests reveal a considerable degree of homology between the two chromosomes (4); a "species barrier" still seems to exist since chromosomal hybrids (recombinants) with other coliform bacteria are rare (35, 36).

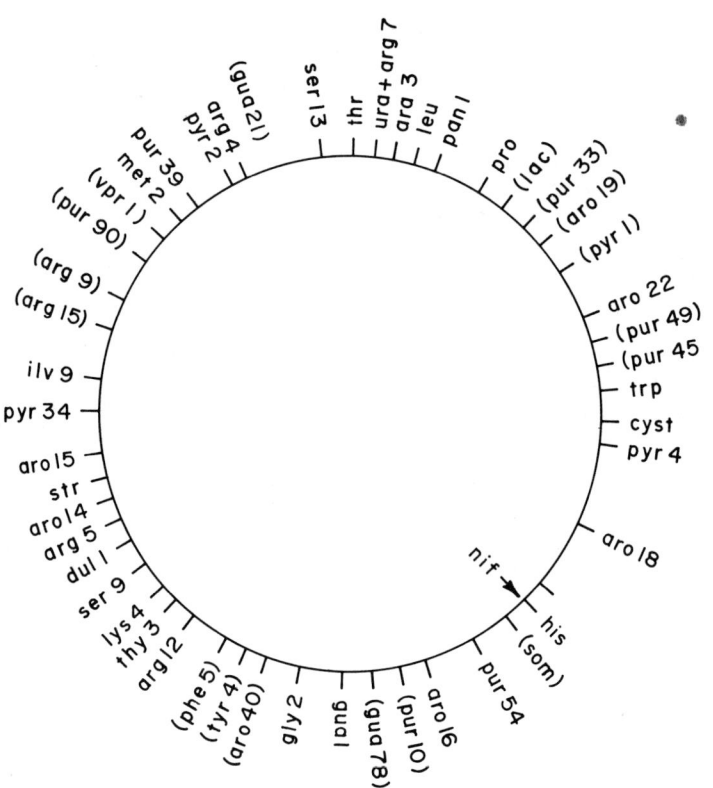

Fig. 13.1. Genetic linkage map of Klebsiella pneumoniae (modified from ref. 58).

nif Genes

Klebsiella are able to fix dinitrogen only under anaerobic conditions. Dinitrogen-fixing strains of Klebsiella have been isolated from leaf nodules of tropical plants (37) and from plant surfaces, soil, and rhizosphere (29, 38). Large numbers occur in the intestines of New Guinea natives whose diet consists mainly of sweet potatoes (39). Although there is no doubt that N_2-fixing strains of K. pneumoniae inhabit the human intestinal tract, the ecological significance of K. pneumoniae as active N_2 fixers in the gut and their role in the nitrogen economy of the individual remain intriguing but as yet unanswered questions.

13.3.2 Dinitrogen-fixing Mutants

Studies on the genetic basis of N_2 fixation in K. pneumoniae depend upon the availability of a wide range of suitably marked mutant strains and means of genetic transfer between strains. Large numbers of genetically stable mutant strains of K. pneumoniae unable to fix dinitrogen are easily obtained (1). Nif-1 through nif-216 plus several hundred other mutants of K. pneumoniae comprise our stock collection. The strain numbers and nif mutation isolation numbers are identical for our collection, thus a specific mutant strain, e.g., nif-1, refers to both the nif mutation and to the strain.

Nif-1 through nif-116 and nif-M1 through nif-M24 isolated in the initial mutant search, were exposed to the mutagen nitrosoguanidine (NTG) for 40 min. Nif-120 through nif-142 were isolated following exposure to NTG for 10 min to reduce the probability of isolating mutant strains carrying multiple mutations.

Dinitrogen fixation (nif⁻) mutants were isolated following chemical mutagenesis and penicillin selection. Mutant and wild-type colonies are easily distinguished based upon colony size and color (Fig. 13.2). Colonies of N_2-fixing strains of K. pneumoniae are darker and much larger than those of mutant colonies. The dark brown color of the N_2-fixing colonies may be due to the synthesis of large amounts of nitrogenase and other brown iron-sulfur proteins. Colonies grown on ammonium or other nitrogen sources are light in color (white to pale yellow) and yield cell-free extracts that are pale yellow. As illustrated in Fig. 13.2, nif⁻ colonies are also light in color. In one mutant selection experiment, numerous

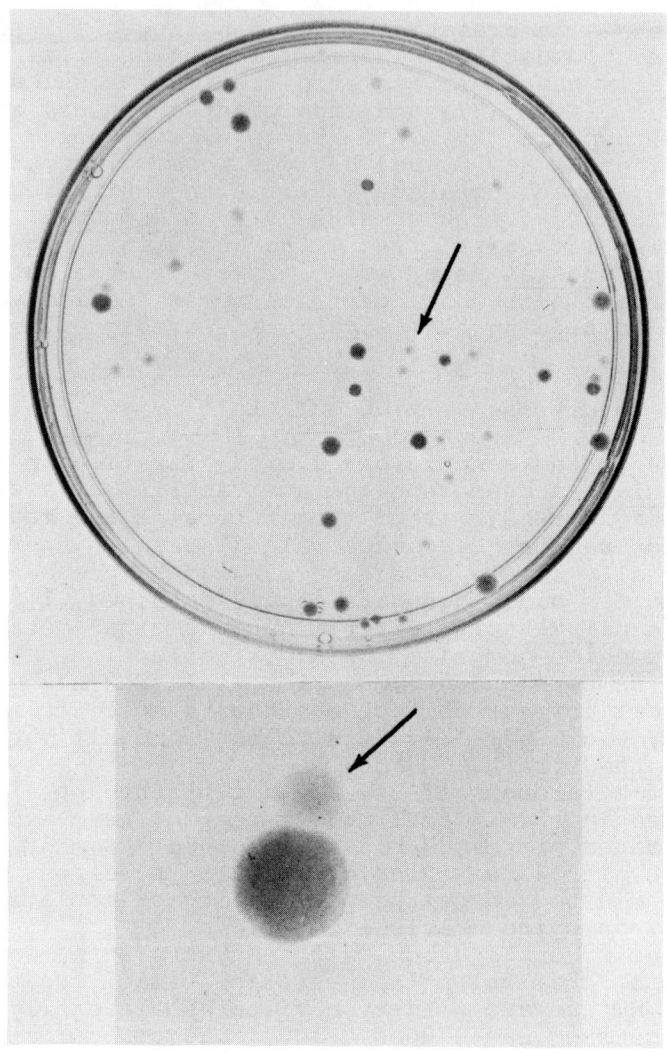

Fig. 13.2. Colonies of wild-type of N fixation mutants of Klebsiella pneumoniae selected by the penicillin technique. Dark colonies are able to fix N , while the small light nif colonies (arrows) grow at the expense of traces of fixed nitrogen present in the agar (from ref. 1).

nif Genes

small but dark colonies were picked along with the tiny white colonies. All dark colonies were either very leaky nif mutants or had defects affecting their growth rates rather than N_2 fixation (40). Initially (1) the minimal plates used to select nif⁻ strains were supplemented with 15 µg/ml ammonium sulfate to allow nif⁻ mutants to form colonies using this low level of fixed nitrogen. It was subsequently found that the ion-agar used in preparing the plates contained sufficient fixed nitrogen to support the growth of tiny colonies.

The genetic stability of nif⁻ mutations was determined by plating large numbers of cells (ca. 10^9) onto nitrogen-free plates (1). The reversion frequency (nif⁻ → nif⁺) was determined as the number of nif⁺ revertant colonies per viable cell plated. Mutant cells were also exposed to chemical mutagens to determine the types of mutations. Most nif⁻ strains spontaneously reverted at frequencies ranging from $1-20 \times 10^{-9}$ revertants per viable cell. A few strains (e.g., nif-1, nif-4, nif-72, and nif-75) were found not to spontaneously revert nor to be reverted by any of the chemical mutagens used (40). It is possible that these strains are deletion mutants, with a segment of the genome containing some nif genes missing.

Nif⁻ strains were screened for the ability to utilize a number of nitrogen sources besides dinitrogen to detect possible pleiotropic mutations. Single colonies picked from storage plates were radially streaked onto minimal agars and a few crystals of each nitrogen source were placed in the center. The growth response was noted after aerobic incubation for 1-3 days at 30°C or anaerobic incubation under argon for 3-5 days at 30°C. The majority of the nif⁻ strains are deficient only in the utilization of N_2 (e.g., nif-6, nif-90, nif-94) (40).

The growth rates of several mutant strains were determined to see if putative N_2 fixation lesions altered cell growth on ammonium. The doubling time for wild-type and most nif⁻ strains are 50-60 min, suggesting that the nif⁻ lesions do not have any deleterious effect upon the utilization of ammonium (1).

A time lag of 8-15 hours between depletion of fixed nitrogen (ammonium) and N_2 fixation is typical for K. pneumoniae M5Al (41, 42) and is presumably due to slow derepression of the N_2 fixation genes. Some strains of K. pneumoniae such as H-26 and H-40 of human origin were found to have shorter derepression periods (40). Yoch and Pengra (42) observed that the utilization of certain nitrogen sources can shorten the growth lag and some eliminate it completely. They demonstrated that when dinitrogen-grown cells were transferred into fresh media containing limiting amounts of fixed nitrogen sources, growth on dinitrogen resumed either shortly after depletion of fixed nitrogen or that N_2 fixation occurred simultaneously with the utilization of the fixed nitrogen source (e.g., aspartate) and no growth lag occurred. They concluded that the utilization of aspartate and a few other amino acids did not repress N_2 fixation. Since aspartate provides nitrogen for cell growth but unlike ammonium does not repress N_2 fixation, it was used as the nitrogen source in media when measuring the nitrogenase activity in whole cells or nif⁻ mutants and for large cell cultures used for the preparation of cell-free extracts.

13.3.3 Search for a Transducing Phage

Transduction in K. aerogenes W-70 mediated by a lysogenic Klebsiella phage, PW-52 has been reported (43). This phage can mediate generalized transduction of auxotrophic markers, but unfortunately the phage has a very narrow host range and does not infect N_2-fixing strains. We examined about 25 natural isolates of Klebsiella for P1 sensitivity. Phage sensitivity was determined by spotting 0.05 ml of a lysate (containing ca. 10^9 pfu/ml) onto a broth agar plate containing a freshly poured suspension (in broth top agar) of the bacteria to be tested. Plates were incubated at 37°C for 1-3 days and examined for zones of lysis. Two Klebsiella strains (M5A1 and UW-1) (29) were clearly sensitive to P1 and three strains [L-3, DK-40 (23), and H-49 (29)] showed faint lysis. Two naturally P1-resistant strains (CDC #2844-64 and 1702-49) were exposed to NTG, and mutants sensitive to P1 were isolated from each (2). Similar phage-sensitive mutants of coliform bacteria have been reported (44), including Salmonella typhimurium LT-2 polysaccharide mutants sensitive to P1. It is probable that many other Klebsiella strains and even other coliform bacteria could be mutated to P1 sensitivity, permitting genetic analysis to be applied.

Two strains of K. pneumoniae (M5Al and UW-1) were found to be sensitive to the lambdoid phage 424 as well as Pl (2). Phage 424 is genetically related to phage λ (45) and might be capable of mediating specialized transduction in K. pneumoniae. Phage 424 was found subject to a restriction-modification system (46) in K. pneumoniae (2). During phage sensitivity experiments it was noticed that 424 (previously grown on strain M5Al) caused a zone of confluent lysis on strain M5Al, while producing only a few plaques on strain UW-1. Starting with the few plaques that did appear, lysates could be prepared from strain UW-1 that were active against this strain but not M5Al. It thus appeared that 424 was restricted and modified by the two K. pneumoniae strains. To quantitate this, 424 lysates were prepared on the two K. pneumoniae strains and on E. coli K-37 (K-12-type restriction/modification) and E. coli 993 (a mutant strain which neither restricts nor modifies). These lysates were used to determine the efficiency of plating of 424 on the four bacterial strains. Both K. pneumoniae strains restrict phage growth on the other. E. coli K-37 restricts phage growth on K. pneumoniae. It cannot be determined from this experiment whether the restriction types are new or unique (47). The point to be stressed here is that nif DNA can indeed be marked and recognized by various donor and recipient strains; this is a feature to be considered in designing genetic crosses of the nif genes.

Two other lambdoid phages, λ and ϕ80, and the transducing phage P-22 did not infect Klebsiella. Strain M5Al is, however, sensitive to the lysogenic coliphage P2.

A summary of the sensitivities of K. pneumoniae M5Al to various transducing and virulent bacteriophages is presented in Table 13.1.

13.4 TRANSDUCTIONAL ANALYSIS

13.4.1 Generalized Transduction

Transduction of the genetic material of K. pneumoniae M5Al mediated by Pl was first investigated using amino acid and vitamin requiring mutant strains (auxotrophs) (1). Pl lysates prepared on wild-type bacteria (donor phage) were used to infect a number of auxotrophs

TABLE 13.1. Sensitivity of *Klebsiella pneumoniae* M5A1 to Various Transducing and Virulent Bacteriophages

Bacteriophages	
Sensitive	Resistant[a]
P1, P2, 424	T4, f2, P22,
T3, T7, If1[b]	MS2, ϕ80, λ,
ϕII[c]	mµ

[a] No clearing occurred in zone containing $\sim 10^7$ phage particles.
[b] Male-specific requiring I-pilus for attachment.
[c] Female-specific, inhibited by presence of R factor.

(recipients). The average transduction frequency is 6.5×10^{-6} prototrophic transuctants/infectious phage, in close agreement with the frequency of transduction for auxotrophic markers in *E. coli* and *Shigella* (48). The data indicate that P1 transduction of *K. pneumoniae* genes is generalized (all genes capable of being transferred) and quantitatively similar to transduction of the genes of *E. coli*. Figure 13.3A shows a cell of *K. pneumoniae* being invaded by phage P1. This transducing phage is shown at higher magnification in the electron micrographs (Fig. 13.3B and C). For further details concerning the morphology and infective properties of P1, see ref. 49.

Fig. 13.3. Electron micrographs of particles of the generalized transducing phage PI. (A) P1 invading a *Klebsiella* cell. (B) P1 particles showing two characteristic head sizes. (C) Contracted particle showing extended tail core. Infected cell stained with 1% uranyl acetate at pH 4.3 and particles (C and B) with 2% potassium phosphotungstate at pH 5.0. Micrographs kindly taken by Alice Taylor (from ref. 40).

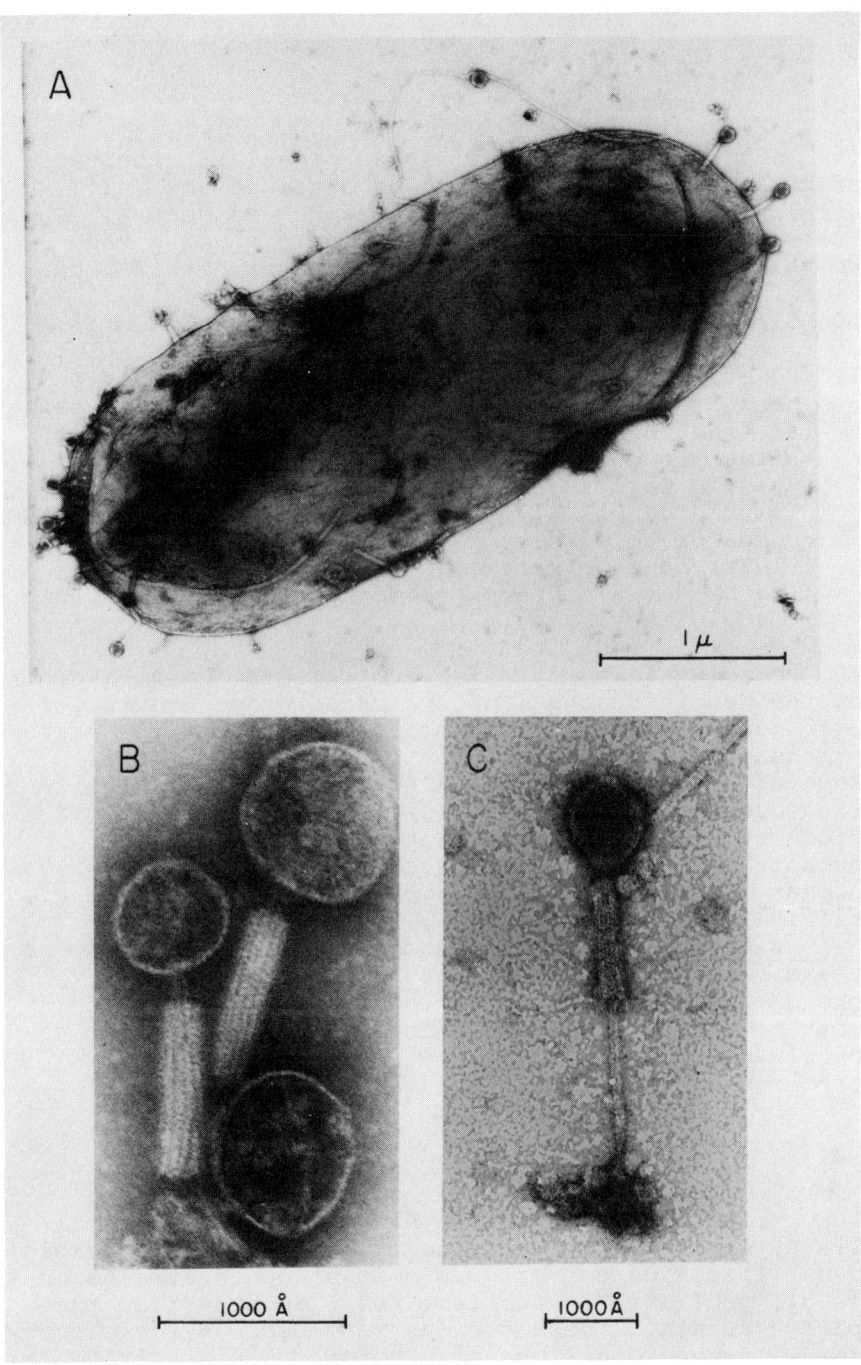

13.4.2 Transduction of nif Genes

The genetic study of N_2 fixation began with analysis of the transduction frequencies obtained when nif⁻ strains were used as both recipients and donors. Such crosses can reveal much information regarding the relative genetic distances within a set of nif⁻ mutations. Earlier work with transducing phages (31, 48, 50, 51) has indicated that:

1. Genetic markers separated by large distances on the chromosome (unlinked genes) give transductants at frequencies comparable to those obtained with wild-type donor phage.
2. Genetic markers that are close to each other (linked genes) give low frequencies of transduction compared to wild-type.
3. P1-transducing particles can transfer a segment of the bacterial chromosome corresponding to approximately 100 genes.

Recipient nif⁻ strains were infected with P1 grown on the donor strains (two cycles of phage growth for nif⁻ donors), plated onto minimal agar, and the nif⁺ transductants recorded after suitable incubation. Several crosses (e.g., nif-88 x nif-95) gave high transduction frequencies indicating that these mutations are in distant but linked genes (1). Other mutants that have been crossed with nif-88 and nif-95 appear to have mutations which are near nif-88 (e.g., nif-104), near nif-95 (e.g., nif-41), or between nif-88 and nif-95 on the chromosome (e.g., nif-23) as these yield transductants at frequencies similar to one of these mutants or of intermediate value with both of them. Therefore these nif lesions appear to be clustered in a relatively small region of the chromosome.

13.4.3 Contransduction of Most nif Genes with his

The chromosomal location of nif mutations was established through the identification of a cotransducible gene (1). A number of double auxotrophic strains (e.g., pro⁻, nif⁻) were isolated with nif-95 as the parent strain. These strains were used as recipients for wild-type donor phage. Phototrophic N_2-fixing colonies arising from such crosses were scored as cotransduction. Of 11 different amino acid and vitamin requirements only his and nif were simultaneously transferred.

To facilitate analysis of the linkage relationship of nif genes with his genes, a set of nif⁻ strains was isolated using a his⁻ parent (hisD2). Strain hisD2 is unable to utilize histidinol as a source of histidine (52) or to complement in vivo with an E. coli episome carrying the histidine operon with a lesion in the hisD gene (53). These data suggest that this K. pneumoniae strain has a defect in a histidine gene analogous to the hisD gene of E. coli and S. typhimuium.

These double auxotrophs (his⁻, nif⁻) were used as recipients for wild-type donor phage to determine the cotransduction frequencies of the various nif lesions with hisD2 (2). As shown in Table 13.2, the cotransduction frequencies range from 78% for nif-213 to 30% for nif-120 and nif-135. Also, note from Table 13.2 that 90% of the nif⁻ strains investigated are of the his-linked genetic type. The cotransduction frequencies of Table 13.2 can be used to make a rough estimate of the length of the DNA of this nif region, i.e., the segment of DNA that lies between the two outside nif markers (nif-213 . . . nif-88). Using the approximations discussed by others (31), we estimate that the nif segment is roughly 15-20 average gene lengths in size, a coding capacity that could more than accommodate the Mo-Fe protein and Fe protein structural genes of nitrogenase. The presence of other nonessential genes spaced between the nif genes in this region is not known nor is the relative orientation of his to nif on the chromosome. Further fine structure analysis must be performed before the apparent clumping of several nif lesions within discreet regions near his can be confirmed.

13.4.4 Rare nif⁻ Loci Unlinked to his

Two mutations, nif-130 and nif-202, appear to be unlinked to his as no nif⁺ colonies were found among the his⁺ transductants (2). Nif-130 and nif-202, both unlinked to his, do not have any phenotypic differences from the majority of nif⁻ mutations that are linked to his. These two strains are likely to represent a new class

TABLE 12.2. Cotransductional Analysis of nif Genes with his (Modified from ref. 2)

Wild-type donor x recipient[a]	Number of his$^+$, nif$^+$ transductants per 200 his$^+$ colonies analyzed	Cotransduction frequency (%)
hisD, nif-213	157	78.5
hisD, nif-131	146	73
hisD, nif-125	145	72.5
hisD, nif-217	141	70.5
hisD, nif-133	138	69
hisD, nif-129	127	63.5
hisD, nif-95	118	59
hisD, nif-132	116	58
hisD, nif-137	115	57.5
hisD, nif-136	112	56
hisD, nif-123	109	54.5
hisD, nif-138	105	52.5
hisD, nif-218	97	48.5
hisD, nif-127	96	48
hisD, nif-209	96	48
hisD, nif-204	95	47.5
hisD, nif-126	94	47
hisD, nif-205	94	47
hisD, nif-215	94	47
hisD, nif-121	93	46.5
hisD, nif-23	92	46
hisD, nif-206	92	46
hisD, nif-214	91	45.5
hisD, nif-122	91	45.5
hisD, nif-216	90	45
hisD, nif-208	89	44.5
hisD, nif-207	88	44
hisD, nif-201	87	43.5
hisD, nif-139	85	42.5
hisD, nif-134	85	42.5
hisD, nif-124	82	41
hisD, nif-128	82	41
hisD, nif-211	81	40.5
hisD, nif-212	80	40

(Continued)

TABLE 12.2. (Continued)

Wild-type donor x recipient[a]	Number of his⁺, nif⁺ transductants per 200 his⁺ colonies analyzed	Cotransduction frequency (%)
hisD, nif-140	80	40
hisD, nif-142	79	39.5
hisD, nif-88	61	30.5
hisD, nif-120	60	30
hisD, nif-135	60	30
hisD, nif-130	0	< 0.5
hisD, nif-202	0	< 0.5
hisD, nif-141	200	100

[a]Histidine-requiring auxotrophy, hisD, was prepared from wild type of nitrosoguanidine treatment and penicillin enrichment (1). The hisD auxotroph was used as parent for selection of various hisD, nif⁻ double auxotrophs.

of nif genes located in different regions of the K. pneumoniae chromosome. An unusual class of nif⁻ mutants was isolated by selecting for simultaneous his⁻, nif⁻ double auxotrophs. As shown in Table 13.2, his⁻, nif⁻ mutants represented by nif-141 display 100% linkage between his and nif markers. A single base mutation is indicated since prototrophic revertants are readily obtained. The nature of this lesion blocking both pathways is unknown but may be regulatory in nature.

Several other mutants, unable to fix dinitrogen whose nif⁻ loci were found to be unlinked to his, have been isolated in the course of our work (40). These strains probably have genetic lesions in various auxiliary pathways supporting dinitrogen fixation. For example, two strains selected for the nif⁻ phenotype on sucrose plates were able to fix dinitrogen with glucose as an energy source (40). An undefined lesion in sucrose metabolism seemed responsible; the strains grew but at reduced rates on sucrose and NH_4^+. Also, a few nif⁻ strains are blocked at the level of ammonia assimilation as described in a later section.

13.4.5 Fine Structure Mapping of nif

The relative order of three genetic loci can be determined through the use of three-factor crosses. A series of such crosses can establish a genetic map. Such analysis has been used to order mutations in the lac operon (54) and the arabinose operon (55). We have employed three-factor crosses to order several nif⁻ loci located near his (2, 56). The experimental procedure for three-factor crosses is similar to cotransduction experiments with nif⁻ strains used as donors rather than wild-type. The mutational order is determined by comparing the results from reciprocal crosses in which each nif⁻ mutation is used as donor or recipient. One reciprocal cross permits the formation of wild-type recombinants with the fewest crossover events. This is reflected in the percentage of his$^+$ transductants regaining N_2-fixing ability. A summary of the order of several nif loci is as follows: hisD nif-213 nif-131 nif-125 nif-217 nif-95 nif-205 nif-138 nif-23 nif-128 nif-208 nif-204 nif-215 nif-120 nif-88.

13.5 DELETIONS OF nif

The biological and physical basis for delections in bacteria and their viruses has been well established. In the simplist sense a deletion represents a shortening of the chromosome by the length of the segment of deleted DNA, a piece varying from a single base pair to tens of thousands of base pairs. Deletions of DNA near the nif region have been isolated in K. pneumoniae using two different methods. The first method takes advantage of a rare excision event associated with the replication of the phage P2. It has been observed in E. coli that the growth of phage P2 occassonally causes the deletion of DNA close to its attachment site H, near the his operon (57). These E. coli deletions are missing the entire his operon and gnd, a gene specifying the enzyme gluconate-6-phosphate dehydrogenase. K. pneumoniae M5A1 is sensitive to phage P2 and we have isolated histidine requiring mutants following the growth of phage P2. These strains cannot utilize N_2 as the sole source of nitrogen and do not produce nitrogenase (2). The eductants (phage P2 induced deletions) are also missing the entire his operon and gnd. Since all the eductants that have been isolated are His⁻ Nif⁻ Gnd⁻, the relative order of these genes could not be determined.

A second method that allows the isolation of strains with deletions of varying lengths has been used to order several genes in the nif region. These deletions are isolated as mutant strains resistant to the Klebsiella phages K3 or K14 that simultaneously require histidine for growth (63). All strains isolated in this manner are missing gluconate-6-phosphate dehydrogenase placing gnd between the gene(s) specifying phage resistance (probably rfb) and the his operon. Several mutant strains are missing only a part of the his operon; they are able to utilize histidinol as a source of histidine. This implies that the hisD gene is present and functional in these strains (hisD is the gene for histidinol dehydrogenase). Complementation analysis of several hisD⁺ strains revealed the presence of several additional his genes. The results suggest that the his operon is oriented on the K. pneumoniae chromosome with the operator end distal to gnd. All strains whose deletions do not extend past the his operon remain Nif⁺. Strains with deletions extending into nif (such as the eductants) are completely missing the his operon and can be divided into two classes based upon the utilization of shikimic acid (shu) as the sole source of carbon and energy. One class can no longer utilize shikimic acid while the other can. Into the latter class fall the eductant strains. These deletion strains allow the establishment of a map of this region of the K. pneumoniae chromosome as follows:

```
                         his
    rfb      gnd     EIFAHBCDGO      nif       shu
```

The strains that remain shu⁺ probably contain deletions that extend only partly through the nif genes. These strains could be used in the determination of the genetic fine structure of the nif genes since they could allow the unambiguous mapping of nif point mutations.

13.6 CONJUGATIONAL ANALYSIS

13.6.1 Circular Linkage Map

Conjugational analysis has recently been employed to map the Klebsiella chromosome (Fig. 13.1); approximately 50 markers have thus far been located (58). In order to select a fertile strain for this work, Matsumoto and Tazaki (30) screened a number of combinations of human isolates of K. pneumoniae for transfer of nutritional markers. The surfaces of petri dishes were flooded with a suspension of a streptomycin-resistant auxotroph as recipient and after the surface dried, loopfuls of potential donor strains were spread on a square patch, 25 per plate, to permit mating in this zone. Growth within the patch indicated prototrophic recombinants had arisen by conjugation. A suitable donor (male) and recipient (female) pair were detected by this simple procedure. Multiple auxotrophs containing up to eight different nutritional defects were constructed and crossed; analysis of the distribution patterns of unselected markers among the recombinants yielded the circular linkage map shown in Fig. 13.1. A striking feature of the Klebsiella map is its similarity with the linkage maps of E. coli and Salmonella, a point discussed by Matsumoto and Tazaki (58). The significance of this map is that it provides a genetic framework of suitable markers for pinpointing the location of the N_2 fixation genes on the Klebsiella chromosome.

13.6.2 Transfer of nif by a "Supermale"

The methods used by Dixon and Postgate (3) for constructing a fertile derivative of the N_2-fixing strain K. pneumoniae M5Al carrying the episome R144drd3 are summarized in Table 13.3. Note that the episome was first "derepressed" while still in E. coli by placing a mutation in a regulatory locus of the episome, which governs infectivity of the episome (59). These strains behave as "supermales" transferring their episomes to females with great efficiency. Strain M5Al without the episome behaves as an excellent recipient (female) for these sexual crosses. Occasionally the episome transfers a segment of bacterial DNA, taken more or less at random from the chromosome (59). The amount of

TABLE 13.3. Genetic Construction of A N₂-fixing Escherichia coli (4)

1. E. coli (episome) $\xrightarrow{\text{chemical mutagen}}$ E. coli (depressed episome)

2. E. coli (derepressed episome) x Klebsiella $\longrightarrow$ Klebsiella (derepressed episome)

3. Klebsiella (derepressed episome) $\xrightarrow{\text{ultraviolet light}}$ Klebsiella ("Hfr-like")

4. Klebsiella ("Hfr-like") x E. coli C $\longrightarrow$ E. coli C (nif⁺)

Klebsiella DNA transferred is not known with precision but probably is in order of several hundreds of genes. F-type transfer, rather than Hfr-type, is indicated by the finding that recombinants receive the functional episome as well as DNA from the donor (4).

Crosses were performed between K. pneumoniae M5Al carrying the episome as donors and nif⁻ mutants of M5Al as recipients. In typical crosses, N_2-fixing recombinant colonies appeared at a frequency of approximately 10^{-5} per donor organism (4). This represents only a small fraction of the recipient cells that receive the episome during mating, suggesting that transfer of chromosomal DNA is a rare event during mating. Dixon and Postgate (3) found that his shows a 95% linkage to their nif-9 and an 85% linkage to nif-2, thus confirming that his and nif are closely linked.

13.7 GENETIC CONSTRUCTION OF N_2-FIXING ESCHERICHIA COLI

The mechanics of this exciting experiment (see Table 13.3) recently carried out by Dixon and Postgate (4) will be described in this section with some important implications stressed in the remaining sections.

Transfer of N_2 fixation genes (nif) within Klebsiella pneumoniae has been accomplished by both transduction and conjugation. In these cases nif was transferred to mutants of the same strain deficient in N_2 fixation (intrastrain crosses). The intergeneric transfer of nif from K. pneumoniae to Escherichia coli, a species that does not naturally fix N_2 has been reported (4). For this experiment a derivative of M5Al was isolated following ultraviolet irradiation and subsequent selection for high-frequency transfer of his. This strain gave rise to his recombinants 20-30 times more frequently than M5Al carrying the depressed episome. The recombination frequencies of a number of other auxotrophic markers were lower than his and diminished as the distance from the histidine operon increased. These observations led Dixon and Postgate (4) to suggest that polarized transfer of chromosomal markers occurred. However, the autonomous sex factor still persisted in the recombinants.

In order to avoid the barrier of host-specific restriction which we have discussed in an earlier

Evidence for a Major nif Operon Near his

section, which often arises in intergeneric bacterial matings, these workers decided to use E. coli strain C as a recipient, since this strain is naturally non-restricting and nonmodifying (46). Transfer of his markers from Klebsiella to an E. coli C his⁻ recipient was followed. His⁺ colonies appeared at the low frequency of approximately 10^{-7} per donor cell. These E. coli his⁺ hybrids were tested for acetylene reduction. In two experiments 10 out of 12 and 2 out of 6 his⁺ hybrids gave positive acetylene reduction. All the hybrids showed reactions typical of E. coli and could easily be distinguished from the donor strain of K. pneumoniae. These strains of E. coli were found capable of N_2 fixation by three separate criteria: acetylene reduction, direct reduction of $^{15}N_2$, and growth on media free of fixed nitrogen.

Most N_2-fixing strains of E. coli C were genetically unstable, readily segregating off nif⁻ derivatives. However, one isolate was stable with all progeny displaying N_2 fixation.

13.8 ADDITIONAL EVIDENCE FOR A MAJOR nif OPERON NEAR his

Klebsiella contains DNA, which codes for the structural and regulatory genes governing nitrogenase and other essential functions of N_2 fixation. These genes might be located in different regions of the chromosome or might be organized as tightly clustered and functionally related groups of genes ("nif operons"). Genetic evidence supporting the concept of a major nif cluster of genes near his has already been cited. Evidence described below supports the idea that this segment of DNA codes for nitrogenase.

13.8.1 Biochemical Evidence

An overwhelming majority of the mutants that we have isolated affect the activity of nitrogenase (1, 2, 40, 56). For example, of 138 nif⁻ mutants of our collection that were assayed for whole cell activity, about two-thirds reduced acetylene at a rate less than 0.5% that of the wild-type strain. Twenty-five mutants gave rates of acetylene reduction from 0.5 to 1% that of wild-type, while 5 strains were 1-2% as active; 17

strains were "leaky" in their growth on N_2 and had acetylene reduction capacity from 2 to 38% that of wild-type. Most of these mutants were found to map near his. It seems clear that most nif⁻ strains that we have isolated are blocked at the level of nitrogenase itself.

At least two genes, corresponding to the Mo-Fe protein and the Fe protein, must code for nitrogenase (II.2). Mutations in either gene might completely destroy nitrogenase activity since neither component can work catalytically without the other. We have attempted to identify separate genetic loci (structural genes) coding for these proteins using two different approaches. In the first experiment, mutant extracts with no nitrogenase activity were tested for the presence of active nitrogenase components by mixing various mutant extracts in combinations of two and testing for nitrogenase activity. As already described, this procedure was first applied to nitrogenase mutants of Azotobacter vinelandii (24) where several in vitro complementation groups were identified. In Klebsiella, functional nitrogenase was reconstituted by mixing certain extracts, which alone were devoid of acetylene reduction capacity. For example, nitrogenase was reconstituted by mixing extracts of nif-23 and nif-95 (56). More than a 100-fold increase was obtained with several of the crosses. Maximum reconstituted levels of enzyme were in the range of 10% that of wild-type. This might be the result of suboptimal conditions for reconstitution or reflect the fact that the levels of components produced by the nif⁻ strains is substantially lower than wild-type. The simplest interpretation of this experiment is that the mutants that show successful "in vitro complementation" each produce an essential portion of the nitrogenase complex defective in the other. Most crosses of extracts from two different mutants do not yeld activity, suggesting that these strains are missing more than one nitrogenase component.

Others (23, 25) were able to pinpoint the defective component(s) of nitrogenase in some of their nif⁻ strains by mixing mutant extracts with purified Fe protein or Mo-Fe protein as described above. These findings prompted us to carry out a similar experiment with mutants nif-95 and nif-88, mapping in separate locations near his. Klebsiella nitrogenase components

Evidence for a Major nif Operon Near his   649

were prepared by DEAE-cellulose chromatography. These components with little nitrogenase activity of their own were mixed with crude extracts of various mutants. The addition of the Mo-Fe protein component, but not the Fe protein of nitrogenase, to extracts of two different nif⁻ strains under study resulted in restoration of a maximum of 20% of the wild-type level of nitrogenase. We concluded from this experiment that both nif-95 and nif-88 synthesized functional Mo-Fe protein but were blocked in production of active Fe protein (56).

Thus, biochemical analysis of nif mutants indicate that the biosynthesis of an active nitrogenase system requires several different gene products. At least three genes located in the his-linked region are involved. Two of these genes, represented by nif-95 and nif-88, are concerned with the Fe protein. Since the Fe protein component is believed to be composed of identical protein subunits, at least one of these genes must be other than a structural gene for the subunit protein. There is currently no evidence to indicate which, if either, is the structural gene. A third gene, mapping between these two, has been identified as a gene for the Mo-Fe component reported to be composed of two dissimilar subunits. Many nif lesions mapping both near and distant from his behave as pleiotropic mutations blocking the activity of both nitrogenase components. Such pleiotropic mutations, though not uncommon in other systems, are often difficult to interpret as in the case of nitrogenase.

13.8.2 Transcription and Translation of the nif Operon(s) by Escherichia coli

As discussed in III.13.7, transfer of the nif cluster of genes near his to E. coli yields N_2-fixing hybrids of E. coli. This experiment provides additional strong evidence that the structural genes of nitrogenase are located in a segment of DNA near his. The possibility that E. coli already harbors the nitrogenase genes in some "repressed" form that requires an activator substance coded for by the Klebsiella DNA seems only a remote possibility. The simplest interpretation of this experiment is that a nif operon(s) coding for nitrogenase subunits is transcribed by E. coli RNA polymerase yielding nif messenger RNA which in turn is translated by the ribosomal protein synthesis machinery of E. coli producing nitrogenase. This experiment helps to pinpoint the location of the structural and regulatory genes of nitrogenase.

13.9 REGULATION OF NITROGEN FIXATION IN K. PNEUMONIAE

The synthesis of nitrogenase in Klebsiella, as well as in other free living nitrogen fixing organisms is severely repressed by the addition of ammonia to the growth medium (41). The synthesis of a number of other enzymes in Klebsiella whose function is to provide the cell with ammonia and/or glutamate by the degradation or reduction of sources of nitrogen (e.g. histidine, proline, nitrate) is also repressed by the presence of ammonia (2, 60, 61). Recently, a model has been proposed that accounts for the repression of enzyme synthesis by ammonia in Klebsiella (see ref. 62 for a more thorough discussion of this model). Control of enzyme synthesis is exerted by activating the transcription of operons whose gene products are involved in the utilization of a nitrogen source (e.g. the hut or nif operons) by the enzyme glutamine synthetase in its biosynthetically active (nonadenylylated) form.

When the growth of Klebsiella is limited by the source of nitrogen, the levels of histidase (the first enzyme in the degradation of histidine) and glutamine synthetase increase approximately ten fold (61, 63). The increased level of glutamine synthetase activity is due both to an increase in the number of enzyme molecules and to a much lower degree of adenylylation than that found in cells grown in the presence of ammonia. There is a specific enzyme system responsive to the nitrogen balance in the cell that can adenylylate or deadenylylate glutamine synthetase. Strains with mutations in glnA (the structural gene for glutamine synthetase) or glnB (a gene for one of the proteins of the adenylylation system) that do not allow the synthesis of active glutamine synthetase are not capable of derepressing the hut operons during nitrogen limited growth (63). glnA mutants of K. pneumoniae (64) also cannot derepress hut expression and additionally cannot synthesize nitrogenase during conditions of nitrogen limitation. In K. aerogenes and K. pneumoniae restoration of glutamine synthetase activity either by transduction to Gln⁺ or by complementation with episomes that contain the glnA⁺ gene of E. coli allows both the hut and nif operons to be normally expressed.

Mutations in glnA can also cause the constitutive synthesis of active glutamine synthetase at high levels (63). Such strains have the GlnC⁻ phenotype and always have derepressed levels of histidase. When the mutations that confer the GlnC⁻ phenotype are present in the same cytoplasm as nif (either by transducing the phenotype from K. aerogenes into K. pneumoniae (64) or by conjugational transfer of F'nif into K. aerogenes) (65) nitrogenase is synthesized even in the presence of 15 mM ammonia. In these GlnC⁻ strains nitrogenase in synthesized anaerobically under all growth conditions. The levels of nitrogenase present in the initial GlnC⁻ strains examined in the presence of ammonia were as high as 30% of the level found in ammonia free cultures. Recently, Shanmugam and Valentine (manuscript in preparation) have isolated GlnC⁻ strains of K. pneumoniae that are fully constitutive for the synthesis of nitrogenase; as much nitrogenase is found in cells grown in the presence or absence of ammonia and is equal to that found in the wild-type strain. These experiments strongly implicate glutamine synthetase as the positive control element for nitrogenase synthesis in K. pneumoniae.

The direct involvement of glutamine synthetase in activating gene expression was demonstrated for the hut operons by Tyler et al. (66). Using a purified in vitro system consisting of λphut DNA, RNA polymerase and nucleoside triphosphates, they observed that the nonadenylylated glutamine synthetase from K. aerogenes stimulated the synthesis of hut messenger RNA. The highly adenylylated form of glutamine synthetase did not stimulate the synthesis of hut m-RNA. Since the genetic and physiological experiments discussed above suggest that nif is controlled by glutamine synthetase in the same manner as hut, it is expected that similar in vitro experiments using λ nif DNA would also show activation of transcription by nonadenylylated glutamine synthetase. Further experiments will be needed to determine the site(s) of action of glutamine synthetase in the nif operons.

13.10 FUTURE OUTLOOK

Future objectives are to obtain genetic and biochemical variants in all key structural and regulatory genes governing N_2 fixation. Nitrogenase is a major target; first the structural and regulatory genes must be conclusively identified. Genetically defined mutations resulting in alterations of primary structure of nitrogenase are of great interest. Conditional lethal mutants (temperature sensitive and nonsense) of nitrogenase represent possible tools for these studies. Ultimately it may be possible to associate different regions of each structural gene with the various catalytic properties of nitrogenase; for example, mutants might be identified with altered substrate affinities (ATP, Mg^{++}, N_2), reductant specificity, enzyme stability, etc. Such mutants may become important biochemical tools for studying the mechanism of action of nitrogenase. In addition genetic studies may yield other valuable information on the ancillary reactions supporting N_2 fixation such as: the metabolism of molybdenum for incorporation into nitrogenase, enzyme assembly, low potential electron transport chains linked to nitrogenase, and repression of nitrogenase. Our current concepts of many of these processes is likely superficial. For example, almost nothing is known about the metabolism of molybdate and its incorporation into nitrogenase. The interesting work of Nason and co-workers (62) suggests that a "molybdenum cofactor" of small molecular weight may be present in nitrogenase. If so, how is this molybdenum compound synthesized? The low-potential electron transport chain linking cellular reducing power to nitrogenase is also of central interest. Induced mutations in low-potential electron carriers such as ferredoxins and flavodoxins have not yet been reported. Fine structure mapping of *nif* genes may eventually pinpoint the regulatory element(s) of DNA essential for expression of nitrogenase, defining the "nitrogenase operon(s)." Finally this information has been and will likely continue to be applied to other N_2-fixing organisms. There remains the virtually unknown area of the genetic basis of symbiotic N_2 fixation in our important leguminous crops.

13.11 ACKNOWLEDGMENT

Original research supported by the National Science Foundation. R. C. V. is the recipient of a Career Development Award from the National Institutes of Health (AI 16595).

13.12 REFERENCES

1. Streicher, S., E. Gurney, and R. C. Valentine, Proc. Nat. Acad. Sci. U. S., 68, 1174 (1971).
2. Streicher, S., E. Gurney, and R. C. Valentine, Nature, 239, 495 (1972).
3. Dixon, R. A., and J. R. Postgate, Nature, 234, 47 (1971).
4. Dixon, R. A., and J. R. Postgate, Nature, 237, 102 (1972).
5. Postgate, J., Symp. Soc. Gen. Microbiol., 21, 287 (1971).
6. Fred, E. B., I. L. Baldwin, and E. McCoy, Root Nodule Bacteria and Leguminous Plants, University of Wisconsin Press, Madison, 1932.
7. Kleczkowska, J., J. Gen. Microbiol., 40, 377 (1965).
8. Schwinghamer, E. A., Antonie van Leeuwenhoek, 33, 121 (1967).
9. Scherrer, A., and J. Dénarié, Plant Soil, Special Vol., 39 (1971).
10. Nutman, P. S., Proc. Roy. Soc. Ser. B, 172, 417 (1969).
11. Dilworth, M. J., Biochim. Biophys. Acta, 184, 432 (1969).
12. Bergersen, F. J., Ann. Rev. Plant Physiol., 22, 121 (1971).
13. Dilworth, M. J., and C. A. Parker, J. Theor. Biol., 25, 208 (1969).
14. Balassa, G., Bacteriol. Rev., 27, 228 (1963).
15. Klein, D. T., and R. M. Klein, J. Bacteriol., 66, 220 (1953).
16. Kern, H., Arch. Mikrobiol., 66, 63 (1969).
17. Heumann, W., Mol. Gen. Genet., 102, 132 (1968).
18. Holsten, R. D., R. C. Burns, R. W. F. Hardy, and R. R. Hebert, Nature, 232, 176 (1971).
19. Karlson, J. L., and H. A. Barker, J. Bacteriol., 56, 671 (1948).
20. Mumford, F. E., J. E. Carnahan, and J. E. Castle, J. Bacteriol., 77, 86 (1959).

21. Benemann, J. R., C. Shew, and R. C. Valentine, Arch. Mikrobiol., 79, 49 (1971).
22. Wyss, D., and M. B. Wyss, J. Bacteriol., 59, 287 (1950).
23. Fisher, R. J., and W. J. Brill, Biochim. Biophys. Acta, 184, 99 (1969).
24. Sorger, G. J., and D. Trofimenkoff, Proc. Nat. Acad. Sci. U. S., 65, 74 (1970).
25. Simon, M. A., and W. J. Brill, J. Bacteriol., 105, 65 (1971).
26. Shestakov, S. V., and N. T. Khyen, Mol. Gen. Genet., 107, 372 (1970).
27. Safferman, R. S., and M. E. Morris, Science, 140, 679 (1963).
28. Asato, Y., and C. E. Folsome, Genetics, 65, 407 (1970).
29. Mahl, M. C., P. W. Wilson, M. A. Fife, and W. H. Ewing, J. Bacteriol., 87, 1482 (1965).
30. Matsumoto, H., and T. Tazaki, Jap. J. Microbiol., 14, 129 (1970).
31. Taylor, A. H., and C. D. Trotter, Bacteriol. Rev., 31, 332 (1967).
32. Sanderson, K. E., Bacteriol. Rev., 31, 354 (1967).
33. Ouellette, C. A., R. H. Burris, and P. W. Wilson, Antonie van Leeuwenhoek, 35, 275 (1969).
34. Natagani, H., M. Shimuzu, and R. C. Valentine, Arch. Mikrobiol., 79, 164 (1971).
35. Baron, L. S., P. Gemski, E. M. Johnson, and J. A. Wohlheiter, Bacteriol. Rev., 82, 362 (1968).
36. Stouthamer, A. H., and Pietersma, Mol. Gen. Genet., 106, 174 (1970).
37. Silver, W. S., Proc. Roy. Soc. London, Ser. B., 172, 389 (1969).
38. Evans, H. J., N. E. R. Campbell, and S. Hill, Can. J. Microbiol., 18, 13 (1972).
39. Bergersen, F. J., and E. H. Hipsley, J. Gen. Microbiol., 60, 61 (1970).
40. Streicher, S., Doctoral dissertation, University of California, Berkeley (1972).
41. Parejko, R. A., and P. W. Wilson, Can. J. Microbiol., 16, 681 (1970).
42. Yoch, D. C., and R. M. Pengra, J. Bacteriol., 92, 618 (1966).
43. MacPhee, D. G., I. W. Sutherland, and J. F. Wilkinson, Nature, 221, 475 (1969).
44. Okada, M., and T. Watanabe, Nature, 217, 854 (1968).
45. Dove, W. F., Ann. Rev. Genet., 2, 305 (1968).

Subject Index

Abscisic acid, 382, 416
Acetylene reduction, assay, 6, 7, 46, 128, 132, 566
   Gunnera, 225
   lichens, 128, 132, 133
   non-legumes, 218, 225
ADH, 89
Alga, associations, 216, 144-148
   nitrogenase activity, effect of C/N ratio on, 99
      effect of combined nitrogen on, 86-89
      effect of dark processes on, 92-93
      location in vegetative cells, 105-106
      respiration and, 104
Amides, 107, 479
Amino acid, auxotrophs, 588
   inhibition, 315
   resistance to, 581, 585, 588
γ-amnio butyric acid, 477
Amino transferase, 476
Ammonia, 4-6, 35, 107, 532-534, 543, 547
   and algal nitrogenase, 86-89
   assimilation by blue-green algae, 86-89
   effect on root nodule infection, 394
   incorporation in blue-green algae, 89
   pool, 110
Anabaena symbiosis, 126, 144
Antagonism, 316-318, 582-584
Antibiotics, 312
   resistance to, 583-586
   streptomycin, 583, 590, 593, 595-601, 603
   viomycin, 583, 584, 593, 605

Antigens, 328-333, 337, 347
   bacteroid, 332
   flagellar, 330
   internal, 334, 347
   somatic, 331
   Vi-like, 334
Ardisia, bacterial endophyte of, 169-172
Argon, 46
Asparagine, 532
Aspartic acid, 532, 547
Aspartic dehydrogenase (ASDH), 89
Assimilate, supply to nodules 481
ATP, and algal nitrogenase, 72, 85-86
ATP, requirement for fixation, 33
Autoplaque, 288
Autotrophy, and algal N_2 fixation, 72-73
Azide, 38, 44
Azolla, symbiosis with *Anabaena*, 126, 144
Azobacter, associations with lichens, 133
Azotophore, 43

Bacteria, oligonitrophilic, 155, 158, 159, 180
Bacteriocins, 318, 325
Bacteriocinogeny, 609
Bacteriophage, rhizobial, 319-325
   host interaction, 321
   morphology, 324
   resistance, 322
   serology, 324
   taxonomy, 346

Bacteroids, 6, 12, 284, 563
  ammonia assimilation in,, 544
  ATP production in, 532, 527, 533-537, 543
  cytochromes in, 538, 540
  development of, 424, 425, 428-431, 436-438, 441
  electron transport in, 536, 540
  ferredoxins in, 542
  flavoproteins in, 538, 541
  membrane envelopes, 520, 530
  metabolism of, 538, 547
  nitrogenase, 542-543
  oxidases in, 537
  suspensions of, 521, 533, 536
Blasia, symbiosis with Nostoc, 142
Blue-green algae, 4, 5, 64-107, 126
  and phosphorus, 84-86
  culture media for, 67-68
  dark fixation by, 71
  effect of desiccation on, 77
  effect of iron on, 81-83
  effect of light on, 69
  effect of oxygen on, 77-81
  effect of pH on, 76-77
  effect of temperature on, 74-76
  growth rates of, 68
  heterotrophy, photoheterotrophy, and N_2-fixation by, 71
  light period and N_2-fixation, 70
  light reactions and N_2-fixation by, 69
  molybdenum requirement of, 83
  N_2-fixation and Emerson enhancement by, 70
  N_2-fixing representatives, 64
  nitrogenase and ATP, 85-86
  production of extracellular N by, 107-110
  pure cultures of, 68

role of oxidative pentose posphate pathway, 72
role of tricarboxylic acid cycle, 72

Calcium, 50, 292, 524
Calvin cycle, 92
Capsule, 282
Carbamyl phosphate, 38, 475
Carbon compounds used by Rhizobium, 289-290
Carbon dioxide, 46
  effect on N_2-fixation by algae, 91
Carbon economy of legume, 485
Carbon monoxide, 45, 234
^{14}C transfer to heterocysts, 100
C/N ratio, 562
  effect on nitrogenase activity of blue-green algae, 99
Carbon substrates for diazotrophic bacteria, 31-33
Cavicularia, symbiosis with Nostoc, 143
Citrulline, 474
Clostridium butylicum, mixed culture, 30
Clostridium pasteurianum, chemostat cultures, 29
Clover, bacteroid development, 425
Cobalt, 49, 83-84, 232, 293, 524
Coccomyxa, in lichens, 128, 139
Colchicine, 382
Collema, N_2-fixation by phycobiont, 130, 136, 140
Combined nitrogen, and algal nitrogenase, 86-89
Competition, in Rhizobium-legume association, 570, 590-592
Complementation, analysis, 643
  in vitro, 628, 648

Subject Index 659

Conjugation, 602-603, 624, 644, 646
Continuous culture, of diazotrophs, 28, 34, 35
Conversion factor, $C_2H_2:N_2$, 221, 226
Copper, 526
Cotransduction, 638-640
Cowpea, bacteroid development in, 428
Coprosma, and foliar associations, 171
Cotledonary reserves, and nodulation, 491
CPS (carbamyl phosphate synthetase), 89
Cross-inoculation groups, 372, 429
 bacteroid development and, 425
 Trema endophyte and, 187
Culture of legumes, 451-454
Cultivar-*Rhizobium* strain interaction, 558, 561, 566, 570-574
Cyariophyceae, N_2-fixing representatives, 64
Cyclic photophosphorylation and heterocysts, 103-104
Cytochromes, 81-82, 301, 541-542
Cytokinins, 156, 168-172, 237, 382, 414

Deletions, 642
Denitrification, 9
Desiccation, 307
 effect of, on blue-green algae, 77
 protectants, 307
2,6-diaminopimelic acid, as cell wall constituent of actinomycete endophyte, 254
3,4-dichlorophenyl-1-1-dimethylurea (DCMU), 259
Diimide, 38

Dinitrogen, Km of, in nitrogenase, 39
Dinitrogen fixation, in *Azolla*, 144
 in bacteria, 21-30
 in blue-green algae, 67-86
 effect of combined nitrogen on, 503-505
 efficiency of, 12, 13
 energy for, 10
 history of, 4, 10, 20
 in lichens, 129-141
 in liverworts, 141-143
 in mosses, 144
 in non-legumes, 187-259
  acetylene reduction, 218-223
  field estimates, 218, 225, 262
  growth experiments, 210-215
  15_N tests, 215-218
 thermodynamics of, 5, 12
Diurnal rhythm, 487
DNA, base ratios, 342, 343
 homology, 342-344, 606, 607, 613
 sharing between bacteria and plant, 611-612
 transformation, 594-600

Effectiveness, loss of, 309, 312, 315, 323
Endocyanoses, 147
Endophyte, blue-green algae as, 64
 leaf-nodule, 171-177
 Gunnera, 258
 root-nodule of non-legume, 247-259
Endospores, 285
Energetic efficiency, 34, 527, 537
Enzymes, aminotransferases, 299
 esterases, 300
 glutamine synthetase, 650-651

Subject Index

glycolytic, 297
β-hydroxybutrate dehydrogenase, 300
ribonucleotide reductase, 300
see also Nitrogenase
Epiphytes, 154-159, 180
Ethane, 46
Ethylene, 156, 381, 441, 445
Exopolysaccharide, 282, 303, 304, 321, 335
Extracellular capsule, 44
Extracellular N production by algae, 107-110

Fahraeus slide technique, 449
Famintzin, 64
Ferredoxin, 27, 82-83
Ferredoxin-NADP oxidoreductase, 93
Fire, 9
Flagella, 282
Fluorescein labelled antibody, 330
Foliar feeding, 483
Fumurate, 538
Fungicides, 310
Fossil distribution, of non-legumes, 208

Genetics of host plant, 558, 561, 562, 563, 568
Genetic variability, 558
Gibberellins, 243, 381
Gibberellin-like substances, 302-311
gln mutants, 650-651
Glutamic acid, 4, 531-533
  dehydrogenase (GDH), 87-89, 532
Glutamine, 532
  synthetase, 88, 650-651
Glycogen, 539
Glycollate, 91
GOGAT (glutamine amide-2-ketoglutarate amino transferase), 88-89

GPT (glutamate:pyruvate amino transferase), 89
Grafts, intercultivar, 566, 568
GTA (glutamate:oxoloacetate amino transferase), 89
Gum, 282, 303, 304, 321, 335
Gunnera, acetylene reduction by, 225
  endophyte of, 258
  growth experiments, 224
  15_{N_2} tests, 225
  N_2-fixation under field conditions, 262
  symbiosis, 260-261

Hatch and Slack pathway, 92
Helium, 46
Hemoglobin, function of, 526-530
  presence of, in root-nodules, 235, 236, 489
Hemoproteins, 301
Herbicides, 311, 381
Heterocysts, in blue-green algae, 65-66
  in lichen phycobionts, 135, 136
  morphology of, 94-98
  nitrogenase in, 102-103
  physiology and function of, 93-106
Heterotrophic growth of blue-green algae, 71
his operon, 640-643
Hormones in infection of legumes, 380-382
Host-controlled modification of phage, 610, 611
hut, 650
Hydrazine, 38
Hydrogen inhibition of N_2-fixation, 26, 45, 225, 233
Hydrogen ions, 295, 301
Hydrogenase, 26, 530-531
Hydroxylamine, 37
Hydroxyl ions, 301

Hyponitrous acid, 37

Immune gel diffusion, 328
Incompatibility in root nodule
 formation in non-legumes,
 252
Indole acetic acid (IAA), 156,
 169, 236, 302, 311, 380
Indole carboxylic acid (ICA),
 237
Ineffective nodules, 431, 444
Infection, in legumes, 370-378,
 382-384, 391-398
 in *Trema*, 252
Infectivity loss in *Rhizobium*,
 315
Interbiosis, 316
Iron, 81-83, 523, 542
Isoenzymes, 348

Leghemoglobin, 520, 526-530,
 535-537
Legumes, culture of, 451-454
 names of, 456-457
 nutritive requirements of,
 522-524
 perennial, 493
 symbiotic cycle in, 489-493
 types of nodulation
  responses, 559-568
Lichens, algal content, 128,
 129
 anatomy and morphology, 129
 cephalodia, 130-134
 culture, 127-129
 environmental effects, 139
 light effects, 139, 141
 moisture content, 140, 141
 rate of N_2-fixation, 134-136
 temperature effects, 139
 thallus form, 129, 130
 translocation of fixed
  nitrogen, 137
Liverworts, 141
"Lucerne fatigue," 323
Lysogeny, 324, 593, 607
Lysogenic conversion, 325, 338

Magnesium, 50, 51, 292
Manganese, 294
Medic, 425
Mereschkowsky, 64
Methane oxidizing bacteria, 25
Methionine sulphoxime, 98-99
Mineral requirements, 228-232,
 292, 495-497, 522
Mixed cultures, 29
Molybdenum, 47, 83, 229-231,
 522, 542, 628, 629, 652
Mosses, 144
Mutants, 339, 627-631
Mutation, 579
 auxotrophic or prototrophic,
  586-588, 590
 radiation induced, 590
 resistance to chemical
  agents, 581-586, 592
Mycorrhiza, 414
Myrsinaceae, leaf-nodule
 symbiosis, 162, 164, 178
Myrica/Casuarina type root-
 nodule, 239

nif, 624-625, 638-643, 647,
 649
 mutants, 627-629, 631-634
Nitramid, 38
Nitrate, 7, 36, 504, 505, 526
 effect on algal nitrogenase,
  86
 as electron acceptor in
  rhizobia, 537
Nitrate reductase, 523
Nitric oxide, 36
Nitrite, 291, 538
Nitrogen, balance, 9, 10
 excretion, 480
 fixation, see Dinitrogen
  fixation
 pool in bacteroids, 532
 in nodules, 532
 in rainfall, 6, 7
 in soil, 10, 260-263
 sources for rhizobia, 291

storage, in blue-green
   algae, 89-90
transfer to host, 228, 478
$^{15}N_2$ methods, 4, 7, 130-144,
   490, 522, 533
Nitrogenase, 5, 6, 299, 625-
   626, 648-649
  bacteroid, 530-538, 542-545
  of blue-green algae, 136
  estimation of, 21, 46, 48
  evolution of, 12, 14
  induction of, 545-546
  interrelationships with photo-
     synthetic and non-photo-
     synthetic processes, 90-93
Nitrous oxide, 36
Nodule, activity parameters,
   489
  Alnus type, 238
  ammonia assimilation in, 520,
     531, 532
  breis, N_2-fixation by, 223,
     533
  callus culture, 255-257
  clover, 526
  development, 399, 410-423,
     441-448
  distribution, 569, 571
  formation in non-legumes,
     239
  ineffective, 431
  intact, 521, 522, 527
  lucerne, 522
  lupin, 526, 534, 538
  mass, 572
  overwintering, 492
  peas, 531, 534
  perennial, 494
  red clover, 527
  respiratory activity of, 485
  serradella, 532, 533, 538
  soybean, 522-538
  structure, 237, 250-251, 520,
     528, 529
  subterranean clover, 522,
     524
  vetch, 531

  see also specific plant
     species
Nodulation, availability of
   oxygen, 500
  effective, 559, 564, 565,
     571
  effect of combined nitrogen,
     503-505
  effects of light, 498
  efficient, 560, 565, 566
  inefficient, 560, 564, 565
  seasonal decline, 494
  specificity, 353
Non-heterocystous blue-green
   algae, 67
Non-legumes, 187-264
Non-nodulated, 490, 560-563
Nostoc, in lichens, 126, 132-
   139
Nutrient solutions, 451, 454

Operon, 650, 652
Oxidases, 541
Oxidative phosphorylation, 35
Oxygen, concentration, 294,
   527, 535, 537
  conformational protection
     against, 43, 44
  effect on blue-green algae,
     77-81
  effect on carbon utiliza-
     tion, 33
  effect on N_2 fixation, 91,
     233, 234
  inactivation by, 533, 534,
     543
  inhibition by, 527
  protection of nitrogenase
     against, 11, 42, 94
  supply in nodules, 521, 526,
     528, 536
Pesticides, 310
pH, 47, 76, 77, 295, 301, 302,
   497
Phages, 320-327, 624, 630,
   634-638
Phaseolus, 216, 218, 428

Subject Index 663

Phloem, 485, 533
Phosphate, 50, 84, 444, 532
Phosphite, 50
Phosphoenol pyruvate carboxylase, 476
Phosphorylation, in blue-green algae, 85-86
Photoheterotrophy, 71
Photorespiration, 91
Photosystem I, role in N_2-fixation, 69
Photosystem II, loss in *Gunnera* endophyte, 259
Photosynthesis, 90-91, 227, 524-526
Phyllosphere, 8, 158-161, 180
Phytoflavin, 82
Plant passage, effect of, 588-589
Plasmalemmosomes, in actinomycete endophyte, 250
Plasmids, 603-605, 613
pO_2, 7, 41-42, 234, 500, 527
Pod removal, 493
Polygalacturonase, 378
Poly-β-hydroxybutyrate, 32, 235 282, 304, 305, 488, 539-540
  dehydrogenase, 540
  depolymerase, 540
Potassium, 294
Proheterocysts, 98
Pseudomonas azotogensis, mixed cultures with, 31
Pseudonodules, 410
Psychotria, N_2-fixation by endophyte, 179
Pyruvate dehydrogenase, 93
Pyruvate:ferredoxin oxidoreductase, 92-93

Replication mapping, 607
Resistance transfer factors, 603, 613
Rhizobitoxine, 567
Rhizobium, cell envelope, 284
  cell surface, 286

cell wall, 286
characteristics, 280, 286, 352
cytochromes, 542, 545
cytology, 281
definition, of genus, 280
  of strain, 280
  of types, 281
fast growers, 280
growth factors, 292, 483
induced chlorosis, 567
lipopolysaccharide, 284, 332
lyophilization, 308
numerical taxonomy, 341
recognition, 351
sensitivity, 289
serogroup, 330, 336, 338
slow growers, 280
survival, 305
symbiosis, in non-legumes, 187, 188, 244-247, 258
type species, 340
variants, 287, 339
Rhizosphere, 7
Rhodopseudomonas capsulata, mixed culture, 30
Rhodospirillum rubrum, continuous culture, 29
Riboflavin, 541, 588, 591
Ribulose-1,5-diphosphate (RuDP), 91, 103
Root, excretions, 480, 501, 561
  hair deformation, 341, 373
  temperature, 498
Rubiaceae, microsymbionts of, 171-179

Seed coat factors, 314
Self-lysis, 289
Serogroups, 563, 564, 570
Serradella nodules, 533, 534, 538, 544
Soil water content, 500
Soybeans, 428, 524-530, 561-570

Stages of growth and nodulation, 571
Stem symbiosis, 198. *See also Gunnera*
Strain acceptance, 569
Strontium, 292
Succinate, 538
Sulfur, 523, 542
Symbiosis, blue-green algae, 126, 141-147
  carbohydrate requirements, 483
  defoliation, 497
  grazing, 498
  growth curves, 491
  light, 497
  pH, 497
Synchronous cultures, 29

Temperature, 47, 74-76, 295, 306, 501
Toxic substances, 308, 310, 314, 567
Transduction, 601-602, 612, 624, 634-638
Transfection, 609

Transfer cells, 478
Transformation, 596-600, 612, 626
Translocation, 136-139
Trema, endophyte, 258
  root nodules, 244-247
Tricarboxylic acid cycle, 72
Tryptophan, 156
Tungsten, 48

Urea, 37
Urease, 37

Vanadium, 48
Vesicles, 247-251
Vitamins, 291
Vitamin B_{12}, 232

Water, 307
  sheath, 160
Waterlogging, 500

Xylem, 525, 532

Zinc, 294

Taxonomic Index

Acacia spp., 409, 420
  Acacia bynoeana, 397
  Acacia cornigera, 397
  Acacia horrida, 409
  Acacia longifolia, 409, 420, 425, 429
  Acacia melanoxylon, 397
Acetobacter spp., 346
Achromobacter spp., 23, 318, 319, 343
Achromobacteriaceae, 161
Actinomyces bovis, 254
Actinomycetales, 187
Aerobacter spp., 157, 319
  Aerobacter aerogenes, 23
Aeschynomene americana, 398
Aeschynomene indica, 398, 399, 411, 421
Agathis spp., 206
Agrobacterium spp., 287, 288, 302, 303, 324, 330, 333, 339-354, 597, 605-607, 612
  Agrobacterium pseudotsugae, 346
  Agrobacterium radiobacter, 291, 346, 377
  Agrobacterium rhizogenes, 345, 346
  Agrobacterium rubi, 346
  Agrobacterium tumefaciens, 303, 304, 330, 345-351, 399, 413, 605, 606, 611-613, 626
Agrostis stolonifera, 111
Albizzia spp., 397
  Albizzia lophantha, 479
Alcaligenes spp., 319
Alnus spp., 189, 191, 195, 200, 201, 204, 210, 211, 218, 220, 227, 231-264, 474, 487
  Alnus glutinosa, 191-193, 200, 210-245, 253, 256, 262

Alnus rubra, 191, 193, 201, 210, 232, 262, 263
Aloe spp., 155
Amblyanthopsis spp., 164
Alysicarpus longifolia, 410
Alysicarpus vaginalis, 410
Amorpha spp., 422
Anabaena spp., 45, 65-70, 81-99, 104-107, 108, 136, 145, 146, 199
  Anabaena ambigua, 65
  Anabaena azollae, 65, 108, 126, 144
  Anabaena cycadeae, 65
  Anabaena cylindrica, 65-70, 74-108
  Anabaena fertilissima, 65
  Anabaena flos-aquae, 65, 84, 86, 99, 100
  Anabaena gelatinosa, 65
  Anabaena humicola, 65
  Anabaena levanderi, 65
  Anabaena naviculoides, 65
  Anabaena variabilis, 65, 92, 108
Anabaenopsis spp., 65, 71
  Anabaenopsis circularis, 65, 71, 84
Anacystis spp., 81
  Anacystis nidulans, 68, 70, 82, 629
Anthericum divaricatum, 13
Anthoceros spp., 111, 126, 142
Anthyllis spp., 290, 341, 347
  Anthyllis vulneraria, 350, 449
Aphanizomenon flos-aquae, 79
Aphanocapsa spp., 93
Arachis, spp., 399
  Arachis erecta, 437

Arachis hypogaea, 377, 398-400, 410-412, 417, 420-425, 436, 437, 444, 479, 501
Arachis nambyguarea, 437
Arachis villosulicarpa, 437
Arctostaphylos spp., 189, 190
Arctostaphylos uva-ursi, 189
Arctotheca nivea, 13
Ardisia spp., 164, 169-173, 178
Ardisia crenata, 164, 174
Ardisia crispa, 162, 164, 169, 174, 177
Ardisia ellipita, 177
Ardisia hortorum, 164, 174
Ardisia humilis, 174
Ardisia punctata, 174
Artemisia spp., 111
Artemisia ludoviciana, 188
Artemisia michauxiana, 188
Arthrobacter spp., 597, 606, 607
Aspergillus spp., 317
Aspergillus niger, 299
Aspergillus terreus, 159
Asterales, 190
Astragalus spp., 302, 347, 437
Astragalus glycyphyllus, 397, 425, 437
Astragalus sinicus, 412
Aulorisa fertilissima, 65
Azomonas agilis, 23
Azomonas insignis, 23
Azomonas macrocytogenes, 23
Azolla spp., 8, 126, 144-148, 199
Azotobacter spp., 4, 5, 7, 11, 12, 20-52, 79, 86, 87, 94, 102, 133, 134, 159, 161, 236, 535, 627
Azotobacter aerogenes, 44
Azotobacter agilis, 20, 23, 32, 36, 50
Azotobacter beijerinckii, 23, 32, 44, 539

Azotobacter chroococcum, 20, 23, 29-51, 160
Azotobacter, galophilium (A. halophilum?), 51
Azotobacter indicum, 23
Azotobacter insignis, 23
Azotobacter macrocytogenes, 23
Azotobacter miscellum, 23
Azotobacter paspali, 23
Azotobacter vinelandii, 20-51, 627-629, 648
Azotococcus agilis, 23
Azotococcus insigne, 23

Bacillus spp., 7, 31, 286, 319
Bacillus circulans, 25
Bacillus foliicola (same as *Bacterium foliicola*), 172, 174, 178
Bacillus indicum, 48
Bacillus macerans, 24
Bacillus polymyxa, 24, 28, 36, 39, 44, 46, 50
Bacillus radicicola, 370, 424
Bacillus subtilis, 88
Bacterium rubiacearum, see *Klebsiella rubiaceamum*
Banksia menziesii, 13
Bdellovibrio spp., 316, 351
Beijerinckia spp., 7, 30-37, 47, 50, 159, 161, 236
Beijerinckia acida, 23
Beijerinckia congensis, 23
Beijerinckia derxii, 23
Beijerinckia fluminensis, 23, 30
Beijerinckia indica, 23, 30, 41, 44, 49
Beijerinckia lacticogenes, 23
Beijerinckia megaterium, 30
Beijerinckia mobilis, 23
Beijerinckia mobilum, 23

Taxonomic Index 667

Betula spp., 191
  Betula nana, 208
Betulaceae, 189
Blasia spp., 142, 143, 148, 199
  Blasia pusilla, 143
Boraginales, 190
Brassica napus, 447
Brassica oleracea, 229
Brevibacterium spp., 319
Bromus inermis, 447
Bryum pendulum, 111

Cactales, 190
Caesalpiniaceae, 14
Cajanus cajan, 396, 397, 409, 421, 423
Cajanus indicus, 421
Cakile maritima, 13
Calamagrostis arundinacea, 190
Calluna vulgaris, 189
Calothrix spp., 71, 74, 108, 132
  Calothrix brevissima, 64
  Calothrix elenkinii, 65
  Calothrix parietina, 65
  Calothrix scopulorum, 65, 74, 108
Canavalia gladiata, 396, 397
Caragana arborescens, 396, 397, 421, 442, 443
Caragana tragacathoides, 397, 437
Carmichaelia australis, 397, 437
Carpinus spp., 191
Carpobrotus aequilateralis, 13
Cassia mimosoides, 398, 412
Casuarina spp., 189-194, 204, 210, 213, 218, 231-239, 248, 254, 260-263
  Casuarina cunninghamiana, 202, 215, 218, 226, 231-239, 253, 254
  Casuarina equisetifolia, 193, 194, 204, 205, 210, 218, 254, 260, 262

Casuarina junghuhniana, 194, 205, 210
Casuarina rumphiana, 194, 210, 220-222
Casuarinales, 5, 189
Cavicularia spp., 161
Ceanothus spp., 189-202, 216, 248-262
  Ceanothus velutinus, 196, 202, 203, 213, 238, 253, 254
Centrosema spp., 493
  Centrosema pubescens, 396, 400, 409, 410, 493, 573
Ceratozamia spp., 126
Cercocarpus spp., 189, 193, 197, 248, 253
  Cercocarpus betuloidies, 202
Chlorobacterium spp., 22
Chlorobium limicola, 22
Chlorobium thiosulfatophilum, 22
Chlorogloea spp., 71, 72
Chlorogloea fritschii, 65, 69-72, 84, 94, 108
Chloropseudomonas, 81
Cicer spp., 348, 614
Cicer arietinum, 396, 412, 417, 421, 436, 442
Chlorospeudomonas ethylicum, 25
Chromatium strain D, 22, 43
Chromatium minutissimum, 22
Chromatium vinosum, 22
Chromobacterium spp., 157, 333, 345, 346
  Chromobacterium lividum, 176, 178
Cladonia spp., 134
  Cladonia impexa, 134
Clostridium spp., 5, 7, 11, 26, 35, 46, 83, 86, 92, 93
  Clostridium aceticum, 24
  Clostridium acetobutyricum, 24
  Clostridium beijerinckii, 24

Taxonomic Index

Clostridium butylicum, 24, 30
Clostridium butyricum, 24, 36
Clostridium felsinium, 24
Clostridium kluyverii, 24
Clostridium lactoacetophilum, 24
Clostridium madisonii, 24
Clostridium pasteurianum, 4, 20, 24, 29-49, 82, 532, 542, 628
Clostridium pectinovorum, 24
Clostridium tetanomorphum, 24
Coccomyxa spp., 128, 139
Collema spp., 74, 130, 137, 139, 140
Collema auriculatum, 132
Collema coccophorus, 132
Collema crispum, 132
Collema fluviatile, 132
Collema furfuraceum, 132
Collema granosum, 130, 132
Collema pulposum, 132
Collema subfuscum, 132
Collema tenax, 130, 136
Collema tuneforme, 132, 136, 140
Colletia spp., 189, 196
Colutea arborescens, 397
Colutea persica, 397
Comptonia spp., 149
Comptonia asplenifolia, 191, 202, 216, 264
Comptonia peregrina, 218
Coprosma spp., 171
Coriaria spp., 189-206, 216, 248, 253, 255, 261
Coriaria japonica, 201, 253
Coriaria myrtifolia, 200, 216, 227, 253
Coriaria ruscifolia, 204
Coriaria thymifolia, 204
Coriariales, 5, 189
Coriariceae, 189
Coronilla glauca, 397

Coronilla iberica, 397
Coronilla scorpioides, 397
Corylus spp., 191
Corynebacterium spp., 157, 319
Corynebacterium autotrophicum, 24
Crotolaria juncea, 396, 421
Crotolaria verrucosa, 396
Crotolaria vespertilio, 398
Cryptococcus laurentii, 155
Cryptostemma calendulaceum, 13
Cyamopsis psoralioides, 396
Cyamposis tetragonolobus, 396, 412, 421, 436
Cyanidium caldarium, 77
Cyanophora paradoxa, 5
Cyanophyceae, 64, 74, 78, 84, 93, 126, 129
Cycadaceae, 8
Cycas spp., 148, 199
Cylindrospermum spp., 108
Cylindrospermum gorakhporense, 66
Cylindrospermum licheniforme, 66
Cylindrospermum maius, 66
Cylindrospermum sphaerica, 66
Cytisus spp., 372, 422
Cytisus capitatus, 398
Cytisus purgans, 398
Cytisus scoparius, 397
Cytisus sessilifolius, 398

Dacrydium spp., 206
Dalbergia lanceolaria, 398
Dalbergia sissoo, 398
Dendriscocaulon umhausense, 132
Derxia spp., 7, 31, 32, 159
Derxia gummosa, 23, 41, 47
Desmodium spp., 493
Desmodium dillenfi, 398
Desmodium fallax, 412
Desmodium gangeticum, 396
Desmodium uncinatum, 398, 493

Desulfotomaculum orientis, 22
Desulfotomaculum ruminis, 22
Desulfovibrio desulfuricans, 22, 34, 35
Desulfovibrio gigas, 22
Desulfovibrio vulgaris, 22, 41
Dichothrix spp., 133
Dioon spp., 199
Dioscorea macoura, 164, 171
Dioscoreaceae, 162, 164
Discaria spp., 196, 248
  Discaria toumatou, 206, 216
Dolichos bifloris, 410
Dolichos lablab, 410
Dolichos sinensis, 372
Doryncium spp., 422
  Doryncium hirsutum, 397
Dryas spp., 189-208, 248, 250, 261
  Dryas drummondii, 197, 202, 209, 238, 262
  Dryas integrifolia, 202, 208
  Dryas octopetala, 193, 200, 204, 208

Eichornia crassippes, 160
Elaeagnaceae, 189
Elaeagnus spp., 189, 195, 201, 202, 250, 253
  Elaeagnus angustifolia, 226, 253
  Elaeagnus commutata, 234, 236
Encephalatos spp., 126
Enterobacter aerogenes, 23
Enterobacter agglomerans, 23
Enterobacter cloacae, 23
Enterobacteriaceae, 161
Enterolobium cyclocarpum, 397
Ephebe lanata, 132
Ervum spp., 372
Erwinia spp., 319
Escherichia coli, 5, 332, 346, 429, 601-605, 613, 624-630, 635-651
Escherichia intermedia, 23
Eucalyptus calophylla, 13

Faba vulgaris, 369
Fagales spp., 5, 189
Fagonia spp., 189, 198
  Fagonia arabica, 187
Fischerella major, 66
Fischerella muscicola, 66
Flavobacterium spp., 157, 176, 319
Frankia spp., 187, 248, 252-257

Galega officinalis, 397
Genista spp., 422
  Genista florida, 398
  Genista hispanica, 398
  Genista pilosa, 398
  Genista siberica, 398
  Genista tinctoria, 398
Geosiphon pyriforme, 147
Glaucocystis spp., 126
  Glaucocystis nostochinearum, 147
Glaux maritima, 111
Gleditschia, 372
Gloeocapsa spp., 45, 67, 68, 80, 81, 94, 106
Glycine spp., 340, 344, 353, 479, 491
  Glycine max, 372-383, 395-400, 409-429, 435-447, 475-484, 490-501, 558
Gossypium hirsutum, 429
Granulobacter pectinovorum, 24
Gunnera spp., 5, 8, 13, 71, 111, 126, 148, 187-204, 224-226, 259-263
  Gunnera arenaria, 226
  Gunnera macrophylla, 205, 225, 258-262
Gunneraceae, 190

Haloragaceae, 189
Hapalosiphon spp., 77, 144
  Hapalosiphon fontinalis, 66
Helianthus spp., 606
  Helianthus annus, 13, 413

Heterodera trifolii, 423
Hippophaë spp., 189, 196
  Hippophaë rhamnoides, 200,
    208, 210, 215, 220, 226,
    232, 233, 236, 254, 262
Hypochoeris glabra, 13

Indigofera spp., 302
  Indigofera atropurpurea, 397
  Indigofera endecaphylla, 396
  Indigofera macrostachya, 397
  Indigofera pseudo-tinctoria,
    412
Inga spp., 409

Klebsiella spp., 5, 7, 41, 88,
  148, 157, 159, 173, 336,
  546, 629-631, 634-636, 643-
  652
Klebsiella aerogenes, 23, 613,
  634, 650, 651
Klebsiella pneumoniae, 23, 36-
  51, 173, 542, 628-652
Klebsiella rubiacearum, 23, 173,
  176, 178

Laburnum anagyroides, 398
Lactobacillus spp., 157
Lathyrus spp., 340, 372, 412,
  422
  Lathyrus aphaca, 397
  Lathyrus montanus, 449
  Lathyrus palustris, 449
  Lathyrus pratensis, 397, 449
  Lathyrus sativus, 396, 397
  Lathyrus sylvestris, 397
Leguminales, 5
Lens spp., 340
  Lens culinaris, 396
Leptogium burgessii, 132
Leptogium lichenoides, 130, 132
Leptogium sinuatum, 132
Leptogium teretiusculum, 132
Lespedeza spp., 422
  Lespedeza stipulacea, 444
Leucaena spp., 302, 348, 349
  Leucaena leucacephala, 398, 410

Libocedrus spp., 206
Lichina spp., 140, 141
  Lichina confinis, 74, 75,
    77, 132, 140
  Lichina pygmaea, 74, 77, 78,
    132
Lipomyces spp., 159
  Lipomyces starkeyii, 159
Lobaria pulmonaria, 132
Lobaria scrobiculata, 132
Lotonomis spp., 287, 289, 291,
  344, 348, 594
Lotus spp., 290, 300, 302,
  340-350, 372, 422, 431,
  437, 451, 650
  Lotus corniculatus, 397,
    412, 415, 425, 435, 441,
    444, 446, 567
  Lotus hispidus, 425
  Lotus pedunculatus, 410, 435,
    441
  Lotus uliginosus, 425
Lupinus spp., 6, 282, 290,
  340-350, 372, 399, 409,
  417-436, 479, 488-501
  Lupinus albus, 398, 412
  Lupinus angustifolius, 398,
    400, 410, 425, 431, 570
  Lupinus arboreus, 400, 494,
    495
  Lupinus digitatus, 57
  Lupinus luteus, 237, 372,
    381, 413, 417, 425, 431,
    436, 492, 570
  Lupinus mutabilis, 369, 398,
    420, 570, 571
  Lupinus perennis, 398
  Lupinus polyphyllus, 398
Lyngbya spp., 67, 108

Macroptilium spp., 350, 353
  Macroptilium atropurpureum,
    236, 375
  Macroptilium lathyroides,
    236
Malpighiales, 189
Massalongia carnosa, 132

Mastigocladus spp., 75
  Mastigocladus laminosus, 66, 76
Medicago spp., 285, 340-354, 373, 400, 410-425, 451, 491, 497, 559
  Medicago arabica, 449
  Medicago arborea, 397
  Medicago denticulata, 391, 396, 397
  Medicago lupulina, 397, 449
  Medicago minimz, 376
  Medicago sativa, 370-381, 391-397, 409, 410, 425-435, 443-446, 496, 559, 565
  Medicago tribuloides, 498
  Medicago truncatula, 376, 395, 397, 419, 425, 443
Melilotus spp., 340, 412
  Melilotus alba, 397, 446, 447, 558
  Melilotus indica, 396, 397
  Melilotus officinalis, 397, 449
Meloidogyne Javanica, 423
Methanobacterium omelianskii, 25
Methylosinus trichosporium, 25
Microcystis flos-aquae, 82
Micromonospora spp., 254
Mimosa spp., 410
  Mimosa bimucronata, 410
  Mimosa caesalpiniaefolia, 410
  Mimosa dysocarpa, 397
  Mimosa invisa, 410
  Mimosa pudica, 410
Minosaceae, 14
Mycobacterium spp., 7, 154, 174, 254
  Mycobacterium flavum, 25, 41
  Mycobacterium lepraemurium, 427
  Mycobacterium rubiacearum, 23, 174, 176
Mycoplana rubra, 174
Myoporum insular, 13

Myrica spp., 189-194, 201-226, 235-253, 261-263
  Myrica cerifera, 215, 221, 223, 226, 233, 237, 239, 253, 262, 533
  Myrica fava, 200
  Myrica gale, 200, 201, 208, 213, 226-235, 253, 475
  Myrica javanica, 200, 205, 211, 212, 263
Myricales, 5
Myrsinaceae, 162, 164, 169, 178
Myrtales, 190

Neorosea androgensis, 164, 169
Nephroma spp., 74, 140
  Nephroma arcticum, 74, 132
  Nephroma laevigatum, 132
Neptunia oleracea, 382, 397
Neurospora crassa, 546
Nicotiana tabacum, 447
Nocardia spp., 254, 319
Nodularia spp., 96
Nostoc spp., 67-74, 82, 96, 98, 108, 111, 126, 132-147, 199, 205, 225, 226, 259, 260, 474
  Nostoc calcicola, 66
  Nostoc commune, 66, 74, 108
  Nostoc cycadae, 66
  Nostoc entophytum, 66, 108
  Nostoc gunnerae, 259
  Nostoc muscorum, 66-72, 76, 81, 86, 87, 98, 102, 108, 155
  Nostoc paludossum, 66
  Nostoc punctiforme, 66, 71, 199, 258, 259
  Nostoc sphaericum, 66, 142, 144, 147
Nostocaceae, 98
Nostochopsis spp., 96

Oenothera drummondii, 13
Onobrychis sativa, 449

Onobrychis viciifolia, 397
Ononis natrix, 397
Ononis repens, 422, 449
Oocystis spp., 126, 147
Ornithopodium tuberosum, 369
Ornithopus spp., 6, 340, 341, 350, 353
  Ornithopus sativus, 375, 396, 397, 410-412, 421, 425, 436, 475
Oryza sativa, 13
Oscillatoria spp., 67

Pannaria rubiginosa, 132
Papilionaceae, 14
Parmeliella atlantica, 132
Parmeliella plumbea, 132
Pavetta spp., 162, 165, 168, 169, 172, 173
  Pavetta gardenifolia, 177
  Pavetta grandiflora, 164, 174, 177
  Pavetta indica, 164
  Pavetta indicum, 174
  Pavetta lanceotota, 177
  Pavetta revoluta, 177
  Pavetta zimmermanniana, 162, 164, 176
Paspalum notatum, 41
Pelodictyon spp., 26
Peltigera spp., 127, 135, 140, 141
  Peltigera aphthosa, 108, 128-140
  Peltigera canina, 77, 129-140
  Peltigera evansiana, 133, 141
  Peltigera horizontalis, 130
  Peltigera polydactyla, 128, 133, 138, 140, 141
  Peltigera praetextata, 133, 134, 135, 141
  Peltigera pruinosa, 133
  Peltigera rufescens, 133, 136, 140
  Peltigera virescens, 130

Penicillium spp., 159, 317
Perlargonium drummondii, 13
Phaseolus spp., 302, 340, 344, 350, 353, 372, 422, 479, 571, 614
  Phaseolus angularis, 412
  Phaseolus atropurpureus, 375
  Phaseolus aureus, 571
  Phaseolus coccineus, 397, 423
  Phaseolus lathyroides, 397
  Phaseolus multiflorus, 423
  Phaseolus oleraceus, 396, 397
  Phaseolus trilobus, 571
  Phaseolus vulgaris, 216, 218, 372, 378, 381, 386, 396, 397, 412-425, 429, 435, 445, 446, 479, 482, 497-501
Phoma spp., 317
Phormidium spp., 67, 82
Phyllobacterium foliicola, 172, 176
Phyllobacterium myrsinaceacum, 174
Phyllocladius spp., 206
Phytomyxa spp., 339
Picea abies, 263
Pinus radiata, 13
Pinus sylvestris, 263
Piptedenia rigida, 398
Pisum spp., 340, 341, 344, 372, 400, 422, 425, 479, 485, 491, 492, 498-505
  Pisum arvense, 413, 505
  Pisum aureus, 571
  Pisum sativum, 369, 382, 386, 391, 395-399, 410-416, 419-427, 433, 478-488, 498-503, 531, 532
Placopsis gelida, 133, 138
Placynthium nigrum, 133
Placynthium pannariellum, 133
Plasmodiophora brassicae, 369
Plectonema spp., 45, 67, 94
  Plectonema boryanum, 67, 99
Plectridium aceticum, 24
  Plectridium pectinovorum, 24

Taxonomic Index   673

Plectridium tetanomorphum, 24
Podocarpus spp., 206
Polychidium muscicola, 133
Pongamia pinnata, 396
Populus spp., 263
Pseudocyphellaria thouarsii, 133
Pseudomonadales, 345
Pseudomonas spp., 157, 318, 319, 346, 351, 597, 603, 605-607, 613
  Pseudomonas aeriginosa, 603, 613
  Pseudomonas azotogensis, 25
  Pseudomonas methanitrificans, 25
Psorelea spp., 302
Psychotria spp., 148, 162, 168, 169, 172, 173, 179
  Psychotria calva, 162, 164, 176, 179
  Psychotria capensis, 176
  Psychotria emetica, 164, 176, 177
  Psychotria faucicola, 162
  Psychotria guerkeana, 162
  Psychotria hirtelli, 176
  Psychotria kirkii, 176
  Psychotria kikwitensis, 162
  Psychotria molleri, 162
  Psychotria mucronata, 164
  Psychotria nairobiensis, 164, 176, 178
  Psychotria punctata (same as *P. bacteriophilia*), 162, 164, 168, 173, 176
  Psychotria schliebenii, 162
Pultanea spp., 302
Purshia spp., 189, 193, 197, 202
  Purshia tridentata, 216

Quercus ilex, 200

Raphidiopsis indica, 67
Rhamnales, 5, 189
Rhamnaceae, 189, 206, 238

Rhizobiaceae, 173
Rhizobium spp., 8, 126, 147, 148, 187, 190, 206, 207, 231-258, 280-354, 370-399, 409-455, 481, 483, 499, 501, 544, 558-573, 578-613, 625-627
  Rhizobium japonicum, 534-545, 558-570, 590-604, 613
  Rhizobium leguminosarum, 531, 581-588, 608
  Rhizobium lupini, 590, 596, 599-603, 605
  Rhizobium meliloti, 558, 559, 565, 581, 585-591, 595-602, 605-609, 613
  Rhizobium phaseoli, 591, 596, 597, 601, 604
  Rhizobium trifolii, 483, 541, 561-563, 565, 568, 569, 572, 582-590, 595-609, 613
Rhodomicrobium vannielii, 22
Rhodospeudomonas capsulata, 22
Rhodospeudomonas gelatinosa, 22
Rhodospeudomonas palustris, 22
Rhodospeudomonas spheroides, 22
Rhodospirillum rubrum, 22
Rhodotorula glutinis, 155
Ricinus communis, 13
Rivulariaceae, 96
Robinia spp., 448, 494
  Robinia pseudoacacia, 397
  Robinia viscoa, 397
Rosaceae, 189, 238
Rosales, 5, 189
Rubiaceae, 161, 162, 164, 165, 169, 178
Rubus fruiticosus, 13

Salix herbaceae, 208
Salix polaris, 208
Salmonella spp., 644
  Salmonella typhimurium, 567, 630, 634, 639
Sanseviera spp., 155
Saprospira grandis, 82

Sarothamnus spp., 494
  Sarothamnus scoparius, 409, 448
Scaevola crassifolia, 13
Scytonema spp., 71, 132, 133
  Scytonema arcangelii, 66
  Scytonema hofmanni, 66
Sesbania spp., 302, 410, 443, 494
  Sesbania bispinosa, 396
  Sesbania grandiflora, 396, 398, 421, 423, 442, 443, 448
Shepherdia spp., 189, 196
  Shepherdia argentea, 201
  Shepherdia canadensis, 201, 216
Shigella spp., 636
Solanum spp., 606
Solorina spp., 74, 140
  Solorina crocea, 74, 133
  Solorina saccata, 133
  Solorina spongiosa, 133
Sophora spp., 448, 494
  Sophora moorcroftiana, 398
  Sophora tomentos, 398
Spartium spp., 442
  Spartium junceum, 398
Sphagnum spp., 77, 144
Stereocaulon spp., 133, 140
  Stereocaulon paschale, 133
Sticta fuliginosa, 133
Sticta limbata, 133
Stigonema spp., 132
  Stigonema dendroideum, 66
Stirlingia latifolia, 13
Stizolobium deeringhianum, 421
Streptomyces spp., 319
Stylosanthes sundaica, 398
Suaeda maritima, 111
Synechococcus spp., 68, 93

Teredora malleolus, 25
Terminosporus kluyverii, 24
Thermopsis spp., 422
Thiocapsa roesopersicina, 22
Tolypothrix tenuis, 66, 70

Trema spp., 8, 187-193, 197, 206, 236, 237, 244-247, 252, 258
  Trema aspera, 187, 206
  Trema cannabina, 187, 193, 206, 207, 222, 246, 431
Tricalysia andongensis, 169
Trichinium manglesii, 13
Trichodesmium spp., 67, 68, 106
Trifolium spp., 307, 314, 340, 341, 344, 353, 372-377, 382-395, 400, 409-412, 417-427, 451, 477, 491, 497, 505, 561, 568
  Trifolium alexandrinum, 396
  Trifolium alpinum, 397
  Trifolium ambiguum, 449, 568
  Trifolium arvense, 396, 449
  Trifolium augustifolium, 396
  Trifolium dubium, 396, 449, 492
  Trifolium fragiferum, 375, 391, 396, 415
  Trifolium glomeratum, 374-376, 379, 381-383, 391-394, 396, 415
  Trifolium hybridum, 449, 568
  Trifolium incarnatum, 376, 396, 397
  Trifolium medium, 449
  Trifolium nigrescens, 396
  Trifolium ornithopodioids, 396
  Trifolium parviflorum, 391-393, 396, 397, 425
  Trifolium pratense, 372-382, 392-397, 413, 425, 433, 435, 440-449, 500, 501, 561, 571, 613
  Trifolium procumbens, 392, 396, 449
  Trifolium repens, 376-381, 394-397, 422-427, 443-449, 479, 483, 493, 497, 499, 568, 572

Trifolium scabrum, 386, 396
Trifolium squamosum, 397
Trifolium subterranueum, 376-381,
　　391-397, 413, 420, 425,
　　426, 435, 440-443, 484,
　　500-505, 524, 564, 565
Trigonella spp., 340
　　Trigonella foenum-graecum, 396,
　　397
Tripascum laxum, 160
Triticum monococcum, 447
Tribulus spp., 189, 198
　　Tribulus alatus, 187, 198
　　Tribulus cistoides, 187, 198

Ulex spp., 424, 494
　　Ulex europaeus, 397, 409, 413,
　　448, 494
Ulmaceae, 187, 189
Umbellales, 190
Urticales, 187, 189

Vicia spp., 340, 353, 372, 400,
　　412, 422, 425, 491, 498
　　Vicia angustifolia, 449, 492
　　Vicia atropurpurea, 425, 498
　　Vicia bengalensis, 531
　　Vicia cracca, 449
　　Vicia faba, 369, 370, 396,
　　　397, 421, 425, 449, 479,
　　　500, 502, 531
　　Vicia hajastana, 447
　　Vicia hirsuta, 374, 382, 386,
　　　392, 396, 397, 425, 449
　　Vicia lathyroides, 449
　　Vicia sativa, 372, 396, 397,
　　　425, 492
　　Vicia sepium, 449
　　Vicia sylvatica, 449
　　Vicia tetrasperma, 397

Vicia villosa, 396, 397
Vigna spp, 344, 353, 429, 437
　　Vigna marina, 400
　　Vigna mungo, 396, 398, 410,
　　　425, 428, 436
　　Vigna owahuensis, 400
　　Vigna radiata, 373, 396, 398,
　　　425, 436
　　Vigna sinensis, 236, 498, 503
　　Vigna unguiculata, 375, 377,
　　　396, 398, 400, 410-413, 417,
　　　421-429, 435, 444
Viminaria spp., 420
　　Viminaria denudata, 421, 423
　　Viminaria juncea, 420, 425, 429
Violales, 190
Vitreoscilla spp., 81

Watsonia bulbillifera, 13
Westiellopsis spp., 107
　　Westiellopsis prolifica, 66,
　　　69, 72, 73, 107
Wisteria spp., 494
　　Wisteria chinensis, 499
　　Wisteria sinensis, 398, 448

Xanthomonas spp., 319, 333
　　Xanthomonas hortoricola, 174

Zamina spp., 199
Zea mays, 160
Zygophyllaceae, 187, 189
Zygophyllum spp., 189, 197
　　Zygophyllum album, 187, 197
　　Zygophyllum coccineum, 197
　　Zygophyllum decumbens, 187,
　　　197
　　Zygophyllum simplex, 187, 197

QR
89.7
T73
V.2